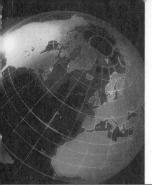

A First
Mathematical
Modeling

4th EDITION

Frank R. Giordano
Naval Postgraduate School

William P. Fox
Naval Postgraduate School

Steven B. Horton
United States Military Academy

Maurice D. Weir
Naval Postgraduate School

BROOKS/COLE
CENGAGE Learning

Australia • Brazil • Japan • Korea • Mexico • Singapore • Spain • United Kingdom • United States

BROOKS/COLE
CENGAGE Learning™

A First Course in Mathematical Modeling, **Fourth Edition**
Frank R. Giordano, William P. Fox, Steven B. Horton, Maurice D. Weir

Publisher: Charlie Van Wagner

Development Editor: Stacy Green

Editorial Assistant: Cynthia Ashton

Technology Project Manager: Sam Subity

Marketing Specialist: Ashley Pickering

Marketing Communications Manager: Linda Yip

Project Manager, Editorial Production:
    Jennifer Risden

Creative Director: Rob Hugel

Art Director: Vernon Boes

Print Buyer: Paula Vang

Permissions Editor: Mardell Glinski-Schultz

Production Service: Matrix Productions

Text Designer: Roy Neuhaus

Copy Editor: Connie Day

Illustrator: Lori Heckelman;
    Macmillan Publishing Solutions

Cover Designer: Roy Neuhaus

Cover/Interior Image: Ralph Mercer/
    Getty Images

Compositor: Macmillan Publishing Solutions

Library of Congress Control Number: 2008921546

International Student Edition:
ISBN-13: 978-0-495-55877-4
ISBN-10: 0-495-55877-X

Cengage Learning International Offices

Asia
cengageasia.com
tel: (65) 6410 1200

Australia/New Zealand
cengage.com.au
tel: (61) 3 9685 4111

Brazil
cengage.com.br
tel: (011) 3665 9900

India
cengage.co.in
tel: (91) 11 30484837/38

Latin America
cengage.com.mx
tel: +52 (55) 1500 6000

UK/Europe/Middle East/Africa
cengage.co.uk
tel: (44) 207 067 2500

Represented in Canada by Nelson Education, Ltd.
tel: (416) 752 9100 / (800) 668 0671
nelson.com

Cengage Learning is a leading provider of customized learning solutions with office locations around the globe, including Singapore, the United Kingdom, Australia, Mexico, Brazil, and Japan. Locate your local office at: **international.cengage.com/region.**

Cengage Learning products are represented in Canada by Nelson Education, Ltd.

For product information: **international.cengage.com**
Visit your local office: **international.cengage.com/regions**
Visit our corporate website: **cengage.com**

Printed in Canada
2 3 4 5 6 7 12 11 10 09

Contents

1 Modeling Change 1

5 Simulation Modeling 177

6 Discrete Probabilistic Modeling 215

9 Dimensional Analysis and Similitude 329

10 Graphs of Functions as Models 371

11 Modeling with a Differential Equation   401

12 Modeling with Systems of Differential Equations   466

13 Optimization of Continuous Models 509

Contents for the Enclosed Compact Disc

1 UMAP Modules (COMAP)

The Undergraduate Applications in Mathematics modules (UMAPs) are developed and produced by the Consortium for Mathematics and Its Applications, Inc. (800-772-6627, **www.comap.com**). UMAPs are particularly suited to supplement the modeling course we propose. The following UMAPs are referenced as projects, further reading, or sources for additional problems and are provided on the CD for easy access.

2 Past Modeling Contest Problems

Past contest problems are excellent sources for modeling projects or sources to design a problem. On the CD we provide links to electronic copies of all contest problems:

Mathematical Contest in Modeling (MCM): *1985–2008*
Interdisciplinary Contest in Modeling (ICM): *1997–2008*
High School Contest in Modeling (HiMCM): *1998–2008*

3 Interdisciplinary Lively Applications Projects (ILAPs)

Interdisciplinary Lively Applications Projects (ILAPs) are developed and produced by the Consortium for Mathematics and Its Applications, Inc., COMAP (800-772-6627, **www.comap.com**). ILAPs are codesigned with a partner discipline to provide in-depth model development and analysis from both a mathematical perspective and that of the partner discipline. We find the following ILAPs to be particularly well suited for the course we propose:

- Car Financing
- Choloform Alert
- Drinking Water
- Electric Power
- Forest Fires
- Game Theory
- Getting the Salt Out
- Health Care
- Health Insurance Premiums
- Hopping Hoop
- Bridge Analysis
- Lagniappe Fund
- Lake Pollution

- Launch the Shuttle
- Pollution Police
- Ramps and Freeways
- Red & Blue CDs
- Drug Poisoning
- Shuttle
- Stocking a Fish Pond
- Survival of Early Americans
- Traffic Lights
- Travel Forecasting
- Tuition Prepayment
- Vehicle Emissions
- Water Purification

4   Technology and Software

Mathematical modeling often requires technology in order to use the techniques discussed in the text, the modules, and ILAPs. We provide extensive examples of technology using spreadsheets (Excel), computer algebra systems (Maple®, Mathematica®, and Matlab®), and the graphing calculator (TI). Application areas include:

- Difference Equations
- Model Fitting
- Empirical Model Construction
- Divided Difference Tables
- Cubic Splines
- Monte Carlo Simulation Models
- Discrete Probabilistic Models
- Reliability Models
- Linear Programming
- Golden Section Search
- Euler's Method for Ordinary Differential Equations
- Euler's Method for Systems of Ordinary Differential Equations
- Nonlinear Optimization

5   Technology Labs

Examples and exercises designed for student use in a laboratory environment are included, addressing the following topics:

- Difference Equations
- Proportionality
- Model Fitting
- Empirical Model Construction
- Monte Carlo Simulation
- Linear Programming
- Discrete Optimization Search Methods
- Ordinary Differential Equations
- Systems of Ordinary Differential Equations
- Continuous Optimization-Search Methods

Preface

To facilitate an early initiation of the modeling experience, the first edition of this text was designed to be taught concurrently or immediately after an introductory business or engineering calculus course. In the second edition, we added chapters treating discrete dynamical systems, linear programming and numerical search methods, and an introduction to probabilistic modeling. Additionally, we expanded our introduction of simulation. In the third edition we included solution methods for some simple dynamical systems to reveal their long-term behavior. We also added basic numerical solution methods to the chapters covering modeling with differential equations. In this edition, we have added a new chapter to address modeling using graph theory. Graph theory is an area of burgeoning interest for modeling contemporary scenarios. Our chapter is intended to introduce graph theory from a modeling perspective and encourage students to pursue the subject in greater detail. We have also added two new sections to the chapter on modeling with a differential equation: discussions of separation of variables and linear equations. Many of our readers had expressed a desire that analytic solutions to first-order differential equations be included as part of their modeling course. The text has been reorganized into two parts: Part One, Discrete Modeling (Chapters 1–9), and Part Two, Continuous Modeling (Chapters 10–13). This organizational structure allows for teaching an entire modeling course that is based on Part One and does not require the calculus. Part Two then addresses continuous models based on optimization and differential equations that can be presented concurrently with freshman calculus. The text gives students an opportunity to cover all phases of the mathematical modeling process. The new CD-ROM accompanying the text contains software, additional modeling scenarios and projects, and a link to past problems from the Mathematical Contest in Modeling. We thank Sol Garfunkel and the COMAP staff for their support of modeling activities that we refer to under Resource Materials below.

Goals and Orientation

The course continues to be a bridge between the study of mathematics and the applications of mathematics to various fields. The course affords the student an early opportunity to see how the pieces of an applied problem fit together. The student investigates meaningful and practical problems chosen from common experiences encompassing many academic disciplines, including the mathematical sciences, operations research, engineering, and the management and life sciences.

This text provides an introduction to the entire modeling process. Students will have opportunities to practice the following facets of modeling and enhance their problem-solving capabilities:

1. *Creative and Empirical Model Construction:* Given a real-world scenario, the student learns to identify a problem, make assumptions and collect data, propose a model, test the assumptions, refine the model as necessary, fit the model to data if appropriate, and analyze the underlying mathematical structure of the model to appraise the sensitivity of the conclusions when the assumptions are not precisely met.

2. *Model Analysis:* Given a model, the student learns to work backward to uncover the implicit underlying assumptions, assess critically how well those assumptions fit the scenario at hand, and estimate the sensitivity of the conclusions when the assumptions are not precisely met.

3. *Model Research:* The student investigates a specific area to gain a deeper understanding of some behavior and learns to use what has already been created or discovered.

Student Background and Course Content

Because our desire is to initiate the modeling experience as early as possible in the student's program, the only prerequisite for Chapters 10, 11, and 12 is a basic understanding of single-variable differential and integral calculus. Although some unfamiliar mathematical ideas are taught as part of the modeling process, the emphasis is on using mathematics that the students already know after completing high school. This is especially true in Part One. The modeling course will then motivate students to study the more advanced courses such as linear algebra, differential equations, optimization and linear programming, numerical analysis, probability, and statistics. The power and utility of these subjects are intimated throughout the text.

Further, the scenarios and problems in the text are not designed for the application of a particular mathematical technique. Instead, they demand thoughtful ingenuity in using fundamental concepts to find reasonable solutions to "open-ended" problems. Certain mathematical techniques (such as Monte Carlo simulation, curve fitting, and dimensional analysis) are presented because often they are not formally covered at the undergraduate level. Instructors should find great flexibility in adapting the text to meet the particular needs of students through the problem assignments and student projects. We have used this material to teach courses to both undergraduate and graduate students—and even as a basis for faculty seminars.

Organization of the Text

The organization of the text is best understood with the aid of Figure 1. The first nine chapters constitute Part One and require only precalculus mathematics as a prerequisite. We begin with the idea of modeling *change* using simple finite difference equations. This approach is quite intuitive to the student and provides us with several concrete models to support

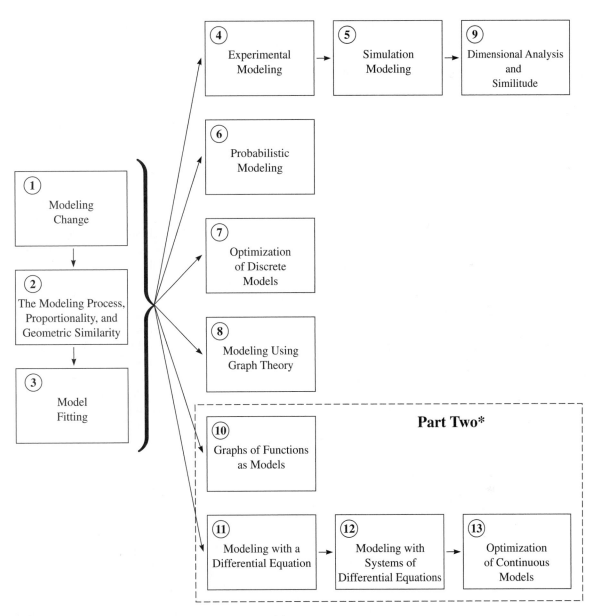

*Part Two requires single-variable calculus as a corequisite.

■ **Figure 1**

Chapter organization and progression

our discussion of the modeling process in Chapter 2. There we classify models, analyze the modeling process, and construct several proportionality models or submodels that are then revisited in the next two chapters. In Chapter 3 the student is presented with three criteria for fitting a specific type of curve to a collected data set, with emphasis on the least-squares criterion. Chapter 4 addresses the problem of capturing the trend of a collected set of data. In this empirical construction process, we begin with fitting simple one-term models approximating collected data sets and then progress to more sophisticated interpolating models, including polynomial smoothing models and cubic splines. Simulation models are discussed in Chapter 5. An empirical model is fit to some collected data, and then Monte Carlo simulation is used to duplicate the behavior being investigated. The presentation motivates the eventual study of probability and statistics.

Chapter 6 provides an introduction to probabilistic modeling. The topics of Markov processes, reliability, and linear regression are introduced, building on scenarios and analysis presented previously. Chapter 7 addresses the issue of finding the best-fitting model using the other two criteria presented in Chapter 3. Linear programming is the method used for finding the "best" model for one of the criteria, and numerical search techniques can be used for the other. The chapter concludes with an introduction to numerical search methods, including the dichotomous and Golden Section methods. Part One ends with Chapter 9, which is devoted to dimensional analysis, a topic of great importance in the physical sciences and engineering.

Part Two is dedicated to the study of continuous models. Chapter 10 treats the construction of continuous graphical models and explores the sensitivity of the models constructed to the assumptions underlying them. In Chapters 11 and 12 we model dynamic (time varying) scenarios. These chapters build on the discrete analysis presented in Chapter 1 by now considering situations where time is varying continuously. Chapter 13 is devoted to the study of continuous optimization. Students get the opportunity to solve continuous optimization problems requiring only the application of elementary calculus and are introduced to constrained optimization problems as well.

Student Projects

Student projects are an essential part of any modeling course. This text includes projects in creative and empirical model construction, model analysis, and model research. Thus we recommend a course consisting of a mixture of projects in all three facets of modeling. These projects are most instructive if they address scenarios that have no unique solution. Some projects should include *real* data that students are either given or can *readily* collect. A combination of individual and group projects can also be valuable. Individual projects are appropriate in those parts of the course in which the instructor wishes to emphasize the development of individual modeling skills. However, the inclusion of a group project early in the course gives students the exhilaration of a "brainstorming" session. A variety of projects is suggested in the text, such as constructing models for various scenarios, completing UMAP[1] modules, or researching a model presented as an example in the text

---

[1]UMAP modules are developed and distributed through COMAP, Inc., 57 Bedford Street, Suite 210, Lexington, MA 02173.

or class. It is valuable for each student to receive a mixture of projects requiring either model construction, model analysis, or model research for variety and confidence building throughout the course. Students might also choose to develop a model in a scenario of particular interest, or analyze a model presented in another course. We recommend five to eight short projects in a typical modeling course. Detailed suggestions on how student projects can be assigned and used are included in the Instructor's Manual that accompany this text.

In terms of the number of scenarios covered throughout the course, as well as the number of homework problems and projects assigned, we have found it better to pursue a few that are developed carefully and completely. We have provided many more problems and projects than can reasonably be assigned to allow for a wide selection covering many different application areas.

Resource Materials

We have found material provided by the Consortium for Mathematics and Its Application (COMAP) to be outstanding and particularly well suited to the course we propose. Individual modules for the undergraduate classroom, UMAP Modules, may be used in a variety of ways. First, they may be used as instructional material to support several lessons. In this mode a student completes the self-study module by working through its exercises (the detailed solutions provided with the module can be conveniently removed before it is issued). Another option is to put together a block of instruction using one or more UMAP modules suggested in the projects sections of the text. The modules also provide excellent sources for "model research," because they cover a wide variety of applications of mathematics in many fields. In this mode, a student is given an appropriate module to research and is asked to complete and report on the module. Finally, the modules are excellent resources for scenarios for which students can practice model construction. In this mode the instructor writes a scenario for a student project based on an application addressed in a particular module and uses the module as background material, perhaps having the student complete the module at a later date. The CD accompanying the text contains most of the UMAPs referenced throughout. Information on the availability of newly developed interdisciplinary projects can be obtained by writing COMAP at the address given previously, calling COMAP at 1-800-772-6627, or electronically: order@comap.com.

A great source of student-group projects are the Mathematical Contest in Modeling (MCM) and the Interdisciplinary Contest in Modeling (ICM). These projects can be taken from the link provided on the CD and tailored by the instructor to meet specific goals for their class. These are also good resources to prepare teams to compete in the MCM and ICM contests. The contest is sponsored by COMAP with funding support from the National Security Agency, the Society of Industrial and Applied Mathematics, the Institute for Operations Research and the Management Sciences, and the Mathematical Association of America. Additional information concerning the contest can be obtained by contacting COMAP, or visiting their website at **www.comap.com**.

The Role of Technology

Technology is an integral part of doing mathematical modeling with this textbook. Technology can be used to support the modeling of solutions in all of the chapters. Rather than incorporating lots of varied technologies into the explanations of the models directly in the text, we decided to include the use of various technology on the enclosed CD. There the student will find templates in Microsoft® Excel®, Maple®, Mathematica®, and Texas Instruments graphing calculators, including the TI-83 and 84 series.

We have chosen to illustrate the use of *Maple* in our discussion of the following topics that are well supported by Maple commands and programming procedures: difference equations, proportionality, model fitting (least squares), empirical models, simulation, linear programming, dimensional analysis, modeling with differential equations, modeling with systems of differential equations, and optimization of continuous models. Maple worksheets for the illustrative examples appearing in the referenced chapters are provided on the CD.

Mathematica was chosen to illustrate difference equations, proportionality, model fitting (least squares), empirical models, simulation, linear programming, graph theory, dimensional analysis, modeling with differential equations, modeling with systems of differential equations, and optimization of continuous models. Mathematica worksheets for illustrative examples in the referenced chapters are provided on the CD.

Excel is a spreadsheet that can be used to obtain numerical solutions and conveniently obtain graphs. Consequently, Excel was chosen to illustrate the iteration process and graphical solutions to difference equations. It was also selected to calculate and graph functions in proportionality, model fitting, empirical modeling (additionally used for divided difference tables and the construction and graphing of cubic splines), Monte Carlo simulation, linear programming (Excel's Solver is illustrated), modeling with differential equations (numerical approximations with both the Euler and the Runge-Kutta methods), modeling with systems of differential equations (numerical solutions), and Optimization of Discrete and Continuous Models (search techniques in single-variable optimization such as the dichotomous and Golden Section searches). Excel worksheets can be found on the website.

The *TI calculator* is a powerful tool for technology as well. Much of this textbook can be covered using the TI calculator. We illustrate the use of TI calculators with difference equations, proportionality, modeling fitting, empirical models (Ladder of Powers and other transformations), simulation, and differential equations (Euler's method to construct numerical solutions).

Acknowledgments

It is always a pleasure to acknowledge individuals who have played a role in the development of a book. We are particularly grateful to Brigadier General (retired) Jack M. Pollin and Dr. Carroll Wilde for stimulating our interest in teaching modeling and for support and guidance in our careers. We're indebted to many colleagues for reading the first edition manuscript and suggesting modifications and problems: Rickey Kolb, John Kenelly, Robert Schmidt, Stan Leja, Bard Mansager, and especially Steve Maddox and Jim McNulty.

We are indebted to a number of individuals who authored or co-authored UMAP materials that support the text: David Cameron, Brindell Horelick, Michael Jaye, Sinan Koont, Stan Leja, Michael Wells, and Carroll Wilde. In addition, we thank Solomon Garfunkel and the entire COMAP staff for their cooperation on this project, especially Roland Cheyney for his help with the production of the CD that accompanies the text. We also thank Tom O'Neil and his students for their contributions to the CD and for helpful suggestions in support of modeling activities. We would like to thank Dr. Amy H. Erickson for her many contributions to the CD and website.

Thanks to the following reviewers: Stephen Alessandrini, Rutgers-Camden; John Cannon, University of Central Florida; Donald Cathcart, Salisbury State University; Catherine Crawford, Elmhurst College; Hajrudin Fejzic, California State University San Bernadino; Michael Frantz, University of La Verne; Larry Hill, Lafayette College; Dick Jardine, Keene State University; Theresa Jeevanjee, Fontbonne University; Andy Keck, Western State College of Colorado; Aprillya Lanz, Clayton State University; Abdelhamid Meziani, Florida International University; Ho-Kuen Ng, San Jose State University; William Paulsen, Arkansas State University; Ken Roblee, Troy State University; Todd Smith, Athens State College; Alexandros Sopasakis, University of Massachusetts; and Ray Toland, Clarkson University.

The production of any mathematics text is a complex process, and we have been especially fortunate in having a superb and creative production staff at Brooks/Cole. In particular, we express our thanks to Craig Barth, our editor for the first edition, Gary Ostedt, the second edition, and Gary Ostedt and Bob Pirtle, our editors for the third edition. We would like to thank everyone from Brooks/Cole who worked with us on this edition, especially Charlie Van Wagner, our Acquisitions Editor, and Stacy Green, our Development Editor. Thanks also to Sara Planck and Matrix Productions for production service.

Frank R. Giordano
William P. Fox
Steven B. Horton
Maurice D. Weir

In memory of a loving wife and friend,
Gwendolyn Galing Fox
(1954–2005)

As the caption of her painting says:

And when our work is done... may it be said, "well done"...

Modeling Change

Introduction

To help us better understand our world, we often describe a particular phenomenon mathematically (by means of a function or an equation, for instance). Such a **mathematical model** is an idealization of the real-world phenomenon and never a completely accurate representation. Although any model has its limitations, a good one can provide valuable results and conclusions. In this chapter we direct our attention to modeling change.

Mathematical Models

In modeling our world, we are often interested in predicting the value of a variable at some time in the future. Perhaps it is a population, a real estate value, or the number of people with a communicative disease. Often a mathematical model can help us understand a behavior better or aid us in planning for the future. Let's think of a mathematical model as a mathematical construct designed to study a particular real-world system or behavior of interest. The model allows us to reach mathematical conclusions about the behavior, as illustrated in Figure 1.1. These conclusions can be interpreted to help a decision maker plan for the future.

Simplification

Most models simplify reality. Generally, models can only approximate real-world behavior. One very powerful simplifying relationship is **proportionality**.

■ **Figure 1.1**

A flow of the modeling process beginning with an examination of real-world data

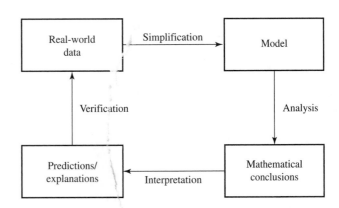

Definition

> Two variables y and x are **proportional** (to each other) if one is always a constant multiple of the other—that is, if
>
> $$y = kx$$
>
> for some nonzero constant k. We write $y \propto x$.

The definition means that the graph of y versus x lies along a straight line through the origin. This graphical observation is useful in testing whether a given data collection reasonably assumes a proportionality relationship. If a proportionality is reasonable, a plot of one variable against the other should approximate a straight line through the origin. Here is an example.

EXAMPLE 1   *Testing for Proportionality*

Table 1.1
Spring–mass system

Mass	Elong.
50	1.000
100	1.875
150	2.750
200	3.250
250	4.375
300	4.875
350	5.675
400	6.500
450	7.250
500	8.000
550	8.750

Consider a spring–mass system, such as the one shown in Figure 1.2. We conduct an experiment to measure the stretch of the spring as a function of the mass (measured as weight) placed on the spring. Consider the data collected for this experiment, displayed in Table 1.1. A scatterplot graph of the stretch or elongation of the spring versus the mass or weight placed on it reveals a straight line passing approximately through the origin (Figure 1.3).

The data appear to follow the proportionality rule that elongation e is proportional to the mass m, or, symbolically, $e \propto m$. The straight line appears to pass through the origin. This geometric understanding allows us to look at the data to determine whether proportionality is a reasonable simplifying assumption and, if so, to estimate the slope k. In this case, the assumption appears valid, so we estimate the constant of proportionality by picking the two points $(200, 3.25)$ and $(300, 4.875)$ as lying along the straight line. We calculate the slope of the line joining these points as

$$slope = \frac{4.875 - 3.25}{300 - 200} = 0.01625$$

■ **Figure 1.2**

Spring–mass system

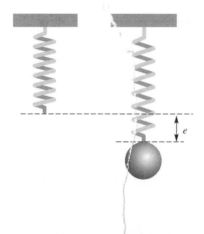

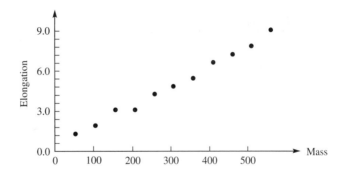

Figure 1.3

Data from spring–mass system

Thus the constant of proportionality is approximately 0.0163, the slope of the line through the origin. We estimate our model as

$$e = 0.0163m$$

We then examine how close our model fits the data by plotting the line it represents superimposed on the scatterplot (Figure 1.4). The graph reveals that the simplifying proportionality model is reasonable because most of the points fall on or very near the line $e = 0.0163m$.

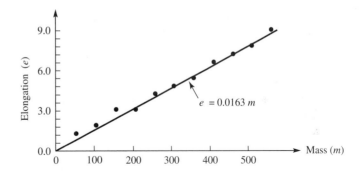

Figure 1.4

Data from spring–mass system with proportionality line

Modeling Change

A powerful paradigm to use in modeling change is

$$future\ value = present\ value + change$$

Often, we wish to predict the future on the basis of what we know now, in the present, and add the change that has been carefully observed. In such cases, we begin by studying the change itself according to the formula

$$change = future\ value - present\ value$$

By collecting data over a period of time and plotting those data, we often can discern patterns to model that capture the trend of the change. If the behavior is taking place over *discrete time periods*, the preceding construct leads to a **difference equation**, which we study in this chapter. If the behavior is taking place *continuously* with respect to time, then the construct leads to a **differential equation** studied in Chapter 11. Both are powerful methodologies for studying change to explain and predict behavior.

1.1   Modeling Change with Difference Equations

In this section we build mathematical models to describe change in an observed behavior. When we observe change, we are often interested in understanding why the change occurs in the way it does, perhaps to analyze the effects of different conditions on the behavior or to predict what will happen in the future. A mathematical model helps us better understand a behavior, while allowing us to experiment mathematically with different conditions affecting it.

Definition

For a sequence of numbers $A = \{a_0, a_1, a_2, a_3, \ldots\}$ the first differences are

$$\Delta a_0 = a_1 - a_0$$
$$\Delta a_1 = a_2 - a_1$$
$$\Delta a_2 = a_3 - a_2$$
$$\Delta a_3 = a_4 - a_3$$

For each positive integer n, the **nth first difference** is

$$\Delta a_n = a_{n+1} - a_n$$

Note from Figure 1.5 that the first difference represents the rise or fall between consecutive values of the sequence—that is, the vertical *change* in the graph of the sequence during one time period.

■ **Figure 1.5**

The first difference of a sequence is the rise in the graph during one time period.

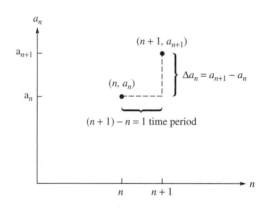

$(n + 1, a_{n+1})$

$\Delta a_n = a_{n+1} - a_n$

(n, a_n)

$(n + 1) - n = 1$ time period

EXAMPLE 1 *A Savings Certificate*

Consider the value of a savings certificate initially worth $1000 that accumulates interest paid each month at 1% per month. The following sequence of numbers represents the value of the certificate month by month.

$$A = (1000, 1010, 1020.10, 1030.30, \ldots)$$

The first differences are as follows:

$$\Delta a_0 = a_1 - a_0 = 1010 - 1000 = 10$$
$$\Delta a_1 = a_2 - a_1 = 1020.10 - 1010 = 10.10$$
$$\Delta a_2 = a_3 - a_2 = 1030.30 - 1020.10 = 10.20$$

Note that the first differences represent the *change in the sequence* during one time period, or the *interest earned* in the case of the savings certificate example.

The first difference is useful for modeling change taking place in discrete intervals. In this example, the change in the value of the certificate from one month to the next is merely the interest paid during that month. If n is the number of months and a_n the value of the certificate after n months, then the change or interest growth in each month is represented by the nth difference

$$\Delta a_n = a_{n+1} - a_n = 0.01a_n$$

This expression can be rewritten as the difference equation

$$a_{n+1} = a_n + 0.01a_n$$

We also know the initial deposit of $1000 (initial value) that then gives the **dynamical system model**

$$a_{n+1} = 1.01a_n, \quad n = 0, 1, 2, 3, \ldots \tag{1.1}$$
$$a_0 = 1000$$

where a_n represents the amount accrued after n months. Because n represents the nonnegative integers $\{0, 1, 2, 3, \ldots\}$, Equation (1.1) represents an *infinite set* of algebraic equations, called a **dynamical system**. Dynamical systems allow us to describe the *change* from one period to the next. The difference equation formula computes the next term, given the immediately previous term in the sequence, but it does not compute the value of a specific term directly (e.g., the savings after 100 periods). We would iterate the sequence to a_{100} to obtain that value.

Because it is change we often observe, we can construct a difference equation by representing or approximating the change from one period to the next. To modify our example, if we were to withdraw $50 from the account each month, the change during a period would be the interest earned during that period minus the monthly withdrawal, or

$$\Delta a_n = a_{n+1} - a_n = 0.01a_n - 50$$

In most examples, mathematically describing the change is not going to be as precise a procedure as illustrated here. Often it is necessary to *plot the change*, *observe a pattern*, and then *describe the change* in mathematical terms. That is, we will be trying to find

$$change = \Delta a_n = some\ function\ f$$

The change may be a function of previous terms in the sequence (as was the case with no monthly withdrawals), or it may also involve some external terms (such as the amount of money withdrawn in the current example or an expression involving the period n). Thus, in constructing models representing change in this chapter, we will be modeling change in discrete intervals, where

$$change = \Delta a_n = a_{n+1} - a_n = f(terms\ in\ the\ sequence,\ external\ terms)$$

Modeling change in this way becomes the art of determining or approximating a function f that represents the change.

Consider a second example in which a difference equation exactly models a behavior in the real world.

EXAMPLE 2   *Mortgaging a Home*

Six years ago your parents purchased a home by financing $80,000 for 20 years, paying monthly payments of $880.87 with a monthly interest of 1%. They have made 72 payments and wish to know how much they owe on the mortgage, which they are considering paying off with an inheritance they received. Or they could be considering refinancing the mortgage with several interest rate options, depending on the length of the payback period. The change in the amount owed each period increases by the amount of interest and decreases by the amount of the payment:

$$\Delta b_n = b_{n+1} - b_n = 0.01b_n - 880.87$$

Solving for b_{n+1} and incorporating the initial condition gives the dynamical system model

$$b_{n+1} = b_n + 0.01b_n - 880.87$$
$$b_0 = 80000$$

where b_n represents the amount owed after n months. Thus,

$$b_1 = 80000 + 0.01(80000) - 880.87 = 79919.13$$
$$b_2 = 79919.13 + 0.01(79919.13) - 880.87 = 79837.45$$

yielding the sequence

$$B = (80000, 79919.13, 79837.45, \ldots)$$

Calculating b_3 from b_2, b_4 from b_3, and so forth in turn, we obtain $b_{72} = \$71{,}523.11$. The sequence is graphed in Figure 1.6.   ■ ■ ■

Months n	Amount owed b_n
0	80000.00
1	79919.13
2	79837.45
3	79754.96
4	79671.64
5	79587.48
6	79502.49
7	79416.64
8	79329.94
9	79242.37
10	79153.92
11	79064.59
12	78974.37

■ **Figure 1.6**

The sequence and graph for Example 2

Let's summarize the important ideas introduced in Examples 1 and 2.

Definition

A **sequence** is a function whose domain is the set of all nonnegative integers and whose range is a subset of the real numbers. A **dynamical system** is a relationship among terms in a sequence. A **numerical solution** is a table of values satisfying the dynamical system.

In the problems for this section we discuss other behaviors in the world that can be modeled exactly by difference equations. In the next section, we use difference equations to approximate observed change. After collecting data for the change and discerning patterns of the behavior, we will use the concept of proportionality to test and fit models that we propose.

1.1 PROBLEMS

SEQUENCES

1. Write out the first five terms a_0–a_4 of the following sequences:

   a. $a_{n+1} = 3a_n, \quad a_0 = 1$

   b. $a_{n+1} = 2a_n + 6, \quad a_0 = 0$

   c. $a_{n+1} = 2a_n(a_n + 3), \quad a_0 = 4$

   d. $a_{n+1} = a_n^2, \quad a_0 = 1$

2. Find a formula for the nth term of the sequence.

   a. $\{3, 3, 3, 3, 3, \ldots\}$

   b. $\{1, 4, 16, 64, 256, \ldots\}$

    c. $\{\frac{1}{2}, \frac{1}{4}, \frac{1}{8}, \frac{1}{16}, \frac{1}{32} \ldots\}$

    d. $\{1, 3, 7, 15, 31, \ldots\}$

DIFFERENCE EQUATIONS

3. By examining the following sequences, write a difference equation to represent the change during the nth interval as a function of the previous term in the sequence.

    a. $\{2, 4, 6, 8, 10, \ldots\}$

    b. $\{2, 4, 16, 256, \ldots\}$

    c. $\{1, 2, 5, 11, 23, \ldots\}$

    d. $\{1, 8, 29, 92, \ldots\}$

4. Write out the first five terms of the sequence satisfying the following difference equations:

    a. $\Delta a_n = \frac{1}{2} a_n, \quad a_0 = 1$

    b. $\Delta b_n = 0.015 b_n, \quad b_0 = 1000$

    c. $\Delta p = 0.001(500 - p_n), \quad p_0 = 10$

    d. $\Delta t_n = 1.5(100 - t_n), \quad t_0 = 200$

DYNAMICAL SYSTEMS

5. By substituting $n = 0, 1, 2, 3$, write out the first four algebraic equations represented by the following dynamical systems:

    a. $a_{n+1} = 3a_n, \quad a_0 = 1$

    b. $a_{n+1} = 2a_n + 6, \quad a_0 = 0$

    c. $a_{n+1} = 2a_n(a_n + 3), \quad a_0 = 4$

    d. $a_{n+1} = a_n^2, \quad a_0 = 1$

6. Name several behaviors you think can be modeled by dynamical systems.

MODELING CHANGE EXACTLY

For Problems 7–10, formulate a dynamical system that models change exactly for the described situation.

7. You currently have $5000 in a savings account that pays 0.5% interest each month. You add another $200 each month.

8. You owe $500 on a credit card that charges 1.5% interest each month. You pay $50 each month and you make no new charges.

9. Your parents are considering a 30-year, $200,000 mortgage that charges 0.5% interest each month. Formulate a model in terms of a monthly payment p that allows the mortgage (loan) to be paid off after 360 payments. *Hint:* If a_n represents the amount owed after n months, what are a_0 and a_{360}?

10. Your grandparents have an annuity. The value of the annuity increases each month by an automatic deposit of 1% interest on the previous month's balance. Your grandparents

withdraw $1000 at the beginning of each month for living expenses. Currently, they have $50,000 in the annuity. Model the annuity with a dynamical system. Will the annuity run out of money? When? *Hint*: What value will a_n have when the annuity is depleted?

11. Repeat Problem 10 with an interest rate of 0.5%.

12. Your current credit card balance is $12,000 with a current rate of 19.9% per year. Interest is charged monthly. Determine what monthly payment p will pay off the card in

a. Two years, assuming no new charges

b. Four years, assuming no new charges

13. Again consider Problem 12 above. Now assume that each month you charge $105. Determine what monthly payment p will pay off the card in

a. Two years

b. Four years

1.1 PROJECTS

1. With the rising price of gas, you wish to buy a new (hybrid) car this year. You narrow your choices to the following hybrids: 2007 Toyota Camry hybrid, 2007 Saturn hybrid, 2007 Honda Civic hybrid, 2007 Nissan Altima hybrid, and 2007 Mercury Mariner hybrid. Each company has offered you its "best deal" as listed below. You are able to allocate approximately $500 for a car payment each month for up to 60 months. Use a dynamical system to determine which of the new hybrid cars you could buy.

Hybrid	"Best Deal"	Cash Down	Interest and Duration
Saturn	$22,045	$1000	5.95% APR for 60 months
Honda Civic	$24,350	$1500	5.5% for 60 months
Toyota Camry	$26,200	$750	6.25% for 60 months
Mariner	$27,515	$1500	6% for 60 months
Altima	$24,990	$1000	5.9% for 60 months

2. You are considering a 30-year mortgage that charges 0.4% interest each month to pay off a $250,000 mortgage.

a. Determine the monthly payment p that allows the loan to be paid off at 360 months.

b. Now assume that you have been paying the mortgage for 8 years and now have an opportunity to refinance the loan. You have a choice between a 20-year loan at 4% per year with interest charged monthly and a 15-year loan at 3.8% per year with interest charged monthly. Each of the loans charges a closing cost of $2500. Determine the monthly payment p for both the 20-year loan and the 15-year loan. Do you think refinancing is the right thing to do? If so, do you prefer the 20-year or the 15-year option?

1.2   Approximating Change with Difference Equations

In most examples, describing the change mathematically will not be as precise a procedure as in the savings certificate and mortgage examples presented in the previous section. Typically, we must plot the change, observe a pattern, and then approximate the change in mathematical terms. In this section we approximate some observed change to complete the expression

$$change = \Delta a_n = some\ function\ f$$

We begin by distinguishing between change that takes place continuously and that which occurs in discrete time intervals.

Discrete Versus Continuous Change

When we construct models involving change, an important distinction is that some change takes place in **discrete** time intervals (such as the depositing of interest in an account), whereas in other cases, the change happens **continuously** (such as the change in the temperature of a cold can of soda on a warm day). Difference equations represent change in the case of discrete time intervals. Later we will see the relationship between discrete change and continuous change (for which calculus was developed). For now, in the several models that follow, we approximate a continuous change by examining data taken at discrete time intervals. Approximating a continuous change by difference equations is an example of model simplification.

EXAMPLE 1   *Growth of a Yeast Culture*

The data in Figure 1.7 were collected from an experiment measuring the growth of a yeast culture. The graph represents the assumption that the change in population is proportional to the current size of the population. That is, $\Delta p_n = (p_{n+1} - p_n) = kp_n$, where p_n represents the size of the population biomass after n hours, and k is a positive constant. The value of k depends on the time measurement.

Although the graph of the data does not lie precisely along a straight line passing exactly through the origin, it can be *approximated* by such a straight line. Placing a ruler over the data to approximate a straight line through the origin, we estimate the slope of the line to be about 0.5. Using the estimate $k = 0.5$ for the slope of the line, we hypothesize the proportionality model

$$\Delta p_n = p_{n+1} - p_n = 0.5p_n$$

yielding the prediction $p_{n+1} = 1.5p_n$. This model predicts a population that increases forever, which is questionable.

Time in hours n	Observed yeast biomass p_n	Change in biomass $p_{n+1}-p_n$
0	9.6	8.7
1	18.3	10.7
2	29.0	18.2
3	47.2	23.9
4	71.1	48.0
5	119.1	55.5
6	174.6	82.7
7	257.3	

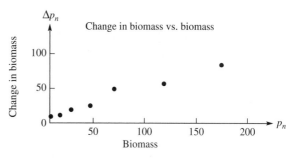

■ **Figure 1.7**

Growth of a yeast culture versus time in hours; data from R. Pearl, "The Growth of Population," *Quart. Rev. Biol.* 2(1927): 532–548

Model Refinement: Modeling Births, Deaths, and Resources

If both births and deaths during a period are proportional to the population, then the change in population should be proportional to the population, as was illustrated in Example 1. However, certain resources (e.g., food) can support only a maximum population level rather than one that increases indefinitely. As these maximum levels are approached, growth should slow.

EXAMPLE 2 *Growth of a Yeast Culture Revisited*

Finding a Model   The data in Figure 1.8 show what actually happens to the yeast culture growing in a restricted area as time increases beyond the eight observations given in Figure 1.7.

From the third column of the data table in Figure 1.8, note that the change in population per hour becomes smaller as the resources become more limited or constrained. From the graph of population versus time, the population appears to be approaching a limiting value, or **carrying capacity**. Based on our graph we estimate the carrying capacity to be 665. (Actually, the graph does not precisely tell us the correct number is 665 and not 664 or 666, for example.) Nevertheless, as p_n approaches 665, the change does slow considerably. Because $665 - p_n$ gets smaller as p_n approaches 665, we propose the model

$$\Delta p_n = p_{n+1} - p_n = k(665 - p_n)p_n$$

which causes the change Δp_n to become increasingly small as p_n approaches 665. Mathematically, this hypothesized model states that the change Δp_n is proportional to the product $(665 - p_n)p_n$. To test the model, plot $(p_{n+1} - p_n)$ versus $(665 - p_n)p_n$ to see if there is a reasonable proportionality. Then estimate the proportionality constant k.

Examining Figure 1.9, we see that the plot does reasonably approximate a straight line projected through the origin. We estimate the slope of the line approximating the data to be

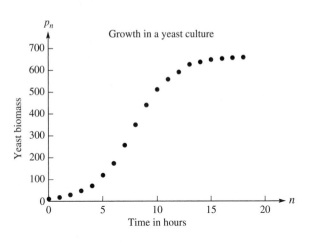

Time in hours n	Yeast biomass p_n	Change/ hour $p_{n+1} - p_n$
0	9.6	8.7
1	18.3	10.7
2	29.0	18.2
3	47.2	23.9
4	71.1	48.0
5	119.1	55.5
6	174.6	82.7
7	257.3	93.4
8	350.7	90.3
9	441.0	72.3
10	513.3	46.4
11	559.7	35.1
12	594.8	34.6
13	629.4	11.4
14	640.8	10.3
15	651.1	4.8
16	655.9	3.7
17	659.6	2.2
18	661.8	

■ **Figure 1.8**

Yeast biomass approaches a limiting population level

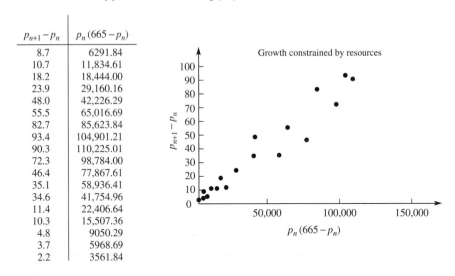

$p_{n+1} - p_n$	$p_n(665 - p_n)$
8.7	6291.84
10.7	11,834.61
18.2	18,444.00
23.9	29,160.16
48.0	42,226.29
55.5	65,016.69
82.7	85,623.84
93.4	104,901.21
90.3	110,225.01
72.3	98,784.00
46.4	77,867.61
35.1	58,936.41
34.6	41,754.96
11.4	22,406.64
10.3	15,507.36
4.8	9050.29
3.7	5968.69
2.2	3561.84

■ **Figure 1.9**

Testing the constrained growth model

about $k \approx 0.00082$, which gives the model

$$p_{n+1} - p_n = 0.00082(665 - p_n)p_n \tag{1.2}$$

Solving the Model Numerically Solving Equation (1.2) for p_{n+1} gives

$$p_{n+1} = p_n + 0.00082(665 - p_n)p_n \tag{1.3}$$

The right side of this equation is a quadratic in p_n. Such dynamical systems are classified as **nonlinear** and generally cannot be solved for analytical solutions. That is, usually we cannot find a formula expressing p_n in terms of n. However, if given that $p_0 = 9.6$, we can substitute in the expression to compute p_1:

$$p_1 = p_0 + 0.00082(665 - p_0)p_0 = 9.6 + 0.00082(665 - 9.6)9.6 = 14.76$$

In a similar manner, we can substitute $p_1 = 14.76$ into Equation (1.3) to compute $p_2 = 22.63$. **Iterating** in this way, we compute a table of values to provide a **numerical solution** to the model. This numerical solution of **model predictions** is presented in Figure 1.10. The predictions and observations are plotted together versus time on the same graph. Note that the model captures fairly well the *trend* of the observed data. ▪ ▪ ▪

Time in hours	Observation	Prediction
0	9.6	9.6
1	18.3	14.8
2	29.0	22.6
3	47.2	34.5
4	71.1	52.4
5	119.1	78.7
6	174.6	116.6
7	257.3	169.0
8	350.7	237.8
9	441.0	321.1
10	513.3	411.6
11	559.7	497.1
12	594.8	565.6
13	629.4	611.7
14	640.8	638.4
15	651.1	652.3
16	655.9	659.1
17	659.6	662.3
18	661.8	663.8

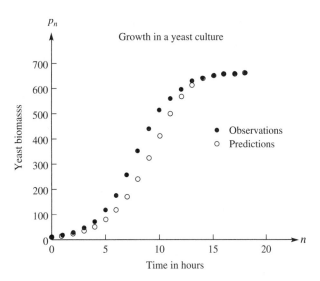

▪ **Figure 1.10**

Model predictions and observations

EXAMPLE 3 *Spread of a Contagious Disease*

Suppose that there are 400 students in a college dormitory and that one or more students has a severe case of the flu. Let i_n represent the number of infected students after n time periods. Assume that some interaction between those infected and those not infected is required to pass on the disease. If all are susceptible to the disease, then $(400 - i_n)$ represents those susceptible but not yet infected. If those infected remain contagious, we can model the change of those infected as a proportionality to the product of those infected by those susceptible but not yet infected, or

$$\Delta i_n = i_{n+1} - i_n = ki_n(400 - i_n) \tag{1.4}$$

In this model the product $i_n(400 - i_n)$ represents the number of possible interactions between those infected and those not infected at time n. A fraction k of these interactions would cause additional infections, represented by Δi_n.

Equation (1.4) has the same form as Equation (1.2), but in the absence of any data we cannot determine a value for the proportionality constant k. Nevertheless, a graph of the predictions determined by Equation (1.4) would have the same S shape as the graph of the yeast population in Figure 1.10.

There are many refinements to this model. For example, we might assume that a segment of the population is not susceptible to the disease, that the infection period is limited, or that infected students are removed from the dorm to prevent interaction with the uninfected. More sophisticated models might even treat the infected and susceptible populations separately.   ■ ■ ■

EXAMPLE 4   *Decay of Digoxin in the Bloodstream*

Digoxin is used in the treatment of heart disease. Doctors must prescribe an amount of medicine that keeps the concentration of digoxin in the bloodstream above an **effective level** without exceeding a **safe level** (there is variation among patients). For an initial dosage of 0.5 mg in the bloodstream, Table 1.2 shows the amount of digoxin a_n remaining in the bloodstream of a particular patient after n days, together with the change Δa_n each day.

Table 1.2   The change a_n in digoxin in a patient's bloodstream

n	0	1	2	3	4	5	6	7	8
a_n	0.500	0.345	0.238	0.164	0.113	0.078	0.054	0.037	0.026
Δa_n	−0.155	−0.107	−0.074	−0.051	−0.035	−0.024	−0.017	−0.011	

A scatterplot of Δa_n versus a_n from Table 1.2 is shown in Figure 1.11. The graph shows that the change Δa_n during a time interval is approximately proportional to the amount of digoxin a_n present in the bloodstream at the beginning of the time interval. The slope of the proportionality line through the origin is approximately $k \approx -0.107/0.345 \approx -0.310$. Since the graph in Figure 1.11 shows the change Δa_n as a linear function of a_n with slope -0.31, we have $\Delta a_n = -0.31 a_n$.

■ **Figure 1.11**

A plot of Δa_n versus a_n from Table 1.2 suggests a straight line through the origin.

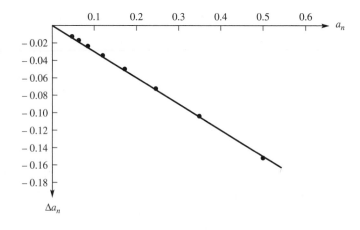

The Model   From Figure 1.11,

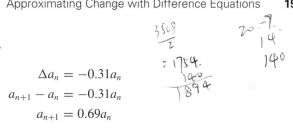

$$\Delta a_n = -0.31a_n$$
$$a_{n+1} - a_n = -0.31a_n$$
$$a_{n+1} = 0.69a_n$$

A difference equation model for the decay of digoxin in the bloodstream, given an initial dosage of 0.5 mg, is

$$a_{n+1} = a_n - 0.31a_n = 0.69a_n,$$
$$a_0 = 0.5$$

■ ■ ■

EXAMPLE 5   *Heating of a Cooled Object*

Now we examine a behavior that is taking place continuously. Suppose a cold can of soda is taken from a refrigerator and placed in a warm classroom and we measure the temperature periodically. The temperature of the soda is initially $40\,°F$ and the room temperature is $72\,°F$. Temperature is a measure of energy per unit volume. Because the volume of soda is small relative to the volume of the room, we would expect the room temperature to remain constant. Furthermore, we assume the entire can of soda is the same temperature, neglecting variation within the can. We might expect the change in temperature per time period to be greater when the difference in temperature between the soda and the room is large and the change in temperature per unit time to be less when the difference in temperature is small. Letting t_n represent the temperature of the soda after n time periods, and letting k be a positive constant of proportionality, we propose

$$\Delta t_n = t_{n+1} - t_n = k(72 - t_n)$$
$$t_0 = 40$$

Many refinements are possible for this model. Although we have assumed k is constant, it actually depends on the shape and conductivity properties of the container, the length of the time period between the measurements, and so on. Also, the temperature of the environment may not be constant in many instances, and it may be necessary to take into account that the temperature is not uniform throughout the can of soda. The temperature of an object may vary in one dimension (as in the case of a thin wire), in two dimensions (such as for a flat plate), or in three dimensions (as in the case of a space capsule reentering the earth's atmosphere). ■ ■ ■

We have presented only a glimpse of the power of difference equations to model change in the world around us. In the next section, we build numerical solutions to some of these models and observe the patterns they exhibit. Having observed certain patterns for various types of difference equations, we will then classify them by their mathematical structures. This will help to determine the long-term behavior of a dynamical system under study.

1.2   PROBLEMS

1. The following data were obtained for the growth of a sheep population introduced into a new environment on the island of Tasmania.[1]

Year	1814	1824	1834	1844	1854	1864
Population	125	275	830	1200	1750	1650

Plot the data. Is there a trend? Plot the change in population versus years elapsed after 1814. Formulate a discrete dynamical system that reasonably approximates the change you have observed.

2. The following data represent the U.S. population from 1790 to 2000. Find a dynamical system model that fits the data fairly well. Test your model by plotting the predictions of the model against the data.

Year	Population	Year	Population	Year	Population
1790	3,929,000	1870	38,558,000	1940	131,669,000
1800	5,308,000	1880	50,156,000	1950	150,697,000
1810	7,240,000	1890	62,948,000	1960	179,323,000
1820	9,638,000	1900	75,995,000	1970	203,212,000
1830	12,866,000	1910	91,972,000	1980	226,505,000
1840	17,069,000	1920	105,711,000	1990	248,710,000
1850	23,192,000	1930	122,755,000	2000	281,416,000
1860	31,443,000				

3. Sociologists recognize a phenomenon called *social diffusion*, which is the spreading of a piece of information, a technological innovation, or a cultural fad among a population. The members of the population can be divided into two classes: those who have the information and those who do not. In a fixed population whose size is known, it is reasonable to assume that the rate of diffusion is proportional to the number who have the information times the number yet to receive it. If a_n denotes the number of people who have the information in a population of N people after n days, formulate a dynamical system to approximate the change in the number of people in the population who have the information.

4. Consider the spreading of a highly communicable disease on an isolated island with population size N. A portion of the population travels abroad and returns to the island infected with the disease. Formulate a dynamical system to approximate the change in the number of people in the population who have the disease.

---

[1] Adapted from J. Davidson, "On the Growth of the Sheep Population in Tasmania," *Trans. R. Soc. S. Australia* 62(1938): 342–346.

5. Assume that we are considering the survival of whales and that if the number of whales falls below a minimum survival level m, the species will become extinct. Assume also that the population is limited by the carrying capacity M of the environment. That is, if the whale population is above M, it will experience a decline because the environment cannot sustain that large a population level. In the following model, a_n represents the whale population after n years. Discuss the model.

$$a_{n+1} - a_n = k(M - a_n)(a_n - m)$$

6. A certain drug is effective in treating a disease if the concentration remains above 100 mg/L. The initial concentration is 640 mg/L. It is known from laboratory experiments that the drug decays at the rate of 20% of the amount present each hour.

   a. Formulate a model representing the concentration at each hour.

   b. Build a table of values and determine when the concentration reaches 100 mg/L.

7. Use the model developed in Problem 6 to prescribe an initial dosage and a maintenance dosage that keeps the concentration above the effective level of 500 ppm but below a safe level of 1000 ppm. Experiment with different values until you have results you think are satisfactory.

8. A humanoid skull is discovered near the remains of an ancient campfire. Archaeologists are convinced the skull is the same age as the original campfire. It is determined from laboratory testing that only 1% of the original amount of carbon-14 remains in the burned wood taken from the campfire. It is known that carbon-14 decays at a rate proportional to the amount remaining and that carbon-14 decays 50% over 5700 years. Formulate a model for carbon-14 dating.

9. The data in the accompanying table show the speed n (in increments of 5 mph) of an automobile and the associated distance a_n in feet required to stop it once the brakes are applied. For instance, $n = 6$ (representing $6 \times 5 = 30$ mph) requires a stopping distance of $a_6 = 47$ ft.

   a. Calculate and plot the change Δa_n versus n. Does the graph reasonably approximate a linear relationship?

   b. Based on your conclusions in part (a), find a difference equation model for the stopping distance data. Test your model by plotting the errors in the predicted values against n. Discuss the appropriateness of the model.

n	1	2	3	4	5	6	7	8	9	10	11	12	13	14	15	16
a_n	3	6	11	21	32	47	65	87	112	140	171	204	241	282	325	376

10. Place a cold can of soda in a room. Measure the temperature of the room, and periodically measure the temperature of the soda. Formulate a model to predict the change in the temperature of the soda. Estimate any constants of proportionality from your data. What are some of the sources of error in your model?

11. Cipro is an antibiotic taken to combat many infections, including anthrax. Cipro is filtered from the blood by the kidneys. Each 24-hour period, the kidneys filter out about one third of the Cipro that was in the blood at the beginning of the 24-hour period.

a. Assume a patient was given only a single 500-mg dose. Use a difference equation to construct a table of values listing the concentration of Cipro in this patient's blood at the end of each day.

b. Now assume that the patient must take an additional 500 mg per day. Use a difference equation to construct a table of values listing the concentration of Cipro at the end of each day.

c. Compare and interpret these two tables.

1.2 PROJECT

1. Complete the UMAP module "The Diffusion of Innovation in Family Planning," by Kathryn N. Harmon, UMAP 303. This module gives an interesting application of finite difference equations to study the process through which public policies are diffused to understand how national governments might adopt family planning policies. (See enclosed CD for the UMAP module.)

1.2 Further Reading

Frauenthal, James C. *Introduction to Population Modeling*. Lexington, MA: COMAP, 1979.

Hutchinson, G. Evelyn. *An Introduction to Population Ecology*. New Haven, CT: Yale University Press, 1978.

Levins, R. "The Strategy of Model Building in Population Biology." *American Scientist* 54 (1966): 421–431.

1.3 Solutions to Dynamical Systems

In this section we build solutions to some dynamical systems, starting with an initial value and iterating a sufficient number of subsequent values to determine the patterns involved. In some cases, we see that the behavior predicted by dynamical systems is characterized by the mathematical structure of the system. In other cases, we see wild variations in the behavior caused by only small changes in the initial values of the dynamical system. We also examine dynamical systems for which small changes in the proportionality constants cause wildly different predictions.

The Method of Conjecture

The **method of conjecture** is a powerful mathematical technique to hypothesize the form of a solution to a dynamical system and then to accept or reject the hypothesis. It is based on exploration and observation from which we attempt to discern a pattern to the solution. Let's begin with an example.

EXAMPLE 1 *A Savings Certificate Revisited*

In the savings certificate example (Example 1, Section 1.1), a savings certificate initially worth $1000 accumulated interest paid each month at 1% of the balance. No deposits or withdrawals occurred in the account, determining the dynamical system

$$a_{n+1} = 1.01a_n \tag{1.5}$$
$$a_0 = 1000$$

Look for a Pattern   We might begin by graphing the balance in the account at the end of each month (after the interest is paid). That is, we graph a_n versus n. The graph is displayed in Figure 1.12.

■ **Figure 1.12**

Growth of the savings certificate balance
$a_{n+1} = 1.01a_n,\ a_0 = 1000$

n	a_n
0	1000.00
1	1010.00
2	1020.10
3	1030.30
4	1040.60
5	1051.01
6	1061.52
7	1072.14
8	1082.86
9	1093.69
10	1104.62
11	1115.67
12	1126.83
13	1138.09
14	1149.47
15	1160.97
16	1172.58
17	1184.30

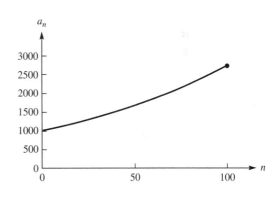

From the graph it appears that the sequence $\{a_0, a_1, a_2, a_3, \ldots\}$ grows without bound. Let's examine the sequence algebraically to gain more insight into the growth pattern.

$$a_1 = 1010.00 = 1.01(1000)$$
$$a_2 = 1020.10 = 1.01(1010) = 1.01(1.01(1000)) = 1.01^2(1000)$$
$$a_3 = 1030.30 = 1.01(1020.10) = 1.01(1.01^2(1000)) = 1.01^3(1000)$$
$$a_4 = 1040.60 = 1.01(1030.30) = 1.01(1.01^3(1000)) = 1.01^4(1000)$$

The pattern of these terms in the sequence suggests that the kth term a_k is the amount 1000 multiplied by $(1.01)^k$.

Conjecture   For $k = 1, 2, 3, \ldots$, the term a_k in the dynamical system (Equation (1.5)) is

$$a_k = (1.01)^k 1000 \tag{1.6}$$

Test the Conjecture   We test the conjecture by examining whether the formula for a_k satisfies the system of Equation (1.5) upon substitution.

$$a_{n+1} = 1.01a_n$$
$$(1.01)^{n+1}1000 = 1.01((1.01)^n 1000) = (1.01)^{n+1}1000$$

Since this last equation is true for every positive integer n, we accept the conjecture in Equation (1.6).

Conclusion   The solution for the term a_k in the dynamical system (1.5) is

$$a_k = (1.01)^k 1000$$

or

$$a_k = (1.01)^k a_0, \quad k = 1, 2, 3, \ldots$$

This solution allows us to compute the balance a_k in the account after k months. For example, after 120 months (or 10 years), the account is worth $a_{120} = (1.01)^{120}(1000) \approx \$3303.90$. After 30 years ($k = 360$), the account is worth \$35,949.64. This calculation is easier to obtain than iterating the dynamical system 360 times, but a formula such as Equation (1.6) can provide even more insight into the long-term behavior of a dynamical system, as we will soon see.   ■ ■ ■

Let's summarize the procedure followed in this example.

The Method of Conjecture

1. Observe a pattern.

2. Conjecture a form of the solution to the dynamical system.

3. Test the conjecture by substitution.

4. Accept or reject the conjecture depending on whether it does or does not satisfy the system after the substitution and algebraic manipulation. For the conjecture to be accepted, the substitution must result in an identity.

Linear Dynamical Systems $a_{n+1} = ra_n$, for r Constant

The dynamical system in Example 1 is of the form $a_{n+1} = ra_n$, where $r = 1.01$. Let's follow the method of conjecture for the more general case in which r is any positive or negative constant, assuming a given initial value a_0.

Look for a Pattern   Examining the terms in the sequence $a_{n+1} = ra_n$, we see that

$$a_1 = ra_0$$
$$a_2 = ra_1 = r(ra_0) = r^2 a_0$$
$$a_3 = ra_2 = r(r^2 a_0) = r^3 a_0$$
$$a_4 = ra_3 = r(r^3 a_0) = r^4 a_0$$

From these terms, we observe a pattern leading to the following conjecture:

Conjecture   For $k = 1, 2, 3, \ldots$, the term a_k in the dynamical system $a_{n+1} = ra_n$ is

$$a_k = r^k a_0 \tag{1.7}$$

Test the Conjecture   We substitute the formula from Equation (1.7) into the dynamical system:

$$a_{n+1} = ra_n$$
$$r^{n+1} a_0 = r(r^n a_0) = r^{n+1} a_0$$

The result is an identity, and we accept the conjecture. Let's summarize our result.

Theorem 1

The solution of the linear dynamical system $a_{n+1} = ra_n$ for r any nonzero constant is

$$a_k = r^k a_0$$

where a_0 is a given initial value.

EXAMPLE 2   *Sewage Treatment*

A sewage treatment plant processes raw sewage to produce usable fertilizer and clean water by removing all other contaminants. The process is such that each hour 12% of remaining contaminants in a processing tank are removed. What percentage of the sewage would remain after 1 day? How long would it take to lower the amount of sewage by half? How long until the level of sewage is down to 10% of the original level?

Solution   Let the initial amount of sewage contaminants be a_0 and let a_n denote the amount after n hours. We then build the model

$$a_{n+1} = a_n - 0.12a_n = 0.88a_n$$

which is a linear dynamical system. From Theorem 1, the solution is

$$a_k = (0.88)^k a_0$$

After 1 day, $k = 24$ hours and the level of sewage remaining is

$$a_{24} = (0.88)^{24} a_0 = 0.0465a_0.$$

That is, the level of contaminants in the sewage has been reduced by more than 95% at the end of the first day.

Half the original contaminants remain when $a_k = 0.5a_0$. Thus,

$$0.5a_0 = a_k = (0.88)^k a_0$$

Solving for k, we find

$$0.5a_0 = (0.88)^k a_0$$
$$(0.88)^k = 0.5,$$
$$k = \frac{\log 0.5}{\log 0.88} = 5.42$$

It takes about 5.42 hours to lower the contaminants to half their original amount.
To reduce the level of contaminants by 90%, we require

$$(0.88)^k a_0 = 0.1a_0$$

so

$$k = \frac{\log 0.1}{\log 0.88} = 18.01$$

It takes 18 hours before the contaminants are reduced to 10% of their original level. ■ ■ ■

Long-Term Behavior of $a_{n+1} = ra_n$, for r Constant

For the linear dynamical system $a_{n+1} = ra_n$, let's consider some significant values of r. If $r = 0$, then all the values of the sequence (except possibly a_0) are zero, so there is no need for further analysis. If $r = 1$, then the sequence becomes $a_{n+1} = a_n$. This is an interesting case because no matter where the sequence starts, it stays there forever (as illustrated in Figure 1.13, in which the sequence is enumerated and graphed for $a_0 = 50$ and also graphed for several other starting values). Values for which a dynamical system remains constant at those values, once reached, are called *equilibrium values* of the system. We define that term more precisely later, but for now note that in Figure 1.13, any starting value would be an equilibrium value.

If $r > 1$, as in Example 1, then the sequence $a_k = r^k a_0$, which solves the linear dynamical system, grows without bound. This growth was illustrated in Figure 1.12.

What happens if r is negative? If we replace 1.01 with -1.01 in Example 1, we obtain the graph in Figure 1.14. Note the oscillation between positive and negative values. Because the negative sign causes the next term in the sequence to be of opposite sign from the previous term, we conclude that, in general, negative values of r cause oscillations in the linear form $a_{n+1} = ra_n$.

■ **Figure 1.13**

Every solution of $a_{n+1} = a_n$ is a constant solution.

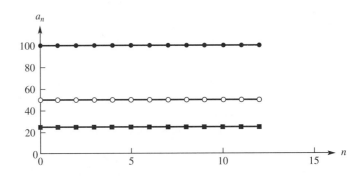

n	a_n
0	1000
1	−1010.00
2	1020.10
3	−1030.30
4	1040.60
5	−1051.01
6	1061.52
7	−1072.14
8	1082.86
9	−1093.69
10	1104.62
11	−1115.67
12	1126.83
13	−1138.09
14	1149.47
15	−1160.97
16	1172.58

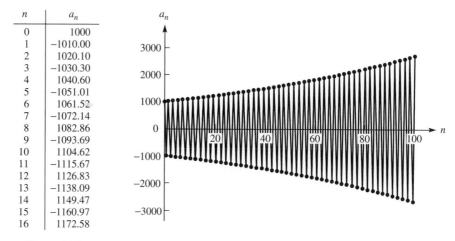

■ **Figure 1.14**

A negative value of r causes oscillation.

What happens if $|r| < 1$? We know what happens if $r = 0$, if $r > 1$, if $r < -1$, and if r is negative in general. If $0 < r < 1$ so that r is a positive fraction less than 1, then r^k approaches 0 as k becomes increasingly large. This means that the sequence $a_k = r^k a_0$ solving the linear system $a_{n+1} = ra_n$ can be made as small as we please once k is large enough. We observed this behavior in Example 2. Another illustration is provided in the digoxin example.

Suppose digoxin decays in the bloodstream such that each day, 69% of the concentration of digoxin remains from the previous day (Example 4 in Section 1.2). If we start with 0.5 mg and a_n represents the amount after n days, we can represent the behavior with the following model (a numerical solution is shown in Figure 1.15).

$$a_{n+1} = 0.69a_n, \quad n = 0, 1, 2, 3, \ldots$$
$$a_0 = 0.5$$

n	a_n
0	0.5
1	0.3450
2	0.23805
3	0.164255
4	0.1133356
5	0.078210157
6	0.053959082
7	0.037231766
8	0.0256899187
9	0.0177726044
10	0.0122309703
11	0.0084393695
12	0.0058231649
13	0.0040179838
14	0.0027724088
15	0.0019296209

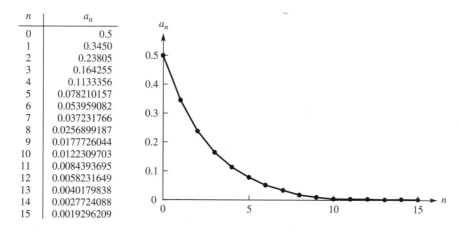

■ **Figure 1.15**

A positive fractional value of r causes decay.

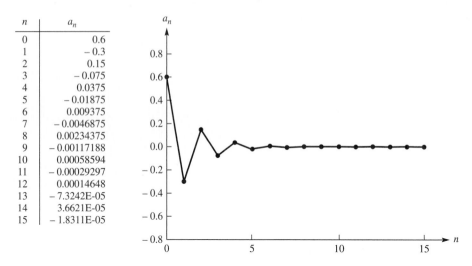

n	a_n
0	0.6
1	-0.3
2	0.15
3	-0.075
4	0.0375
5	-0.01875
6	0.009375
7	-0.0046875
8	0.00234375
9	-0.00117188
10	0.00058594
11	-0.00029297
12	0.00014648
13	-7.3242E-05
14	3.6621E-05
15	-1.8311E-05

■ **Figure 1.16**
A negative fractional value of r causes decay with oscillation about 0.

The behavior for $-1 < r < 0$, where r is a negative fraction, also captures decay to the value 0. However, in this case the sequence $a_k = r^k a_0$ alternates in sign as it approaches 0. This is illustrated in Figure 1.16 for the linear dynamical system

$$a_{n+1} = -0.5a_n,$$
$$a_0 = 0.6$$

Let's summarize our observations (Figure 1.17):

Long-term behavior for $a_{n+1} = ra_n$

$r = 0$	Constant solution and equilibrium value at 0
$r = 1$	All initial values are constant solutions
$r < 0$	Oscillation
$|r| < 1$	Decay to limiting value of 0
$|r| > 1$	Growth without bound

Dynamical Systems of the Form $a_{n+1} = ra_n + b$, Where r and b Are Constants

Now let's add a constant b to the dynamical system we previously studied. Again, we want to classify the nature of the long-term behavior for all possible cases. We begin with a definition.

■ **Figure 1.17**

Long-term behaviors for
$a_{n+1} = ra_n$, $r \neq 0$, $|r| > 1$
and $|r| < 1$

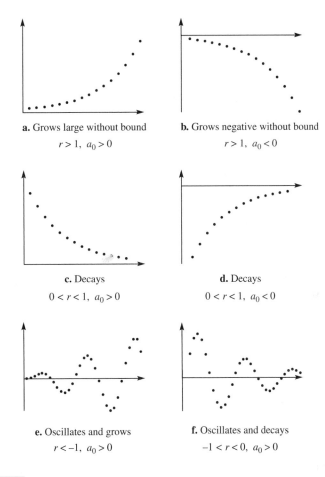

a. Grows large without bound
$r > 1$, $a_0 > 0$

b. Grows negative without bound
$r > 1$, $a_0 < 0$

c. Decays
$0 < r < 1$, $a_0 > 0$

d. Decays
$0 < r < 1$, $a_0 < 0$

e. Oscillates and grows
$r < -1$, $a_0 > 0$

f. Oscillates and decays
$-1 < r < 0$, $a_0 > 0$

Definition

A number a is called an **equilibrium value** or **fixed point** of a dynamical system
$a_{n+1} = f(a_n)$ if $a_k = a$ for all $k = 1, 2, 3, \ldots$ when $a_0 = a$. That is, $a_k = a$ is a
constant solution to the dynamical system.

A consequence of the definition is that a is an equilibrium value for $a_{n+1} = f(a_n)$ if
and only if $a = f(a)$ when $a_0 = a$. This result is shown by simple substitution. Equilibrium
values can be helpful in understanding the long-term behavior of a dynamical system such
as $a_{n+1} = ra_n + b$. Let's consider three examples to gain insight into its behavior.

EXAMPLE 3 *Prescription for Digoxin*

Consider again the digoxin problem. Recall that digoxin is used in the treatment of heart
patients. The objective of the problem is to consider the decay of digoxin in the bloodstream
to prescribe a dosage that keeps the concentration between acceptable levels (so that it is

both safe and effective). Suppose we prescribe a daily drug dosage of 0.1 mg and know that half the digoxin remains in the system at the end of each dosage period. This results in the dynamical system

$$a_{n+1} = 0.5a_n + 0.1$$

Now consider three starting values, or initial doses:

$$A: a_0 = 0.1$$
$$B: a_0 = 0.2$$
$$C: a_0 = 0.3$$

In Figure 1.18 we compute the numerical solutions for each case.

n	A a_n	B a_n	C a_n
0	0.1	0.2	0.3
1	0.15	0.2	0.25
2	0.175	0.2	0.225
3	0.1875	0.2	0.2125
4	0.19375	0.2	0.20625
5	0.196875	0.2	0.203125
6	0.1984375	0.2	0.2015625
7	0.19921875	0.2	0.20078125
8	0.19960938	0.2	0.20039063
9	0.19980469	0.2	0.20019531
10	0.19990234	0.2	0.20009766
11	0.19995117	0.2	0.20004883
12	0.19997559	0.2	0.20002441
13	0.19998779	0.2	0.20001221
14	0.1999939	0.2	0.2000061
15	0.19999695	0.2	0.20000305

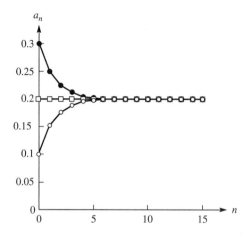

■ **Figure 1.18**

Three initial digoxin doses

Note that the value 0.2 is an equilibrium value, because once that value is reached, the system remains at 0.2 forever. Furthermore, if we start below the equilibrium (as in Case A) or above the equilibrium (as in Case C), apparently we approach the equilibrium value as a *limit*. In the problems, you are asked to compute solutions for starting values even closer to 0.2, lending evidence that 0.2 is a *stable* equilibrium value.

When digoxin is prescribed, the concentration level must stay above an effective level for a period of time without exceeding a safe level. In the problems, you are asked to find initial and subsequent doses that are both *safe* and *effective*.   ▪ ▪ ▪

EXAMPLE 4   *An Investment Annuity*

Return now to the savings account problem and consider an *annuity*. Annuities are often planned for retirement purposes. They are basically savings accounts that pay interest on the amount present and allow the investor to withdraw a fixed amount each month until the account is depleted. An interesting issue (posed in the problems) is to determine the amount one must save monthly to build an annuity allowing for withdrawals, beginning at a certain age with a specified amount for a desired number of years, before the account's depletion. For now, consider 1% as the monthly interest rate and a monthly withdrawal of $1000. This gives the dynamical system

$$a_{n+1} = 1.01a_n - 1000$$

Now suppose we made the following initial investments:

$$\text{A: } a_0 = 90{,}000$$
$$\text{B: } a_0 = 100{,}000$$
$$\text{C: } a_0 = 110{,}000$$

The numerical solutions for each case are graphed in Figure 1.19.

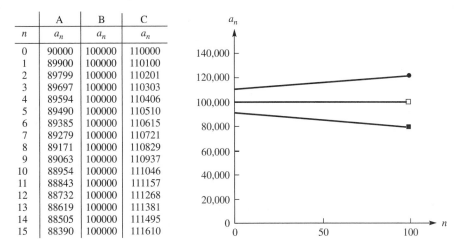

	A	B	C
n	a_n	a_n	a_n
0	90000	100000	110000
1	89900	100000	110100
2	89799	100000	110201
3	89697	100000	110303
4	89594	100000	110406
5	89490	100000	110510
6	89385	100000	110615
7	89279	100000	110721
8	89171	100000	110829
9	89063	100000	110937
10	88954	100000	111046
11	88843	100000	111157
12	88732	100000	111268
13	88619	100000	111381
14	88505	100000	111495
15	88390	100000	111610

■ **Figure 1.19**

An annuity with three initial investments

Notice that the value 100,000 is an equilibrium value; once it is reached, the system remains there for all subsequent values. But if we start above that equilibrium, there is growth without bound. (Try plotting with $a_0 = \$100{,}000.01$.) On the other hand, if we start below $100,000, the savings are used up at an increasing rate. (Try $99,999.99.) Note how drastically different the long-term behaviors are even though the starting values differ by only $0.02. In this situation, we say that the equilibrium value 100,000 is *unstable*: If we start close to the value (even within a penny) we do not remain close. Look at the dramatic differences displayed by Figures 1.18 and 1.19. Both systems show equilibrium values, but the first is stable and the second is unstable. ▪ ▪ ▪

In Examples 3 and 4 we considered cases in which $|r| < 1$ and $|r| > 1$. Let's see what happens if $r = 1$.

EXAMPLE 5 *A Checking Account*

Most students cannot keep enough cash in their checking accounts to earn any interest. Suppose you have an account that pays no interest and that each month you pay only your dorm rent of $300, giving the dynamical system

$$a_{n+1} = a_n - 300$$

The result is probably obvious from the application, but when compared with Figures 1.18 and 1.19 the graph is revealing. In Figure 1.20 we plot the numerical solution for a starting value of 3000. Do you see how drastically the graph of the solution differs from those of the previous examples? Can you find an equilibrium, as in Examples 3 and 4? ∎ ∎ ∎

■ Figure 1.20

A checking account to pay dorm costs

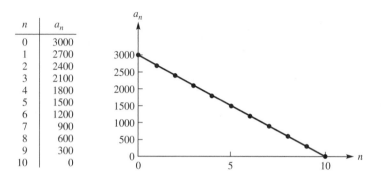

n	a_n
0	3000
1	2700
2	2400
3	2100
4	1800
5	1500
6	1200
7	900
8	600
9	300
10	0

Now let's collect our observations thus far, classifying the three examples and their long-term behavior according to the value of the constant r.

Finding and Classifying Equilibrium Values

Determining if equilibrium values exist, and classifying them as stable or unstable, assists us immensely in analyzing the long-term behavior of a dynamical system. Consider again Examples 3 and 4. In Example 3, how did we know that a starting value of 0.2 would result in a constant solution or equilibrium value? Similarly, how do we know the answer to the same question for an investment of $100,000 in Example 4? For a dynamical system of the form

$$a_{n+1} = ra_n + b \tag{1.8}$$

denote by a the equilibrium value, if one exists. From the definition of an equilibrium value, if we start at a we must remain there for all n; that is, $a_{n+1} = a_n = a$ for all n. Substituting a for a_{n+1} and a_n in Equation (1.8) yields

$$a = ra + b$$

and solving for a we find

$$a = \frac{b}{1-r}, \quad \text{if } r \neq 1$$

If $r = 1$ and $b = 0$, then every initial value results in a constant solution (as shown in Figure 1.13). Hence, every initial value is an equilibrium value. The following theorem summarizes our observations.

Theorem 2

The equilibrium value for the dynamical system

$$a_{n+1} = ra_n + b, \quad r \neq 1$$

is

$$a = \frac{b}{1-r}$$

If $r = 1$ and $b = 0$, every number is an equilibrium value. If $r = 1$ and $b \neq 0$, no equilibrium value exits.

Applying Theorem 2 to Example 3, we find the equilibrium value is

$$a = \frac{0.1}{1 - 0.5} = 0.2$$

For Example 4, we compute the equilibrium as

$$a = \frac{-1000}{1 - 1.01} = 100{,}000$$

In Example 5, $r = 1$ and no equilibrium exists because $b = -300$. Also, the graph of the solution is a line. These examples also provide the following insights into the nature of the long-term behavior we observed according to the value of the constant r.

Dynamical system $a_{n+1} = ra_n + b, \ b \neq 0$

Value of r	Long-term behavior observed
$|r| < 1$	Stable equilibrium
$|r| > 1$	Unstable equilibrium
$r = 1$	Graph is a line with no equilibrium

In Example 3, we saw from Figure 1.18 that the term a_k approaches the equilibrium value 0.2 as k becomes increasingly large. Because $r^k = (0.5)^k$ tends to 0 for large k, it seems reasonable to conjecture that the solution is of the form $a_k = (0.5)^k c + 0.2$ for some constant c that depends on the initial condition.

Let's test the conjecture that the form

$$a_k = r^k c + \frac{b}{1 - r}$$

solves the system $a_{n+1} = ra_n + b$, where $r \neq 1$.

Substituting into the system, we have

$$a_{n+1} = ra_n + b$$

$$r^{n+1}c + \frac{b}{1 - r} = r\left(r^n c + \frac{b}{1 - r}\right) + b$$

$$r^{n+1}c + \frac{b}{1 - r} = r^{n+1}c + \frac{rb}{1 - r} + b$$

$$\frac{b}{1 - r} = \frac{rb}{1 - r} + b$$

$$b = rb + b(1 - r)$$

Since this last equation is an identity, we accept the conjecture. Let's summarize our result.

Theorem 3

The solution of the dynamical system $a_{n+1} = ra_n + b$, $r \neq 1$ is

$$a_k = r^k c + \frac{b}{1 - r}$$

for some constant c (which depends on the initial condition).

EXAMPLE 6   An Investment Annuity Revisited

For the annuity modeled in Example 4, how much of an initial investment do we need to deplete the annuity in 20 years (or 240 months)?

Solution   The equilibrium value of the system $a_{n+1} = 1.01a_n - 1000$ is 100,000 and we want $a_{240} = 0$. From Theorem 3 we have

$$a_{240} = 0 = (1.01)^{240}c + 100{,}000$$

and solving this equation gives $c = -100{,}000/(1.01)^{240} = -9180.58$ (to the nearest cent). To find the initial investment a_0, we again use Theorem 3:

$$a_0 = (1.01)^0 c + 100{,}000 = -9180.58 + 100{,}000 = 90{,}819.42$$

Thus, an initial investment of $90,819.42 allows us to withdraw $1000 per month from the account for 20 years (a total withdrawal of $240,000). At the end of 20 years the account is depleted.

Nonlinear Systems

An important advantage of discrete systems is that numerical solutions can be constructed for any dynamical system when an initial value is given. We have seen that long-term behavior can be sensitive to the starting value and to the values of the parameter r. Recall the model for the yeast biomass from Section 1.2:

$$p_{n+1} = p_n + 0.00082(665 - p_n)p_n$$

After some algebraic manipulation, this dynamical system can be rewritten in the simpler form

$$a_{n+1} = r(1 - a_n)a_n \tag{1.9}$$

where $a_n = 0.0005306p_n$, and $r = 1.546$. The behavior of the sequence determined by Equation (1.9) is very sensitive to the value of the parameter r. In Figure 1.21 (see page 32) we plot a numerical solution for various values of r, beginning with $a_0 = 0.1$.

Note how remarkably different the behavior is in each of the six cases. In Figure 1.21a we see that for $r = 1.546$ the behavior approaches a limit of about 0.35 directly from the starting value 0.10. In Figure 1.21b, we see that for $r = 2.75$ the behavior approaches a limit of about 0.65 but oscillates across it during the approach. In Figure 1.21c, where $r = 3.25$, we see that the behavior now approaches two values (0.5 and 0.8), again in an oscillatory fashion, although this is not apparent from the scatterplot. We call such behavior periodic with a two-cycle. In Figure 1.21d for $r = 3.525$ we see a four-cycle, and in Figure 1.21e for $r = 3.555$ (just slightly larger than before) we observe an eight-cycle. Finally, in Figure 1.21f for $r = 3.75$ sufficiently large, there is no pattern whatsoever, and it is impossible to predict the long-term behavior of the model. Notice how radically the behaviors change with very small changes in the parameter r. The behavior exhibited in Figure 1.21f is called *chaotic behavior*. Chaotic systems demonstrate sensitivity to the constant parameters of the system. A given chaotic system can be quite sensitive to initial conditions.

1.3 PROBLEMS

1. Find the solution to the difference equations in the following problems:

   a. $a_{n+1} = 3a_n, \quad a_0 = 1$

   b. $a_{n+1} = 5a_n, \quad a_0 = 10$

   c. $a_{n+1} = 3a_n/4, \quad a_0 = 64$

   d. $a_{n+1} = 2a_n - 1, \quad a_0 = 3$

   e. $a_{n+1} = -a_n + 2, \quad a_0 = -1$

   f. $a_{n+1} = 0.1a_n + 3.2, \quad a_0 = 1.3$

2. For the following problems, find an equilibrium value if one exists. Classify the equilibrium value as stable or unstable.

   a. $a_{n+1} = 1.1a_n$

   b. $a_{n+1} = 0.9a_n$

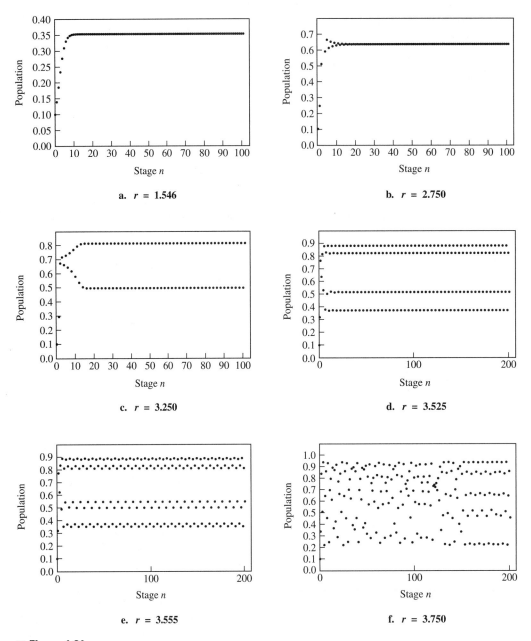

Figure 1.21

Long-term behavior exhibited by numerical solutions to the equation $a_{n+1} = r(1 - a_n)a_n$ for six values of the parameter r

c. $a_{n+1} = -0.9a_n$

d. $a_{n+1} = a_n$

e. $a_{n+1} = -1.2a_n + 50$

f. $a_{n+1} = 1.2a_n - 50$

g. $a_{n+1} = 0.8a_n + 100$

h. $a_{n+1} = 0.8a_n - 100$

i. $a_{n+1} = -0.8a_n + 100$

j. $a_{n+1} = a_n - 100$

k. $a_{n+1} = a_n + 100$

3. Build a numerical solution for the following *initial value problems*. Plot your data to observe patterns in the solution. Is there an equilibrium solution? Is it stable or unstable?

   a. $a_{n+1} = -1.2a + 50,$ $a_0 = 1000$

   b. $a_{n+1} = 0.8a_n - 100,$ $a_0 = 500$

   c. $a_{n+1} = 0.8a_n - 100,$ $a_0 = -500$

   d. $a_{n+1} = -0.8a_n + 100,$ $a_0 = 1000$

   e. $a_{n+1} = a_n - 100,$ $a_0 = 1000$

4. For the following problems, find the solution to the difference equation and the equilibrium value if one exists. Discuss the long-term behavior of the solution for various initial values. Classify the equilibrium values as stable or unstable.

   a. $a_{n+1} = -a_n + 2,$ $a_0 = 1$

   b. $a_{n+1} = a_n + 2,$ $a_0 = -1$

   c. $a_{n+1} = a_n + 3.2,$ $a_0 = 1.3$

   d. $a_{n+1} = -3a_n + 4,$ $a_0 = 5$

5. You currently have $5000 in a savings account that pays 0.5% interest each month. You add $200 each month. Build a numerical solution to determine when the account reaches $20,000.

6. You owe $500 on a credit card that charges 1.5% interest each month. You can pay $50 each month with no new charges. What is the equilibrium value? What does the equilibrium value mean in terms of the credit card? Build a numerical solution. When will the account be paid off? How much is the last payment?

7. Your parents are considering a 30-year, $100,000 mortgage that charges 0.5% interest each month. Formulate a model in terms of a monthly payment p that allows the mortgage (loan) to be paid off after 360 payments. *Hint:* If a_n represents the amount owed after n months, what are a_0 and a_{360}? Experiment by building numerical solutions to find a value of p that works.

8. Your parents are considering a 30-year mortgage that charges 0.5% interest each month. Formulate a model in terms of a monthly payment p that allows the mortgage (loan) to be paid off after 360 payments. Your parents can afford a monthly payment of $1500. Experiment to determine the maximum amount of money they can borrow. *Hint:* If a_n represents the amount owed after n months, what are a_0 and a_{360}?

9. Your grandparents have an annuity. The value of the annuity increases each month as 1% interest on the previous month's balance is deposited. Your grandparents withdraw $1000 each month for living expenses. Currently, they have $50,000 in the annuity. Model the annuity with a dynamical system. Find the equilibrium value. What does the equilibrium value represent for this problem? Build a numerical solution to determine when the annuity is depleted.

10. *Continuation of Example 4, Section 1.2*: Find the equilibrium value of the digoxin model. What is the significance of the equilibrium value?

11. *Continuation of Problem 6, Section 1.2*: Experiment with different initial and maintenance doses. Find a combination that is convenient, considering the time between doses and the amount that will be taken as measures of convenience.

12. *Continuation of Problem 8, Section 1.2*: Determine the age of the humanoid skull found near the remains of an ancient campfire.

13. Consider the spreading of a rumor through a company of 1000 employees, all working in the same building. We assume that the spreading of a rumor is similar to the spreading of a contagious disease (see Example 3, Section 1.2) in that the number of people hearing the rumor each day is proportional to the product of the number who have heard the rumor previously and the number who have not heard the rumor. This is given by

$$r_{n+1} = r_n + 1000kr_n(1000 - r_n)$$

where k is a parameter that depends on how fast the rumor spreads and n is the number of days. Assume $k = 0.001$ and further assume that four people initially have heard the rumor. How soon will all 1000 employees have heard the rumor?

14. Consider modeling the contagious disease Ebola. (You may be interested in researching on the Internet just how deadly this virus is.) An animal research laboratory is located in Restin, Virginia, a suburb of Washington, D.C. with a population of 856,900 people. A monkey with the Ebola virus has escaped from its captivity in the laboratory and infected one employee. This employee reports to a Restin hospital later with Ebola symptoms. The Infectious Disease Center (IDC) in Atlanta gets a call and begins to model the spread of the disease. Build a model for the IDC with the following growth rates to determine the number of people who will be infected in Restin after 2 weeks.

   a. $k = 0.25$

   b. $k = 0.025$

   c. $k = 0.0025$

   d. $k = 0.00025$

15. Again consider the spreading of a rumor (see Problem 13 in this section), but now assume a company with 2000 employees. The rumor concerns the number of mandatory terminations that this company must absorb. Based on the model presented in Problem 13, build a model for the company with the following rumor growth rates to determine the number who have heard the rumor after 1 week.

   a. $k = 0.25$

   b. $k = 0.025$

c. $k = 0.0025$

d. $k = 0.00025$

e. List some ways of controlling the growth rate.

1.3 PROJECTS

1. You plan to invest part of your paycheck to finance your children's education. You want to have enough in the account to draw $1000 a month every month for 8 years beginning 20 years from now. The account pays 0.5% interest each month.

   a. How much money will you need 20 years from now to accomplish the financial objective? Assume you stop investing when your first child begins college—a safe assumption.

   b. How much must you deposit each month during the next 20 years?

2. Assume we are considering the survival of whales and that if the number of whales falls below a minimum survival level m, the species will become extinct. Assume also that the population is limited by the carrying capacity M of the environment. That is, if the whale population is above M, then it will experience a decline because the environment cannot sustain that large a population level. In the following model, a_n represents the whale population after n years. Build a numerical solution for $M = 5000$, $m = 100$, $k = 0.0001$, and $a_0 = 4000$.

$$a_{n+1} - a_n = k(M - a_n)(a_n - m)$$

   Now experiment with different values for M, m, and k. Try several starting values for a_0. What does your model predict?

3. *A Killer Virus*—You have volunteered for the Peace Corps and have been sent to Rwanda to help in humanitarian aid. You meet with the World Health Organization (WHO) and find out about a new killer virus, Hanta. If just one copy of the virus enters the human body, it can start reproducing very rapidly. In fact, the virus doubles its numbers in 1 hour. The human immune system can be quite effective, but this virus hides in normal cells. As a result, the human immune response does not begin until the virus has 1 million copies floating within the body. One of the first actions of the immune system is to raise the body temperature, which in turn lowers the virus replication rate to 150% per hour. The fever and then flu-like symptoms are usually the first indication of the illness. Some people with the virus assume that they have only a flu or a bad cold. This assumption leads to deadly consequences because the immune response alone is not enough to combat this deadly virus. At maximum reaction, the immune systems alone can kill only 200,000 copies of the virus per hour. Model this initial phase of the illness (before antibiotics) for a volunteer infected with 1 copy of the virus.

   a. How long will it take for the immune response to begin?

   b. If the number of copies of the virus reaches 1 billion, the virus cannot be stopped. Determine when this happens.

c. When the number of copies of the virus reaches 1 trillion, the person will die. Determine when this occurs.

To combat this virus fully, the infected person needs to receive an injection and hourly doses of an antibiotic. The antibiotic does not affect the replication rate of the virus (the fever keeps it at 150%), but the immune system and the antibiotics together kill 500,000,000 copies of the virus per hour.

d. Model the second phase of the virus (after the antibiotics are taken). Determine the latest time at which you can start administering the antibiotic in order to save the person. Analyze your model and discuss its strengths and weaknesses. (See enclosed CD for the UMAP module.)

4. *Mercury in Fish*—Public officials are worried about the elevated levels of toxic mercury pollution in the reservoirs that provide the drinking water to your city. They have asked for your assistance in analyzing the severity of the problem. Scientists have known about the adverse affects of mercury on the health of humans for more than a century. The term *mad as a hatter* stems from the nineteenth-century use of mercuric nitrate in the making of felt hats. Human activities are responsible for most mercury emitted into the environment. For example, mercury, a by-product of coal, comes from the smokestack emissions of old, coal-fired power plants in the Midwest and South and is disseminated by acid rain. Its particles rise on the smokestack plumes and hitch a ride on prevailing winds, which often blow northeast. After colliding with mountains, the particles drop to earth. Once in the ecosystem, microorganisms in the soil and reservoir sediment break down the mercury and produce a very toxic chemical known as methyl mercury.

Mercury undergoes a process known as bioaccumulation. This occurs when organisms take in contaminants more rapidly than their bodies can eliminate them. Therefore, the amount of mercury in their bodies accumulates over time. Humans can eliminate mercury from their system at a rate proportional to the amount remaining. Methyl mercury decays 50% every 65 to 75 days (known as the half-life of mercury) if no further mercury is ingested during that time.

Officials in your city have collected and tested 2425 samples of largemouth bass from the reservoirs and provided the following data. All fish were contaminated. The mean value of the methyl mercury in the fish samples was 0.43 μg (microgram) per gram. The average weight of the fish was 0.817 kg.

a. Assume the average adult person (70 kg) eats one fish (0.817 kg) per day. Construct a difference equation to model the accumulation of methyl mercury in the average adult. Assume the half-life is approximately 70 days. Use your model to determine the maximum amount of methyl mercury that the average adult human will accumulate in her or his lifetime.

b. You find out that there is a lethal limit to the amount of mercury in the body; it is 50 mg/kg. What is the maximum number of fish per month that can be eaten without exceeding this lethal limit?

5. Complete the UMAP module "Difference Equations with Applications," by Donald R. Sherbert, UMAP 322. This module presents a good introduction to solving first- and second-order linear difference equations, including the method of undetermined coefficients for nonhomogeneous equations. Applications to problems in population and economic modeling are included.

1.4 Systems of Difference Equations

In this section we consider systems of difference equations. For selected starting values, we build numerical solutions to get an indication of the long-term behavior of the system. As we saw in Section 1.3, equilibrium values are values of the dependent variable(s) for which there is no change in the system once the equilibrium values are obtained. For the systems considered in this section, we find the equilibrium values and then explore starting values in their vicinity. If we start close to an equilibrium value, we want to know whether the system will

a. Remain close
b. Approach the equilibrium value
c. Not remain close

What happens near equilibrium values provides insight concerning the long-term behavior of the system. Does the system demonstrate periodic behavior? Are there oscillations? Does the long-term behavior described by the numerical solution appear to be sensitive to

a. The initial conditions?
b. Small changes in the constants of proportionality used to model the behavior under study?

Although our approach in this section is numerical, we revisit several scenarios described in this section in Chapter 12 when we treat systems of differential equations. Our goal now is to model several behaviors with difference equations and explore numerically the behavior predicted by the model.

EXAMPLE 1 *A Car Rental Company*

A car rental company has distributorships in Orlando and Tampa. The company specializes in catering to travel agents who want to arrange tourist activities in both cities. Consequently, a traveler will rent a car in one city and drop the car off in the second city. Travelers may begin their itinerary in either city. The company wants to determine how much to charge for this drop-off convenience. Because cars are dropped off in both cities, will a sufficient number of cars end up in each city to satisfy the demand for cars in that city? If not, how many cars must the company transport from Orlando to Tampa or from Tampa to Orlando? The answers to these questions will help the company figure out its expected costs.

The historical records reveal that 60% of the cars rented in Orlando are returned to Orlando, whereas 40% end up in Tampa. Of the cars rented from the Tampa office, 70% are returned to Tampa, whereas 30% end up in Orlando. Figure 1.22 is helpful in summarizing the situation.

■ **Figure 1.22**

Car rental offices in
Orlando and Tampa

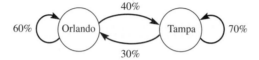

Dynamical Systems Model   Let's develop a model of the system. Let n represent the number of business days. Now define

$$O_n = \text{the number of cars in Orlando at the end of day } n$$
$$T_n = \text{the number of cars in Tampa at the end of day } n$$

Thus the historical records reveal the system

$$O_{n+1} = 0.6O_n + 0.3T_n$$
$$T_{n+1} = 0.4O_n + 0.7T_n$$

Equilibrium Values   The equilibrium values for the system are those values of O_n and T_n for which no change in the system takes place. Let's call the equilibrium values, if they exist, O and T, respectively. Then $O = O_{n+1} = O_n$ and $T = T_{n+1} = T_n$ simultaneously. Substitution in our model yields the following requirements for the equilibrium values:

$$O = 0.6O + 0.3T$$
$$T = 0.4O + 0.7T$$

This system is satisfied whenever $O = \frac{3}{4}T$. For example, if the company owns 7000 cars and starts with 3000 in Orlando and 4000 in Tampa, then our model predicts that

$$O_1 = 0.6(3000) + 0.3(4000) = 3000$$
$$T_1 = 0.4(3000) + 0.7(4000) = 4000$$

Thus this system remains at $(O, T) = (3000, 4000)$ if we start there.

Next let's explore what happens if we start at values other than the equilibrium values. Let's iterate the *system* for the following four initial conditions:

Four starting values for the car
rental problem

	Orlando	Tampa
Case 1	7000	0
Case 2	5000	2000
Case 3	2000	5000
Case 4	0	7000

A numerical solution, or table of values, for each of the starting values is graphed in Figure 1.23.

■ **Figure 1.23**

The rental car problem

n	Orlando	Tampa
0	7000	0
1	4200	2800
2	3360	3640
3	3108	3892
4	3032.4	3967.6
5	3009.72	3990.28
6	3002.916	3997.084
7	3000.875	3999.125

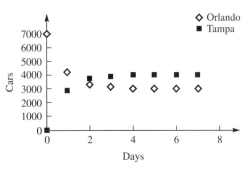

a. Case 1

n	Orlando	Tampa
0	5000	2000
1	3600	3400
2	3180	3820
3	3054	3946
4	3016.2	3983.8
5	3004.86	3995.14
6	3001.458	3998.542
7	3000.437	3999.563

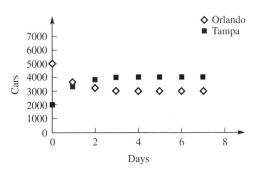

b. Case 2

n	Orlando	Tampa
0	2000	5000
1	2700	4300
2	2910	4090
3	2973	4027
4	2991.9	4008.1
5	2997.57	4002.43
6	2999.271	4000.729
7	2999.781	4000.219

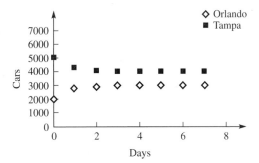

c. Case 3

n	Orlando	Tampa
0	0	7000
1	2100	4900
2	2730	4270
3	2919	4081
4	2975.7	4024.3
5	2992.71	4007.29
6	2997.813	4002.187
7	2999.344	4000.656

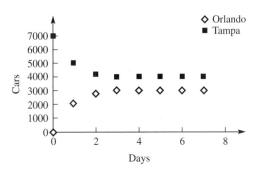

d. Case 4

Sensitivity to Initial Conditions and Long-Term Behavior   In each of the four cases, within a week the system is very close to the equilibrium value (3000, 4000) even in the absence of any cars at one of the two sites. Our results suggest that the equilibrium value is stable and insensitive to the starting values. Based on these explorations, we are inclined to predict that the system approaches the equilibrium where $\frac{3}{7}$ of the fleet ends up in Orlando and the remaining $\frac{4}{7}$ in Tampa. This information is helpful to the company. Knowing the demand patterns in each city, the company can estimate how many cars it needs to ship. In the problem set, we ask you to explore the system to determine whether it is sensitive to the coefficients in the equations for O_{n+1} and T_{n+1}.

EXAMPLE 2   *The Battle of Trafalgar*

In the battle of Trafalgar in 1805, a combined French and Spanish naval force under Napoleon fought a British naval force under Admiral Nelson. Initially, the French–Spanish force had 33 ships, and the British had 27 ships. During an encounter, each side suffers a loss equal to 10% of the number of ships of the opposing force. Fractional values are meaningful and indicate that one or more ships are not at full capacity.

Dynamical Systems Model   Let n denote the encounter stage during the course of the battle and define

$$B_n = \text{the number of British ships at stage } n$$
$$F_n = \text{the number of French–Spanish ships at stage } n$$

Then, after an encounter at stage n, the number of ships remaining on each side is

$$B_{n+1} = B_n - 0.1F_n$$
$$F_{n+1} = F_n - 0.1B_n$$

Figure 1.24 shows the numerical solution of the battle for the starting values $B_0 = 27$ and $F_0 = 33$. For full-force engagements we see that the British force would be roundly defeated after 11 encounters, with only 3 ships remaining and with at least 1 of these badly damaged. At the end of the battle, after 11 encounter stages, the French–Spanish fleet would still have approximately 18 ships.

Lord Nelson's Divide-and-Conquer Strategy   Napoleon's force of 33 ships was arranged essentially along a line separated into three groups as shown in Figure 1.25. Lord Nelson's strategy was to engage force A with 13 British ships (holding 14 in reserve). He then planned to combine those ships that survived the skirmish against force A with the 14 ships in reserve to engage force B. Finally, after the battle with force B, he planned to use all remaining ships to engage force C.

Assuming each side loses 5% of the number of ships of the opposing force for each of the three battles (to enhance the graphs), Figure 1.26 (see pages 42 and 43) shows the numerical solution of each battle. In battle A, we see that the British defeat the French, with only 1 British ship damaged; the French have (approximately) 1 of their 3 ships remaining. Joining forces, battle B begins with a force of better than 26 ships for the British and 18 ships

Stage	Brittish force	French force
1	27.0000	33.0000
2	23.7000	30.3000
3	20.6700	27.9300
4	17.8770	25.8630
5	15.2907	24.0753
6	12.8832	22.5462
7	10.6285	21.2579
8	8.5028	20.1951
9	6.4832	19.3448
10	4.5488	18.6965
11	2.6791	18.2416

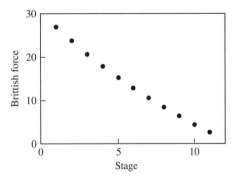

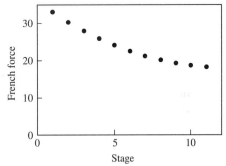

■ **Figure 1.24**

Numerical solution of the battle between British and French–Spanish ships in 1805 under full-force engagements

Force B = 17   Force A = 3   Force C = 13

■ **Figure 1.25**

Configuration of Napoleon's fleet

for the French (1 ship from battle A joining the French force B). The second battle results in the French losing all but 1 of their force B ships and 1 heavily damaged ship; the British have 19 ships intact plus a heavily damaged ship.

Entering battle C with the French again, and combining Britain's force C with its remaining ships from battle B, we see that Nelson wins the day. At the end of the final battle, the French have but 1 ship remaining intact, whereas the British have 12 intact ships. Each side has a damaged ship as well.

The results predicted by our model using the divide-and-conquer strategy are similar to what actually happened historically. The British fleet under Lord Nelson did win the battle of Trafalgar, although the French disengaged during the third battle and returned to France with approximately 13 ships. Unfortunately, Lord Nelson was killed during the battle, but his strategy was brilliant. Without it, the British would have lost their fleet. ■ ■ ■

Battle A

Stage	British force	French force
1	13.0000	3.00000
2	12.8500	2.35000
3	12.7325	1.70750
4	12.6471	1.07088

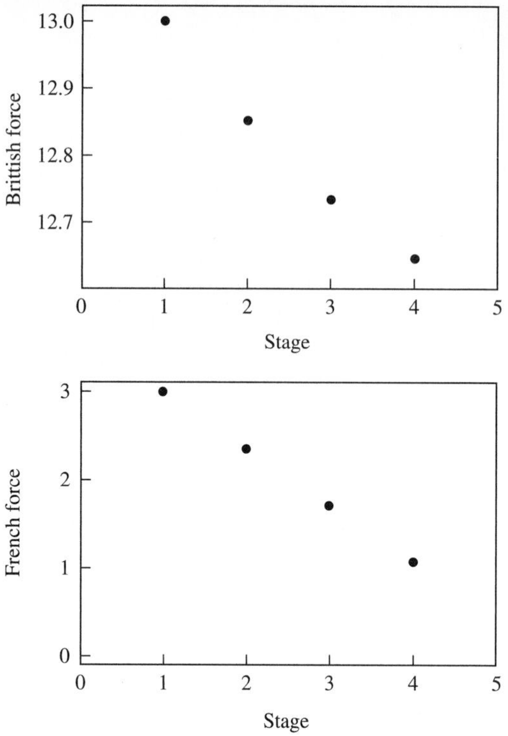

Battle B

Stage	British force	French force
1	26.6471	18.0709
2	25.7436	16.7385
3	24.9066	15.4513
4	24.1341	14.2060
5	23.4238	12.9993
6	22.7738	11.8281
7	22.1824	10.6894
8	21.6479	9.5803
9	21.1689	8.4979
10	20.7440	7.4395
11	20.3720	6.4023
12	20.0519	5.3837
13	19.7827	4.3811
14	19.5637	3.3919
15	19.3941	2.4138
16	19.2734	1.4441

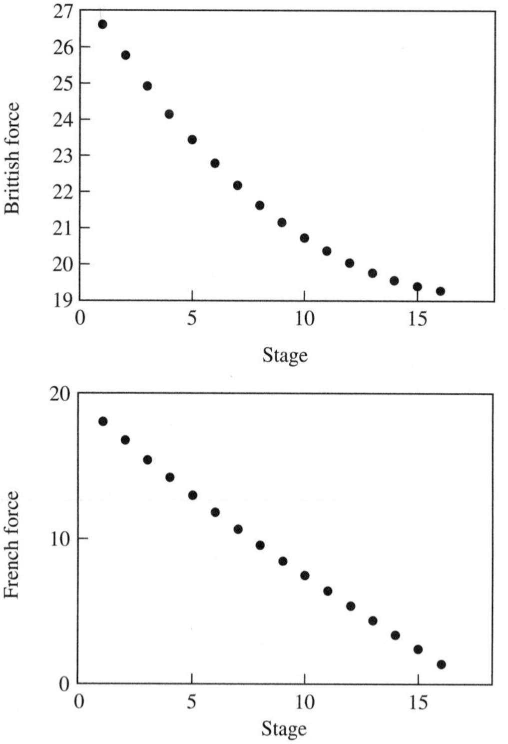

Figure 1.26

Lord Nelson's divide-and-conquer strategy against the French–Spanish force surprised them, giving an advantage to the British.

Battle C

Stage	British force	French force
1	19.2734	14.4441
2	18.5512	13.4804
3	17.8772	12.5529
4	17.2495	11.6590
5	16.6666	10.7965
6	16.1268	9.9632
7	15.6286	9.1569
8	15.1707	8.3754
9	14.7520	7.6169
10	14.3711	6.8793
11	14.0272	6.1607
12	13.7191	5.4594
13	13.4462	4.7734
14	13.2075	4.1011
15	13.0024	3.4407
16	12.8304	2.7906
17	12.6909	2.1491
18	12.5834	1.5146

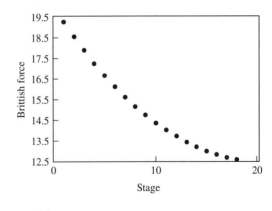

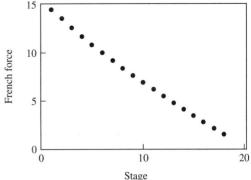

■ **Figure 1.26**

Continued

EXAMPLE 3 *Competitive Hunter Model—Spotted Owls and Hawks*

Suppose a species of spotted owls competes for survival in a habitat that also supports hawks. Suppose also that in the absence of the other species, each individual species exhibits unconstrained growth in which the change in the population during an interval of time (such as 1 day) is proportional to the population size at the beginning of the interval. If O_n represents the size of the spotted owl population at the end of day n and H_n represents the competing hawk population, then

$$\Delta O_n = k_1 O_n \quad \text{and} \quad \Delta H_n = k_2 H_n$$

Here, k_1 and k_2 are the constant positive growth rates. The effect of the presence of the second species is to diminish the growth rate of the other species, and vice versa. Although there are many ways to model the mutually detrimental interaction of the two species, we will assume that this decrease is approximately proportional to the number of possible interactions between the two species. Therefore, one submodel is to assume that the decrease is proportional to the product of O_n and H_n. These considerations are modeled by the equations

$$\Delta O_n = k_1 O_n - k_3 O_n H_n$$
$$\Delta H_n = k_2 H_n - k_4 O_n H_n$$

Solving each equation for the $n + 1$st term gives

$$O_{n+1} = (1 + k_1)O_n - k_3 O_n H_n$$
$$H_{n+1} = (1 + k_2)H_n - k_4 O_n H_n$$

where k_1–k_4 are positive constants. Now, let's choose specific values for the constants of proportionality and consider the system:

$$O_{n+1} = 1.2O_n - 0.001 O_n H_n \tag{1.10}$$
$$H_{n+1} = 1.3H_n - 0.002 O_n H_n$$

Equilibrium Values   If we call the equilibrium values (O, H), then we must have $O = O_{n+1} = O_n$ and $H = H_{n+1} = H_n$ simultaneously. Substituting into the system yields

$$O = 1.2O - 0.001 OH$$
$$H = 1.3H - 0.002 OH$$

or

$$0 = 0.2O - 0.001 OH = O(0.2 - 0.001 H)$$
$$0 = 0.3H - 0.002 OH = H(0.3 - 0.002 O)$$

The first equation indicates that there is no change in the owl population if $O = 0$ or $H = 0.2/0.001 = 200$. The second equation indicates there is no change in the hawk population if $H = 0$ or $O = 0.3/0.002 = 150$, as depicted in Figure 1.27. Note that equilibrium values exist at $(O, H) = (0, 0)$ and $(O, H) = (150, 200)$ because *neither* population will change at those points. (Substitute the equilibrium values into Equation (1.10) to check that the system indeed remains at $(0, 0)$ and $(150, 200)$ if either of those points represents the starting values.)

■ **Figure 1.27**

If the owl population begins at 150 and the hawk population begins at 200, the two populations remain at their starting values.

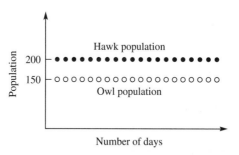

Now let's analyze what happens in the vicinity of the equilibrium values we have found. Let's build numerical solutions for the three starting populations given here. Note that the first two values are close to the equilibrium value $(150, 200)$, whereas the third is near the origin.

	Owls	Hawks
Case 1	151	199
Case 2	149	201
Case 3	10	10

Iterating Equation (1.10), beginning with the starting values given, results in the numerical solutions shown in Figure 1.28. Note that in each case, one of the two species eventually drives the other to extinction.

Sensitivity to Initial Conditions and Long-Term Behavior   Suppose 350 owls and hawks are to be placed in a habitat modeled by Equation (1.10). If 150 of the birds are owls, our model predicts the owls will remain at 150 forever. If 1 owl is removed from the habitat (leaving 149), then the model predicts that the owl population will die out. If 151 owls are placed in the habitat, however, the model predicts that the owls will grow without bound and the hawks will disappear. This model is extremely sensitive to the initial conditions. The equilibrium values are unstable in the sense that if we start close to either equilibrium value, we do not remain close. Note how the model predicts that coexistence of the two species in a single habitat is highly unlikely because one of the two species will eventually dominate the habitat. In the problem set, you are asked to explore this system further by examining other starting points and by changing the coefficients of the model.   ■ ■ ■

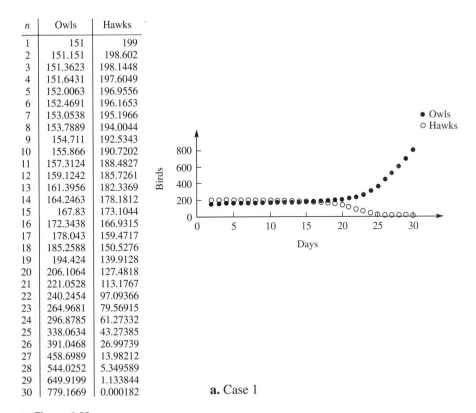

n	Owls	Hawks
1	151	199
2	151.151	198.602
3	151.3623	198.1448
4	151.6431	197.6049
5	152.0063	196.9556
6	152.4691	196.1653
7	153.0538	195.1966
8	153.7889	194.0044
9	154.711	192.5343
10	155.866	190.7202
11	157.3124	188.4827
12	159.1242	185.7261
13	161.3956	182.3369
14	164.2463	178.1812
15	167.83	173.1044
16	172.3438	166.9315
17	178.043	159.4717
18	185.2588	150.5276
19	194.424	139.9128
20	206.1064	127.4818
21	221.0528	113.1767
22	240.2454	97.09366
23	264.9681	79.56915
24	296.8785	61.27332
25	338.0634	43.27385
26	391.0468	26.99739
27	458.6989	13.98212
28	544.0252	5.349589
29	649.9199	1.133844
30	779.1669	0.000182

a. Case 1

■ **Figure 1.28**

Either owls or hawks dominate the competition. (*Continues*)

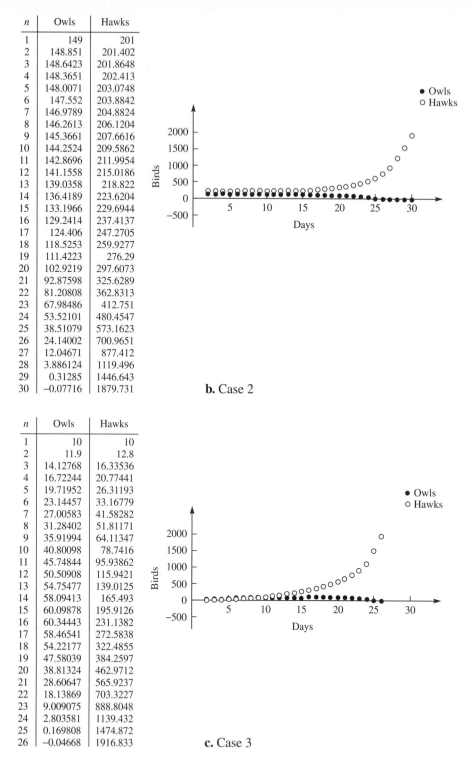

n	Owls	Hawks
1	149	201
2	148.851	201.402
3	148.6423	201.8648
4	148.3651	202.413
5	148.0071	203.0748
6	147.552	203.8842
7	146.9789	204.8824
8	146.2613	206.1204
9	145.3661	207.6616
10	144.2524	209.5862
11	142.8696	211.9954
12	141.1558	215.0186
13	139.0358	218.822
14	136.4189	223.6204
15	133.1966	229.6944
16	129.2414	237.4137
17	124.406	247.2705
18	118.5253	259.9277
19	111.4223	276.29
20	102.9219	297.6073
21	92.87598	325.6289
22	81.20808	362.8313
23	67.98486	412.751
24	53.52101	480.4547
25	38.51079	573.1623
26	24.14002	700.9651
27	12.04671	877.412
28	3.886124	1119.496
29	0.31285	1446.643
30	−0.07716	1879.731

b. Case 2

n	Owls	Hawks
1	10	10
2	11.9	12.8
3	14.12768	16.33536
4	16.72244	20.77441
5	19.71952	26.31193
6	23.14457	33.16779
7	27.00583	41.58282
8	31.28402	51.81171
9	35.91994	64.11347
10	40.80098	78.7416
11	45.74844	95.93862
12	50.50908	115.9421
13	54.75477	139.0125
14	58.09413	165.493
15	60.09878	195.9126
16	60.34443	231.1382
17	58.46541	272.5838
18	54.22177	322.4855
19	47.58039	384.2597
20	38.81324	462.9712
21	28.60647	565.9237
22	18.13869	703.3227
23	9.009075	888.8048
24	2.803581	1139.432
25	0.169808	1474.872
26	−0.04668	1916.833

c. Case 3

■ **Figure 1.28**

Continued

EXAMPLE 4 *Voting Tendencies of the Political Parties*

Consider a three-party system with Republicans, Democrats, and Independents. Assume that in the next election, 75% of the those who voted Republican again vote Republican, 5% vote Democrat, and 20% vote Independent. Of those who voted Democrat before, 20% vote Republican, 60% again vote Democrat, and 20% vote Independent. Of those who voted Independent, 40% vote Republican, 20% vote Democrat, and 40% again vote Independent. Assume that these tendencies continue from election to election and that no additional voters enter or leave the system. These tendencies are depicted in Figure 1.29.

■ **Figure 1.29**

Voting tendencies among Republicans, Democrats, and Independents

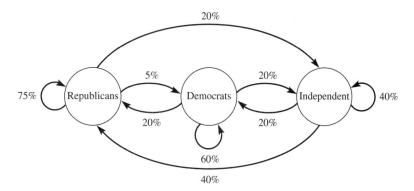

To formulate a system of difference equations, let n represent the nth election and define

$$R_n = \text{the number of Republican voters in the } n\text{th election}$$
$$D_n = \text{the number of Democrat voters in the } n\text{th election}$$
$$I_n = \text{the number of Independent voters in the } n\text{th election}$$

Formulating the system of difference equations, we have the following dynamical system:

$$R_{n+1} = 0.75R_n + 0.20D_n + 0.40I_n$$
$$D_{n+1} = 0.05R_n + 0.60D_n + 0.20I_n$$
$$I_{n+1} = 0.20R_n + 0.20D_n + 0.40I_n$$

Equilibrium Values If we call the equilibrium values (R, D, I), then we must have $R = R_{n+1} = R_n$, $D = D_{n+1} = D_n$, and $I = I_{n+1} = I_n$ simultaneously. Substituting into the dynamical system yields

$$-0.25R + 0.20D + 0.40I = 0$$
$$0.05R - 0.40D + 0.20I = 0$$
$$0.20R + 0.20D - 0.60I = 0$$

There are an infinite number of solutions to this system of equations. Letting $I = 1$, the system is satisfied if $R = 2.2221$ and $D = 0.7777694$ (approximately). Suppose the system has 399,998 voters. Then $R = 222,221$, $D = 77,777$, and $I = 100,000$ voters should approximate the equilibrium values. Let's use a spreadsheet to check the equilibrium values and several other values. The total voters in the system is 399,998, with the initial voting as follows:

	Republicans	Democrats	Independents
Case 1	222,221	77,777	100,000
Case 2	227,221	82,777	90,000
Case 3	100,000	100,000	199,998
Case 4	0	0	399,998

The numerical solutions for the starting values are graphed in Figure 1.30.

Sensitivity to Initial Conditions and Long-Term Behavior   Suppose initially there are 399,998 voters in the system and all remain in the system. At least for the starting values we investigated, the system approaches the same result, even if initially there are no Republicans or Democrats in the system. The equilibrium investigated appears to be stable. Starting values in the vicinity of the equilibrium appear to approach the equilibrium. What about the origin? Is it stable? In the problem set you are asked to explore this system further

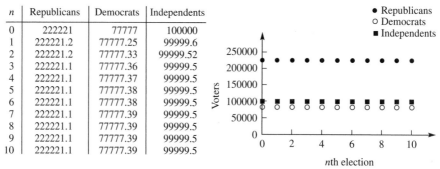

n	Republicans	Democrats	Independents
0	222221	77777	100000
1	222221.2	77777.25	99999.6
2	222221.2	77777.33	99999.52
3	222221.1	77777.36	99999.5
4	222221.1	77777.37	99999.5
5	222221.1	77777.38	99999.5
6	222221.1	77777.38	99999.5
7	222221.1	77777.39	99999.5
8	222221.1	77777.39	99999.5
9	222221.1	77777.39	99999.5
10	222221.1	77777.39	99999.5

a. Case 1

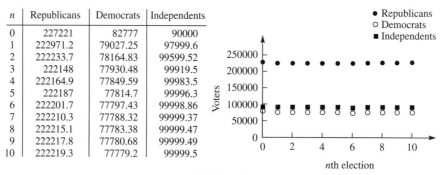

n	Republicans	Democrats	Independents
0	227221	82777	90000
1	222971.2	79027.25	97999.6
2	222233.7	78164.83	99599.52
3	222148	77930.48	99919.5
4	222164.9	77849.59	99983.5
5	222187	77814.7	99996.3
6	222201.7	77797.43	99998.86
7	222210.3	77788.32	99999.37
8	222215.1	77783.38	99999.47
9	222217.8	77780.68	99999.49
10	222219.3	77779.2	99999.5

b. Case 2

■ **Figure 1.30**

Voting tendencies

n	Republicans	Democrats	Independents
0	100000	100000	199998
1	174999.2	104999.6	119999.2
2	200249	95749.56	103999.4
3	210936.4	88262.07	100799.5
4	216174.5	83663.96	100159.5
5	218927.5	81039	100031.5
6	220416	79576.08	100005.9
7	221229.6	78767.63	100000.8
8	221676	78322.21	99999.76
9	221921.4	78077.08	99999.55
10	222056.3	77942.23	99999.51

c. Case 3

n	Republicans	Democrats	Independents
0	0	0	399998
1	159999.2	79999.6	159999.2
2	199999	87999.56	111999.4
3	212398.9	85199.57	102399.5
4	217298.9	82219.59	100479.5
5	219609.9	80292.6	100095.5
6	220804.1	79175.15	100018.7
7	221445.6	78549.04	100003.3
8	221795.4	78202.37	100000.3
9	221987.1	78011.25	99999.65
10	222092.4	77906.03	99999.53

d. Case 4

■ **Figure 1.30**

Continued

by examining other starting points and changing the coefficients of the model. Additionally, you are asked to investigate systems in which voters enter and leave the system. ■ ■ ■

1.4 PROBLEMS

1. Consider Example 1, A Car Rental Company. Experiment with different values for the coefficients. Iterate the resulting dynamical system for the given initial values. Then experiment with different starting values. Do your experimental results indicate that the model is sensitive

   a. to the coefficients?

   b. to the starting values?

2. Consider Example 3, Competitive Hunter Model—Spotted Owls and Hawks. Experiment with different values for the coefficients using the starting values given. Then try different starting values. What is the long-term behavior? Do your experimental results indicate that the model is sensitive

   a. to the coefficients?

   b. to the starting values?

3. In analyzing the battle of Trafalgar in 1805, we saw that if the two forces simply engaged head-on, the British lost the battle and approximately 24 ships, whereas the French–Spanish force lost approximately 15 ships. We also saw that Lord Nelson could overcome a superior force by employing a divide-and-conquer strategy. An alternative strategy for overcoming a superior force is to increase the technology employed by the inferior force. Suppose that the British ships were equipped with superior weaponry, and that the French–Spanish losses equaled 15% of the number of ships of the opposing force, whereas the British suffered casualties equal to 5% of the opposing force.

   a. Formulate a system of difference equations to model the number of ships possessed by each force. Assume the French–Spanish force starts with 33 ships and the British start with 27 ships.

   b. Build a numerical solution to determine who wins under the new assumption in a head-on engagement.

   c. Build a numerical solution for the three battles employing Nelson's divide-and-conquer strategy coupled with the superior weaponry of the British ships.

4. Suppose the spotted owls' primary food source is a single prey: mice. An ecologist wishes to predict the population levels of spotted owls and mice in a wildlife sanctuary. Letting M_n represent the mouse population after n years and O_n the predator owl population, the ecologist has suggested the model

$$M_{n+1} = 1.2M_n - 0.0010O_n M_n$$
$$O_{n+1} = 0.7O_n + 0.002O_n M_n$$

   The ecologist wants to know whether the two species can coexist in the habitat and whether the outcome is sensitive to the starting populations.

   a. Compare the signs of the coefficients of the preceding model with the signs of the coefficients of the owls–hawks model in Example 3. Explain the sign of each of the four coefficients $1.2, -0.001, 0.7$, and 0.002 in terms of the predator–prey relationship being modeled.

   b. Test the initial populations in the following table and predict the long-term outcome:

	Owls	Mice
Case A	150	200
Case B	150	300
Case C	100	200
Case D	10	20

   c. Now experiment with different values for the coefficients using the starting values given. Then try different starting values. What is the long-term behavior? Do your experimental results indicate that the model is sensitive to the coefficients? Is it sensitive to the starting values?

5. In Example 4, Voting Tendencies of the Political Parties, experiment with starting values near the origin. Does the origin appear to be a stable equilibrium? Explain. Experiment with different values for the coefficients using the starting values given.

Then try different starting values. What is the long-term behavior? Do your experimental results indicate that the model is sensitive to the coefficients? To the starting values?

Now assume that each party recruits new party members. Initially assume that the total number of voters increases as each party recruits unregistered citizens. Experiment with different values for new party members. What is the long-term behavior? Does it seem to be sensitive to the recruiting rates? How would you adjust your model to reflect that the total number of citizens in the voting district is constant? Adjust the model to reflect what you think is happening in your voting district. What do you think will happen in your district over the long haul?

6. An economist is interested in the variation of the price of a single product. It is observed that a high price for the product in the market attracts more suppliers. However, increasing the quantity of the product supplied tends to drive the price down. Over time, there is an interaction between price and supply. The economist has proposed the following model, where P_n represents the price of the product at year n, and Q_n represents the quantity.

$$P_{n+1} = P_n - 0.1(Q_n - 500)$$
$$Q_{n+1} = Q_n + 0.2(P_n - 100)$$

a. Does the model make sense intuitively? What is the significance of the constants 100 and 500? Explain the significance of the signs of the constants -0.1 and 0.2.

b. Test the initial conditions in the following table and predict the long-term behavior.

	Price	Quantity
Case A	100	500
Case B	200	500
Case C	100	600
Case D	100	400

7. In 1868, the accidental introduction into the United States of the cottony-cushion insect (*Icerya purchasi*) from Australia threatened to destroy the American citrus industry. To counteract this situation, a natural Australian predator, a ladybird beetle (*Novius cardinalis*), was imported. The beetles kept the insects to a relatively low level. When DDT (an insecticide) was discovered to kill scale insects, farmers applied it in the hope of reducing the scale insect population even further. However, DDT turned out to be fatal to the beetle as well, and the overall effect of using the insecticide was to increase the numbers of the scale insect. Let C_n and B_n represent the cottony-cushion insect and ladybird beetle population levels, respectively, after n days. Generalizing the model in Problem 4, we have

$$C_{n+1} = C_n + k_1 C_n - k_2 B_n C_n$$
$$B_{n+1} = B_n - k_3 B_n + k_4 B_n C_n$$

where the k_i are positive constants.

a. Discuss the meaning of each k_i in the predator–prey model.

b. What assumptions are implicitly being made about the growth of each species in the absence of the other species?

c. Pick values for your coefficients and try several starting values. What is the long-term behavior predicted by your model? Vary the coefficients. Do your experimental results indicate that the model is sensitive to the coefficients? To the starting values?

d. Modify the predator–prey model to reflect a predator–prey system in which farmers apply (on a regular basis) an insecticide that destroys both the insect predator and the insect prey at a rate proportional to the numbers present.

1.4 PROJECTS (See enclosed CD for UMAP modules.)

1. Complete the requirements of the UMAP module "Graphical Analysis of Some Difference Equations in Biology," by Martin Eisen, UMAP 553. The growth of many biological populations can be modeled by difference equations. This module shows how the behavior of the solutions to certain equations can be predicted by graphical techniques.

2. Prepare a summary of the paper by May et al. listed among the Further Reading titles for this section.

3. Complete the modules "The Growth of Partisan Support I: Model and Estimation" (UMAP 304) and "The Growth of Partisan Support II: Model Analytics" (UMAP 305), by Carol Weitzel Kohfeld. UMAP 304 presents a simple model of political mobilization, refined to include the interaction between supporters of a particular party and recruitable nonsupporters. UMAP 305 investigates the mathematical properties of the first-order quadratic-difference equation model. The model is tested using data from three U.S. counties.

1.4 Further Reading

Clark, Colin W. *Mathematical Bioeconomics*: *The Optimal Management of Renewable Resources*. New York: Wiley, 1976.

May, R. M. *Stability and Complexity in Model Ecosystems*, Monographs in Population Biology VI. Princeton, NJ: Princeton University Press, 2001.

May, R.M., ed. *Theoretical Ecology*: *Principles and Applications*. Philadelphia: Saunders, 1976.

May, R. M., J. R. Beddington, C. W. Clark, S. J. Holt, & R. M. Lewis, "Management of Multispecies Fisheries." *Science* 205 (July 1979): 267–277.

Schom, A., & M. Joseph. *Trafalgar, Countdown to Battle, 1803–1805*. London: Simon & Shuster, 1990.

Shubik, M., ed. *Mathematics of Conflict*. Amsterdam: Elsevier Science, 1983.

Tuchinsky, P. M. *Man in Competition with the Spruce Budworm*, UMAP Expository Monograph. The population of tiny caterpillars periodically explodes in the evergreen forests of eastern Canada and Maine. They devour the trees' needles and cause great damage to forests that are central to the economy of the region. The province of New Brunswick is using mathematical models of the budworm/forest interaction in an effort to plan for and control the damage. The monograph surveys the ecological situation and examines the computer simulation and models currently in use.

The Modeling Process, Proportionality, and Geometric Similarity

Introduction

In Chapter 1 we presented graphical models representing population size, drug concentration in the bloodstream, various financial investments, and the distribution of cars between two cities for a rental company. Now we examine more closely the process of mathematical modeling. To gain an understanding of the processes involved in mathematical modeling, consider the two worlds depicted in Figure 2.1. Suppose we want to understand some behavior or phenomenon in the real world. We may wish to make predictions about that behavior in the future and analyze the effects that various situations have on it.

■ **Figure 2.1**

The real and mathematical worlds

Real-world systems	Mathematical world
Observed behavior or phenomenon	Models Mathematical operations and rules Mathematical conclusions

For example, when studying the populations of two interacting species, we may wish to know if the species can coexist within their environment or if one species will eventually dominate and drive the other to extinction. In the case of the administration of a drug to a person, it is important to know the correct dosage and the time between doses to maintain a safe and effective level of the drug in the bloodstream.

How can we construct and use models in the mathematical world to help us better understand real-world systems? Before discussing how we link the two worlds together, let's consider what we mean by a real-world system and why we would be interested in constructing a mathematical model for a system in the first place.

A **system** is an assemblage of objects joined in some regular interaction or interdependence. The modeler is interested in understanding how a particular system works, what causes changes in the system, and how sensitive the system is to certain changes. He or she is also interested in predicting what changes might occur and when they occur. How might such information be obtained?

For instance, suppose the goal is to draw conclusions about an observed phenomenon in the real world. One procedure would be to conduct some real-world behavior trials or experiments and observe their effect on the real-world behavior. This is depicted on the left side of Figure 2.2. Although such a procedure might minimize the loss in fidelity incurred by a less direct approach, there are many situations in which we would not want to follow such a course of action. For instance, there may be prohibitive financial and human costs

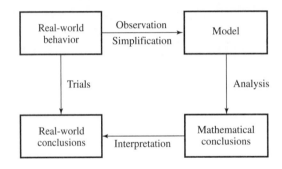

for conducting even a single experiment, such as determining the level of concentration at
which a drug proves to be fatal or studying the radiation effects of a failure in a nuclear
power plant near a major population area. Or we may not be willing to accept even a
single experimental failure, such as when investigating different designs for a heat shield
for a spacecraft carrying astronauts. Moreover, it may not even be possible to produce a
trial, as in the case of investigating specific change in the composition of the ionosphere
and its corresponding effect on the polar ice cap. Furthermore, we may be interested in
generalizing the conclusions beyond the specific conditions set by one trial (such as a cloudy
day in New York with temperature 82 °F, wind 15–20 miles per hour, humidity 42%, and
so on). Finally, even though we succeed in predicting the real-world behavior under some
very specific conditions, we have not necessarily *explained* why the particular behavior
occurred. (Although the abilities to predict and explain are often closely related, the ability
to predict a behavior does not necessarily imply an understanding of it. In Chapter 3 we
study techniques specifically designed to help us make predictions even though we cannot
explain satisfactorily all aspects of the behavior.) The preceding discussion underscores the
need to develop indirect methods for studying real-world systems.

An examination of Figure 2.2 suggests an alternative way of reaching conclusions about
the real world. First, we make specific observations about the behavior being studied and
identify the factors that seem to be involved. Usually we cannot consider, or even identify,
all the factors involved in the behavior, so we make simplifying assumptions that eliminate
some factors. For instance, we may choose to neglect the humidity in New York City, at least
initially, when studying radioactive effects from the failure of a nuclear power plant. Next,
we conjecture tentative relationships among the factors we have selected, thereby creating
a rough model of the behavior. Having constructed a model, we then apply appropriate
mathematical analysis leading to conclusions about the model. Note that these conclusions
pertain only to the model, not to the actual real-world system under investigation. Because
we made some simplifications in constructing the model, and because the observations
on which the model is based invariably contain errors and limitations, we must carefully
account for these anomalies before drawing any inferences about the real-world behavior.
In summary, we have the following rough modeling procedure:

1. Through observation, identify the primary factors involved in the real-world behavior, possibly making simplifications.

2. Conjecture tentative relationships among the factors.

3. Apply mathematical analysis to the resultant model.

4. Interpret mathematical conclusions in terms of the real-world problem.

■ **Figure 2.3**

The modeling process as a closed system

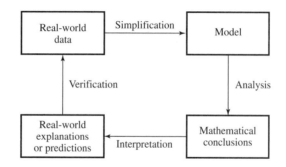

We portrayed this flow of the modeling process in the introduction to Chapter 1 and show it again in Figure 2.3 as a closed system. Given some real-world system, we gather sufficient data to formulate a model. Next we analyze the model and reach mathematical conclusions about it. Then we interpret the model and make predictions or offer explanations. Finally, we test our conclusions about the real-world system against new observations and data. We may then find we need to go back and refine the model to improve its predictive or descriptive capabilities. Or perhaps we will discover that the model really does not fit the real world accurately, so we must formulate a new model. We will study the various components of this modeling process in detail throughout the book.

2.1 Mathematical Models

For our purposes we define a **mathematical model** as a mathematical construct designed to study a particular real-world system or phenomenon. We include graphical, symbolic, simulation, and experimental constructs. Mathematical models can be differentiated further. There are existing mathematical models that can be identified with some particular real-world phenomenon and used to study it. Then there are those mathematical models that we construct specifically to study a special phenomenon. Figure 2.4 depicts this differentiation between models. Starting from some real-world phenomenon, we can represent it mathematically by constructing a new model or selecting an existing model. On the other hand, we can replicate the phenomenon experimentally or with some kind of simulation.

Regarding the question of constructing a mathematical model, a variety of conditions can cause us to abandon hope of achieving any success. The mathematics involved may be so complex and intractable that there is little hope of analyzing or solving the model, thereby defeating its utility.

This complexity can occur, for example, when attempting to use a model given by a system of partial differential equations or a system of nonlinear algebraic equations. Or the problem may be so large (in terms of the number of factors involved) that it is impossible to

■ **Figure 2.4**

The nature of the model

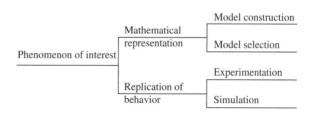

capture all the necessary information in a single mathematical model. Predicting the global effects of the interactions of a population, the use of resources, and pollution is an example of such an impossible situation. In such cases we may attempt to replicate the behavior *directly* by conducting various experimental trials. Then we collect data from these trials and analyze the data in some way, possibly using statistical techniques or curve-fitting procedures. From the analysis, we can reach certain conclusions.

In other cases, we may attempt to replicate the behavior *indirectly*. We might use an analog device such as an electrical current to model a mechanical system. We might use a scaled-down model such as a scaled model of a jet aircraft in a wind tunnel. Or we might attempt to replicate a behavior on a digital computer—for instance, simulating the global effects of the interactions of population, use of resources, and pollution or simulating the operation of an elevator system during morning rush hour.

The distinction between the various model types as depicted in Figure 2.4 is made solely for ease of discussion. For example, the distinction between experiments and simulations is based on whether the observations are obtained directly (experiments) or indirectly (simulations). In practical models this distinction is not nearly so sharp; one master model may employ several models as submodels, including selections from existing models, simulations, and experiments. Nevertheless, it is informative to contrast these types of models and compare their various capabilities for portraying the real world.

To that end, consider the following properties of a model.

Fidelity: The preciseness of a model's representation of reality

Costs: The total cost of the modeling process

Flexibility: The ability to change and control conditions affecting the model as required data are gathered

It is useful to know the degree to which a given model possesses each of these characteristics. However, since specific models vary greatly, even within the classes identified in Figure 2.4, the best we can hope for is a comparison of the relative performance between the classes of models for each of the characteristics. The comparisons are depicted in Figure 2.5, where the ordinate axis denotes the degree of effectiveness of each class.

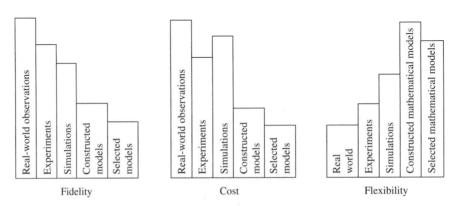

■ **Figure 2.5**

Comparisons among the model types

Let's summarize the results shown in Figure 2.5. First, consider the characteristic of fidelity. We would expect observations made directly in the real world to demonstrate the greatest fidelity, even though some testing bias and measurement error may be present. We would expect experimental models to show the next greatest fidelity because behavior is observed directly in a more controlled environment such as a laboratory. Because simulations incorporate indirect observations, they suffer a further loss in fidelity. Whenever a mathematical model is constructed, real-world conditions are simplified, resulting in more loss of fidelity. Finally, any selected model is based on additional simplifications that are not even tailored to the specific problem, and these simplifications imply still further loss in fidelity.

Next, consider cost. Generally, we would expect any selected mathematical model to be the least expensive. Constructed mathematical models bear an additional cost of tailoring the simplifications to the phenomenon being studied. Experiments are usually expensive to set up and operate. Likewise, simulations use indirect devices that are often expensive to develop, and simulations commonly involve large amounts of computer space, time, and maintenance.

Finally, consider flexibility. Constructed mathematical models are generally the most flexible because different assumptions and conditions can be chosen relatively easily. Selected models are less flexible because they are developed under specific assumptions; nevertheless, specific conditions can often be varied over wide ranges. Simulations usually entail the development of some other indirect device to alter assumptions and conditions appreciably. Experiments are even less flexible because some factors are very difficult to control beyond specific ranges. Observations of real-word behavior have little flexibility because the observer is limited to the specific conditions that pertain at the time of the observation. Moreover, other conditions might be highly improbable, or impossible, to create. It is important to understand that our discussion is only qualitative in nature and that there are many exceptions to these generalizations.

Construction of Models

In the preceding discussion we viewed modeling as a process and briefly considered the form of the model. Now let's focus on the construction of mathematical models. We begin by presenting an outline of a procedure that is helpful in constructing models. In the next section, we illustrate the various steps in the procedure by discussing several real-world examples.

STEP 1 **IDENTIFY THE PROBLEM.** What is the problem you would like to explore? Typically this is a difficult step because in real-life situations no one simply hands you a mathematical problem to solve. Usually you have to sort through large amounts of data and identify some particular aspect of the situation to study. Moreover, it is imperative to be sufficiently precise (ultimately) in the formulation of the problem to allow for translation of the verbal statements describing the problem into mathematical symbols. This translation is accomplished through the next steps. It is important to realize that the answer to the question posed might not lead directly to a usable problem identification.

STEP 2 **MAKE ASSUMPTIONS.** Generally, we cannot hope to capture in a usable mathematical model all the factors influencing the identified problem. The task is simplified by reducing the number of factors under consideration. Then, relationships among the remaining variables must be determined. Again, by assuming relatively simple relationships, we can reduce the complexity of the problem. Thus the assumptions fall into two main activities:

a. CLASSIFY THE VARIABLES. What things influence the behavior of the problem identified in Step 1? List these things as variables. The variables the model seeks to explain are the dependent variables, and there may be several of these. The remaining variables are the independent variables. Each variable is classified as dependent, independent, or neither.

You may choose to neglect some of the independent variables for either of two reasons. First, the effect of the variable may be relatively small compared to other factors involved in the behavior. Second, a factor that affects the various alternatives in about the same way may be neglected, even though it may have a very important influence on the behavior under investigation. For example, consider the problem of determining the optimal shape for a lecture hall, where readability of a chalkboard or overhead projection is a dominant criterion. Lighting is certainly a crucial factor, but it would affect all possible shapes in about the same way. By neglecting such a variable, and perhaps incorporating it later in a separate, more refined model, the analysis can be simplified considerably.

b. DETERMINE INTERRELATIONSHIPS AMONG THE VARIABLES SELECTED FOR STUDY. Before we can hypothesize a relationship among the variables, we generally must make some additional simplifications. The problem may be so complex that we cannot see a relationship among all the variables initially. In such cases it may be possible to study submodels. That is, we study one or more of the independent variables separately. Eventually we will connect the submodels together. Studying various techniques, such as proportionality, will aid in hypothesizing relationships among the variables.

STEP 3 **SOLVE OR INTERPRET THE MODEL.** Now put together all the submodels to see what the model is telling us. In some cases the model may consist of mathematical equations or inequalities that must be solved to find the information we are seeking. Often, a problem statement requires a best solution, or *optimal solution*, to the model. Models of this type are discussed later.

Often, we will find that we are not quite ready to complete this step, or we may end up with a model so unwieldy we cannot solve or interpret it. In such situations we might return to Step 2 and make additional simplifying assumptions. Sometimes we will even want to return to Step 1 to redefine the problem. This point will be amplified in the following discussion.

STEP 4 **VERIFY THE MODEL.** Before we can use the model, we must test it out. There are several questions to ask before designing these tests and collecting data—a process that can be expensive and time-consuming. First, does the model answer the problem identified in Step 1, or did it stray from the key issue as we constructed the model? Second, is the model usable in a practical sense? That is, can we really gather the data necessary to operate the model? Third, does the model make common sense?

Once the commonsense tests are passed, we will want to test many models using actual data obtained from empirical observations. We need to be careful to design the test in such a way as to include observations over the *same range* of values of the various independent variables we expect to encounter when actually using the model. The assumptions made in Step 2 may be reasonable over a restricted range of the independent variables but very poor outside of those values. For instance, a frequently used interpretation of Newton's second law states that the net force acting on a body is equal to the mass of the body times its acceleration. This law is a reasonable model until the speed of the object approaches the speed of light.

Be careful about the conclusions you draw from any tests. Just as we cannot prove a theorem simply by demonstrating many cases that support the theorem, we cannot extrapolate broad generalizations from the particular evidence we gather about our model. A model does not become a law just because it is verified repeatedly in some specific instances. Rather, we *corroborate the reasonableness* of our model through the data we collect.

STEP 5   IMPLEMENT THE MODEL. Of course, our model is of no use just sitting in a filing cabinet. We will want to explain our model in terms that the decision makers and users can understand if it is ever to be of use to anyone. Furthermore, unless the model is placed in a user-friendly mode, it will quickly fall into disuse. Expensive computer programs sometimes suffer such a demise. Often the inclusion of an additional step to facilitate the collection and input of the data necessary to operate the model determines its success or failure.

STEP 6   MAINTAIN THE MODEL. Remember that the model is derived from a specific problem identified in Step 1 and from the assumptions made in Step 2. Has the original problem changed in any way, or have some previously neglected factors become important? Does one of the submodels need to be adjusted?

We summarize the steps for constructing mathematical models in Figure 2.6. We should not be too enamored of our work. Like any model, our procedure is an approximation process and therefore has its limitations. For example, the procedure seems to consist of discrete steps leading to a usable result, but that is rarely the case in practice. Before offering an alternative procedure that emphasizes the iterative nature of the modeling process, let's discuss the advantages of the methodology depicted in Figure 2.6.

The process shown in Figure 2.6 provides a methodology for progressively focusing on those aspects of the problem we wish to study. Furthermore, it demonstrates a curious blend of creativity with the scientific method used in the modeling process. The first two steps are more artistic or original in nature. They involve abstracting the essential features of the problem under study, neglecting any factors judged to be unimportant, and postulating relationships precise enough to help answer the questions posed by the problem. However, these relationships must be simple enough to permit the completion of the remaining steps.

■ **Figure 2.6**

Construction of a mathematical model

Step 1. Identify the problem.
Step 2. Make assumptions.
      a. Identify and classify the variables.
      b. Determine interrelationships between the variables and submodels.
Step 3. Solve the model.
Step 4. Verify the model.
      a. Does it address the problem?
      b. Does it make common sense?
      c. Test it with real-world data.
Step 5. Implement the model.
Step 6. Maintain the model.

Although these steps admittedly involve a degree of craftsmanship, we will learn some scientific techniques we can apply to appraise the importance of a particular variable and the preciseness of an assumed relationship. Nevertheless, when generating numbers in Steps 3 and 4, remember that the process has been largely inexact and intuitive.

EXAMPLE 1 *Vehicular Stopping Distance*

Scenario Consider the following rule often given in driver education classes:

Allow one car length for every 10 miles of speed under normal driving conditions, but more distance in adverse weather or road conditions. One way to accomplish this is to use the 2-second rule for measuring the correct following distance no matter what your speed. To obtain that distance, watch the vehicle ahead of you pass some definite point on the highway, like a tar strip or overpass shadow. Then count to yourself "one thousand and one, one thousand and two;" that is 2 seconds. If you reach the mark before you finish saying those words, then you are following too close behind.

The preceding rule is implemented easily enough, but how good is it?

Problem Identification Our ultimate goal is to test this rule and suggest another rule if it fails. However, the statement of the problem—How good is the rule?—is vague. We need to be more specific and spell out a problem, or ask a question, whose solution or answer will help us accomplish our goal while permitting a more exact mathematical analysis. Consider the following problem statement: *Predict the vehicle's total stopping distance as a function of its speed.*

Assumptions We begin our analysis with a rather obvious model for total stopping distance:

$$\text{total stopping distance} = \text{reaction distance} + \text{braking distance}$$

By *reaction distance*, we mean the distance the vehicle travels from the instant the driver perceives a need to stop to the instant when the brakes are actually applied. *Braking distance* is the distance required for the brakes to bring the vehicle to a complete stop.

First let's develop a submodel for reaction distance. The reaction distance is a function of many variables, and we start by listing just two of them:

$$\text{reaction distance} = f(\text{response time, speed})$$

We could continue developing the submodel with as much detail as we like. For instance, response time is influenced by both individual driving factors and the vehicle operating system. System time is the time from when the driver touches the brake pedal until the brakes are mechanically applied. For modern cars we would probably neglect the influence of the system because it is quite small in comparison to the human factors. The portion of the response time determined by the driver depends on many things, such as reflexes, alertness, and visibility. Because we are developing only a general rule, we could just incorporate average values and conditions for these latter variables. Once all the variables deemed important to the submodel have been identified, we can begin to determine interrelationships among them. We suggest a submodel for reaction distance in the next section.

Next consider the braking distance. The weight and speed of the vehicle are certainly important factors to be taken into account. The efficiency of the brakes, type and condition of the tires, road surface, and weather conditions are other legitimate factors. As before, we would most likely assume average values and conditions for these latter factors. Thus, our initial submodels give braking distance as a function of vehicular weight and speed:

$$\text{braking distance} = h(\text{weight, speed})$$

In the next section we also suggest and analyze a submodel for braking distance.

Finally, let's discuss briefly the last three steps in the modeling process for this problem. We would want to test our model against real-world data. Do the predictions afforded by the model agree with real driving situations? If not, we would want to assess some of our assumptions and perhaps restructure one (or both) of our submodels. If the model does predict real driving situations accurately, then does the rule stated in the opening discussion agree with the model? The answer gives an objective basis for answering, How good is the rule? Whatever rule we come up with (to implement the model), it must be easy to understand and easy to use if it is going to be effective. In this example, maintenance of the model does not seem to be a particular issue. Nevertheless, we would want to be sensitive to the effects on the model of such changes as power brakes or disc brakes, a fundamental change in tire design, and so on. ▨ ▨ ▩

Let's contrast the modeling process presented in Figure 2.6 with the scientific method. One version of the **scientific method** is as follows:

STEP 1 Make some general observations of a phenomenon.

STEP 2 Formulate a hypothesis about the phenomenon.

STEP 3 Develop a method to test that hypothesis.

STEP 4 Gather data to use in the test.

STEP 5 Test the hypothesis using the data.

STEP 6 Confirm or deny the hypothesis.

By design, the mathematical modeling process and scientific method have similarities. For instance, both processes involve making assumptions or hypotheses, gathering real-world data, and testing or verification using that data. These similarities should not be surprising; though recognizing that part of the modeling process is an art, we do attempt to be scientific and objective whenever possible.

There are also subtle differences between the two processes. One difference lies in the primary goal of the two processes. In the modeling process, assumptions are made in selecting which variables to include or neglect and in postulating the interrelationships among the included variables. The goal in the modeling process is to *hypothesize a model*, and, as with the scientific method, evidence is gathered to corroborate that model. Unlike the scientific method, however, the objective is not to *confirm* or *deny* the model (we already know it is not precisely correct because of the simplifying assumptions we have made) but rather to test its *reasonableness*. We may decide that the model is quite satisfactory and useful, and elect to accept it. Or we may decide that the model needs to be refined or simplified. In extreme cases we may even need to redefine the problem, in a sense rejecting the model altogether. We will see in subsequent chapters that this decision process really constitutes the heart of mathematical modeling.

Iterative Nature of Model Construction

Model construction is an iterative process. We begin by examining some system and iden-tifying the particular behavior we wish to predict or explain. Next we identify the variables and simplifying assumptions, and then we generate a model. We will generally start with a rather simple model, progress through the modeling process, and then refine the model as the results of our validation procedure dictate. If we cannot come up with a model or solve the one we have, we must *simplify* it (Figure 2.7). This is done by treating some variables as constants, by neglecting or aggregating some variables, by assuming simple relationships (such as linearity) in any submodel, or by further restricting the problem under investigation. On the other hand, if our results are not precise enough, we must refine the model (Figure 2.7).

Refinement is generally achieved in the opposite way to simplification: We introduce ad-ditional variables, assume more sophisticated relationships among the variables, or expand the scope of the problem. By simplification and refinement, we determine the generality, realism, and precision of our model. This process cannot be overemphasized and constitutes the art of modeling. These ideas are summarized in Table 2.1.

We complete this section by introducing several terms that are useful in describing models. A model is said to be **robust** when its conclusions do not depend on the precise satisfaction of the assumptions. A model is **fragile** if its conclusions do depend on the precise satisfaction of some sort of conditions. The term **sensitivity** refers to the degree of change in a model's conclusions as some condition on which they depend is varied; the greater the change, the more sensitive is the model to that condition.

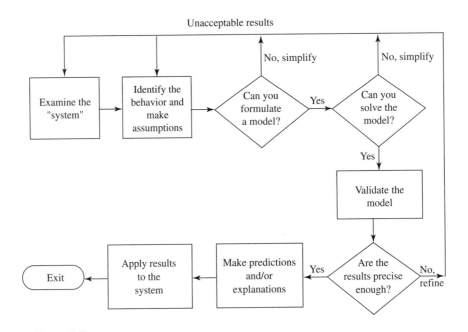

■ **Figure 2.7**

The iterative nature of model construction

Table 2.1   The art of mathematical modeling: simplifying or refining the model as required

Model simplification	Model refinement
1. Restrict problem identification.	1. Expand the problem.
2. Neglect variables.	2. Consider additional variables.
3. Conglomerate effects of several variables.	3. Consider each variable in detail.
4. Set some variables to be constant.	4. Allow variation in the variables.
5. Assume simple (linear) relationships.	5. Consider nonlinear relationships.
6. Incorporate more assumptions.	6. Reduce the number of assumptions.

2.1   PROBLEMS

In Problems 1–8, the scenarios are vaguely stated. From these vague scenarios, identify a problem you would like to study. Which variables affect the behavior you have identified in the problem identification? Which variables are the most important? Remember, there are really no right answers.

1. The population growth of a single species.

2. A retail store intends to construct a new parking lot. How should the lot be illuminated?

3. A farmer wants to maximize the yield of a certain crop of food grown on his land. Has the farmer identified the correct problem? Discuss alternative objectives.

4. How would you design a lecture hall for a large class?

5. An object is to be dropped from a great height. When and how hard will it hit the ground?

6. How should a manufacturer of some product decide how many units of that product should be manufactured each year and how much to charge for each unit?

7. The United States Food and Drug Administration is interested in knowing if a new drug is effective in the control of a certain disease in the population.

8. How fast can a skier ski down a mountain slope?

For the scenarios presented in Problems 9–17, identify a problem worth studying and list the variables that affect the behavior you have identified. Which variables would be neglected completely? Which might be considered as constants initially? Can you identify any submodels you would want to study in detail? Identify any data you would want collected.

9. A botanist is interested in studying the shapes of leaves and the forces that mold them. She clips some leaves from the bottom of a white oak tree and finds the leaves to be rather broad and not very deeply indented. When she goes to the top of the tree, she finds very deeply indented leaves with hardly any broad expanse of blade.

10. Animals of different sizes work differently. Small ones have squeaky voices, their hearts beat faster, and they breathe more often than larger ones. On the other hand, the skeleton of a larger animal is more robustly built than that of a small animal. The ratio of the

diameter to the length is greater in a larger animal than it is in a smaller one. Thus there are regular distortions in the proportions of animals as the size increases from small to large.

11. A physicist is interested in studying properties of light. He wants to understand the path of a ray of light as it travels through the air into a smooth lake, particularly at the interface of the two different media.

12. A company with a fleet of trucks faces increasing maintenance costs as the age and mileage of the trucks increase.

13. People are fixated by speed. Which computer systems offer the most speed?

14. How can we improve our ability to sign up for the best classes each term?

15. How should we save a portion of our earnings?

16. Consider a new company that is just getting started in producing a single product in a competitive market situation. Discuss some of the short-term and long-term goals the company might have as it enters into business. How do these goals affect employee job assignments? Would the company necessarily decide to maximize profits in the short run?

17. Discuss the differences between using a model to predict versus using one to explain a real-world system. Think of some situations in which you would like to explain a system. Likewise, imagine other situations in which you would want to predict a system.

2.1 PROJECTS

1. Consider the taste of brewed coffee. What are some of the variables affecting taste? Which variables might be neglected initially? Suppose you hold all variables fixed except water temperature. Most coffeepots use boiled water in some manner to extract the flavor from the ground coffee. Do you think boiled water is optimal for producing the best flavor? How would you test this submodel? What data would you collect and how would you gather them?

2. A transportation company is considering transporting people between skyscrapers in New York City via helicopter. You are hired as a consultant to determine the number of helicopters needed. Identify an appropriate problem precisely. Use the model-building process to identify the data you would like to have to determine the relationships between the variables you select. You may want to redefine your problem as you proceed.

3. Consider wine making. Suggest some objectives a commercial producer might have. Consider taste as a submodel. What are some of the variables affecting taste? Which variables might be neglected initially? How would you relate the remaining variables? What data would be useful to determine the relationships?

4. Should a couple buy or rent a home? As the cost of a mortgage rises, intuitively, it would seem that there is a point where it no longer pays to buy a house. What variables determine the total cost of a mortgage?

5. Consider the operation of a medical office. Records have to be kept on individual patients, and accounting procedures are a daily task. Should the office buy or lease a small computer system? Suggest objectives that might be considered. What variables would you consider? How would you relate the variables? What data would you like to have to determine the relationships between the variables you select? Why might solutions to this problem differ from office to office?

6. When should a person replace his or her vehicle? What factors should affect the decision? Which variables might be neglected initially? Identify the data you would like to have to determine the relationships among the variables you select.

7. How far can a person long jump? In the 1968 Olympic Games in Mexico City, Bob Beamon of the United States increased the record by a remarkable 10%, a record that stood through the 1996 Olympics. List the variables that affect the length of the jump. Do you think the low air density of Mexico City accounts for the 10% difference?

8. Is college a financially sound investment? Income is forfeited for 4 years, and the cost of college is extremely high. What factors determine the total cost of a college education? How would you determine the circumstances necessary for the investment to be profitable?

2.2 Modeling Using Proportionality

We introduced the concept of proportionality in Chapter 1 to model change. Recall that

$$y \propto x \text{ if and only if } y = kx \text{ for some constant } k > 0 \qquad (2.1)$$

Of course, if $y \propto x$, then $x \propto y$ because the constant k in Equation (2.1) is greater than zero and then $x = (\frac{1}{k})y$. The following are other examples of proportionality relationships:

$$y \propto x^2 \text{ if and only if } y = k_1 x^2 \text{ for } k_1 \text{ a constant} \qquad (2.2)$$
$$y \propto \ln x \text{ if and only if } y = k_2 \ln x \text{ for } k_2 \text{ a constant} \qquad (2.3)$$
$$y \propto e^x \text{ if and only if } y = k_3 e^x \text{ for } k_3 \text{ a constant} \qquad (2.4)$$

In Equation (2.2), $y = kx^2, k > 0$, so we also have $x \propto y^{1/2}$ because $x = (\frac{1}{\sqrt{k}})y^{1/2}$. This leads us to consider how to link proportionalities together, a transitive rule for proportionality:

$$y \propto x \quad \text{and} \quad x \propto z, \text{ then} \quad y \propto z$$

Thus, any variables proportional to the same variables are proportional to one another.

Now let's explore a geometric interpretation of proportionality. In Equation (2.1), $y = kx$ yields $k = y/x$. Thus, k may be interpreted as the tangent of the angle θ depicted in Figure 2.8, and the relation $y \propto x$ defines a set of points along a line in the plane with angle of inclination θ.

Comparing the general form of a proportionality relationship $y = kx$ with the equation for a straight line $y = mx + b$, we can see that the graph of a proportionality relationship is a

■ **Figure 2.8**

Geometric interpretation of
$y \propto x$

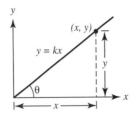

■ **Figure 2.9**

Geometric interpretation of
Models (a) (2.2), (b) (2.3),
and (c) (2.4)

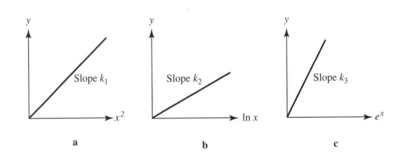

line (possibly extended) passing through the origin. If we plot the proportionality variables for Models (2.2)–(2.4), we obtain the straight-line graphs presented in Figure 2.9.

It is important to note that not just any straight line represents a proportionality relationship: The y-intercept *must* be zero so that the line passes through the origin. Failure to recognize this point can lead to erroneous results when using our model. For example, suppose we are interested in predicting the volume of water displaced by a boat as it is loaded with cargo. Because a floating object displaces a volume of water equal to its weight, we might be tempted to assume that the total volume y of displaced water is proportional to the weight x of the added cargo. However, there is a flaw with that assumption because the unloaded boat already displaces a volume of water equal to its weight. Although the graph of total volume of displaced water versus weight of added cargo is given by a straight line, it is not given by a line passing through the origin (Figure 2.10), so the proportionality assumption is incorrect.

■ **Figure 2.10**

A straight-line relationship
exists between displaced
volume and total weight,
but it is not a proportionality
because the line fails to
pass through the origin.

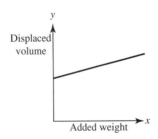

A proportionality relationship may, however, be a reasonable *simplifying assumption,* depending on the size of the y-intercept and the slope of the line. The domain of the independent variable can also be significant since the relative error

$$\frac{y_a - y_p}{y_a}$$

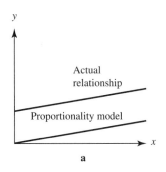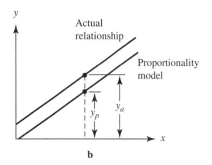

■ Figure 2.11

Proportionality as a simplifying assumption

is greater for small values of x. These features are depicted in Figure 2.11. If the slope is nearly zero, proportionality may be a poor assumption because the initial displacement dwarfs the effect of the added weight. For example, there would be virtually no effect in placing 400 lb on an aircraft carrier already weighing many tons. On the other hand, if the initial displacement is relatively small and the slope is large, the effect of the initial displacement is dwarfed quickly, and proportionality is a good simplifying assumption.

EXAMPLE 1 *Kepler's Third Law*

To assist in further understanding the idea of proportionality, let's examine one of the famous proportionalities from Table 2.2, Kepler's third law. In 1601, the German astronomer Johannes Kepler became director of the Prague Observatory. Kepler had been helping Tycho Brahe in collecting 13 years of observations on the relative motion of the planet Mars. By 1609, Kepler had formulated his first two laws:

1. Each planet moves along an ellipse with the sun at one focus.
2. For each planet, the line from the sun to the planet sweeps out equal areas in equal times.

Kepler spent many years verifying these laws and formulating the third law given in Table 2.2, which relates the orbital periods and mean distances of the planets from the sun.

Table 2.2 Famous proportionalities

Hooke's law: $F = kS$, where F is the restoring force in a spring stretched or compressed a distance S.

Newton's law: $F = ma$ or $a = \frac{1}{m}F$, where a is the acceleration of a mass m subjected to a net external force F.

Ohm's law: $V = iR$, where i is the current induced by a voltage V across a resistance R.

Boyle's law: $V = \frac{k}{p}$, where under a constant temperature k, the volume V is inversely proportional to the pressure p.

Einstein's theory of relativity: $E = c^2M$, where under the constant speed of light squared c^2, the energy E is proportional to the mass M of the object.

Kepler's third law: $T = cR^{\frac{3}{2}}$, where T is the period (days) and R is the mean distance to the sun.

Table 2.3   Orbital periods and mean distances of planets from the sun

Planet	Period (days)	Mean distance (millions of miles)
Mercury	88.0	36
Venus	224.7	67.25
Earth	365.3	93
Mars	687.0	141.75
Jupiter	4331.8	483.80
Saturn	10,760.0	887.97
Uranus	30,684.0	1764.50
Neptune	60,188.3	2791.05
Pluto	90,466.8	3653.90

The data shown in Table 2.3 are from the 1993 *World Almanac*.

In Figure 2.12, we plot the period versus the mean distance to the $\frac{3}{2}$ power. The plot approximates a line that projects through the origin. We can easily estimate the slope (constant of proportionality) by picking any two points that lie on the line passing through the origin:

$$slope = \frac{90,466.8 - 88}{220,869.1 - 216} \approx 0.410$$

We estimate the model to be $T = 0.410R^{3/2}$. ▪ ▪ ■

■ **Figure 2.12**

Kepler's third law as a proportionality

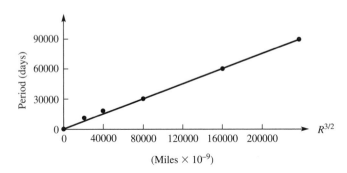

(Miles $\times 10^{-9}$)

Modeling Vehicular Stopping Distance

Consider again the scenario posed in Example 1 of Section 2.1. Recall the general rule that allows one car length for every 10 mph of speed. It was also stated that this rule is the same as allowing for 2 seconds between cars. The rules are in fact different from one another (at least for most cars). For the rules to be the same, at 10 mph both should allow one car length:

$$1 \text{ car length} = \text{distance} = \left(\frac{\text{speed in ft}}{\text{sec}}\right)(2 \text{ sec})$$

$$= \left(\frac{10 \text{ miles}}{\text{hr}}\right)\left(\frac{5280 \text{ ft}}{\text{mi}}\right)\left(\frac{1 \text{ hr}}{3600 \text{ sec}}\right)(2 \text{ sec}) = 29.33 \text{ ft}$$

This is an unreasonable result for an average car length of 15 ft, so the rules are *not* the same.

■ Figure 2.13

Geometric interpretation of
the one-car-length rule

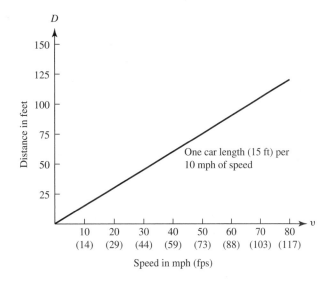

Let's interpret the one-car-length rule geometrically. If we assume a car length of 15 ft and plot this rule, we obtain the graph shown in Figure 2.13, which shows that the distance allowed by the rule is proportional to the speed. In fact, if we plot the speed in feet per second, the constant of proportionality has the units seconds and represents the total time for the equation $D = kv$ to make sense. Moreover, in the case of a 15-ft car, we obtain a constant of proportionality as follows:

$$k = \frac{15 \text{ ft}}{10 \text{ mph}} = \frac{15 \text{ ft}}{52{,}800 \text{ ft}/3600 \text{ sec}} = \frac{90}{88} \text{ sec}$$

In our previous discussion of this problem, we presented the model

$$\text{total stopping distance} = \text{reaction distance} + \text{braking distance}$$

Let's consider the submodels for reaction distance and braking distance.
Recall from Example 1 of Section 2.1 that

$$\text{reaction distance} = f(\text{response time, speed})$$

Now assume that the vehicle continues at constant speed from the time the driver determines the need to stop until the brakes are applied. Under this assumption, reaction distance d_r is simply the product of response time t_r and velocity v:

$$d_r = t_r v \qquad (2.5)$$

To test Submodel (2.5), plot measured reaction distance versus velocity. If the resultant graph approximated a straight line through the origin, we could estimate the slope t_r and feel fairly confident in the submodel. Alternatively, we could test a group of drivers representative of the assumptions made in the example in Section 2.1 and estimate t_r directly.
Next, consider the braking distance:

$$\text{braking distance} = h(\text{weight, speed})$$

Suppose that there is a panic stop and the maximum brake force F is applied throughout the stop. The brakes are basically an energy-dissipating device; that is, the brakes do work on the vehicle producing a change in the velocity that results in a loss of kinetic energy. Now, the work done is the force F times the braking distance d_b. This work must equal the change in kinetic energy, which, in this situation, is simply $0.5\ mv^2$. Thus, we have

$$\text{work done} = Fd_b = 0.5mv^2 \tag{2.6}$$

Next, we consider how the force F relates to the mass of the car. A reasonable design criterion would be to build cars in such a way that the maximum deceleration is constant when the maximum brake force is applied, regardless of the mass of the car. Otherwise, the passengers and driver would experience an unsafe jerk during the braking to a complete stop. This assumption means that the panic deceleration of a larger car, such as a Cadillac, is the same as that of a small car, such as a Honda, owing to the design of the braking system. Moreover, constant deceleration occurs throughout the panic stop. From Newton's second law, $F = ma$, it follows that the force F is proportional to the mass. Combining this result with Equation (2.6) gives the proportionality relation

$$d_b \propto v^2$$

At this point we might want to design a test for the two submodels, or we could test the submodels against the data provided by the U.S. Bureau of Public Roads given in Table 2.4.

Figure 2.14 depicts the plot of driver reaction distance against velocity using the data in Table 2.4. The graph is a straight line of approximate slope 1.1 passing through the origin; our results are too good! Because we always expect some deviation in experimental results, we should be suspicious. In fact, the results of Table 2.4 are based on Submodel (2.5), where an average response time of 3/4 sec was obtained independently. Thus we might later decide to design another test for the submodel.

To test the submodel for braking distance, we plot the observed braking distance recorded in Table 2.4 against v^2, as shown in Figure 2.15. Proportionality seems to be

Table 2.4   Observed reaction and braking distances

Speed (mph)	Driver reaction distance (ft)	Braking distance* (ft)		Total stopping distance (ft)	
20	22	18–22	(20)	40–44	(42)
25	28	25–31	(28)	53–59	(56)
30	33	36–45	(40.5)	69–78	(73.5)
35	39	47–58	(52.5)	86–97	(91.5)
40	44	64–80	(72)	108–124	(116)
45	50	82–103	(92.5)	132–153	(142.5)
50	55	105–131	(118)	160–186	(173)
55	61	132–165	(148.5)	193–226	(209.5)
60	66	162–202	(182)	228–268	(248)
65	72	196–245	(220.5)	268–317	(292.5)
70	77	237–295	(266)	314–372	(343)
75	83	283–353	(318)	366–436	(401)
80	88	334–418	(376)	422–506	(464)

*Interval given includes 85% of the observations based on tests conducted by the U.S. Bureau of Public Roads. Figures in parentheses represent average values.

■ Figure 2.14

Proportionality of reaction distance and speed

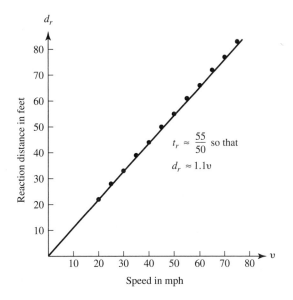

a reasonable assumption at the lower speeds, although it does seem to be less convincing at the higher speeds. By graphically fitting a straight line to the data, we estimate the slope and obtain the submodel:

$$d_b = 0.054v^2 \qquad (2.7)$$

We will learn how to fit the model to the data analytically in Chapter 3.

■ Figure 2.15

Proportionality of braking distance and the speed squared

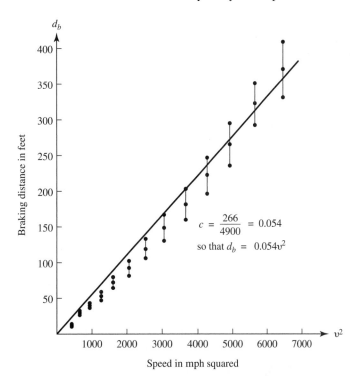

■ **Figure 2.16**

Total stopping distance

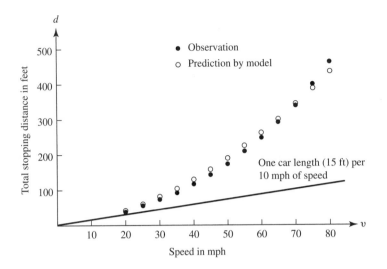

Summing Submodels (2.6) and (2.7), we obtain the following model for the total stopping distance d:

$$d = 1.1v + 0.054v^2 \tag{2.8}$$

The predictions of Model (2.8) and the actual observed stopping distance recorded in Table 2.4 are plotted in Figure 2.16. Considering the grossness of the assumptions and the inaccuracies of the data, the model seems to agree fairly reasonably with the observations up to 70 mph. The rule of thumb of one 15-ft car length for every 10 mph of speed is also plotted in Figure 2.16. We can see that the rule significantly underestimates the total stopping distance at speeds exceeding 40 mph.

Let's suggest an alternative rule of thumb that is easy to understand and use. Assume the driver of the trailing vehicle must be fully stopped by the time he or she reaches the point occupied by the lead vehicle at the exact time of the observation. Thus, the driver must trail the lead vehicle by the total stopping distance, based either on Model (2.8) or on the observed data in Table 2.4. The maximum stopping distance can readily be converted to a trailing time. The results of these computations for the observed distances, in which 85% of the drivers were able to stop, are given in Table 2.5. These computations suggest the following general rule:

Speed (mph)	Guideline (sec)
0–10	1
10–40	2
40–60	3
60–75	4

This alternative rule is plotted in Figure 2.17. An alternative to using such a rule might be to convince manufacturers to modify existing speedometers to compute stopping distance and time for the car's speed v based on Equation (2.8). We will revisit the braking distance problem in Section 11.3 with a model based on the derivative.

Table 2.5 Time required to allow the proper stopping distance

Speed				Trailing time required for maximum stopping
(mph)	(fps)	Stopping distance* (ft)		distance (sec)
20	(29.3)	42	(44)†	1.5
25	(36.7)	56	(59)	1.6
30	(44.0)	73.5	(78)	1.8
35	(51.3)	91.5	(97)	1.9
40	(58.7)	116	(124)	2.1
45	(66.0)	142.5	(153)	2.3
50	(73.3)	173	(186)	2.5
55	(80.7)	209.5	(226)	2.8
60	(88.0)	248	(268)	3.0
65	(95.3)	292.5	(317)	3.3
70	(102.7)	343	(372)	3.6
75	(110.0)	401	(436)	4.0
80	(117.3)	464	(506)	4.3

*Includes 85% of the observations based on tests conducted by the U.S. Bureau of Public Roads.
†Figures in parentheses under stopping distance represent maximum values and are used to calculate trailing times.

■ **Figure 2.17**

Total stopping distance and alternative general rule. The plotted observations are the maximum values from Table 2.4.

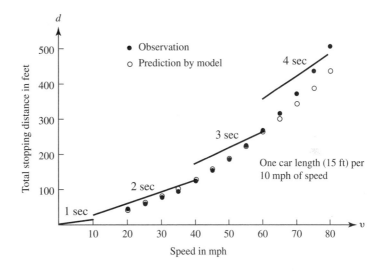

2.2 PROBLEMS

1. Show graphically the meaning of the proportionality $y \propto u/v$.

2. Explain the meaning of the following proportionality argument and illustrate it graphically.

$$w \propto \left(\frac{f}{u}\right)^{-2}$$

3. If a spring is stretched 0.37 in. by a 14-lb force, what stretch will be produced by a 9-lb force? By a 22-lb force? Assume Hooke's law, which asserts that the distance stretched is proportional to the force applied.

4. If an architectural drawing is scaled so that 0.75 in. represents 4 ft, what length represents 27 ft?

5. A road map has scale 1 inch = 6 miles. You measure the distance from home to the ski resort you plan to go visit as 11.75 inches. How many miles will you be traveling? What assumptions are you making?

6. Determine whether the following data support a proportionality argument for $y \propto z^{1/2}$. If so, estimate the slope.

y	3.5	5	6	7	8
z	3	6	9	12	15

In Problems 7–12, determine whether the data set supports the stated proportionality model.

7. Force $\propto$ Stretch

Force	10	20	30	40	50	60	70	80	90
Stretch	19	57	94	134	173	216	256	297	343

8. $y \propto x^3$

y	19	25	32	51	57	71	113	141	123	187	192	205	252	259	294
x	17	19	20	22	23	25	28	31	32	33	36	37	38	39	41

9. $d \propto v^2$

d	22	28	33	39	44	50	55	61	66	72	77
v	20	25	30	35	40	45	50	55	60	65	70

10. $y \propto x^2$

y	4	11	22	35	56	80	107	140	175	215
x	1	2	3	4	5	6	7	8	9	10

11. $y \propto x^3$

y	0	1	2	6	14	24	37	58	82	114
x	1	2	3	4	5	6	7	8	9	10

12. $y \propto e^x$

y	6	15	42	114	311	845	2300	6250	17000	46255
x	1	2	3	4	5	6	7	8	9	10

13. A new planet is discovered beyond Pluto at a mean distance to the sun of 4004 million miles. Using Kepler's third law, determine an estimate for the time T to travel around the sun in an orbit.

14. For the vehicular stopping distance model, design a test to determine the average response time. Design a test to determine average reaction distance. Discuss the difference between the two statistics. If you were going to use the results of these tests to predict the total stopping distance, would you want to use the average reaction distance? Explain your reasoning.

15. For the submodel concerning braking distance in the vehicular stopping distance model, how would you design a brake system so that the maximum deceleration is constant for all vehicles regardless of their mass? Consider the surface area of the brake pads and the capacity of the hydraulic system to apply a force.

2.2 PROJECTS

1. Consider an automobile suspension system. Build a model that relates the stretch (or compression) of the spring to the mass it supports. If possible, obtain a car spring and collect data by measuring the change in spring size to the mass supported by the spring. Graphically test your proportionality argument. If it is reasonable, find the constant of proportionality.

2. Research and prepare a 10-minute report on Hooke's law.

2.3 Modeling Using Geometric Similarity

Geometric similarity is a concept related to proportionality and can be useful to simplify the mathematical modeling process.

Definition

Two objects are said to be **geometrically similar** if there is a one-to-one correspondence between points of the objects such that the ratio of distances between corresponding points is constant for all possible pairs of points.

For example, consider the two boxes depicted in Figure 2.18. Let l denote the distance between the points A and B in Figure 2.18a, and let l' be the distance between the corresponding points A' and B' in Figure 2.18b. Other corresponding points in the two figures, and the associated distances between the points, are marked the same way. For the boxes to be geometrically similar, it must be true that

$$\frac{l}{l'} = \frac{w}{w'} = \frac{h}{h'} = k, \quad \text{for some constant } k > 0$$

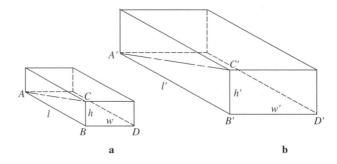

Let's interpret the last result geometrically. In Figure 2.18, consider the triangles ABC and $A'B'C'$. If the two boxes are geometrically similar, these two triangles must be similar. The same argument can be applied to any corresponding pair of triangles, such as CBD and $C'B'D'$. Thus, *corresponding angles are equal for objects that are geometrically similar.* In other words, the shape is the same for two geometrically similar objects, and one object is simply an enlarged copy of the other. We can think of geometrically similar objects as scaled replicas of one another, as in an architectural drawing in which all the dimensions are simply scaled by some constant factor k.

One advantage that results when two objects are geometrically similar is a simplification in certain computations, such as volume and surface area. For the boxes depicted in Figure 2.18, consider the following argument for the ratio of the volumes V and V':

$$\frac{V}{V'} = \frac{lwh}{l'w'h'} = k^3 \tag{2.9}$$

Similarly, the ratio of their total surface areas S and S' is given by

$$\frac{S}{S'} = \frac{2lh + 2wh + 2wl}{2l'h' + 2w'h' + 2w'l'} = k^2 \tag{2.10}$$

Not only are these ratios immediately known once the scaling factor k has been specified, but also the surface area and volume may be expressed as a proportionality in terms of some selected **characteristic dimension**. Let's select the length l as the characteristic dimension. Then with $l/l' = k$, we have

$$\frac{S}{S'} = k^2 = \frac{l^2}{l'^2}$$

Therefore,

$$\frac{S}{l^2} = \frac{S'}{l'^2} = \text{constant}$$

holds for any two geometrically similar objects. That is, surface area is always proportional to the square of the characteristic dimension length:

$$S \propto l^2$$

Likewise, volume is proportional to the length cubed:

$$V \propto l^3$$

Thus, if we are interested in some function depending on an object's length, surface area, and volume, for example,

$$y = f(l, S, V)$$

we could express all the function arguments in terms of some selected characteristic dimension, such as length, giving

$$y = g(l, l^2, l^3)$$

Geometric similarity is a powerful simplifying assumption.

EXAMPLE 1 *Raindrops from a Motionless Cloud*

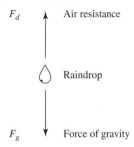

Suppose we are interested in the terminal velocity of a raindrop from a motionless cloud. Examining the free-body diagram, we note that the only forces acting on the raindrop are gravity and drag. Assume that the atmospheric drag on the raindrop is proportional to its surface area S times the square of its speed v. The mass m of the raindrop is proportional to the weight of the raindrop (assuming constant gravity in Newton's second law):

$$F = F_g - F_d = ma$$

Under terminal velocity (assuming v equals v_t), we have $a = 0$, so Newton's second law is reduced to

$$F_g - F_d = 0$$

or

$$F_g = F_d$$

We are assuming that $F_d \propto Sv^2$ and that F_g is proportional to weight w. Since $m \propto w$, we have $F_g \propto m$.

Next we assume all the raindrops are geometrically similar. This assumption allows us to relate area and volume so that

$$S \propto l^2 \quad \text{and} \quad V \propto l^3$$

for any characteristic dimension l. Thus $l \propto S^{1/2} \propto V^{1/3}$, which implies

$$S \propto V^{2/3}$$

Because weight and mass are proportional to volume, the transitive rule for proportionality gives

$$S \propto m^{2/3}$$

From the equation $F_g = F_d$, we now have $m \propto m^{2/3} v_t^2$. Solving for the terminal velocity, we have

$$m^{1/3} \propto v_t^2 \quad \text{or} \quad m^{1/6} \propto v_t$$

Therefore, the terminal velocity of the raindrop is proportional to its mass raised to the one-sixth power.   ■ ■ ■

Testing Geometric Similarity

The principle of geometric similarity suggests a convenient method for testing to determine whether it holds among a collection of objects. Because the definition requires that the ratio of distances between corresponding pairs of points be the same for all pairs of points, we can test that requirement to see if the objects in a given collection are geometrically similar.

For example, we know that circles are geometrically similar (because all circles have the same shape, possibly varying only in size). If c denotes the circumference of a circle, d its diameter, and s the length of arc along the circle subtended by a given (fixed) angle θ, then we know from geometry that

$$c = \pi d \quad \text{and} \quad s = \left(\frac{d}{2}\right)\theta$$

Thus, for any two circles,

$$\frac{c_1}{c_2} = \frac{\pi d_1}{\pi d_2} = \frac{d_1}{d_2}$$

and

$$\frac{s_1}{s_2} = \frac{(d_1/2)\theta}{(d_2/2)\theta} = \frac{d_1}{d_2}$$

That is, the ratio of distances between corresponding points (points that have the same fixed angle) as we go around any two circles is always the ratio of their diameters. This observation supports the reasonableness of the geometric similarity argument for the circles.

EXAMPLE 2   *Modeling a Bass Fishing Derby*

For conservation purposes, a sport fishing club wishes to encourage its members to release their fish immediately after catching them. The club also wishes to grant awards based on the total weight of fish caught: honorary membership in the 100 Pound Club, Greatest Total

Weight Caught during a Derby Award, and so forth. How does someone fishing determine the weight of a fish he or she has caught? You might suggest that each individual carry a small portable scale. However, portable scales tend to be inconvenient and inaccurate, especially for smaller fish.

Problem Identification   We can identify the problem as follows: *Predict the weight of a fish in terms of some easily measurable dimensions.*

Assumptions   Many factors that affect the weight of a fish can easily be identified. Different species have different shapes and different average weights per unit volume (weight density) based on the proportions and densities of meat, bone, and so on. Gender also plays an important role, especially during spawning season. The various seasons probably have a considerable effect on weight.

Since a general rule for sport fishing is sought, let's initially restrict attention to a single species of fish, say bass, and assume that within the species the average weight density is constant. Later, it may be desirable to refine our model if the results prove unsatisfactory or if it is determined that considerable variability in density does exist. Furthermore, let's neglect gender and season. Thus, initially we will predict weight as a function of size (volume) and constant average weight density.

Assuming that all bass are geometrically similar, the volume of any bass is proportional to the cube of some characteristic dimension. Note that we are not assuming any particular shape, but only that the bass are scaled models of one another. The basic shape can be quite irregular as long as the ratio between corresponding pairs of points in two distinct bass remains constant for all possible pairs of points. This idea is illustrated in Figure 2.19.

■ **Figure 2.19**

Fish that are geometrically similar are simply scaled models of one another.

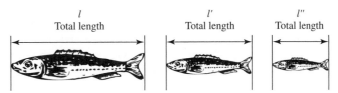

l
Total length

l'
Total length

l''
Total length

Now choose the length l of the fish as the characteristic dimension. This choice is depicted in Figure 2.19. Thus, the volume of a bass satisfies the proportionality

$$V \propto l^3$$

Because weight W is volume times average weight density and we are assuming a constant average density, it follows immediately that

$$W \propto l^3$$

Model Verification   Let's test our model. Consider the following data collected during a fishing derby.

Length, l (in.)	14.5	12.5	17.25	14.5	12.625	17.75	14.125	12.625
Weight, W (oz)	27	17	41	26	17	49	23	16

■ **Figure 2.20**

If the model is valid, the graph of W versus l^3 should be a straight line passing through the origin.

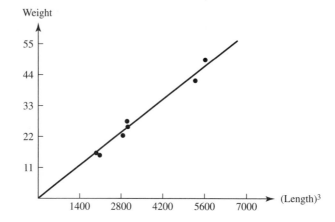

If our model is correct, then the graph of W versus l^3 should be a straight line passing through the origin. The graph showing an approximating straight line is presented in Figure 2.20. (Note that the judgment here is qualitative. In Chapter 3 we develop analytic methods to determine a best-fitting model for collected data.)

Let's accept the model, at least for further testing, based on the small amount of data presented so far. Because the data point $(14.5^3, 26)$ lies along the line we have drawn in Figure 2.20, we can estimate the slope of the line as $26/3049 = 0.00853$, yielding the model

$$W = 0.00853l^3 \tag{2.11}$$

Of course, if we had drawn our line a little differently, we would have obtained a slightly different slope. In Chapter 3 you will be asked to show analytically that the coefficients that minimize the sum of squared deviations between the model $W = kl^3$ and the given data points is $k = 0.008437$. A graph of Model (2.11) is presented in Figure 2.21, showing also a plot of the original data points.

Model (2.11) provides a convenient general rule. For example, from Figure 2.21 we might estimate that a 12-in. bass weighs approximately 1 lb. This means that an 18-in. bass

■ **Figure 2.21**

Graph of the model $W = 0.00853\,l^3$

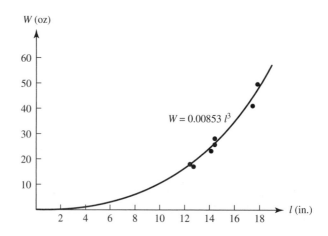

should weigh approximately $(1.5)^3 = 3.4$ lb and a 24-in. bass approximately $2^3 = 8$ lb. For the fishing derby, a card converting the length of a caught fish to its weight in ounces or pounds could be given to each angler, or a cloth tape or a retractable metal tape could be marked with a conversion scale if the use of the rule becomes popular enough. A conversion scale for Model (2.11) is as follows:

Length (in.)	12	13	14	15	16	17	18	19	20	21	22	23	24	25	26
Weight (oz)	15	19	23	29	35	42	50	59	68	79	91	104	118	133	150
Weight (lb)	0.9	1.2	1.5	1.8	2.2	2.6	3.1	3.7	4.3	4.9	5.7	6.5	7.4	8.3	9.4

Even though our rule seems reasonable based on the limited data we have obtained, an angler may not like the rule because it does not reward catching a fat fish: The model treats fat and skinny fish alike. Let's address this dissatisfaction. Instead of assuming that the fish are geometrically similar, assume that only their cross-sectional areas are similar. This does not imply any particular shape for the cross section, only that the definition of geometric similarity is satisfied. We choose the characteristic dimension to be the girth, g, defined subsequently.

Now assume that the major portion of the weight of the fish is from the main body. Thus, the head and tail contribute relatively little to the total weight. Constant terms can be added later if our model proves worthy of refinement. Next assume that the main body is of varying cross-sectional area. Then the volume can be found by multiplying the average cross-sectional area A_{avg} by the effective length l_{eff}:

$$V \approx l_{eff}(A_{avg})$$

How shall the effective length l_{eff} and the average cross-sectional area A_{avg} be measured? Have the fishing contestants measure the length of the fish l as before, and assume the proportionality $l_{eff} \propto l$. To estimate the average cross-sectional area, have each angler take a cloth measuring tape and measure the circumference of the fish at its widest point. Call this measurement the girth, g. Assume the average cross-sectional area is proportional to the square of the girth. Combining these two proportionality assumptions gives

$$V \propto l g^2$$

Finally, assuming constant density, $W \propto V$, as before, so

$$W = k l g^2 \tag{2.12}$$

for some positive constant k.

Several assumptions have been made, so let's get an initial test of our model. Consider again the following data.

Length, l (in.)	14.5	12.5	17.25	14.5	12.625	17.75	14.125	12.625
Girth, g (in.)		9.75	8.375	11.0	9.75	8.5	12.5	9.0	8.5
Weight, W (oz)	27	17	41	26	17	49	23	16

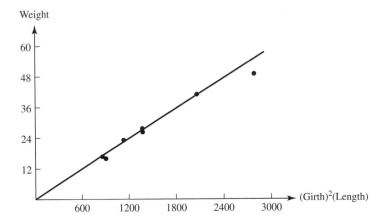

Because our model suggests a proportionality between W and lg^2, we consider a plot of W versus lg^2. This plot is depicted in Figure 2.22. The plotted data lie approximately along a straight line passing through the origin, so the proportionality assumption seems reasonable. Now the point corresponding to the 41-oz fish happens to lie along the line shown in Figure 2.22, so the slope can be estimated as

$$\frac{41}{(17.25)(11)^2} \approx 0.0196$$

This computation leads to the model

$$W = 0.0196lg^2 \tag{2.13}$$

In Chapter 3 we show analytically that choosing the slope in such a way that the sum of the squared deviations from the given data points is minimized leads to the model

$$W = 0.0187lg^2$$

An angler would probably be happier with the new rule (2.13), because doubling the girth leads to a fourfold increase in the weight of the fish. However, the model appears more inconvenient to apply. Because $1/0.0196 \approx 50.9$, we could round this coefficient to 50 and have an angler apply one of the simple rules:

$$W = \frac{lg^2}{50} \text{ for } W \text{ in ounces and } l, g \text{ measured in inches}$$

$$W = \frac{lg^2}{800} \text{ for } W \text{ in pounds and } l, g \text{ measured in inches}$$

However, the application of either of the preceding rules would probably require the contestant to record the length and the girth of each fish and then compute its weight on a four-function calculator. Or perhaps he or she could be given a two-dimensional card showing the correct weights for different values of length and girth. The competitors probably would prefer a simple plastic disk on which girth and length measurements could be entered in such a way that the weight of the bass appeared in a window. You are asked to design such a disk in the following problem section. ■ ■ ■

EXAMPLE 3 *Modeling the Size of the "Terror Bird"*

South America and Africa began to drift apart along the mid-oceanic ridge approximately 80 million years ago. Because of the great tectonic force, South America drifted to the west, away from Africa. During the next 75 million years, South America's flora and fauna evolved in isolation and produced many different plant and animal forms. About 7 million years ago, the Isthmus of Panama arose, connecting South America with North America. This allowed for the interchange of formerly separated species. The interchange enabled animals such as camels, deer, elephants, horses, cats, and dogs to travel south, and the North American fauna received an infusion of mammals such as giant ground sloths, anteaters, and aquatic capybaras. They were accompanied by groups of birds called terror birds.

The terror birds were giant, flightless, predatory birds. The terror bird known as *Titanis walleri* was a fleet hunter that would lie in ambush and attack from the tall grasses. These birds killed with their beaks, pinning down their prey with an inner toe claw 4 to 5 in. long and shredding their prey. These birds had arms (not wings) that were most like those of a bipedal dinosaur. Figure 2.23 shows an artist's rendering of the terror bird. It is the largest predatory bird known to have existed, and paleontologists believe these flightless predatory birds evolved as the dominant carnivores on land.

■ **Figure 2.23**

Artist's rendering of
Titanis walleri

The various terror birds ranged in size from 5 ft to 9 ft tall, *Titanis* being the largest. Because very little fossil material of *Titanis* has been discovered, its exact size is unclear. According to Dr. Chandler, who uncovered the fossil in 1994, estimates range from 6 to 7 ft tall. Let's see whether we can use modeling to learn something more about the terror bird.

Problem Identification   Predict the weight of the terror bird as a function of the circumference of its femur.

Assumptions and Variables   We assume that the terror birds are geometrically similar to other large birds of today. With this assumption of geometric similarity, we have that the volume of the bird is proportional to any characteristic dimension cubed.

$$V \propto l^3$$

If we assume a constant weight density, then the terror bird's volume is proportional to its weight, $V \propto W$, and we have

$$V \propto W \propto l^3$$

Let the characteristic dimension l be the circumference of the femur, which is chosen because it supported the body weight, giving the model

$$W = kl^3, k > 0$$

Testing the Model   We use the data set for various bird sizes in Table 2.6 for our model. First, we have a data set from birds of various sizes. These data are appropriate because the terror bird was a bird.

Figure 2.24 shows a scatterplot for Table 2.6 and reveals that the trend is concave up and increasing.

Table 2.6   Femur circumference and body weight of birds

Femur circumference (cm)	Body weight (kg)
0.7943	0.0832
0.7079	0.0912
1.000	0.1413
1.1220	0.1479
1.6982	0.2455
1.2023	0.2818
1.9953	0.7943
2.2387	2.5119
2.5119	1.4125
2.5119	0.8913
3.1623	1.9953
3.5481	4.2658
4.4668	6.3096
5.8884	11.2202
6.7608	19.95
15.136	141.25
15.85	158.4893

Because our proposed model is $W = kl^3, k > 0$, we plot W versus l^3 and obtain approximately a straight line through the origin (see Figure 2.25). Thus our proposed model is reasonably accurate. The slope of the line projected through the origin is approximately 0.0398, giving

$$W = 0.0398l^3$$

■ Figure 2.24

Scatterplot of bird data

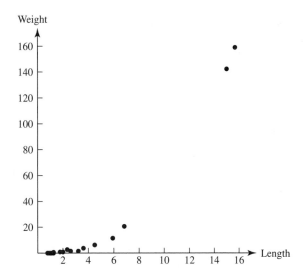

We next plot this model with the original data to visually examine how close the model comes to fitting the trend of the data (see Figure 2.26).

Predicting the terror bird's weight using the model $W = 0.0398l^3$ and the circumference of the femur of the terror bird measured as 21 cm, we find that the weight was approximately 368.58 kg.

Problem 7 in this section gives dinosaur data relating the size (weight) of prehistoric animals to the circumference of their femurs. Completing that problem will enable you to compare the model obtained in Figure 2.26 with a second model based on dinosaur data. ■ ■ ■

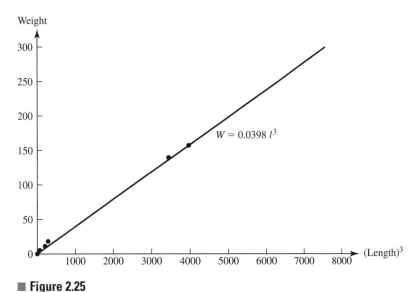

■ Figure 2.25

Plot of weight versus length^3

■ **Figure 2.26**

The model and the data

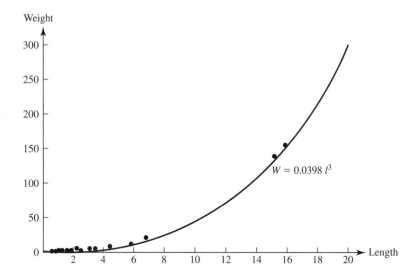

PROBLEMS

1. Suppose two maps of the same country on different scales are drawn on tracing paper and superimposed, with possibly one of the maps being turned over before being super-imposed on the other. Show that there is just one place that is represented by the same spot on both maps.

2. Consider a 20-lb pink flamingo that stands 3 ft in height and has legs that are 2 ft in length. Model the height and leg length of a 100-lb flamingo. What assumptions are necessary? Are they reasonable assumptions?

3. An object is sliding down a ramp inclined at an angle of θ radians and attains a terminal velocity before reaching the bottom. Assume that the drag force caused by the air is proportional to Sv^2, where S is the cross-sectional area perpendicular to the direction of motion and v is the speed. Further assume that the sliding friction between the object and the ramp is proportional to the normal weight of the object. Determine the relationship between the terminal velocity and the mass of the object. If two different boxes, weighing 600 and 800 lb, are pushed down the ramp, find the relationship between their terminal velocities.

4. Assume that under certain conditions the heat loss of an object is proportional to the exposed surface area. Relate the heat loss of a cubic object with side length 6 in. to one with a side length of 12 in. Now, consider two irregularly shaped objects, such as two submarines. Relate the heat loss of a 70-ft submarine to that of a 7-ft scale model. Suppose you are interested in the amount of energy needed to maintain a constant internal temperature in the submarine. Relate the energy needed in the actual submarine to that required by the scaled model. Specify the assumptions you have made.

5. Consider the situation of two warm-blooded adult animals essentially at rest and under the same conditions (as in a zoo). Assume that the animals maintain the same body temperature and that the energy available to maintain this temperature is proportional to the amount of food provided to them. Challenge this assumption. If you are willing to assume that the animals are geometrically similar, relate the amounts of food necessary to maintain their body temperatures to their lengths and volumes. (*Hint:* See Problem 4.) List any assumptions you have made. What additional assumptions are necessary to relate the amount of food necessary to maintain their body weight?

6. Design a plastic disk to perform the calculations given by Model (2.13).

7. Consider Models (2.11) and (2.13). Which do you think is better? Why? Discuss the models qualitatively. In Chapter 3 you will be asked to compare the two models analytically.

8. In what circumstances, if any, will Models (2.11) and (2.13) coincide? Explain fully.

9. Consider the models $W \propto l^2 g$ and $W \propto g^3$. Interpret each of these models geometrically. Explain how these two models differ from Models (2.11) and (2.13), respectively. In what circumstances, if any, would the four models coincide? Which model do you think would do the best job of predicting W? Why? In Chapter 3 you will be asked to compare the four models analytically.

   a. Let $A(x)$ denote a typical cross-sectional area of a bass, $0 \leq x \leq l$, where l denotes the length of the fish. Use the mean value theorem from calculus to show that the volume V of the fish is given by

   $$V = l \cdot \overline{A}$$

   where $\overline{A}$ is the average value of $A(x)$.

   b. Assuming that $\overline{A}$ is proportional to the square of the girth g and that weight density for the bass is constant, establish that

   $$W \propto l g^2$$

2.3 PROJECTS

1. *Superstars*—In the TV show *Superstars* the top athletes from various sports compete against one another in a variety of events. The athletes vary considerably in height and weight. To compensate for this in the weight-lifting competition, the body weight of the athlete is subtracted from his lift. What kind of relationship does this suggest? Use the following table, which displays the winning lifts at the 1996 Olympic Games, to show this relationship.

Winning lifts from the 2000 Olympic Games

Max participant weight (kg)	Lift (kg)	Male (M) or female (F)
<48	185	F
48–53	225	F
58–63	242.50	F
75	245	F
>75	300	F
56	305	M
62	325	M
69	357.50	M
77	376.50	M
85	390	M
94	405	M
105	425	M
>105	472.5	M

Physiological arguments have been proposed that suggest that the strength of a muscle is proportional to its cross-sectional area. Using this submodel for strength, construct a model relating lifting ability and body weight. List all assumptions. Do you have to assume that all weight lifters are geometrically similar? Test your model with the data provided.

Now consider a refinement to the previous model. Suppose there is a certain amount of body weight that is independent of size in adults. Suggest a model that incorporates this refinement, and test it against the data provided.

Criticize the use of the preceding data. What data would you really like to have to handicap the weight lifters? Who is the best weight lifter according to your models? Suggest a rule of thumb for the *Superstars* show to handicap the weight lifters.

2. *Heart Rate of Birds*—Warm-blooded animals use large quantities of energy to maintain body temperature because of the heat loss through the body surface. In fact, biologists believe that the primary energy drain on a resting warm-blooded animal is maintenance of body temperature.

Bird	Body weight (g)	Pulse rate (beats/min)
Canary	20	1000
Pigeon	300	185
Crow	341	378
Buzzard	658	300
Duck	1100	190
Hen	2000	312
Goose	2300	240
Turkey	8750	193
Ostrich	71000	60–70

Data from A. J. Clark, *Comparative Physiology of the Heart* (New York: Macmillan, 1977), p. 99.

a. Construct a model relating blood flow through the heart to body weight. Assume that the amount of energy available is proportional to the blood flow through the lungs, which is the source of oxygen. Assuming the least amount of blood needed to circulate, the amount of available energy will equal the amount of energy used to maintain the body temperature.

b. The following data relate weights of some birds to their heart rate measured in beats per minute. Construct a model that relates heart rate to body weight. Discuss the assumptions of your model. Use the data to check your model.

3. *Heart Rate of Mammals*—The following data relate the weights of some mammals to their heart rate in beats per minute. Based on the discussion relating blood flow through the heart to body weight, as presented in Project 2, construct a model that relates heart rate to body weight. Discuss the assumptions of your model. Use the data to check your model.

Mammal	Body weight (g)	Pulse rate (beats/min)
Vespergo pipistrellas	4	660
Mouse	25	670
Rat	200	420
Guinea pig	300	300
Rabbit	2000	205
Little dog	5000	120
Big dog	30000	85
Sheep	50000	70
Man	70000	72
Horse	450000	38
Ox	500000	40
Elephant	3000000	48

Data from A. J. Clark, *Comparative Physiology of the Heart* (New York: Macmillan, 1977), p. 99.

4. *Lumber Cutters*—Lumber cutters wish to use readily available measurements to estimate the number of board feet of lumber in a tree. Assume they measure the diameter of the tree in inches at waist height. Develop a model that predicts board feet as a function of diameter in inches.

Use the following data for your test:

x	17	19	20	23	25	28	32	38	39	41
y	19	25	32	57	71	113	123	252	259	294

The variable x is the diameter of a ponderosa pine in inches, and y is the number of board feet divided by 10.

a. Consider two separate assumptions, allowing each to lead to a model. Completely analyze each model.

i. Assume that all trees are right-circular cylinders and are approximately the same height.

   ii. Assume that all trees are right-circular cylinders and that the height of the tree is proportional to the diameter.

b. Which model appears to be better? Why? Justify your conclusions.

5. *Racing Shells*—If you have been to a rowing regatta, you may have observed that the more oarsmen there are in a boat, the faster the boat travels. Investigate whether there is a mathematical relationship between the speed of a boat and the number of crew members. Consider the following assumptions (partial list) in formulating a model:

   a. The total force exerted by the crew is constant for a particular crew throughout the race.

   b. The drag force experienced by the boat as it moves through the water is proportional to the square of the velocity times the wet surface area of the hull.

   c. Work is defined as force times distance. Power is defined as work per unit time.

Crew	Distance (m)	Race 1 (sec)	Race 2 (sec)	Race 3 (sec)	Race 4 (sec)	Race 5 (sec)	Race 6 (sec)
1	2500	20:53	22:21	22:49	26:52		
2	2500	19:11	19:17	20:02			
4	2500	16:05	16:42	16:43	16:47	16:51	17:25
8	2500	9:19	9:29	9:49	9:51	10:21	10:33

   Hint: When additional oarsmen are added to a shell, it is not obvious whether the amount of *force* is proportional to the number in the crew or the amount of *power* is proportional to the number in the crew. Which assumption appears the most reasonable? Which yields a more accurate model?

6. *Scaling a Braking System*—Suppose that after years of experience, your auto company has designed an optimum braking system for its prestigious full-sized car. That is, the distance required to brake the car is the best in its weight class, and the occupants feel the system is very smooth. Your firm has decided to build cars in the lighter weight classes. Discuss how you would scale the braking system of your current car to have the same performance in the smaller versions. Be sure to consider the hydraulic system and the size of the brake pads. Would a simple geometric similarity suffice? Let's suppose that the wheels are scaled in such a manner that the pressure (at rest) on the tires is constant in all car models. Would the brake pads seem proportionally larger or smaller in the scaled-down cars?

7. Data have been collected on numerous dinosaurs during the prehistoric period. Using proportionality and geometric similarity, build a mathematical model to relate the weight of the terror bird to its femur circumference. Recall that the femur circumference of the terror bird in Example 3 was 21 cm. Compare the weight found using this new model to the weight found in Example 3. Which model would you prefer? Give reasons justifying your preference.

Dinosaur data

Name	Femur circumference (mm)	Weight (kg)
Hypsilophodonitdae	103	55
Ornithomimdae	136	115
Thescelosauridae	201	311
Ceratosauridae	267	640
Allosauridae	348	1230
Hadrosauridae-1	400	1818
Hadrosauridae-2	504	3300
Hadrosauridae-3	512	3500
Tyrannosauridae	534	4000

2.4 Automobile Gasoline Mileage

Scenario   During periods of concern when oil shortages and embargoes create an energy crisis, there is always interest in how fuel economy varies with vehicular speed. We suspect that, when driven at low speeds and in low gears, automobiles convert power relatively inefficiently and that, when they are driven at high speeds, drag forces on the vehicle increase rapidly. It seems reasonable, then, to expect that automobiles have one or more speeds that yield optimum fuel mileage (the most miles per gallon of fuel). If this is so, fuel mileage would decrease beyond that optimal speed, but it would be beneficial to know just how this decrease takes place. Moreover, is the decrease significant? Consider the following excerpt from a newspaper article (written when a national 55-mph speed limit existed):

> *Observe the 55-mile-an-hour national highway speed limit. For every 5 miles an hour over 50, there is a loss of 1 mile to the gallon. Insisting that drivers stay at the 55-mile-an-hour mark has cut fuel consumption 12 percent for Ryder Truck Lines of Jacksonville, Florida—a savings of 631,000 gallons of fuel a year. The most fuel-efficient range for driving generally is considered to be between 35 and 45 miles an hour.*[1]

Note especially the suggestion that there is a loss of 1 mile to the gallon for every 5 miles an hour over 50 mph. How good is this general rule?

Problem Identification   *What is the relationship between the speed of a vehicle and its fuel mileage?* By answering this question, we can assess the accuracy of this rule.

Assumptions   Let's consider the factors influencing fuel mileage. First, there are propulsion forces that drive the vehicle forward. These forces depend on the power available from the type of fuel being burned, the engine's efficiency in converting that potential power, gear ratios, air temperature, and many other factors, including vehicular velocity. Next, there are drag forces that tend to retard the vehicle's forward motion. The drag forces include frictional

---

[1]"Boost Fuel Economy," *Monterey Peninsula Herald*, May 16, 1982.

effects that depend on the vehicle's weight, the type and condition of the tires, and the condition of the road surface. Air resistance is another drag force and depends on the vehicular speed, vehicular surface area and shape, the wind, and air density. Another factor influencing fuel mileage is related to the driving habits of the driver. Does he or she drive at constant speeds or constantly accelerate? Does he or she drive on level or mountainous terrain? Thus, fuel mileage is a function of several factors, summarized in the following equation:

$$\text{fuel mileage} = f(\text{propulsion forces, drag forces, driving habits, and so on})$$

It is clear that the answer to the original problem will be quite detailed considering all the possible combinations of car types, drivers, and road conditions. Because such a study is far too ambitious to be undertaken here, we restrict the problem we are willing to address.

Restricted Problem Identification   *For a particular driver, driving his or her car on a given day on a level highway at constant highway speeds near the optimal speed for fuel economy, provide a qualitative explanation of how fuel economy varies with small increases in speed.*

Under this restricted problem, environmental conditions (such as air temperature, air density, and road conditions) can be considered as constant. Because we have specified that the driver is driving his or her car, we have fixed the tire conditions, the shape and surface of the vehicle, and fuel type. By restricting the highway driving speeds to be near the optimal speed, we obtain the simplifying assumptions of constant engine efficiency and constant gear ratio over small changes in vehicular velocity. Restricting a problem as originally posed is a very powerful technique for obtaining a manageable model.

The newspaper article from which this problem was derived also gave the rule of thumb that for every 5 mph over 50 mph, there is a loss of 1 mile to the gallon. Let's graph this rule. If you plot the loss in miles per gallon against speed minus 50, the graph is a straight line passing through the origin as depicted in Figure 2.27. Let's see if this linear graph is qualitatively correct.

Since the automobile is to be driven at constant speeds, the acceleration is zero. We know from Newton's second law that the resultant force must be zero, or that the forces of propulsion and resistance must be in equilibrium. That is,

$$F_p = F_r$$

where F_p represents the propulsion force and F_r the resisting force.

■ **Figure 2.27**

For every 5 mph over
50 mph, there is a loss of
1 mile to the gallon.

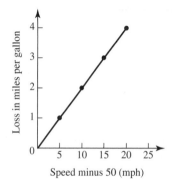

Speed minus 50 (mph)

First consider the force of propulsion. Each gallon of gasoline contains an amount of energy, say K. If C_r represents the amount of fuel burned per unit time, then $C_r K$ represents the power available to the car. Assuming a constant rate of power conversion, it follows that the converted power is proportional to $C_r K$. Because power is the product of force and velocity for a constant force, this argument yields the proportionality relation

$$F_p \propto \frac{C_r K}{v}$$

When we further assume a constant fuel rating K, this last proportionality simplifies to

$$F_p \propto \frac{C_r}{v} \tag{2.14}$$

Next consider the resisting force. Because we restricted the problem to highway speeds, it is reasonable to assume that frictional forces are small when compared to the drag forces caused by air resistance. At highway speeds, one sensible submodel for these drag forces is

$$F_r \propto S v^2$$

where S is the cross-sectional area perpendicular to the direction of the moving car. (This assumption is commonly used in engineering for moderate sizes of S.) Because S is constant in our restricted problem, it follows that

$$F_r \propto v^2$$

Application of the condition $F_p = F_r$ and the proportionality (2.14) then yields

$$\frac{C_r}{v} \propto v^2$$

or

$$C_r \propto v^3 \tag{2.15}$$

The proportionality (2.15) gives the qualitative information that the fuel consumption rate should increase as the cube of the velocity. However, fuel consumption rate is not an especially good indicator of fuel efficiency: Although proportionality (2.15) says that the car is using more fuel per unit time at higher speeds, the car is also traveling farther. Therefore, we define gasoline mileage as follows:

$$\text{mileage} = \frac{\text{distance}}{\text{consumption}}$$

Substitution of vt for distance and $C_r t$ for consumption then gives the proportionality

$$\text{mileage} = \frac{v}{C_r} \propto v^{-2} \tag{2.16}$$

Thus, gasoline mileage is inversely proportional to the square of the velocity.

Model (2.16) provides some useful qualitative information to assist us in explaining automobile fuel consumption. First, we should be very suspicious of the rule of thumb suggesting a linear graph as depicted in Figure 2.27. We should be careful about the conclusions we do draw. Although the power relationship in Equation (2.16) appears impressive, it is valid only over a restricted range of speeds. Over that restricted range the relationship could be nearly linear, depending on the size of the constant of proportionality. Moreover, do not forget that we have ignored many factors in our analysis and have assumed that several important factors are constant. Thus, our model is quite fragile, and its use is limited to a qualitative explanation over a restricted range of speeds. But then, that is precisely how we identified the problem.

2.4   PROBLEMS

1. In the automobile gasoline mileage example, suppose you plot miles per gallon against speed as a graphical representation of the general rule (instead of the graph depicted in Figure 2.27). Explain why it would be difficult to deduce a proportionality relationship from that graph.

2. In the automobile gasoline mileage example, assume the drag forces are proportional to Sv, where S is the cross-sectional area perpendicular to the direction of the moving car and v is its speed. What conclusions can you draw? Discuss the factors that might influence the choice of Sv^2 over Sv for the drag forces submodel. How could you test the submodel?

3. Discuss several factors that were completely ignored in our analysis of the gasoline mileage problem.

2.5   Body Weight and Height, Strength and Agility

Body Weight and Height

A question of interest to almost all Americans is, How much should I weigh? A rule often given to people desiring to run a marathon is 2 lb of body weight per inch of height, but shorter marathon runners seem to have a much easier time meeting this rule than taller ones. Tables have been designed to suggest weights for different purposes. Doctors are concerned about a reasonable weight for health purposes, and Americans seek weight standards based on physical appearance. Moreover, some organizations, such as the Army, are concerned about physical conditioning and define an upper weight allowance for acceptability. Quite often these weight tables are categorized in some manner. For example, consider Table 2.7, which gives upper weight limits of acceptability for males between the ages of 17 and 21. (The table has no further delineators such as bone structure.)

    If the differences between successive weight entries in Table 2.7 are computed to determine how much weight is allowed for each additional inch of height, it will be seen that throughout a large portion of the table, a constant 5 lb per inch is allowed (with some fours and sixes appearing at the lower and upper ends of the scales, respectively). Certainly this last rule is more liberal than the rule recommended for marathoners (2 lb per inch), but

Table 2.7 Weight versus height for males aged 17–21

Height (in.)	Weight (lb)	Height (in.)	Weight (lb)
60	132	71	185
61	136	72	190
62	141	73	195
63	145	74	201
64	150	75	206
65	155	76	212
66	160	77	218
67	165	78	223
68	170	79	229
69	175	80	234
70	180		

just how reasonable a constant-weight-per-height rule is remains to be seen. In this section we examine qualitatively how weight and height should vary.

Body weight depends on a number of factors, some of which we have mentioned. In addition to height, bone density could be a factor. Is there a significant variation in bone density, or is it essentially constant? What about the relative volume occupied by the bones? Is the volume essentially constant, or are there heavy, medium, and light bone structures? And what about a body density factor? How can differences in the densities of bone, muscle, and fat be accounted for? Do these densities vary? Is body density typically a function of age and gender in the sense that the relative composition of muscle, bone, and fat varies as a person becomes older? Are there different compositions of muscle and fat between males and females of the same age?

Let's define the problem so that bone density is considered constant (by accepting an upper limit) and predict weight as a function of height, gender, age, and body density. The purposes or basis of the weight table must also be specified, so we base the table on physical appearance.

Problem Identification We identify the problem as follows: *For various heights, genders, and age groups, determine upper weight limits that represent maximum levels of acceptability based on physical appearance.*

Assumptions Assumptions about body density are needed if we are to successfully predict weight as a function of height. As one simplifying assumption, suppose that some parts of the body are composed of an inner core of a different density. Assume too that the inner core is composed primarily of bones and muscle material and that the outer core is primarily a fatty material, giving rise to the different densities (Figure 2.28). We next construct submodels to explain how the weight of each core might vary with height.

How does body weight vary with height? To begin, assume that for adults, certain parts of the body, such as the head, have the same volume and density for different people. Thus the weight of an adult is given by

$$W = k_1 + W_{in} + W_{out} \tag{2.17}$$

■ Figure 2.28

Assume that parts of the body are composed of an inner and outer core of distinct densities.

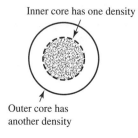

Inner core has one density

Outer core has another density

where $k_1 > 0$ is the constant weight of those parts having the same volume and density for different individuals, and W_{in} and W_{out} are the weights of the inner and outer cores, respectively.

Next, let's turn our attention to the inner core. How does the volume of the extremities and trunk vary with height? We know that people are not geometrically similar because they do not appear to be scaled models of one another, having different shapes and different relative proportions of the trunk and extremities. By definition of the problem at hand, however, we are concerned with an upper weight limit based on physical appearance. Even though this may be somewhat subjective, it would seem reasonable that whatever image might be visualized as an upper limit standard of acceptability for a 74-in. person would be a scaled image of a 65-in. person. Thus, for purposes of our problem, geometric similarity of individuals is a reasonable assumption. Note that no particular shape is being assumed, but only that the ratios of distances between corresponding points in individuals are the same. Under this assumption, the volume of each component we are considering is proportional to the cube of a characteristic dimension, which we select to be height h. Hence, the sum of the components must be proportional to the cube of the height, or

$$V_{in} \propto h^3 \tag{2.18}$$

Now, what should be assumed about the average weight density of the inner core? Assuming that the inner core is composed of muscle and bone, each of which has a different density, what percentage of the total volume of the inner core is occupied by the bones? If bone diameter is assumed to be proportional to the height, then the total volume occupied by the bones is proportional to the cube of the height. This implies that the percentage of the total volume of the inner core occupied by the bones in geometrically similar individuals is constant. It follows from the ensuing argument that the average weight density ρ_{in} is constant as well. For example, consider the average weight density ρ_{avg} of a volume V consisting of two components V_1 and V_2, each with a density ρ_1 and ρ_2. Then

$$V = V_1 + V_2$$

and

$$\rho_{avg} V = W = \rho_1 V_1 + \rho_2 V_2$$

yield

$$\rho_{avg} = \rho_1 \frac{V_1}{V} + \rho_2 \frac{V_2}{V}$$

Therefore, as long as the ratios V_1/V and V_2/V do not change, the average weight density ρ_{avg} is constant. Application of this result to the inner core implies that the average weight density ρ_{in} is constant, yielding

$$W_{in} = V_{in}\rho_{in} \propto h^3$$

or

$$W_{in} = k_2 h^3 \quad \text{for} \quad k_2 > 0 \tag{2.19}$$

Note that the preceding submodel includes any case of materials with densities different from that of muscles and bone (such as tendons, ligaments, and organs) as long as the percentage of the total volume of the inner core occupied by those materials is constant.

Now consider the outer core of fatty material. Because the table is to be based on personal appearance, it can be argued that the thickness of the outer core should be constant regardless of the height (see Problem 3). If τ represents this thickness, then the weight of the outer core is

$$W_{out} = \tau \rho_{out} S_{out}$$

where S_{out} is the surface area of the outer core and ρ_{out} is the density of the outer core. Again assuming that the subjects are geometrically similar, it follows that the surface area is proportional to the square of the height. If the density of the outer core of fatty material is assumed to be constant for all individuals, then we have

$$W_{out} \propto h^2$$

It may be argued, however, that taller people can carry a greater thickness for the fatty layer. If it is assumed that the thickness of the outer core is proportional to the height, then

$$W_{out} \propto h^3$$

Allowing both these assumptions to reside in a single submodel gives

$$W_{out} = k_3 h^2 + k_4 h^3, \quad \text{where } k_3, k_4 \geq 0 \tag{2.20}$$

Here the constants k_3 and k_4 are allowed to assume a zero value.

Summing the submodels represented by Equations (2.17), (2.19), and (2.20) to determine a model for weight yields

$$W = k_1 + k_3 h^2 + k_5 h^3 \quad \text{for} \quad k_1, k_5 > 0 \quad \text{and} \quad k_3 \geq 0 \tag{2.21}$$

where $k_5 = k_2 + k_4$. Note that Model (2.21) suggests variations in weight of a higher order than the first power of h. If the model is valid, then taller people will indeed have a difficult time satisfying the linear rules given earlier. At the moment, however, our judgment can be only qualitative because we have not verified our submodels. Some ideas on how to test the model are discussed in the problem set. Furthermore, we have not placed any relative significance on the higher-order terms occurring in Equation (2.21). In the study of statistics, regression techniques are given to provide insight into the significance of each term in the model.

Model Interpretation Let's interpret the general rules given earlier, which allowed a constant weight increase for each additional inch of height, in terms of our submodel. Consider the amount of weight attributable to an increase in the length of the trunk. Because the total allowable weight increase per inch is assumed constant by the given rules, the portion allowed for the trunk increase may also be assumed constant. To allow a constant weight increase, the trunk must increase in length while maintaining *the same cross-sectional area*. This implies, for example, that the waist size remains constant. Let's suppose that in the age 17–21 category, a 30-in. waist is judged the upper limit acceptable for the sake of personal appearance in a male with a height of 66 in. The 2-lb-per-inch rule would allow a 30-in. waist for a male with a height of 72 in. as well. On the other hand, the model based on geometric similarity suggests that all distances between corresponding points should increase by the same ratio. Thus, the male with a height of 72 in. should have a waist of 30(72/66), or approximately 32.7 in., to be proportioned similarly. Comparing the two models, we get the following data:

Height (in.)	Linear models (in., waist measure)	Geometric similarity model (in., waist measure)
66	30	30.0
72	30	32.7
78	30	35.5
84	30	38.2

Now we can see why tall marathoners who follow the 2-lb-per-inch rule appear *very thin*.

Strength and Agility

Consider a competitive sports contest in which men or women of various sizes compete in events emphasizing strength (such as weight lifting) or agility (such as running an obstacle course). How would you handicap such events? Let's define a problem as follows.

Problem Identification *For various heights, weights, genders, and age groups, determine their relationship with agility in competitive sports.*

Assumptions Let's initially neglect gender and age. We assume that agility is proportional to the ratio strength/weight. We further assume that strength is proportional to the size of the muscles being used in the event, and we measure this size in terms of the muscles' cross-sectional area. (See Project 1 in Section 2.3.) Recall too that weight is proportional to volume (assuming constant weight density). If we assume all participants are geometrically similar, we have

$$\text{agility} \propto \frac{\text{strength}}{\text{weight}} \propto \frac{l^2}{l^3} \propto \frac{1}{l}$$

This is clearly a *nonlinear relationship* between agility and the characteristic dimension, l. We also see that under these assumptions, agility has a *nonlinear relationship* with weight. How would you collect data for your model? How would you test and verify your model?

2.5 PROBLEMS

1. Describe in detail the data you would like to obtain to test the various submodels supporting Model (2.21). How would you go about collecting the data?

2. Tests exist to measure the percentage of body fat. Assume that such tests are accurate and that a great many carefully collected data are available. You may specify any other statistics, such as waist size and height, that you would like collected. Explain how the data could be arranged to check the assumptions underlying the submodels in this section. For example, suppose the data for males between ages 17 and 21 with constant body fat and height are examined. Explain how the assumption of constant density of the inner core could be checked.

3. A popular measure of physical condition and personal appearance is the pinch test. To administer this test, you measure the thickness of the outer core at selected locations on the body by pinching. Where and how should the pinch be made? What thickness of pinch should be allowed? Should the pinch thickness be allowed to vary with height?

4. It has been said that gymnastics is a sport that requires great agility. Use the model and assumptions developed for agility in this section to argue why there are few tall gymnasts.

2.5 PROJECT

1. Consider an endurance test that measures only aerobic fitness. This test could be a swimming test, running test, or bike test. Assume that we want all competitors to do an equal amount of work. Build a mathematical model that relates work done by the competitor to some measurable characteristic, such as height or weight. Next consider a refinement using kinetic energy in your model. Collect some data for one of these aerobic tests and determine the reasonableness of these models.

3

Model Fitting

Introduction

In the mathematical modeling process we encounter situations that cause us to analyze data for different purposes. We have already seen how our assumptions can lead to a model of a particular type. For example, in Chapter 2 when we analyzed the distance required to bring a car to a safe stop once the brakes are applied, our assumptions led to a submodel of the form

$$d_b = Cv^2$$

where d_b is the distance required to stop the car, v is the velocity of the car at the time the brakes are applied, and C is some arbitrary constant of proportionality. At this point we can collect and analyze sufficient data to determine whether the assumptions are reasonable. If they are, we want to determine the constant C that selects the particular member from the family $y = Cv^2$ corresponding to the braking distance submodel.

We may encounter situations in which there are different assumptions leading to different submodels. For example, when studying the motion of a projectile through a medium such as air, we can make different assumptions about the nature of a drag force, such as the drag force being proportional to v or v^2. We might even choose to neglect the drag force completely. As another example, when we are determining how fuel consumption varies with automobile velocity, our different assumptions about the drag force can lead to models that predict that mileage varies as $C_1 v^{-1}$ or as $C_2 v^{-2}$. The resulting problem can be thought of in the following way: First, use some collected data to choose C_1 and C_2 in a way that selects the curve from each family that best fits the data, and then choose whichever resultant model is more appropriate for the particular situation under investigation.

A different case arises when the problem is so complex as to prevent the formulation of a model explaining the situation. For instance, if the submodels involve partial differential equations that are not solvable in closed form, there is little hope for constructing a master model that can be solved and analyzed without the aid of a computer. Or there may be so many significant variables involved that one would not even attempt to construct an explicative model. In such cases, experiments may have to be conducted to investigate the behavior of the independent variable(s) within the range of the data points.

The preceding discussion identifies three possible tasks when we are analyzing a collection of data points:

1. Fitting a selected model type or types to the data.

2. Choosing the most appropriate model from competing types that have been fitted. For example, we may need to determine whether the best-fitting exponential model is a better model than the best-fitting polynomial model.

3. Making predictions from the collected data.

In the first two tasks a model or competing models exist that seem to *explain* the observed behavior. We address these two cases in this chapter under the general heading of *model fitting*. However, in the third case, a model does not exist to explain the observed behavior. Rather, there exists a collection of data points that can be used to *predict* the behavior within some range of interest. In essence, we wish to construct an *empirical model* based on the collected data. In Chapter 4 we study such empirical model construction under the general heading of *interpolation*. It is important to understand both the philosophical and the mathematical distinctions between model fitting and interpolation.

Relationship Between Model Fitting and Interpolation

Let's analyze the three tasks identified in the preceding paragraph to determine what must be done in each case. In Task 1 the precise meaning of *best* model must be identified and the resulting mathematical problem resolved. In Task 2 a criterion is needed for comparing models of different types. In Task 3 a criterion must be established for determining how to make predictions in between the observed data points.

Note the difference in the modeler's attitude in each of these situations. In the two model-fitting tasks a relationship of a particular type is strongly suspected, and the modeler is willing to accept some deviation between the model and the collected data points to have a model that satisfactorily *explains* the situation under investigation. In fact, the modeler expects errors to be present in both the model and the data. On the other hand, when interpolating, the modeler is strongly guided by the data that have been carefully collected and analyzed, and a curve is sought that captures the trend of the data to *predict* in between the data points. Thus, the modeler generally attaches little explicative significance to the interpolating curves. In all situations the modeler may ultimately want to make predictions from the model. However, the modeler tends to emphasize the proposed *models* over the data when model fitting, whereas when interpolating, he or she places greater confidence in the *collected data* and attaches less significance to the form of the model. In a sense, explicative models are *theory* driven, whereas predictive models are *data* driven.

Let's illustrate the preceding ideas with an example. Suppose we are attempting to relate two variables, y and x, and have gathered the data plotted in Figure 3.1. If the

■ **Figure 3.1**

Observations relating the variables y and x

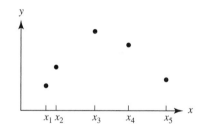

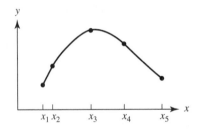

modeler is going to make predictions based solely on the data in the figure, he or she might
use a technique such as *spline interpolation* (which we study in Chapter 4) to pass a smooth
polynomial through the points. Note that in Figure 3.2 the interpolating curve passes through
the data points and captures the trend of the behavior over the range of observations.

Suppose that in studying the particular behavior depicted in Figure 3.1, the modeler
makes assumptions leading to the expectation of a quadratic model, or parabola, of the form
$y = C_1 x^2 + C_2 x + C_3$. In this case the data of Figure 3.1 would be used to determine the
arbitrary constants C_1, C_2, and C_3 to select the best parabola (Figure 3.3). The fact that
the parabola may deviate from some or all of the data points would be of no concern. Note
the difference in the values of the predictions made by the curves in Figures 3.2 and 3.3
in the vicinity of the values x_1 and x_5.

A modeler may find it necessary to fit a model and also to interpolate in the same problem. Using the best-fitting model of a given type may prove unwieldy, or even impossible,
for subsequent analysis involving operations such as integration or differentiation. In such
situations the model may be replaced with an interpolating curve (such as a polynomial)
that is more readily differentiated or integrated. For example, a step function used to model
a square wave might be replaced by a trigonometric approximation to facilitate subsequent
analysis. In these instances the modeler desires the interpolating curve to approximate
closely the essential characteristics of the function it replaces. This type of interpolation is
usually called **approximation** and is typically addressed in introductory numerical analysis
courses.

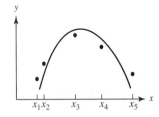

Sources of Error in the Modeling Process

Before discussing criteria on which to base curve-fitting and interpolation decisions, we
need to examine the modeling process to ascertain where errors can arise. If error considerations are neglected, undue confidence may be placed in intermediate results, causing faulty
decisions in subsequent steps. Our goal is to ensure that all parts of the modeling process
are computationally compatible and to consider the effects of cumulative errors likely to
exist from previous steps.

For purposes of easy reference, we classify errors under the following category scheme:

1. Formulation error
2. Truncation error
3. Round-off error
4. Measurement error

Formulation errors result from the assumption that certain variables are negligible or from simplifications in describing interrelationships among the variables in the various submodels. For example, when we determined a submodel for braking distance in Chapter 2, we completely neglected road friction, and we assumed a very simple relationship for the nature of the drag force due to air resistance. Formulation errors are present in even the best models.

Truncation errors are attributable to the numerical method used to solve a mathematical problem. For example, we may find it necessary to approximate $\sin x$ with a polynomial representation obtained from the power series

$$\sin x = x - \frac{x^3}{3!} + \frac{x^5}{5!} - \cdots$$

An error will be introduced when the series is truncated to produce the polynomial.

Round-off errors are caused by using a finite digit machine for computation. Because all numbers cannot be represented exactly using only finite representations, we must always expect round-off errors to be present. For example, consider a calculator or computer that uses 8-digit arithmetic. Then the number $\frac{1}{3}$ is represented by .33333333 so that 3 times $\frac{1}{3}$ is the number .99999999 rather than the actual value 1. The error 10^{-8} is due to round-off. The ideal real number $\frac{1}{3}$ is an *infinite* string of decimal digits .3333..., but any calculator or computer can do arithmetic only with numbers having finite precision. When many arithmetic operations are performed in succession, each with its own round-off, the accumulated effect of round-off can significantly alter the numbers that are supposed to be the answer. Round-off is just one of the things we have to live with—*and be aware of*—when we use computing machines.

Measurement errors are caused by imprecision in the data collection. This imprecision may include such diverse things as human errors in recording or reporting the data or the actual physical limitations of the laboratory equipment. For example, considerable measurement error would be expected in the data reflecting the reaction distance and the braking distance in the braking distance problem.

3.1 Fitting Models to Data Graphically

Assume the modeler has made certain assumptions leading to a model of a particular type. The model generally contains one or more parameters, and sufficient data must be gathered to determine them. Let's consider the problem of data collection.

The determination of how many data points to collect involves a trade-off between the cost of obtaining them and the accuracy required of the model. As a minimum, the modeler needs at least as many data points as there are arbitrary constants in the model curve. Additional points are required to determine any arbitrary constants involved with the requirements of the best-fit method we are using. The *range* over which the model is to be used determines the endpoints of the interval for the independent variable(s).

The *spacing* of the data points within that interval is also important because any part of the interval over which the model must fit particularly well can be weighted by using unequal spacing. We may choose to take more data points where maximum use of the model is expected, or we may collect more data points where we anticipate abrupt changes in the dependent variable(s).

Even if the experiment has been carefully designed and the trials meticulously conducted, the modeler needs to appraise the accuracy of the data before attempting to fit the model. How were the data collected? What is the accuracy of the measuring devices used in the collection process? Do any points appear suspicious? Following such an appraisal and elimination (or replacement) of spurious data, it is useful to think of each data point as an interval of relative confidence rather than as a single point. This idea is shown in Figure 3.4. The length of each interval should be commensurate with the appraisal of the errors present in the data collection process.

■ **Figure 3.4**

Each data point is thought of as an interval of confidence.

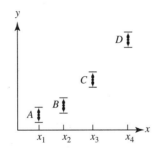

Visual Model Fitting with the Original Data

Suppose we want to fit the model $y = ax + b$ to the data shown in Figure 3.4. How might we choose the constants a and b to determine the line that best fits the data? Generally, when more than two data points exist, not all of them can be expected to lie exactly along a single straight line, even if such a line accurately models the relationship between the two variables x and y. Ordinarily, there will be some vertical discrepancy between a few of the data points and any particular line under consideration. We refer to these vertical discrepancies as **absolute deviations** (Figure 3.5). For the best-fitting line, we might try to

■ **Figure 3.5**

Minimizing the sum of the absolute deviations from the fitted line

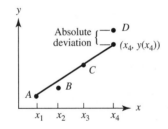

minimize the sum of these absolute deviations, leading to the model depicted in Figure 3.5. Although success may be achieved in minimizing the sum of the absolute deviations, the absolute deviation from individual points may be quite large. For example, consider point D in Figure 3.5. If the modeler has confidence in the accuracy of this data point, there will be concern for the predictions made from the fitted line near the point. As an alternative, suppose a line is selected that minimizes the largest deviation from any point. Applying this criterion to the data points might give the line shown in Figure 3.6.

■ Figure 3.6

Minimizing the largest absolute deviation from the fitted line

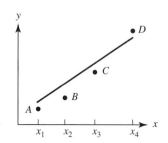

Although these visual methods for fitting a line to data points may appear imprecise, the methods are often quite *compatible* with the accuracy of the modeling process. The grossness of the assumptions and the imprecision involved in the data collection may not warrant a more sophisticated analysis. In such situations, the blind application of one of the analytic methods to be presented in Section 3.2 may lead to models far less appropriate than one obtained graphically. Furthermore, a visual inspection of the model fitted graphically to the data immediately gives an impression of *how good* the fit is and *where* it appears to fit well. Unfortunately, these important considerations are often overlooked in problems with large amounts of data analytically fitted via computer codes. Because the model-fitting portion of the modeling process seems to be more precise and analytic than some of the other steps, there is a tendency to place undue faith in the numerical computations.

Transforming the Data

Most of us are limited visually to fitting only lines. So how can we graphically fit curves as models? Suppose, for example, that a relationship of the form $y = Ce^x$ is suspected for some submodel and the data shown in Table 3.1 have been collected.

The model states that y is proportional to e^x. Thus, if we plot y versus e^x, we should obtain approximately a straight line. The situation is depicted in Figure 3.7. Because the plotted data points do lie approximately along a line that projects through the origin, we conclude that the assumed proportionality is reasonable. From the figure, the slope of the line is approximated as

$$C = \frac{165 - 60.1}{54.6 - 20.1} \approx 3.0$$

Table 3.1 Collected data

x	1	2	3	4
y	8.1	22.1	60.1	165

■ **Figure 3.7**

Plot of y versus e^x for the data given in Table 3.1

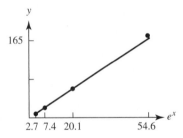

Now let's consider an alternative technique that is useful in a variety of problems. Take the logarithm of each side of the equation $y = Ce^x$ to obtain

$$\ln y = \ln C + x$$

Note that this expression is an equation of a line in the variables $\ln y$ and x. The number $\ln C$ is the intercept when $x = 0$. The transformed data are shown in Table 3.2 and plotted in Figure 3.8. Semilog paper or a computer is useful when plotting large amounts of data.

Table 3.2 The transformed data from Table 3.1

x	1	2	3	4
$\ln y$	2.1	3.1	4.1	5.1

■ **Figure 3.8**

Plot of $\ln y$ versus x using Table 3.2

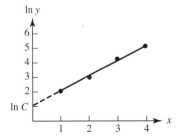

From Figure 3.8, we can determine that the intercept $\ln C$ is approximately 1.1, giving $C = e^{1.1} \approx 3.0$ as before.

A similar transformation can be performed on a variety of other curves to produce linear relationships among the resulting transformed variables. For example, if $y = x^a$, then

$$\ln y = a \ln x$$

is a linear relationship in the transformed variables $\ln y$ and $\ln x$. Here, log–log paper or a computer is useful when plotting large amounts of data.

Let's pause and make an important observation. Suppose we do invoke a transformation and plot $\ln y$ versus x, as in Figure 3.8, and find the line that successfully minimizes the sum of the absolute deviations of the transformed data points. The line then determines $\ln C$, which in turn produces the proportionality constant C. Although it is not obvious, the resulting model $y = Ce^x$ is not the member of the family of exponential curves of the form ke^x that minimizes the sum of the absolute deviations from the original data points (when we plot y versus x). This important idea will be demonstrated both graphically and analytically in the ensuing discussion. When transformations of the form $y = \ln x$ are made, the distance concept is distorted. Although a fit that is compatible with the inherent limitations of a graphical analysis may be obtained, the modeler must be aware of this distortion and *verify the model using the graph from which it is intended to make predictions or conclusions—namely the y versus x graph in the original data rather than the graph of the transformed variables.*

We now present an example illustrating how a transformation may distort distance in the xy-plane. Consider the data plotted in Figure 3.9 and assume the data are expected to fit a model of the form $y = Ce^{1/x}$. Using a logarithmic transformation as before, we find

$$\ln y = \frac{1}{x} + \ln C$$

■ **Figure 3.9**

A plot of some collected data points

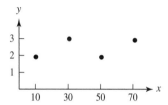

■ **Figure 3.10**

A plot of the transformed data points

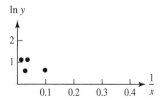

A plot of the points $\ln y$ versus $1/x$ based on the original data is shown in Figure 3.10. Note from the figure how the transformation distorts the distances between the original data points and squeezes them all together. Consequently, if a straight line is made to fit the transformed data plotted in Figure 3.10, the absolute deviations appear relatively small (i.e., small computed on the Figure 3.10 scale rather than on the Figure 3.9 scale). If we were to plot the fitted model $y = Ce^{1/x}$ to the data in Figure 3.9, we would see that it fits the data relatively poorly, as shown in Figure 3.11.

From the preceding example, it can be seen that if a modeler is not careful when using transformations, he or she can be tricked into selecting a relatively poor model. This realization becomes especially important when comparing alternative models. Very serious errors can be introduced when selecting the best model unless all comparisons are made with the original data (plotted in Figure 3.9 in our example). Otherwise, the choice of best model may be determined by a peculiarity of the transformation rather than on the

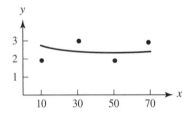

■ Figure 3.11

A plot of the curve $y = Ce^{1/x}$ based on the value $\ln C = 0.9$ from Figure 3.10

merits of the model and how well it fits the original data. Although the danger of making transformations is evident in this graphical illustration, a modeler may be fooled if he or she is not especially observant because many computer codes fit models by first making a transformation. If the modeler intends to use indicators such as the sum of the absolute deviations to make decisions about the adequacy of a particular submodel or choose among competing submodels, the modeler must first ascertain how those indicators were computed.

3.1 PROBLEMS

1. The model in Figure 3.2 would normally be used to predict behavior between x_1 and x_5. What would be the danger of using the model to predict y for values of x less than x_1 or greater than x_5? Suppose we are modeling the trajectory of a thrown baseball.

2. The following table gives the elongation e in inches per inch (in./in.) for a given stress S on a steel wire measured in pounds per square inch (lb/in.^{2}). Test the model $e = c_1 S$ by plotting the data. Estimate c_1 graphically.

$S \ (\times 10^{-3})$	5	10	20	30	40	50	60	70	80	90	100
$e \ (\times 10^5)$	0	19	57	94	134	173	216	256	297	343	390

3. In the following data, x is the diameter of a ponderosa pine in inches measured at breast height and y is a measure of volume—number of board feet divided by 10. Test the model $y = ax^b$ by plotting the transformed data. If the model seems reasonable, estimate the parameters a and b of the model graphically.

x	17	19	20	22	23	25	28	31	32	33	36	37	38	39	41
y	19	25	32	51	57	71	113	141	123	187	192	205	252	259	294

4. In the following data, V represents a mean walking velocity and P represents the population size. We wish to know if we can predict the population size P by observing how fast people walk. Plot the data. What kind of a relationship is suggested? Test the following models by plotting the appropriate transformed data.

   a. $P = aV^b$

   b. $P = a \ln V$

V	2.27	2.76	3.27	3.31	3.70	3.85	4.31	4.39	4.42
P	2500	365	23700	5491	14000	78200	70700	138000	304500

V	4.81	4.90	5.05	5.21	5.62	5.88
P	341948	49375	260200	867023	1340000	1092759

5. The following data represent the growth of a population of fruit flies over a 6-week period. Test the following models by plotting an appropriate set of data. Estimate the parameters of the following models.

   a. $P = c_1 t$

   b. $P = ae^{bt}$

t (days)	7	14	21	28	35	42
P (number of observed flies)	8	41	133	250	280	297

6. The following data represent (hypothetical) energy consumption normalized to the year 1900. Plot the data. Test the model $Q = ae^{bx}$ by plotting the transformed data. Estimate the parameters of the model graphically.

x	Year	Consumption Q
0	1900	1.00
10	1910	2.01
20	1920	4.06
30	1930	8.17
40	1940	16.44
50	1950	33.12
60	1960	66.69
70	1970	134.29
80	1980	270.43
90	1990	544.57
100	2000	1096.63

7. In 1601 the German astronomer Johannes Kepler became director of the Prague Observatory. Kepler had been helping Tycho Brahe in collecting 13 years of observations on the relative motion of the planet Mars. By 1609 Kepler had formulated his first two laws:

   i. Each planet moves on an ellipse with the sun at one focus.

   ii. For each planet, the line from the sun to the planet sweeps out equal areas in equal times.

Kepler spent many years verifying these laws and formulating a third law, which relates the planets' orbital periods and mean distances from the sun.

a. Plot the period time T versus the mean distance r using the following updated observational data.

Planet	Period (days)	Mean distance from the sun (millions of kilometers)
Mercury	88	57.9
Venus	225	108.2
Earth	365	149.6
Mars	687	227.9
Jupiter	4,329	778.1
Saturn	10,753	1428.2
Uranus	30,660	2837.9
Neptune	60,150	4488.9
Pluto	90,670	5876.7

b. Assuming a relationship of the form

$$T = Cr^a$$

determine the parameters C and a by plotting $\ln T$ versus $\ln r$. Does the model seem reasonable? Try to formulate Kepler's third law.

3.2   Analytic Methods of Model Fitting

In this section we investigate several criteria for fitting curves to a collection of data points. Each criterion suggests a method for selecting the best curve from a given family so that according to the criterion, the curve most accurately represents the data. We also discuss how the various criteria are related.

Chebyshev Approximation Criterion

In the preceding section we graphically fit lines to a given collection of data points. One of the best-fit criteria used was to minimize the largest distance from the line to any corresponding data point. Let's analyze this geometric construction. Given a collection of m data points (x_i, y_i), $i = 1, 2, \ldots, m$, fit the collection to the line $y = mx + b$, determined by the parameters a and b, that minimizes the distance between any data point (x_i, y_i) and its corresponding data point on the line $(x_i, ax_i + b)$. That is, minimize the largest absolute deviation $|y_i - y(x_i)|$ over the entire collection of data points. Now let's generalize this criterion.

Given some function type $y = f(x)$ and a collection of m data points (x_i, y_i), minimize the largest absolute deviation $|y_i - f(x_i)|$ over the entire collection. That is, determine the parameters of the function type $y = f(x)$ that minimizes the number

$$\text{Maximum } |y_i - f(x_i)| \qquad i = 1, 2, \ldots, m \qquad (3.1)$$

This important criterion is often called the **Chebyshev approximation criterion**. The difficulty with the Chebyshev criterion is that it is often complicated to apply in practice, at least using only elementary calculus. The optimization problems that result from applying the criterion may require advanced mathematical procedures or numerical algorithms necessitating the use of a computer.

■ **Figure 3.12**

The line segment AC is divided into two segments, AB and BC.

For example, suppose you want to measure the line segments AB, BC, and AC represented in Figure 3.12. Assume your measurements yield the estimates $AB = 13$, $BC = 7$, and $AC = 19$. As you should expect in any physical measuring process, discrepancy results. In this situation, the values of AB and BC add up to 20 rather than the estimated $AC = 19$. Let's resolve the discrepancy of 1 unit using the Chebyshev criterion. That is, we will assign values to the three line segments in such a way that the largest absolute deviation between any corresponding pair of assigned and observed values is minimized. Assume the same degree of confidence in each measurement so that each measurement has equal weight. In that case, the discrepancy should be distributed equally across each segment, resulting in the predictions $AB = 12\frac{2}{3}$, $BC = 6\frac{2}{3}$, and $AC = 19\frac{1}{3}$. Thus, each absolute deviation is $\frac{1}{3}$. Convince yourself that reducing any of these deviations causes one of the other deviations to increase. (Remember that $AB + BC$ must equal AC.) Let's formulate the problem symbolically.

Let x_1 represent the true value of the length of the segment AB and x_2 the true value for BC. For ease of our presentation, let r_1, r_2, and r_3 represent the discrepancies between the true and measured values as follows:

$$x_1 - 13 = r_1 \text{ (line segment } AB)$$
$$x_2 - 7 = r_2 \text{ (line segment } BC)$$
$$x_1 + x_2 - 19 = r_3 \text{ (line segment } AC)$$

The numbers r_1, r_2, and r_3 are called **residuals**. Note that residuals can be positive or negative, whereas absolute deviations are always positive.

If the Chebyshev approximation criterion is applied, values are assigned to x_1 and x_2 in such a way as to minimize the largest of the three numbers $|r_1|$, $|r_2|$, $|r_3|$. If we call that largest number r, then we want to

$$\text{Minimize } r$$

subject to the three conditions:

$$|r_1| \le r \quad \text{or} \quad -r \le r_1 \le r$$
$$|r_2| \le r \quad \text{or} \quad -r \le r_2 \le r$$
$$|r_3| \le r \quad \text{or} \quad -r \le r_3 \le r$$

Each of these conditions can be replaced by two inequalities. For example, $|r_1| \le r$ can be replaced by $r - r_1 \ge 0$ and $r + r_1 \ge 0$. If this is done for each condition, the problem can be stated in the form of a classical mathematical problem:

$$\text{Minimize } r$$

Subject to

$$
\begin{array}{lll}
r - x_1 & + 13 \ge 0 & (r - r_1 \ge 0) \\
r + x_1 & - 13 \ge 0 & (r + r_1 \ge 0) \\
r & - x_2 + 7 \ge 0 & (r - r_2 \ge 0) \\
r & + x_2 - 7 \ge 0 & (r + r_2 \ge 0) \\
r - x_1 & - x_2 + 19 \ge 0 & (r - r_3 \ge 0) \\
r + x_1 & + x_2 - 19 \ge 0 & (r + r_3 \ge 0)
\end{array}
$$

This problem is called a **linear program**. We will discuss linear programs further in Chapter 7. Even large linear programs can be solved by computer implementation of an algorithm known as the **Simplex Method**. In the preceding line segment example, the Simplex Method yields a minimum value of $r = \frac{1}{3}$, and $x_1 = 12\frac{2}{3}$ and $x_2 = 6\frac{2}{3}$.

We now generalize this procedure. Given some function type $y = f(x)$, whose parameters are to be determined, and a collection of m data points (x_i, y_i), define the residuals $r_i = y_i - f(x_i)$. If r represents the largest absolute value of these residuals, then the problem is

$$\text{Minimize } r$$

subject to

$$
\left.
\begin{array}{l}
r - r_i \ge 0 \\
r + r_i \ge 0
\end{array}
\right\} \quad \text{for } i = 1, 2, \ldots, m
$$

Although we discuss linear programs in Chapter 7, we should note here that the model resulting from this procedure is not always a linear program; for example, consider fitting the function $f(x) = \sin kx$. Also note that many computer codes of the Simplex algorithm require using variables that are allowed to assume only nonnegative values. This requirement can be accomplished with simple substitution (see Problem 5).

As we will see, alternative criteria lead to optimization problems that often can be resolved more conveniently. Primarily for this reason, the Chebyshev criterion is not used often for fitting a curve to a finite collection of data points. However, its application should be considered whenever minimizing the largest absolute deviation is important. (We consider several applications of the criterion in Chapter 7.) Furthermore, the principle underlying the Chebyshev criterion is extremely important when one is replacing a function defined over an interval by another function and the largest difference between the two functions over the interval must be minimized. This topic is studied in approximation theory and is typically covered in introductory numerical analysis.

Minimizing the Sum of the Absolute Deviations

When we were graphically fitting lines to the data in Section 3.1, one of our criteria minimized the total sum of the absolute deviations between the data points and their corresponding points on the fitted line. This criterion can be generalized: Given some function type $y = f(x)$ and a collection of m data points (x_i, y_i), minimize the sum of the absolute deviations $|y_i - f(x_i)|$. That is, determine the parameters of the function type $y = f(x)$ to minimize

$$\sum_{i=1}^{m} |y_i - f(x_i)| \tag{3.2}$$

If we let $R_i = |y_i - f(x_i)|$, $i = 1, 2, \ldots, m$ represent each absolute deviation, then the preceding criterion (3.2) can be interpreted as minimizing the length of the line formed by adding together the numbers R_i. This is illustrated for the case $m = 2$ in Figure 3.13.

Figure 3.13

A geometric interpretation of minimizing the sum of the absolute deviations

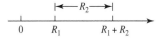

Although we geometrically applied this criterion in Section 3.1 when the function type $y = f(x)$ was a line, the general criterion presents severe problems. To solve this optimization problem using the calculus, we need to differentiate the sum (3.2) with respect to the parameters of $f(x)$ to find the critical points. However, the various derivatives of the sum fail to be continuous because of the presence of the absolute values, so we will not pursue this criterion further. In Chapter 7 we consider other applications of the criterion and present techniques for approximating solutions numerically.

Least-Squares Criterion

Currently, the most frequently used curve-fitting criterion is the **least-squares criterion**. If we use the same notation shown earlier, the problem is to determine the parameters of the function type $y = f(x)$ to minimize the sum

$$\sum_{i=1}^{m} |y_i - f(x_i)|^2 \tag{3.3}$$

Part of the popularity of this criterion stems from the ease with which the resulting optimization problem can be solved using only the calculus of several variables. However, relatively recent advances in mathematical programming techniques (such as the Simplex Method for solving many applications of the Chebyshev criterion) and advances in numerical methods for approximating solutions to the criterion (3.2) promise to dissipate this advantage. The justification for the use of the least-squares method increases when we consider probabilistic arguments that assume the errors are distributed randomly. However, we will not discuss such probabilistic arguments until later in the text.

We now give a geometric interpretation of the least-squares criterion. Consider the case of three data points and let $R_i = |y_i - f(x_i)|$ denote the absolute deviation between the observed and predicted values for $i = 1, 2, 3$. Think of the R_i as the scalar components of a deviation vector, as depicted in Figure 3.14. Thus the vector $\mathbf{R} = R_1\mathbf{i} + R_2\mathbf{j} + R_3\mathbf{k}$ represents the resultant deviation between the observed and predicted values. The magnitude

A geometric interpretation
of the least-squares
criterion

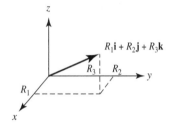

of the deviation vector is given by

$$|\mathbf{R}| = \sqrt{R_1^2 + R_2^2 + R_3^2}$$

To minimize $|\mathbf{R}|$ we can minimize $|\mathbf{R}|^2$ (see Problem 1). Thus, the least-squares problem is to determine the parameters of the function type $y = f(x)$ such that

$$|\mathbf{R}|^2 = \sum_{i=1}^{3} R_i^2 = \sum_{i=1}^{3} |y_i - f(x_i)|^2$$

is minimized. That is, we may interpret the least-squares criterion as minimizing the magnitude of the vector whose coordinates represent the absolute deviation between the observed and predicted values.

Relating the Criteria

The geometric interpretations of the three curve-fitting criteria help in providing a qualitative description comparing the criteria. Minimizing the sum of the absolute deviations tends to treat each data point with equal weight and to average the deviations. The Chebyshev criterion gives more weight to a single point potentially having a large deviation. The least-squares criterion is somewhere in between as far as weighting individual points with significant deviations is concerned. But let's be more precise. Because the Chebyshev and least-squares criteria are the most convenient to apply analytically, we now derive a method for relating the deviations resulting from using these two criteria.

Suppose the Chebyshev criterion is applied and the resulting optimization problem solved to yield the function $f_1(x)$. The absolute deviations resulting from the fit are defined as follows:

$$|y_i - f_1(x_i)| = c_i, \quad i = 1, 2, \ldots, m$$

Now define $c_{\max}$ as the largest of the absolute deviations c_i. There is a special significance attached to $c_{\max}$. Because the parameters of the function $f_1(x)$ are determined so as to minimize the value of $c_{\max}$, it is the minimal largest absolute deviation obtainable.

On the other hand, suppose the least-squares criterion is applied and the resulting optimization problem solved to yield the function $f_2(x)$. The absolute deviations resulting from the fit are then given by

$$|y_i - f_2(x_i)| = d_i, \quad i = 1, 2, \ldots, m$$

Define d_{max} as the largest of the absolute deviations d_i. At this point it can only be said that d_{max} is at least as large as c_{max} because of the special significance of the latter as previously discussed. However, let's attempt to relate d_{max} and c_{max} more precisely.

The special significance the least-squares criterion attaches to the d_i is that the sum of their squares is the smallest such sum obtainable. Thus, it must be true that

$$d_1^2 + d_2^2 + \cdots + d_m^2 \leq c_1^2 + c_2^2 + \cdots + c_m^2$$

Because $c_i \leq c_{max}$ for every i, these inequalities imply

$$d_1^2 + d_2^2 + \cdots + d_m^2 \leq mc_{max}^2$$

or

$$\sqrt{\frac{d_1^2 + d_2^2 + \cdots + d_m^2}{m}} \leq c_{max}$$

For ease of discussion, define

$$D = \sqrt{\frac{d_1^2 + d_2^2 + \cdots + d_m^2}{m}}$$

Thus,

$$D \leq c_{max} \leq d_{max}$$

This last relationship is very revealing. Suppose it is more convenient to apply the least-squares criterion in a particular situation, but there is concern about the largest absolute deviation c_{max} that may result. If we compute D, a lower bound on c_{max} is obtained, and d_{max} gives an upper bound. Thus, if there is considerable difference between D and d_{max}, the modeler should consider applying the Chebyshev criterion.

3.2 PROBLEMS

1. Using elementary calculus, show that the minimum and maximum points for $y = f(x)$ occur among the minimum and maximum points for $y = f^2(x)$. Assuming $f(x) \geq 0$, why can we minimize $f(x)$ by minimizing $f^2(x)$?

2. For each of the following data sets, formulate the mathematical model that minimizes the largest deviation between the data and the line $y = ax + b$. If a computer is available, solve for the estimates of a and b.

a.

x	1.0	2.3	3.7	4.2	6.1	7.0
y	3.6	3.0	3.2	5.1	5.3	6.8

b.

x	29.1	48.2	72.7	92.0	118	140	165	199
y	0.0493	0.0821	0.123	0.154	0.197	0.234	0.274	0.328

c.

x	2.5	3.0	3.5	4.0	4.5	5.0	5.5
y	4.32	4.83	5.27	5.74	6.26	6.79	7.23

3. For the following data, formulate the mathematical model that minimizes the largest deviation between the data and the model $y = c_1 x^2 + c_2 x + c_3$. If a computer code is available, solve for the estimates of c_1, c_2, and c_3.

x	0.1	0.2	0.3	0.4	0.5
y	0.06	0.12	0.36	0.65	0.95

4. For the following data, formulate the mathematical model that minimizes the largest deviation between the data and the model $P = ae^{bt}$. If a computer code is available, solve for the estimates of a and b.

t	7	14	21	28	35	42
P	8	41	133	250	280	297

5. Suppose the variable x_1 can assume any real value. Show that the following substitution using nonnegative variables x_2 and x_3 permits x_1 to assume any real value.

$$x_1 = x_2 - x_3, \qquad \text{where } x_1 \text{ is unconstrained}$$

and

$$x_2 \geq 0 \qquad \text{and} \qquad x_3 \geq 0$$

Thus, if a computer code allows only nonnegative variables, the substitution allows for solving the linear program in the variables x_2 and x_3 and then recovering the value of the variable x_1.

3.3   Applying the Least-Squares Criterion

Suppose that our assumptions lead us to expect a model of a certain type and that data have been collected and analyzed. In this section the least-squares criterion is applied to estimate the parameters for several types of curves.

Fitting a Straight Line

Suppose a model of the form $y = Ax + B$ is expected and it has been decided to use the m data points $(x_i, y_i), i = 1, 2, \ldots, m$ to estimate A and B. Denote the least-squares estimate of $y = Ax + B$ by $y = ax + b$. Applying the least-squares criterion (3.3) to this situation requires the minimization of

$$S = \sum_{i=1}^{m} [y_i - f(x_i)]^2 = \sum_{i=1}^{m} (y_i - ax_i - b)^2$$

A necessary condition for optimality is that the two partial derivatives $\partial S/\partial a$ and $\partial S/\partial b$ equal zero, yielding the equations

$$\frac{\partial S}{\partial a} = -2 \sum_{i=1}^{m} (y_i - ax_i - b)x_i = 0$$

$$\frac{\partial S}{\partial b} = -2 \sum_{i=1}^{m} (y_i - ax_i - b) = 0$$

These equations can be rewritten to give

$$\left. \begin{array}{c} a \sum_{i=1}^{m} x_i^2 + b \sum_{i=1}^{m} x_i = \sum_{i=1}^{m} x_i y_i \\ a \sum_{i=1}^{m} x_i + mb = \sum_{i=1}^{m} y_i \end{array} \right\} \qquad (3.4)$$

The preceding equations can be solved for a and b once all the values for x_i and y_i are substituted into them. The solutions (see Problem 1 at the end of this section) for the parameters a and b are easily obtained by elimination and are found to be

$$a = \frac{m \sum x_i y_i - \sum x_i \sum y_i}{m \sum x_i^2 - \left(\sum x_i \right)^2}, \qquad \text{the \textbf{slope}} \qquad (3.5)$$

and

$$b = \frac{\sum x_i^2 \sum y_i - \sum x_i y_i \sum x_i}{m \sum x_i^2 - \left(\sum x_i \right)^2}, \qquad \text{the \textbf{intercept}} \qquad (3.6)$$

Computer codes are easily written to compute these values for a and b for any collection of data points. Equations (3.4) are called the **normal equations**.

Fitting a Power Curve

Now let's use the least-squares criterion to fit a curve of the form $y = Ax^n$, where n is fixed, to a given collection of data points. Call the least-squares estimate of the model $f(x) = ax^n$. Application of the criterion then requires minimization of

$$S = \sum_{i=1}^{m} [y_i - f(x_i)]^2 = \sum_{i=1}^{m} [y_i - ax_i^n]^2$$

A necessary condition for optimality is that the derivative ds/da equal zero, giving the equation

$$\frac{dS}{da} = -2 \sum_{i=1}^{m} x_i^n [y_i - ax_i^n] = 0$$

Solving the equation for a yields

$$a = \frac{\sum x_i^n y_i}{\sum x_i^{2n}} \tag{3.7}$$

Remember, the number n is *fixed* in Equation (3.7).

The least-squares criterion can be applied to other models as well. The limitation in applying the method lies in calculating the various derivatives required in the optimization process, setting these derivatives to zero, and solving the resulting equations for the parameters in the model type.

For example, let's fit $y = Ax^2$ to the data shown in Table 3.3 and predict the value of y when $x = 2.25$.

Table 3.3 Data collected to fit $y = Ax^2$

x	0.5	1.0	1.5	2.0	2.5
y	0.7	3.4	7.2	12.4	20.1

In this case, the least-squares estimate a is given by

$$a = \frac{\sum x_i^2 y_i}{\sum x_i^4}$$

We compute $\sum x_i^4 = 61.1875$, $\sum x_i^2 y_i = 195.0$ to yield $a = 3.1869$ (to four decimal places). This computation gives the least-squares approximate model

$$y = 3.1869x^2$$

When $x = 2.25$, the predicted value for y is 16.1337.

Transformed Least-Squares Fit

Although the least-squares criterion appears easy to apply in theory, in practice it may be difficult. For example, consider fitting the model $y = Ae^{Bx}$ using the least-squares criterion. Call the least-squares estimate of the model $f(x) = ae^{bx}$. Application of the criterion then requires the minimization of

$$S = \sum_{i=1}^{m} [y_i - f(x_i)]^2 = \sum_{i=1}^{m} \left[y_i - ae^{bx_i} \right]^2$$

A necessary condition for optimality is that $\partial S / \partial a = \partial S / \partial b = 0$. Formulate the conditions and convince yourself that solving the resulting system of nonlinear equations would not be easy. Many simple models result in derivatives that are very complex or in systems of equations that are difficult to solve. For this reason, we use transformations that allow us to *approximate* the least-squares model.

In graphically fitting lines to data in Section 3.1, we often found it convenient to transform the data first and then fit a line to the transformed points. For example, in graphically fitting $y = Ce^x$, we found it convenient to plot $\ln y$ versus x and then fit a line to the transformed data. The same idea can be used with the least-squares criterion to simplify the computational aspects of the process. In particular, if a convenient substitution can be found so that the problem takes the form $Y = AX + B$ in the transformed variables X and Y, then Equation (3.4) can be used to fit a line to the transformed variables. We illustrate the technique with the example that we just worked out.

Suppose we wish to fit the power curve $y = Ax^N$ to a collection of data points. Let's denote the estimate of A by α and the estimate of N by n. Taking the logarithm of both sides of the equation $y = \alpha x^n$ yields

$$\ln y = \ln \alpha + n \ln x \tag{3.8}$$

When the variables $\ln y$ versus $\ln x$ are plotted, Equation (3.8) yields a straight line. On that graph, $\ln \alpha$ is the intercept when $\ln x = 0$ and the slope of the line is n. Using Equations (3.5) and (3.6) to solve for the slope n and intercept $\ln \alpha$ with the transformed variables and $m = 5$ data points, we have

$$n = \frac{5 \sum (\ln x_i)(\ln y_i) - \left(\sum \ln x_i\right)\left(\sum \ln y_i\right)}{5 \sum (\ln x_i)^2 - \left(\sum \ln x_i\right)^2}$$

$$\ln \alpha = \frac{\sum (\ln x_i)^2 (\ln y_i) - \left(\sum \ln x_i\right)(\ln y_i) \sum \ln x_i}{5 \sum (\ln x_i)^2 - \left(\sum \ln x_i\right)^2}$$

For the data displayed in Table 3.3 we get $\sum \ln x_i = 1.3217558$, $\sum \ln y_i = 8.359597801$, $\sum (\ln x_i)^2 = 1.9648967$, $\sum (\ln x_i)(\ln y_i) = 5.542315175$, yielding $n = 2.062809314$ and $\ln \alpha = 1.126613508$, or $\alpha = 3.085190815$. Thus, our least-squares best fit of Equation (3.8) is (rounded to four decimal places)

$$y = 3.0852x^{2.0628}$$

This model predicts $y = 16.4348$ when $x = 2.25$. Note, however, that this model fails to be a quadratic like the one we fit previously.

Suppose we still wish to fit a *quadratic* $y = Ax^2$ to the collection of data. Denote the estimate of A by a_1 to distinguish this constant from the constants a and α computed previously. Taking the logarithm of both sides of the equation $y = a_1 x^2$ yields

$$\ln y = \ln a_1 + 2 \ln x$$

In this situation the graph of $\ln y$ versus $\ln x$ is a straight line of slope 2 and intercept $\ln a_1$. Using the second equation in (3.4) to compute the intercept, we have

$$2 \sum \ln x_i + 5 \ln a_1 = \sum \ln y_i$$

For the data displayed in Table 3.3, we get $\sum \ln x_i = 1.3217558$ and $\sum \ln y_i = 8.359597801$. Therefore, this last equation gives $\ln a_1 = 1.14321724$, or $a_1 = 3.136844129$, yielding the least-squares best fit (rounded to four decimal places)

$$y = 3.1368x^2$$

The model predicts $y = 15.8801$ when $x = 2.25$, which differs significantly from the value 16.1337 predicted by the first quadratic $y = 3.1869x^2$ obtained as the least-squares best fit of $y = Ax^2$ without transforming the data. We compare these two quadratic models (as well as a third model) in the next section.

The preceding example illustrates two facts. First, if an equation can be transformed to yield an equation of a straight line in the transformed variables, Equation (3.4) can be used directly to solve for the slope and intercept of the transformed graph. Second, the least-squares best fit to the transformed equations *does not* coincide with the least-squares best fit of the original equations. The reason for this discrepancy is that the resulting optimization problems are different. In the case of the original problem, we are finding the curve that minimizes the sum of the squares of the deviations using the original data, whereas in the case of the transformed problem, we are minimizing the sum of the squares of the deviations using the *transformed* variables.

3.3   PROBLEMS

1. Solve the two equations given by (3.4) to obtain the values of the parameters given by Equations (3.5) and (3.6), respectively.

2. Use Equations (3.5) and (3.6) to estimate the coefficients of the line $y = ax + b$ such that the sum of the squared deviations between the line and the following data points is minimized.

a.

x	1.0	2.3	3.7	4.2	6.1	7.0
y	3.6	3.0	3.2	5.1	5.3	6.8

b.

x	29.1	48.2	72.7	92.0	118	140	165	199
y	0.0493	0.0821	0.123	0.154	0.197	0.234	0.274	0.328

c.

x	2.5	3.0	3.5	4.0	4.5	5.0	5.5
y	4.32	4.83	5.27	5.74	6.26	6.79	7.23

For each problem, compute D and d_{max} to bound c_{max}. Compare the results to your solutions to Problem 2 in Section 3.2.

3. Derive the equations that minimize the sum of the squared deviations between a set of data points and the quadratic model $y = c_1 x^2 + c_2 x + c_3$. Use the equations to find estimates of c_1, c_2, and c_3 for the following set of data.

x	0.1	0.2	0.3	0.4	0.5
y	0.06	0.12	0.36	0.65	0.95

Compute D and d_{max} to bound c_{max}. Compare the results with your solution to Problem 3 in Section 3.2.

4. Make an appropriate transformation to fit the model $P = ae^{bt}$ using Equation (3.4). Estimate a and b.

t	7	14	21	28	35	42
P	8	41	133	250	280	297

5. Examine closely the system of equations that result when you fit the quadratic in Problem 3. Suppose $c_2 = 0$. What would be the corresponding system of equations? Repeat for the cases $c_1 = 0$ and $c_3 = 0$. Suggest a system of equations for a cubic. Check your result. Explain how you would generalize the system of Equation (3.4) to fit any polynomial. Explain what you would do if one or more of the coefficients in the polynomial were zero.

6. A general rule for computing a person's weight is as follows: For a female, multiply the height in inches by 3.5 and subtract 108; for a male, multiply the height in inches by 4.0 and subtract 128. If the person is small bone-structured, adjust this computation by subtracting 10%; for a large bone-structured person, add 10%. No adjustment is made for an average-size person. Gather data on the weight versus height of people of differing age, size, and gender. Using Equation (3.4), fit a straight line to your data for males and another straight line to your data for females. What are the slopes and intercepts of those lines? How do the results compare with the general rule?

In Problems 7–10, fit the data with the models given, using least squares.

7.

x	1	2	3	4	5
y	1	1	2	2	4

 a. $y = b + ax$
 b. $y = ax^2$

8. Data for stretch of a spring

$x(\times 10^{-3})$	5	10	20	30	40	50	60	70	80	90	100
$y(\times 10^{-5})$	0	19	57	94	134	173	216	256	297	343	390

 a. $y = ax$
 b. $y = b + ax$
 c. $y = ax^2$

9. Data for the ponderosa pine

x	17	19	20	22	23	25	28	31	32	33	36	37	39	42
y	19	25	32	51	57	71	113	140	153	187	192	205	250	260

 a. $y = ax + b$
 b. $y = ax^2$

   c. $y = ax^3$

   d. $y = ax^3 + bx^2 + c$

10. Data for planets

Body	Period (sec)	Distance from sun (m)
Mercury	7.60×10^6	5.79×10^{10}
Venus	1.94×10^7	1.08×10^{11}
Earth	3.16×10^7	1.5×10^{11}
Mars	5.94×10^7	2.28×10^{11}
Jupiter	3.74×10^8	7.79×10^{11}
Saturn	9.35×10^8	1.43×10^{12}
Uranus	2.64×10^9	2.87×10^{12}
Neptune	5.22×10^9	4.5×10^{12}
Pluto	7.82×10^9	5.91×10^{12}

Fit the model $y = ax^{3/2}$.

3.3 PROJECTS

1. Compete the requirements of the module "Curve Fitting via the Criterion of Least Squares," by John W. Alexander, Jr., UMAP 321. (See enclosed CD for UMAP module.) This unit provides an easy introduction to correlations, scatter diagrams (polynomial, log-arithmic, and exponential scatters), and lines and curves of regression. Students construct scatter diagrams, choose appropriate functions to fit specific data, and use a computer program to fit curves. Recommended for students who wish an introduction to statistical measures of correlation.

2. Select a project from Projects 1–7 in Section 2.3 and use least squares to fit your proposed proportionality model. Compare your least-squares results with the model used from Section 2.3. Find the bounds on the Chebyshev criterion and interpret the results.

3.3 Further Reading

Burden, Richard L., & J. Douglas Faires. *Numerical Analysis*, 7th ed. Pacific Grove, CA: Brooks/Cole, 2001.

Cheney, E. Ward, & David Kincaid. *Numerical Mathematics and Computing.* Monterey, CA: Brooks/Cole, 1984.

Cheney, E. Ward, & David Kincaid. *Numerical Analysis*, 4th ed. Pacific Grove, CA: Brooks/Cole, 1999.

Hamming, R. W. *Numerical Methods for Scientists and Engineers.* New York: McGraw-Hill, 1973.

Stiefel, Edward L. *An Introduction to Numerical Mathematics.* New York: Academic Press, 1963.

3.4 Choosing a Best Model

Let's consider the adequacy of the various models of the form $y = Ax^2$ that we fit using the least-squares and transformed least-squares criteria in the previous section. Using the least-squares criterion, we obtained the model $y = 3.1869x^2$. One way of evaluating how well the model fits the data is to compute the deviations between the model and the actual data. If we compute the sum of the squares of the deviations, we can bound c_{max} as well. For the model $y = 3.1869x^2$ and the data given in Table 3.3, we compute the deviations shown in Table 3.4.

Table 3.4 Deviations between the data in Table 3.3 and the fitted model $y = 3.1869x^2$

x_i	0.5	1.0	1.5	2.0	2.5
y_i	0.7	3.4	7.2	12.4	20.1
$y_i - y(x_i)$	−0.0967	0.2131	0.02998	−0.3476	0.181875

From Table 3.4 we compute the sum of the squares of the deviations as 0.20954, so $D = (0.20954/5)^{1/2} = 0.204714$. Because the largest absolute deviation is 0.3476 when $x = 2.0$, c_{max} can be bounded as follows:

$$D = 0.204714 \le c_{max} \le 0.3476 = d_{max}$$

Let's find c_{max}. Because there are five data points, the mathematical problem is to minimize the largest of the five numbers $|r_i| = |y_i - y(x_i)|$. Calling that largest number r, we want to minimize r subject to $r \ge r_i$ and $r \ge -r_i$ for each $i = 1, 2, 3, 4, 5$. Denote our model by $y(x) = a_2 x^2$. Then, substitution of the observed data points in Table 3.3 into the inequalities $r \ge r_i$ and $r \ge -r_i$ for each $i = 1, 2, 3, 4, 5$ yields the following linear program:

Minimize r

subject to

$$r - r_1 = r - (0.7 - 0.25a_2) \ge 0$$
$$r + r_1 = r + (0.7 - 0.25a_2) \ge 0$$
$$r - r_2 = \quad r - (3.4 - a_2) \quad \ge 0$$
$$r + r_2 = \quad r + (3.4 - a_2) \quad \ge 0$$
$$r - r_3 = r - (7.2 - 2.25a_2) \ge 0$$
$$r + r_3 = r + (7.2 - 2.25a_2) \ge 0$$
$$r - r_4 = \quad r - (12.4 - 4a_2) \quad \ge 0$$
$$r + r_4 = \quad r + (12.4 - 4a_2) \quad \ge 0$$
$$r - r_5 = r - (20.1 - 6.25a_2) \ge 0$$
$$r + r_5 = r + (20.1 - 6.25a_2) \ge 0$$

In Chapter 7 we show that the solution of the preceding linear program yields $r = 0.28293$ and $a_2 = 3.17073$. Thus, we have reduced our largest deviation from $d_{max} = 0.3476$ to $c_{max} = 0.28293$. Note that we can reduce the largest deviation no further than 0.28293 for the model type $y = Ax^2$.

We have now determined three estimates of the parameter A for the model type $y = Ax^2$. Which estimate is best? For each model we can readily compute the deviations from each data point as recorded in Table 3.5.

Table 3.5   Summary of the deviations for each model $y = Ax^2$

x_i	y_i	$y_i - 3.1869x_i^2$	$y_i - 3.1368x_i^2$	$y_i - 3.17073x_i^2$
0.5	0.7	−0.0967	−0.0842	−0.0927
1.0	3.4	0.2131	0.2632	0.2293
1.5	7.2	0.029475	0.1422	0.0659
2.0	12.4	−0.3476	−0.1472	−0.2829
2.5	20.1	0.181875	0.4950	0.28293

For each of the three models we can compute the sum of the squares of the deviations and the maximum absolute deviation. The results are shown in Table 3.6.

As we would expect, each model has something to commend it. However, notice the increase in the sum of the squares of the deviations in the transformed least-squares model. It is tempting to apply a simple rule, such as choose the model with the smallest absolute deviation. (Other statistical indicators of goodness of fit exist as well. For example, see *Probability and Statistics in Engineering and Management Science*, by William W. Hines and Douglas C. Montgomery, New York: Wiley, 1972.) These indicators are useful for eliminating obviously poor models, but there is no easy answer to the question, Which model is best? The model with the smallest absolute deviation or the smallest sum of squares may fit very poorly over the range where you intend to use it most. Furthermore, as you will see in Chapter 4, models can easily be constructed that pass through each data point, thereby yielding a zero sum of squares and zero maximum deviation. So we need to answer the question of which model is best on a case-by-case basis, taking into account such things as the purpose of the model, the precision demanded by the scenario, the accuracy of the data, and the range of values for the independent variable over which the model will be used.

When choosing among models or judging the adequacy of a model, we may find it tempting to rely on the value of the best-fit criterion being used. For example, it is tempting to choose the model that has the smallest sum of squared deviations for the given data set or to conclude that a sum of squared deviations less than a predetermined value indicates

Table 3.6   Summary of the results for the three models

Criterion	Model	$\sum [y_i - y(x_i)]^2$	Max $|y_i - y(x_i)|$
Least-squares	$y = 3.1869x^2$	0.2095	0.3476
Transformed least-squares	$y = 3.1368x^2$	0.3633	0.4950
Chebyshev	$y = 3.17073x^2$	0.2256	0.28293

■ Figure 3.15

In all of these graphs, the model $y = x$ has the same sum of squared deviations.

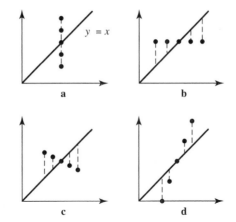

a good fit. However, in isolation these indicators may be very misleading. For example, consider the data displayed in Figure 3.15. In all of the four cases, the model $y = x$ results in exactly the same sum of squared deviations. Without the benefit of the graphs, therefore, we might conclude that in each case the model fits the data about the same. However, as the graphs show, there is a significant variation in each model's ability to capture the trend of the data. The following examples illustrate how the various indicators may be used to help in reaching a decision on the adequacy of a particular model. Normally, a graphical plot is of great benefit.

EXAMPLE 1 *Vehicular Stopping Distance*

Let's reconsider the problem of predicting a motor vehicle's stopping distance as a function of its speed. (This problem was addressed in Sections 2.2 and 3.3.) In Section 3.3 the submodel in which reaction distance d_r was proportional to the velocity v was tested graphically, and the constant of proportionality was estimated to be 1.1. Similarly, the submodel predicting a proportionality between braking distance d_b and the square of the velocity was tested. We found reasonable agreement with the submodel and estimated the proportionality constant to be 0.054. Hence, the model for stopping distance was given by

$$d = 1.1v + 0.054v^2 \tag{3.9}$$

We now fit these submodels analytically and compare the various fits.

To fit the model using the least-squares criterion, we use the formula from Equation (3.7):

$$A = \frac{\sum x_i y_i}{\sum x_i^2}$$

where y_i denotes the driver reaction distance and x_i denotes the speed at each data point. For the 13 data points given in Table 2.3, we compute $\sum x_i y_i = 40905$ and $\sum x_i^2 = 37050$, giving $A = 1.104049$.

For the model type $d_b = Bv^2$, we use the formula

$$B = \frac{\sum x_i^2 y_i}{\sum x_i^4}$$

where y_i denotes the average braking distance and x_i denotes the speed at each data point. For the 13 data points given in Table 2.4, we compute $\sum x_i^2 y_i = 8258350$ and $\sum x_i^4 = 152343750$, giving $B = 0.054209$. Because the data are relatively imprecise and the modeling is done qualitatively, we round the coefficients to obtain the model

$$d = 1.104v + 0.0542v^2 \tag{3.10}$$

Model (3.10) does not differ significantly from that obtained graphically in Chapter 3. Next, let's analyze how well the model fits. We can readily compute the deviations between the observed data points in Table 2.3 and the values predicted by Models (3.9) and (3.10). The deviations are summarized in Table 3.7. The fits of both models are very similar. The largest absolute deviation for Model (3.9) is 30.4 and for Model (3.10) it is 28.8. Note that both models overestimate the stopping distance up to 70 mph, and then they begin to underestimate the stopping distance. A better-fitting model would be obtained by directly fitting the data for total stopping distance to

$$d = k_1 v + k_2 v^2$$

instead of fitting the submodels individually as we did. The advantage of fitting the submodels individually and then testing each submodel is that we can measure how well they explain the behavior.

A plot of the proposed model(s) and the observed data points is useful to determine how well the model fits the data. Model (3.10) and the observations are plotted in Figure 3.16. It

Table 3.7 Deviations from the observed data points and Models (3.9) and (3.10)

Speed	Graphical model (3.9)	Least-squares model (3.10)
20	1.6	1.76
25	5.25	5.475
30	8.1	8.4
35	13.15	13.535
40	14.4	14.88
45	16.35	16.935
50	17	17.7
55	14.35	15.175
60	12.4	13.36
65	7.15	8.255
70	−1.4	−0.14
75	−14.75	−13.325
80	−30.4	−28.8

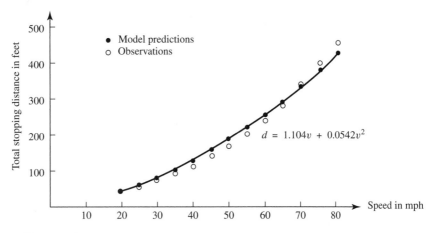

Figure 3.16

A plot of the proposed model and the observed data points provides a visual check on the adequacy of the model.

is evident from the figure that a definite trend exists in the data and that Model (3.10) does a reasonable job of capturing that trend, especially at the lower speeds.

A powerful technique for quickly determining where the model is breaking down is to plot the deviations (residuals) as a function of the independent variable(s). For Model (3.10), a plot of the deviations is given in Figure 3.17 showing that the model is indeed reasonable up to 70 mph. Beyond 70 mph there is a breakdown in the model's ability to predict the observed behavior.

Let's examine Figure 3.17 more closely. Note that although the deviations up to 70 mph are relatively small, they are all positive. If the model fully explains the behavior, not only

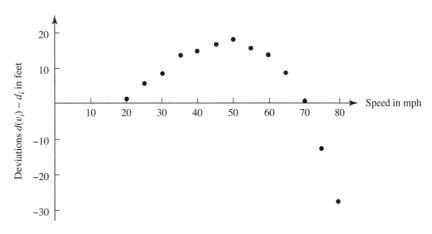

Figure 3.17

A plot of the deviations (residuals) reveals those regions where the model does not fit well.

should the deviations be small, but some should be positive and some negative. Why? In Figure 3.17 we note a definite pattern in the nature of the deviations, which might cause us to reexamine the model and/or the data. The nature of the pattern in the deviations can give us clues on how to refine the model further. In this case, the imprecision in the data collection process probably does not warrant further model refinement.   ■ ■ ■

3.4 PROBLEMS

For Problems 1–6, find a model using the least-squares criterion either on the data or on the transformed data (as appropriate). Compare your results with the graphical fits obtained in the problem set 3.1 by computing the deviations, the maximum absolute deviation, and the sum of the squared deviations for each model. Find a bound on c_{max} if the model was fit using the least-squares criterion.

1. Problem 3 in Section 3.1

2. Problem 4a in Section 3.1

3. Problem 4b in Section 3.1

4. Problem 5a in Section 3.1

5. Problem 2 in Section 3.1

6. Problem 6 in Section 3.1

7. **a.** In the following data, W represents the weight of a fish (bass) and l represents its length. Fit the model $W = kl^3$ to the data using the least-squares criterion.

Length, l (in.)	14.5	12.5	17.25	14.5	12.625	17.75	14.125	12.625
Weight, W (oz)	27	17	41	26	17	49	23	16

   b. In the following data, g represents the girth of a fish. Fit the model $W = klg^2$ to the data using the least-squares criterion.

Length, l (in.)	14.5	12.5	17.25	14.5	12.625	17.75	14.125	12.625
Girth, g (in.)	9.75	8.375	11.0	9.75	8.5	12.5	9.0	8.5
Weight, W (oz)	27	17	41	26	17	49	23	16

   c. Which of the two models fits the data better? Justify fully. Which model do you prefer? Why?

8. Use the data presented in Problem 7b to fit the models $W = cg^3$ and $W = kgl^2$. Interpret these models. Compute appropriate indicators and determine which model is best. Explain.

3.4 PROJECTS

1. Write a computer program that finds the least-squares estimates of the coefficients in the following models.

   a. $y = ax^2 + bx + c$

   b. $y = ax^n$

2. Write a computer program that computes the deviation from the data points and any model that the user enters. Assuming that the model was fitted using the least-squares criterion, compute D and d_{max}. Output each data point, the deviation from each data point, D, d_{max}, and the sum of the squared deviations.

3. Write a computer program that uses Equations (3.4) and the appropriate transformed data to estimate the parameters of the following models.

   a. $y = bx^n$

   b. $y = be^{ax}$

   c. $y = a \ln x + b$

   d. $y = ax^2$

   e. $y = ax^3$

4 Experimental Modeling

Introduction

In Chapter 3 we discussed the philosophical differences between curve fitting and interpolation. When fitting a curve, the modeler is using some assumptions that select a particular type of model that explains the behavior under observation. If collected data then corroborate the reasonableness of those assumptions, the modeler's task is to choose the parameters of the selected curve that best fits the data according to some criterion (such as least squares). In this situation the modeler expects, and willingly accepts, some deviations between the fitted model and the collected data to obtain a model explaining the behavior. The problem with this approach is that in many cases the modeler is unable to construct a tractable model form that satisfactorily explains the behavior. Thus, the modeler does not know what kind of curve actually describes the behavior. If it is necessary to predict the behavior nevertheless, the modeler may conduct experiments (or otherwise gather data) to investigate the behavior of the dependent variable(s) for selected values of the independent variable(s) within some range. In essence, the modeler desires to construct an **empirical model** *based on the collected data* rather than select a model based on certain assumptions. In such cases the modeler is strongly influenced by the data that have been carefully collected and analyzed, so he or she seeks a curve that captures the trend of the data to *predict* in between the data points.

For example, consider the data shown in Figure 4.1a. If the modeler's assumptions lead to the expectation of a quadratic model, a parabola will fit the data points, as illustrated in Figure 4.1b. However, if the modeler has no reason to expect a particular type of model, a smooth curve may pass through the data points instead, as illustrated in Figure 4.1c.

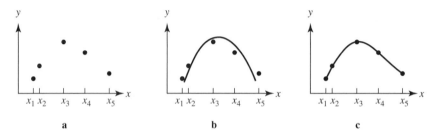

■ **Figure 4.1**

If the modeler expects a quadratic relationship, a parabola may be fit to the data, as in b. Otherwise, a smooth curve may be passed through the points, as in c.

In this chapter, we address the construction of empirical models. In Section 4.1, we study the selection process for simple one-term models that capture the trend of the data. A few scenarios are addressed for which few modelers would attempt to construct an explicative model—namely, predicting the harvest of sea life in the Chesapeake Bay. In Section 4.2, we discuss the construction of higher-order polynomials that pass through the collected data points. In Section 4.3, we investigate smoothing of data using low-order polynomials. Finally, in Section 4.4, we present the technique of cubic spline interpolation, where a distinct cubic polynomial is used across successive pairs of data points.

4.1   Harvesting in the Chesapeake Bay and Other One-Term Models

Let's consider a situation in which a modeler has collected some data but is unable to construct an explicative model. In 1992, the *Daily Press* (a newspaper in Virginia) reported some observations (data) collected during the past 50 years on harvesting sea life in the Chesapeake Bay. We will examine several scenarios using observations from (a) harvesting bluefish and (b) harvesting blue crabs by the commercial industry of the Chesapeake Bay. Table 4.1 shows the data we will use in our one-term models.

A scatterplot of harvesting bluefish versus time is shown in Figure 4.2, and a scatterplot of harvesting blue crabs is shown in Figure 4.3. Figure 4.2 clearly shows a tendency to harvest more bluefish over time, indicating or suggesting the availability of bluefish. A more precise description is not so obvious. In Figure 4.3, the tendency is for the increase of harvesting of blue crabs. Again, a precise model is not so obvious.

In the rest of this section, we suggest how we might begin to predict the availability of bluefish over time. Our strategy will be to transform the data of Table 4.1 in such a way that the resulting graph approximates a line, thus achieving a working model. But how do

Table 4.1   Harvesting the bay, 1940–1990

Year	Bluefish (lb)	Blue crabs (lb)
1940	15,000	100,000
1945	150,000	850,000
1950	250,000	1,330,000
1955	275,000	2,500,000
1960	270,000	3,000,000
1965	280,000	3,700,000
1970	290,000	4,400,000
1975	650,000	4,660,000
1980	1,200,000	4,800,000
1985	1,500,000	4,420,000
1990	2,750,000	5,000,000

■ **Figure 4.2**

Scatterplot of harvesting
bluefish versus base year
(5-year periods from 1940
to 1990)

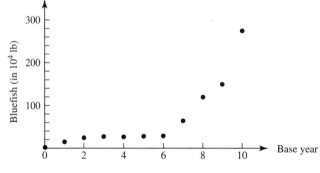

■ **Figure 4.3**

Scatterplot of harvesting
blue crabs versus base year
(5-year periods from 1940
to 1990)

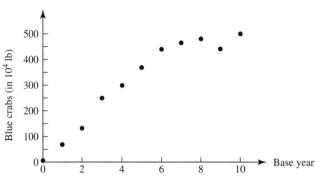

Ladder of
powers

$$\vdots$$
$$z^2$$
$$z$$
$$\sqrt{z}$$
$$\log z$$
$$\frac{1}{\sqrt{z}}$$
$$\frac{1}{z}$$
$$\frac{1}{z^2}$$
$$\vdots$$

we determine the transformation? We will use the ladder of powers[1] of a variable z to help in the selection of the appropriate linearizing transformation.

Figure 4.4 shows a set of five data points (x, y) together with the line $y = x$, for $x > 1$. Suppose we change the y value of each point to $\sqrt{y}$. This procedure yields a new relation $y = \sqrt{x}$ whose y values are closer together over the domain in question. Note that all the y values are reduced, but the larger values are reduced more than the smaller ones.

Changing the y value of each point to $\log y$ has a similar but more pronounced effect, and each additional step down the ladder produces a stronger version of the same effect.

We started in Figure 4.4 in the simplest way—with a linear function. However, that was only for convenience. If we take a concave up, positive-valued function, such as $y = f(x)$, $x > 1$,

then some transformation in the ladder below y—changing the y values to $\sqrt{y}$, or $\log y$, or a more drastic change—squeezes the right-hand tail downward and increases the likelihood of generating a new function more linear than the original. Which transformation should be used is a matter of trial and error and experience. Another possibility is to stretch the right-hand tail to the right (try changing the x values to x^2, x^3 values, etc.).

---

[1] See Paul F. Velleman and David C. Hoaglin, *Applications, Basics, and Computing of Exploratory Data Analysis* (Boston: Duxbury Press, 1981), p. 49.

■ Figure 4.4

Relative effects of three transformations

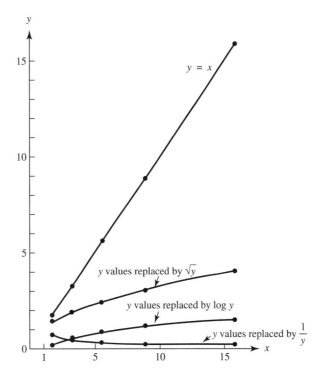

Table 4.2

Ladder of Transformations

$$\vdots$$
$$z^3$$
$$z^2$$
z (no change)

$$*\begin{cases} \sqrt{z} \\ \log z \\ \frac{-1}{\sqrt{z}} \\ \frac{-1}{z} \\ \frac{-1}{z^2} \end{cases}$$

$$\vdots$$

---

*The transformations most often used.

If we take a concave down, positive-valued increasing function, $y = f(x)$, $x > 1$, such as

we can hope to linearize it further by stretching the right-hand tail upward (try changing y values to y^2, y^3 values, etc.). Another possibility is to squeeze the right-hand tail to the left (try changing the x values to $\sqrt{x}$, or $\log x$, or by a more drastic choice from the ladder).

Note that although replacing z by $1/z$ or $1/z^2$ and so on may sometimes have a desirable effect, such replacements also have an undesirable one—an increasing function is converted into a decreasing one. As a result, when using the transformation in the ladder of powers below $\log z$, data analysts generally use a negative sign to keep the transformation data in the same order as the original data. Table 4.2 shows the ladder of transformations as it is generally used.

With this information on the ladder of transformations, let's return to the Chesapeake Bay harvesting data.

EXAMPLE 1   *Harvesting Bluefish*

Recall from the scatterplot in Figure 4.2 that the trend of the data appears to be increasing and concave up. Using the ladder of powers to squeeze the right-hand tail downward, we can change y values by replacing y with $\log y$ or other transformations down the ladder. Another choice would be to replace x values with x^2 or x^3 values or other powers up the

Table 4.3 Harvesting the bay: Bluefish, 1940–1990

Year	Base year	Bluefish (lb)
	x	y
1940	0	15,000
1945	1	150,000
1950	2	250,000
1955	3	275,000
1960	4	270,000
1965	5	280,000
1970	6	290,000
1975	7	650,000
1980	8	1,200,000
1985	9	1,550,000
1990	10	2,750,000

ladder. We will use the data displayed in Table 4.3, where 1940 is the base year of $x = 0$ for numerical convenience, with each base year representing a 5-year period.

We begin by squeezing the right-hand tail rightward by changing x to various values going up the ladder (x^2, x^3, etc.). None of these transformations results in a linear graph. We next change y to values of $\sqrt{y}$ and log y, going down the ladder. Both $\sqrt{y}$ and log y plots versus x appeared more linear than the transformations to the x variable. (The plots of y versus x^2 and x versus $\sqrt{y}$ were identical in their linearity.) We choose the log y versus x model. We fit with least squares the model of the form

$$\log y = mx + b$$

and obtain the following estimated curve:

$$\log y = 0.7231 + 0.1654x$$

where x is the base year and log y is to the base 10 and y is measured in 10^4 pounds. (See Figure 4.5.)

Using the property that $y = \log n$ if and only if $10^y = n$, we can rewrite this equation (with the aid of a calculator) as

$$y = 5.2857(1.4635)^x \qquad (4.1)$$

Figure 4.5

Superimposed data and model $y = 5.2857(1.4635)^x$

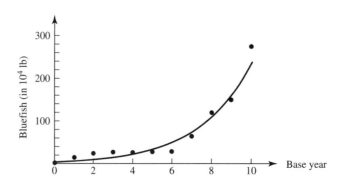

where y is measured in 10^4 pounds of bluefish and x is the base year. The plot of the model appears to fit the data reasonably well. Figure 4.5 shows the graph of this curve superimposed on the scatterplot. We will accept some error to have a simple one-term model. ▪ ▪ ▪

EXAMPLE 2   *Harvesting Blue Crabs*

Recall from our original scatterplot, Figure 4.3, that the trend of the data is increasing and concave down. With this information, we can utilize the ladder of transformations. We will use the data in Table 4.4, modified by making 1940 (year $x = 0$) the base year, with each base year representing a 5-year period.

Table 4.4   Harvesting the bay: Blue crabs, 1940–1990

Year	Base year	Blue crabs (lb)
	x	y
1940	0	100,000
1945	1	850,000
1950	2	1,330,000
1955	3	2,500,000
1960	4	3,000,000
1965	5	3,700,000
1970	6	4,400,000
1975	7	4,660,000
1980	8	4,800,000
1985	9	4,420,000
1990	10	5,000,000

As previously stated, we can attempt to linearize these data by changing y values to y^2 or y^3 values or to others moving up the ladder. After several experiments, we chose to replace the x values with $\sqrt{x}$. This squeezes the right-hand tail to the left. We provide a plot of y versus $\sqrt{x}$ (Figure 4.6). In Figure 4.7, we superimpose a line $y = k\sqrt{x}$ projected through the origin (no y-intercept). We use least squares from Chapter 3 to find k, yielding

$$y = 158.344\sqrt{x} \tag{4.2}$$

where y is measured in 10^4 pounds of blue crabs and x is the base year.

▪ **Figure 4.6**

Blue crabs (in 10^4 lb) versus $\sqrt{x}$

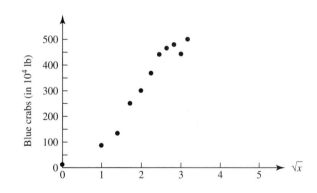

Figure 4.7

The line $y = 158.344\sqrt{x}$

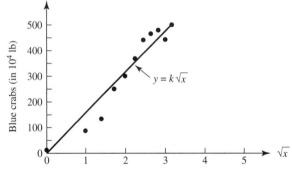

Figure 4.8

Superimposed data and
model $y = 158.344\sqrt{x}$

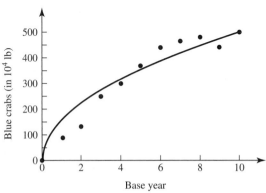

Base year

Figure 4.8 shows the graph of the preceding model superimposed on the scatterplot. Note that the curve seems to be reasonable because it fits the data about as well as expected for a simple one-term model. ∎ ∎ ∎

Verifying the Models

How good are the predictions based on the models? Part of the answer lies in comparing the observed values with those predicted. We can calculate the residuals and the relative errors for each pair of data. In many cases, the modeler is asked to predict or extrapolate to the future. How would these models hold up in predicting the amounts of harvests from the bay in the year 2010?

The following result for the bluefish may be larger than one might predict, whereas the result for the blue crabs may be a little lower than one might predict.

$$\text{Bluefish } y = 5.2857(1.4635)^{14} = 1092.95 \ (10^4 \text{ lb}) \approx 10.9 \text{ million lb}$$
$$\text{Blue crabs } y = 158.344\sqrt{14} = 592.469 \ (10^4 \text{ lb}) \approx 5.92 \text{ million lb}$$

These simple, one-term models should be used for interpolation and not extrapolation. We discuss modeling population growth more thoroughly in Chapters 11 and 12.

Let's summarize the ideas of this section. When we are constructing an empirical model, we always begin with a careful analysis of the collected data. Do the data suggest the existence of a trend? Are there data points that obviously lie outside the trend? If such outliers do exist, it may be desirable to discard them or, if they were obtained experimentally, to repeat the experiment as a check for a data collection error. When it is clear that a trend exists, we next attempt to find a function that will transform the data into a straight line (approximately). In addition to trying the functions listed in the ladder of transformations presented in this section, we can also attempt the transformations discussed in Chapter 3. Thus, if the model $y = ax^b$ were selected, we would plot $\ln y$ versus $\ln x$ to see if a straight line results. Likewise, when investigating the appropriateness of the model $y = ae^{bx}$, we would plot $\ln y$ versus x to see if a straight line results. Keep in mind our discussion in Chapter 3 about how the use of transformations may be deceiving, especially if the data points are squeezed together. Our judgment is strictly qualitative; the idea is to determine if a particular model type appears promising. When we are satisfied that a certain model type does seem to capture the trend of the data, we can estimate the parameters of the model graphically or using the analytical techniques discussed in Chapter 3. Eventually, we must analyze the goodness of fit using the indicators discussed in Chapter 3. Remember to graph the proposed model against the original data points, not the transformed data. If we are dissatisfied with the fit, we can investigate other one-term models. Because of their inherent simplicity, however, one-term models cannot fit all data sets. In such situations, other techniques can be used; we discuss these methods in the next several sections.

4.1 PROBLEMS

In 1976, Marc and Helen Bornstein studied the pace of life.[2] To see if life becomes more hectic as the size of the city becomes larger, they systematically observed the mean time required for pedestrians to walk 50 feet on the main streets of their cities and towns. In Table 4.5, we present some of the data they collected. The variable P represents the population of the town or city, and the variable V represents the mean velocity of pedestrians walking the 50 feet. Problems 1–5 are based on the data in Table 4.5.

1. Fit the model $V = CP^a$ to the "pace of life" data in Table 4.5. Use the transformation $\log V = a \log P + \log C$. Plot $\log V$ versus $\log P$. Does the relationship seem reasonable?

   a. Make a table of $\log P$ versus $\log V$.

   b. Construct a scatterplot of your log–log data.

   c. Eyeball a line l onto your scatterplot.

   d. Estimate the slope and the intercept.

   e. Find the linear equation that relates $\log V$ and $\log P$.

   f. Find the equation of the form $V = CP^a$ that expresses V in terms of P.

2. Graph the equation you found in Problem 1f superimposed on the original scatterplot.

---

[2]Bornstein, Marc H., and Helen G. Bornstein, "The Pace of Life." *Nature* 259 (19 February 1976): 557–559.

Table 4.5 Population and mean velocity over a 50-foot course, for 15 locations*

Location	Population P	Mean velocity V (ft/sec)
(1) Brno, Czechoslovakia	341,948	4.81
(2) Prague, Czechoslovakia	1,092,759	5.88
(3) Corte, Corsica	5,491	3.31
(4) Bastia, France	49,375	4.90
(5) Munich, Germany	1,340,000	5.62
(6) Psychro, Crete	365	2.76
(7) Itea, Greece	2,500	2.27
(8) Iraklion, Greece	78,200	3.85
(9) Athens, Greece	867,023	5.21
(10) Safed, Israel	14,000	3.70
(11) Dimona, Israel	23,700	3.27
(12) Netanya, Israel	70,700	4.31
(13) Jerusalem, Israel	304,500	4.42
(14) New Haven, U.S.A.	138,000	4.39
(15) Brooklyn, U.S.A.	2,602,000	5.05

*Bornstein data.

3. Using the data, a calculator, and the model you determined for V (Problem 1f), complete Table 4.6.

4. From the data in Table 4.6, calculate the mean (i.e., the average) of the Bornstein errors $|V_{observed} - V_{predicted}|$. What do the results suggest about the merit of the model?

Table 4.6 Observed mean velocity for 15 locations

Location*	Observed velocity V	Predicted velocities
1	4.81	
2	5.88	
3	3.31	
4	4.90	
5	5.62	
6	2.76	
7	2.27	
8	3.85	
9	5.21	
10	3.70	
11	3.27	
12	4.31	
13	4.42	
14	4.39	
15	5.05	

*For location names, see Table 4.5.

Table 4.7   Oysters in the bay

Year	Oysters harvested (bushels)
1940	3,750,000
1945	3,250,000
1950	2,800,000
1955	2,550,000
1960	2,650,000
1965	1,850,000
1970	1,500,000
1975	1,000,000
1980	1,100,000
1985	750,000
1990	330,000

5. Solve Problems 1–4 with the model $V = m(\log P) + b$. Compare the errors with those computed in Problem 4. Compare the two models. Which is better?

6. Table 4.7 and Figure 4.9 present data representing the commercial harvesting of oysters in Chesapeake Bay. Fit a simple, one-term model to the data. How well does the best one-term model you find fit the data? What is the largest error? The average error?

■ **Figure 4.9**

Oysters (millions of pounds) versus base year

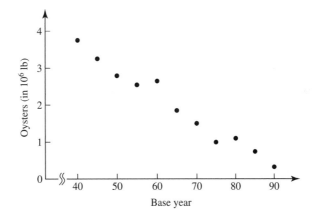

7. In Table 4.8, X is the Fahrenheit temperature, and Y is the number of times a cricket chirps in 1 minute. Fit a model to these data. Analyze how well it fits.

8. Fit a model to Table 4.9. Do you recognize the data? What relationship can be inferred from them?

9. The following data measure two characteristics of a ponderosa pine. The variable X is the diameter of the tree, in inches, measured at breast height; Y is a measure of volume—the number of board feet divided by 10. Fit a model to the data. Then express Y in terms of X.

Diameter and volume for 20 ponderosa pine trees

Observation number	X	Y	Observation number	X	Y
1	36	192	11	31	141
2	28	113	12	20	32
3	28	88	13	25	86
4	41	294	14	19	21
5	19	28	15	39	231
6	32	123	16	33	187
7	22	51	17	17	22
8	38	252	18	37	205
9	25	56	19	23	57
10	17	16	20	39	265

Data reported in Croxton, Cowden, and Klein, *Applied General Statistics*, p. 421.

Table 4.8 Temperature and chirps per minute for 20 crickets

Observation number	X	Y	Observation number	X	Y
1	46	40	11	61	96
2	49	50	12	62	88
3	51	55	13	63	99
4	52	63	14	64	110
5	54	72	15	66	113
6	56	70	16	67	120
7	57	77	17	68	127
8	58	73	18	71	137
9	59	90	19	72	132
10	60	93	20	71	137

Data inferred from a scatterplot in Frederick E. Croxton, Dudley J. Cowden, and Sidney Klein, *Applied General Statistics*, 3rd ed. (Englewood Cliffs, NJ: Prentice-Hall, 1967), p. 390.

Table 4.9

Observation number	X	Y
1	35.97	0.241
2	67.21	0.615
3	92.96	1.000
4	141.70	1.881
5	483.70	11.860
6	886.70	29.460
7	1783.00	84.020
8	2794.00	164.800
9	3666.00	248.400

10. The following data represent the length and weight of a set of fish (bass). Model weight as a function of the length of the fish.

Length (in.)	12.5	12.625	14.125	14.5	17.25	17.75
Weight (oz)	17	16.5	23	26.5	41	49

11. The following data give the population of the United States from 1800 to 2000. Model the population (in thousands) as a function of the year. How well does your model fit? Is a one-term model appropriate for these data? Why?

Year	1800	1820	1840	1860	1880	1900	1920
Population (thousands)	5308	9638	17,069	31,443	50,156	75,995	105,711

Year	1940	1960	1980	1990	2000
Population (thousands)	131,669	179,323	226,505	248,710	281,416

4.1 PROJECT

1. Compete the requirements of UMAP 551, "The Pace of Life, An Introduction to Model Fitting," by Bruce King. Prepare a short summary for classroom discussion.

4.1 Further Reading

Bornstein, Marc H., & Helen G. Bornstein. "The Pace of Life." *Nature* 259 (19 February 1976): 557–559.

Croxton, Fredrick E., Dudley J. Crowden, & Sidney Klein. *Applied General Statistics*, 7th ed. Englewood Cliffs, NJ: Prentice-Hall, 1985.

Neter, John, & William Wassermann. *Applied Linear Statistical Models*, 4th ed. Boston: McGraw-Hill, 1996.

Vellman, Paul F., & David C. Hoaglin. *Applications, Basics, and Computing of Exploratory Data Analysis*. Boston: Duxbury Press, 1984.

Yule, G. Udny. "Why Do We Sometimes Get Nonsense-Correlations between Time Series? A Study in Sampling and the Nature of Time Series." *Journal of the Royal Statistical Society* 89 (1926): 1–69.

4.2 High–Order Polynomial Models

In Section 4.1, we investigated the possibility of finding a simple, one-term model that captures the trend of the collected data. Because of their inherent simplicity, one-term models facilitate model analysis, including sensitivity analysis, optimization, and estimation of rates of change and area under the curve. Because of their mathematical simplicity, however, one-term models are limited in their ability to capture the trend of any collection of data. In some cases, models with more than one term must be considered. The remainder of this chapter considers one type of multiterm model—namely, the polynomial. Because polynomials are easy to integrate and to differentiate, they are especially popular. However, polynomials also have their disadvantages. For example, it is far more appropriate to approximate a data set having a vertical asymptote using a quotient of polynomials $p(x)/q(x)$ rather than a single polynomial.

Let's begin by studying polynomials that pass through each point in a data set that includes only one observation for each value of the independent variable.

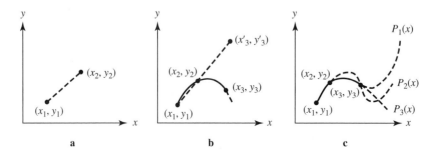

■ **Figure 4.10**

A unique polynomial of at most degree 2 can be passed through three data points (**a** and **b**), but an infinite number of polynomials of degree greater than 2 can be passed through three data points (**c**)

Consider the data in Figure 4.10a. Through the two given data points, a unique line $y = a_0 + a_1 x$ can pass. Determine the constants a_0 and a_1 by the conditions that the line passes through the points (x_1, y_1) and (x_2, y_2). Thus,

$$y_1 = a_0 + a_1 x_1$$

and

$$y_2 = a_0 + a_1 x_2$$

In a similar manner, a *unique* polynomial function of (at most) degree 2, $y = a_0 + a_1 x + a_2 x^2$, can be passed through three distinct points, as shown in Figure 4.10b. Determine the constants a_0, a_1, and a_2 by solving the following system of linear equations:

$$y_1 = a_0 + a_1 x_1 + a_2 x_1^2$$
$$y_2 = a_0 + a_1 x_2 + a_2 x_2^2$$
$$y_3 = a_0 + a_1 x_3 + a_2 x_3^2$$

Let's explain why the qualifier "at most" is needed with this polynomial function. Note that if the three points in Figure 4.10b happen to lie along a straight line, then the unique polynomial function of at most degree 2 passing through the points would necessarily be a straight line (a polynomial of degree 1) rather than a quadratic function, as generally would be expected. The descriptor *unique* is also important. There are an infinite number of polynomials of degree greater than 2 that pass through the three points depicted in Figure 4.10b. (Convince yourself of this fact before proceeding by using Figure 4.10c.)

There is, however, only one polynomial of degree 2 or less. Although this fact may not be obvious, we later state a theorem in its support. For now, remember from high school geometry that a unique circle, which is also represented by an algebraic equation of degree 2, is determined by three points in a plane. Next we illustrate these ideas in an applied problem; we then discuss the advantages and disadvantages of the procedure.

EXAMPLE 1 *Elapsed Time of a Tape Recorder*

We collected data relating the counter on a particular tape recorder with its elapsed playing time. Suppose we are unable to build an explicative model of this system but are still interested in predicting what may occur. How can we resolve this difficulty? As an example, let's construct an empirical model to predict the amount of elapsed time of a tape recorder as a function of its counter reading.

Thus, let c_i represent the counter reading and t_i (sec) the corresponding amount of elapsed time. Consider the following data:

c_i	100	200	300	400	500	600	700	800
t_i (sec)	205	430	677	945	1233	1542	1872	2224

One empirical model is a polynomial that passes through each of the data points. Because we have eight data points, a unique polynomial of at most degree 7 is expected. Denote the polynomial symbolically by

$$P_7(c) = a_0 + a_1 c + a_2 c^2 + a_3 c^3 + a_4 c^4 + a_5 c^5 + a_6 c^6 + a_7 c^7$$

The eight data points require that the constants a_i satisfy the following system of linear algebraic equations:

$$205 = a_0 + 1 a_1 + 1^2 a_2 + 1^3 a_3 + 1^4 a_4 + 1^5 a_5 + 1^6 a_6 + 1^7 a_7$$
$$430 = a_0 + 2 a_1 + 2^2 a_2 + 2^3 a_3 + 2^4 a_4 + 2^5 a_5 + 2^6 a_6 + 2^7 a_7$$
$$\vdots$$
$$2224 = a_0 + 8 a_1 + 8^2 a_2 + 8^3 a_3 + 8^4 a_4 + 8^5 a_5 + 8^6 a_6 + 8^7 a_7$$

Large systems of linear equations can be difficult to solve with great numerical precision. In the preceding illustration, we divided each counter reading by 100 to lessen the numerical difficulties. Because the counter data values are being raised to the seventh power, it is easy to generate numbers differing by several orders of magnitude. It is important to have as much accuracy as possible in the coefficients a_i because each is being multiplied by a number raised to a power as high as 7. For instance, a small a_7 may become significant as c becomes large. This observation suggests why there may be dangers in using even good polynomial functions that capture the trend of the data when we are beyond the range of the observations. The following solution to this system was obtained with the aid of a handheld calculator program:

$$a_0 = -13.9999923 \qquad a_4 = -5.354166491$$
$$a_1 = 232.9119031 \qquad a_5 = 0.8013888621$$
$$a_2 = -29.08333188 \qquad a_6 = -0.0624999978$$
$$a_3 = 19.78472156 \qquad a_7 = 0.0019841269$$

Let's see how well the empirical model fits the data. Denoting the polynomial predictions by $P_7(c_i)$, we find

c_i	100	200	300	400	500	600	700	800
t_i	205	430	677	945	1233	1542	1872	2224
$P_7(c_i)$	205	430	677	945	1233	1542	1872	2224

Rounding the predictions for $P_7(c_i)$ to four decimal places gives complete agreement with the observed data (as would be expected) and results in zero absolute deviations. Now we can see the folly of applying any of the criteria of best-fit studies in Chapter 3 as the sole judge for the best model. Can we really consider this model to be better than other models we could propose?

Let's see how well this new model $P_7(c_i)$ captures the trend of the data. The model is graphed in Figure 4.11.

▧ Figure 4.11

An empirical model for predicting the elapsed time of a tape recorder

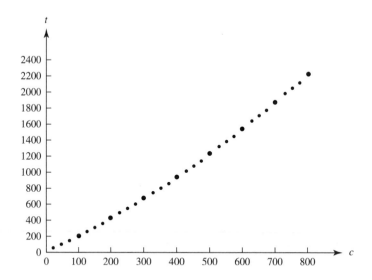

Lagrangian Form of the Polynomial

From the preceding discussion we might expect that given $(n+1)$ distinct data points, there is a unique polynomial of at most degree n that passes through all the data points. Because there are the same number of coefficients in the polynomial as there are data points, intuitively we would think that only one such polynomial exits. This hypothesis is indeed the case, although we will not prove that fact here. Rather, we present the Lagrangian form of the cubic polynomial followed by a brief discussion of how the coefficients may be found for higher-order polynomials.

Suppose the following data have been collected:

x	x_1	x_2	x_3	x_4
y	y_1	y_2	y_3	y_4

Consider the following cubic polynomial:

$$P_3(x) = \frac{(x - x_2)(x - x_3)(x - x_4)}{(x_1 - x_2)(x_1 - x_3)(x_1 - x_4)} y_1 + \frac{(x - x_1)(x - x_3)(x - x_4)}{(x_2 - x_1)(x_2 - x_3)(x_2 - x_4)} y_2$$

$$+ \frac{(x - x_1)(x - x_2)(x - x_4)}{(x_3 - x_1)(x_3 - x_2)(x_3 - x_4)} y_3 + \frac{(x - x_1)(x - x_2)(x - x_3)}{(x_4 - x_1)(x_4 - x_2)(x_4 - x_3)} y_4$$

Convince yourself that the polynomial is indeed cubic and agrees with the value y_i when $x = x_i$. Notice that the x_i values must all be different to avoid division by zero. Observe the pattern for forming the numerator and the denominator for the coefficient of each y_i. This same pattern is followed when forming polynomials of any desired degree. The procedure is justified by the following result.

Theorem 1

If $x_0, x_1, \ldots, x_n$ are $(n + 1)$ distinct points and $y_0, y_1, \ldots, y_n$ are corresponding observations at these points, then there exists a unique polynomial $P(x)$, of at most degree n, with the property that

$$y_k = P(x_k) \quad \text{for each } k = 0, 1, \ldots, n$$

This polynomial is given by

$$P(x) = y_0 L_0(x) + \cdots + y_n L_n(x) \tag{4.3}$$

where

$$L_k(x) = \frac{(x - x_0)(x - x_1) \cdots (x - x_{k-1})(x - x_{k+1}) \cdots (x - x_n)}{(x_k - x_0)(x_k - x_1) \cdots (x_k - x_{k-1})(x_k - x_{k+1}) \cdots (x_k - x_n)}$$

Because the polynomial (4.3) passes through each of the data points, the resultant sum of absolute deviations is zero. Considering the various criteria of best fit presented in Chapter 3, we are tempted to use high-order polynomials to fit larger sets of data. After all, the fit is precise. Let's examine both the advantages and the disadvantages of using high-order polynomials.

Advantages and Disadvantages of High-Order Polynomials

As we have seen on several occasions in previous chapters, it may be of interest to determine the area under the curve representing our model or its rate of change at a particular point.

Polynomial functions have the distinct advantage of being easily integrated and differentiated. If a polynomial can be found that reasonably represents the underlying behavior, it will be easy to approximate the integral and the derivative of the unknown true model as well. Now consider some of the disadvantages of higher-order polynomials. For the 17 data points presented in Table 4.10, it is clear that the trend of the data is $y = 0$ for all x over the interval $-8 \leq x \leq 8$.

Table 4.10

x_i	-8	-7	-6	-5	-4	-3	-2	-1	0	1	2	3	4	5	6	7	8
y_i	0	0	0	0	0	0	0	0	0	0	0	0	0	0	0	0	0

Suppose Equation (4.3) is used to determine a polynomial that passes through the points. Because there are 17 distinct data points, it is possible to pass a unique polynomial of degree at most 16 through the given points. The graph of a polynomial passing through the data points is depicted in Figure 4.12.

◼ Figure 4.12

Fitting a higher-order polynomial through the data points in Table 4.10

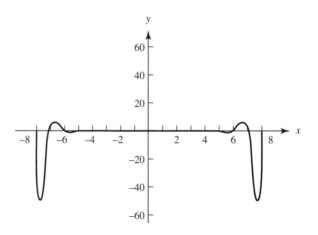

Note that although the polynomial does pass through the data points (within tolerances of computer round-off error), there is severe oscillation of the polynomial near each end of the interval. Thus, there would be gross error in estimating y between the data points near $+8$ or -8. Likewise, consider the error in using the derivative of the polynomial to estimate the rate of change of the data or in using the area under the polynomial to estimate the area trapped by the data. This tendency of high-order polynomials to oscillate severely near the endpoints of the interval is a serious disadvantage to using them.

Here is another example of fitting a higher-order polynomial to data. The scatterplot for the data suggests a smooth, increasing, concave up curve (see Figure 4.13).

x	0.55	1.2	2	4	6.5	12	16
y	0.13	0.64	5.8	102	210	2030	3900

■ Figure 4.13

Scatterplot of data

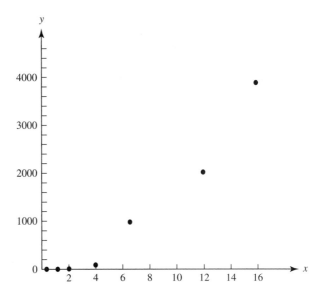

We fit a 6th-order polynomial to the data (where we have seven pairs, $n = 7$) and then plot the polynomial in Figure 4.14.

$$y = -0.0138x^6 + 0.5084x^5 - 6.4279x^4 + 34.8575x^3 - 73.9916x^2 + 64.3128x - 18.0951$$

Thus, although the higher-order $(n - 1)$st-order polynomial gives a perfect fit, its change from increasing to decreasing at the endpoint makes prediction questionable beyond the range of the data. Moreover within the range of the data, the polynomial changes from increasing to decreasing, also making interpolation questionable.

Let's illustrate another disadvantage of high-order polynomials. Consider the three sets of data presented in Table 4.11.

Let's discuss the three cases. Case 2 is merely Case 1 with less precise data: There is one less significant digit for each of the observed data points. Case 3 is Case 1 with an error

■ Figure 4.14

The plot of 6th-order polynomial fit superimposed on the scatterplot

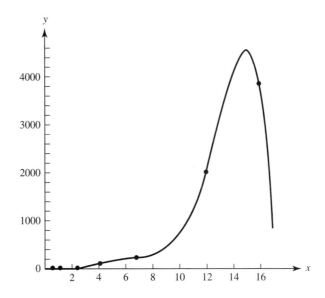

Table 4.11

x_i	0.2	0.3	0.4	0.6	0.9
Case 1: y_i	2.7536	3.2411	3.8016	5.1536	7.8671
Case 2: y_i	2.754	3.241	3.802	5.154	7.867
Case 3: y_i	2.7536	3.2411	3.8916	5.1536	7.8671

introduced in the observation corresponding to $x = 0.4$. Note that the error occurs in the third significant digit (3.8916 instead of 3.8016). Intuitively, we would think that the three interpolating polynomials would be similar because the trends of the data are all similar. Let's determine the interpolating polynomials and see if that is really the situation.

Because there are five distinct data points in each case, a unique polynomial of at most degree 4 can be passed through each set of data. Denote the fourth-degree polynomial symbolically as follows:

$$P_4(x) = a_0 + a_1x + a_2x^2 + a_3x^3 + a_4x^4$$

Table 4.12 tabulates the coefficients a_0, a_1, a_2, a_3, a_4 (to four decimal places) determined by fitting the data points in each of the three cases. Note how sensitive the values of the coefficients are to the data. Nevertheless, the *graphs* of the polynomials are nearly the same over the interval of observations $(0.2, 0.9)$.

Table 4.12

	a_0	a_1	a_2	a_3	a_4
Case 1	2	3	4	-1	1
Case 2	2.0123	2.8781	4.4159	-1.5714	1.2698
Case 3	3.4580	-13.2000	64.7500	-91.0000	46.0000

The graphs of the three fourth-degree polynomials representing each case are presented in Figure 4.15. This example illustrates the sensitivity of the coefficients of high-order polynomials to small changes in the data. Because we do expect measurement error to occur, the tendency of high-order polynomials to oscillate, as well as the sensitivity of their coefficients to small changes in the data, is a disadvantage that restricts their usefulness in modeling. In the next two sections we consider techniques that address the deficiencies noted in this section.

4.2 PROBLEMS

1. For the tape recorder problem in this section, give a system of equations determining the coefficients of a polynomial that passes through each of the data points. If a computer is available, determine and sketch the polynomial. Does it represent the trend of the data?

2. Consider the "pace of life" data from Problem 1, Section 4.1. Consider fitting a 14th-order polynomial to the data. Discuss the disadvantages of using the polynomial to make predictions. If a computer is available, determine and graph the polynomial.

■ Figure 4.15

Small measurement errors can cause huge differences in the coefficients of the higher-order polynomials that result; note that the polynomials diverge outside the range of observations.

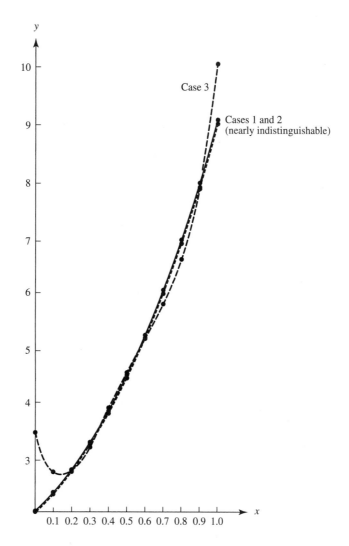

3. In the following data, X is the Fahrenheit temperature and Y is the number of times a cricket chirps in 1 minute (see Problem 7, Section 4.1). Make a scatterplot of the data and discuss the appropriateness of using an 18th-degree polynomial that passes through the data points as an empirical model. If you have a computer available, fit a polynomial to the data and plot the results.

X	46	49	51	52	54	56	57	58	59	60
Y	40	50	55	63	72	70	77	73	90	93

X	61	62	63	64	66	67	68	71	72
Y	96	88	99	110	113	120	127	137	132

4. In the following data, X represents the diameter of a ponderosa pine measured at breast height, and Y is a measure of volume—number of board feet divided by 10. Make a scatterplot of the data. Discuss the appropriateness of using a 13th-degree polynomial that passes through the data points as an empirical model. If you have a computer available, fit a polynomial to the data and graph the results.

X	17	19	20	22	23	25	31	32	33	36	37	38	39	41
Y	19	25	32	51	57	71	141	123	187	192	205	252	248	294

4.3   Smoothing: Low-Order Polynomial Models

We seek methods that retain many of the conveniences found in high-order polynomials without incorporating their disadvantages. One popular technique is to choose a low-order polynomial regardless of the number of data points. This choice normally results in a situation in which the number of data points exceeds the number of constants necessary to determine the polynomial. Because there are fewer constants to determine than there are data points, the low-order polynomial generally will not pass through all the data points. For example, suppose it is decided to fit a quadratic to a set of 10 data points. Because it is generally impossible to force a quadratic to pass through 10 data points, it must be decided which quadratic best fits the data (according to some criterion, as discussed in Chapter 3). This process, which is called **smoothing**, is illustrated in Figure 4.16. The combination of using a low-order polynomial and not requiring that it pass through each data point reduces both the tendency of the polynomial to oscillate and its sensitivity to small changes in the data. This quadratic function smooths the data because it is not required to pass through all the data points.

The process of smoothing requires two decisions. First, the order of the interpolating polynomial must be selected. Second, the coefficients of the polynomial must be determined according to some criterion for the best-fitting polynomial. The problem that results is an optimization problem of the form addressed in Chapter 3. For example, it may be decided to fit a quadratic model to 10 data points using the least-squares best-fitting criterion. We will review the process of fitting a polynomial to a set of data points using the least-squares criterion and then later return to the more difficult question of how to best choose the order of the interpolating polynomial.

■ **Figure 4.16**

The quadratic function smooths the data because it is not required to pass through all the data points.

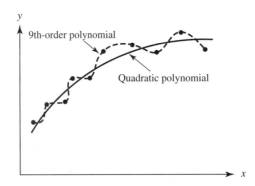

EXAMPLE 1 *Elapsed Time of a Tape Recorder Revisited*

Consider again the tape recorder problem modeled in the previous section. For a particular cassette deck or tape recorder equipped with a counter, relate the counter to the amount of playing time that has elapsed. If we are interested in predicting the elapsed time but are unable to construct an explicative model, it may be possible to construct an empirical model instead. Let's fit a second-order polynomial of the following form to the data:

$$P_2(c) = a + bc + dc^2$$

where c is the counter reading, $P_2(c)$ is the elapsed time, and a, b, and d are constants to be determined. Consider the collected data for the tape recorder problem in the previous section, shown in Table 4.13.

Table 4.13 Data collected for the tape recorder problem

c_i	100	200	300	400	500	600	700	800
t_i (sec)	205	430	677	945	1233	1542	1872	2224

Our problem is to determine the constants a, b, and d so that the resultant quadratic model best fits the data. Although other criteria might be used, we will find the quadratic that minimizes the sum of the squared deviations. Mathematically, the problem is

$$\text{Minimize } S = \sum_{i=1}^{m} \left[t_i - \left(a + bc_i + dc_i^2 \right) \right]^2$$

The necessary conditions for a minimum to exist ($\partial S/\partial a = \partial S/\partial b = \partial S/\partial d = 0$) yield the following equations:

$$ma + \left(\sum c_i \right) b + \left(\sum c_i^2 \right) d = \sum t_i$$

$$\left(\sum c_i \right) a + \left(\sum c_i^2 \right) b + \left(\sum c_i^3 \right) d = \sum c_i t_i$$

$$\left(\sum c_i^2 \right) a + \left(\sum c_i^3 \right) b + \left(\sum c_i^4 \right) d = \sum c_i^2 t_i$$

For the data given in Table 4.13, the preceding system of equations becomes

$$8a + 3600b + 2{,}040{,}000d = 9128$$
$$3600a + 2{,}040{,}000b + 1{,}296{,}000{,}000d = 5{,}318{,}900$$
$$2{,}040{,}000a + 1{,}296{,}000{,}000b + 8.772 \times 10^{11}d = 3{,}435{,}390{,}000$$

Solution of the preceding system yields the values $a = 0.14286$, $b = 1.94226$, and $d = 0.00105$, giving the quadratic

$$P_2(c) = 0.14286 + 1.94226c + 0.00105c^2$$

We can compute the deviation between the observations and the predictions made by the model $P_2(c)$:

c_i	100	200	300	400	500	600	700	800
t_i	205	430	677	945	1233	1542	1872	2224
$t_i - P_2(c_i)$	0.167	−0.452	0.000	0.524	0.119	−0.214	−0.476	0.333

Note that the deviations are very small compared to the order of magnitude of the times.

■ ■ ■

When we are considering the use of a low-order polynomial for smoothing, two issues come to mind:

1. Should a polynomial be used?
2. If so, what order of polynomial would be appropriate?

The derivative concept can help in answering these two questions.

Divided Differences

Notice that a quadratic function is characterized by the properties that its second derivative is constant and its third derivative is zero. That is, given

$$P(x) = a + bx + cx^2$$

we have

$$P'(x) = b + 2cx$$
$$P''(x) = 2c$$
$$P'''(x) = 0$$

However, the only information available is a set of discrete data points. How can these points be used to estimate the various derivatives? Refer to Figure 4.17, and recall the definition of the derivative:

$$\frac{dy}{dx} = \lim_{\Delta x \to 0} \frac{\Delta y}{\Delta x}$$

■ **Figure 4.17**

The derivative of $y = f(x)$ at $x = x_1$ is the limit of the slope of the secant line.

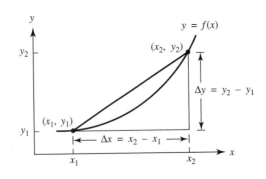

Because dy/dx at $x = x_1$ can be interpreted geometrically as the slope of the line tangent to the curve there, we see from Figure 4.17 that unless Δx is small, the ratio $\Delta y/\Delta x$ is probably not a good estimate of dy/dx. Nevertheless, if dy/dx is to be zero everywhere, then Δy must go to zero. Thus, we can compute the *differences* $y_{i+1} - y_i = \Delta y$ between successive function values in our tabled data to gain insight into what the first derivative is doing.

Likewise, because the first derivative is a function, the process can be repeated to estimate the second derivative. That is, the differences between successive estimates of the first derivative can be computed to approximate the second derivative. Before describing the entire process, we illustrate this idea with a simple example.

We know that the curve $y = x^2$ passes through the points $(0, 0)$, $(2, 4)$, $(4, 16)$, $(6, 36)$, and $(8, 64)$. Suppose the data displayed in Table 4.14 have been collected. Using the data in Table 4.14, we can construct a *difference table*, as shown in Table 4.15.

Table 4.14 A hypothetical set of collected data

x_i	0	2	4	6	8
y_i	0	4	16	36	64

Table 4.15 A difference table for the data of Table 4.14

Data		Differences			
x_i	y_i	Δ	Δ^2	Δ^3	Δ^4
0	0				
		4			
2	4		8		
		12		0	
4	16		8		0
		20		0	
6	36		8		
		28			
8	64				

The first differences, denoted by Δ, are constructed by computing $y_{i+1} - y_i$ for $i = 1, 2, 3, 4$. The second differences, denoted by Δ^2, are computed by finding the difference between successive first differences from the Δ column. The process can be continued, column by column, until Δ^{n-1} is computed for n data points. Note from Table 4.15 that the second differences in our example are constant and the third differences are zero. These results are consistent with the fact that a quadratic function has a constant second derivative and a zero third derivative.

Even if the data are essentially quadratic in nature, we would not expect the differences to go to zero precisely because of the various errors present in the modeling and data collection processes. We might, however, expect the data to become small. Our judgment of the significance of small can be improved by computing **divided differences**. Note that the differences computed in Table 4.15 are estimates of the numerator of each of the various order derivatives. These estimates can be improved by dividing the numerator by the corresponding estimate of the denominator.

Consider the three data points and corresponding estimates of the first and second derivative, called the first and second divided differences, respectively, in Table 4.16. The first divided difference follows immediately from the ratio of $\Delta y/\Delta x$. Because the second

Table 4.16   The first and second divided differences estimate the first and second derivatives, respectively

Data		First divided difference	Second divided difference
x_1	y_1		
		$\dfrac{y_2 - y_1}{x_2 - x_1}$	
			$\dfrac{\dfrac{y_3 - y_2}{x_3 - x_2} - \dfrac{y_2 - y_1}{x_2 - x_1}}{x_3 - x_1}$
x_2	y_2		
		$\dfrac{y_3 - y_2}{x_3 - x_2}$	
x_3	y_3		

■ **Figure 4.18**

The second divided difference may be interpreted as the difference between the adjacent slopes (first divided differences) divided by the length of the interval over which the change has taken place.

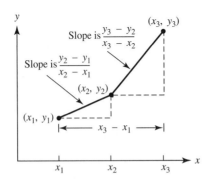

derivative represents the rate of change of the first derivative, we can estimate how much the first derivative changes between x_1 and x_3. That is, we can compute the differences between the adjacent first divided differences and divide by the length of the interval over which that change takes place ($x_3 - x_1$ in this case). Refer to Figure 4.18 for a geometric interpretation of the second divided difference.

In practice it is easy to construct a divided difference table. We generate the next-higher-order divided difference by taking differences between adjacent current-order divided differences and then dividing them by the length of the interval over which the change has taken place. Using Δ^n to denote the nth divided difference, a divided difference table for the data of Table 4.14 is displayed in Table 4.17.

Table 4.17   A divided difference table for the data of Table 4.14

Data		Divided differences		
x_i	y_i	Δ	Δ^2	Δ^3
0	0			
2	4	4/2 = 2		
4	16	12/2 = 6	4/4 = 1	0/6 = 0
6	36	20/2 = 10	4/4 = 1	0/6 = 0
8	64	28/2 = 14	4/4 = 1	

$\Delta x = 6$

It is easy to remember what the numerator should be in each divided difference of the table. To remember what the denominator should be for a given divided difference, we can construct diagonal lines back to y_i of the original data entries and compute the differences in the corresponding x_i. This is illustrated for a third-order divided difference in Table 4.17. This construction becomes more critical when the x_i are unequally spaced.

EXAMPLE 2 *Elapsed Time of a Tape Recorder Revisited Again*

Returning now to our construction of an empirical model for the elapsed time for a tape recorder, how might the order of the smoothing polynomial be chosen? Let's begin by constructing the divided difference table for the given data from Table 4.13. The divided differences are displayed in Table 4.18.

Table 4.18 A divided difference table for the tape recorder data

Data		Divided differences			
x_i	y_i	Δ	Δ^2	Δ^3	Δ^4
100	205				
200	430	2.2500	0.0011		
300	677	2.4700	0.0011	0.0000	0.0000
400	945	2.6800	0.0010	0.0000	0.0000
500	1233	2.8800	0.0011	0.0000	0.0000
600	1542	3.0900	0.0011	0.0000	0.0000
700	1872	3.3000	0.0011	0.0000	
800	2224	3.5200			

Note from Table 4.18 that the second divided differences are essentially constant and that the third divided differences equal zero to four decimal places. The table suggests the data are essentially quadratic, which supports the use of a quadratic polynomial as an empirical model. The modeler may now want to reinvestigate the assumptions to determine whether a quadratic relationship seems reasonable. ■ ■ ■

Observations on Difference Tables

Several observations about divided difference tables are in order. First, the x_i must be distinct and listed in increasing order. It is important to be sensitive to x_i that are close together because division by a small number can cause numerical difficulties. The scales used to measure both the x_i and the y_i must also be considered. For example, suppose the x_i represent distances and are currently measured in miles. If the units are changed to feet, the denominators become much larger, resulting in divided differences that are much smaller. Thus, judgment on what is small is relative and qualitative. Remember, however, that we are trying to decide whether a low-order polynomial is worth further investigation. Before accepting the model, we would want to graph it and analyze the goodness of fit.

When using a divided difference table, we must be sensitive to errors and irregularities that occur in the data. Measurement errors can propagate themselves throughout the table and even magnify themselves. For example, consider the following difference table:

Δ^{n-1}	Δ^n	Δ^{n+1}
6.01		
	−0.01	
6.00		0.02
	0.01	
6.01		

Let's suppose that the Δ^{n-1} column is actually constant except for the presence of a relatively small measurement error. Note how the measurement error gives rise to the negative sign in the Δ^n column. Note also that the magnitudes of the numbers in the Δ^{n+1} column are larger than those of the numbers in the Δ^n column, even though the Δ^{n-1} column is essentially constant. The effect of these errors not only is present in subsequent columns but also spreads to other rows. Because errors and irregularities are normally present in collected data, it is important to be sensitive to the effects they can cause in difference tables.

Historically, divided difference tables were used to determine various forms of interpolating polynomials that passed through a chosen subset of the data points. Today other interpolating techniques, such as smoothing and cubic splines, are more popular. Nevertheless, difference tables are easily constructed on a computer and, like the derivatives that they approximate, provide an inexpensive source of useful information about the data.

EXAMPLE 3 *Vehicular Stopping Distance*

The following problem was presented in Section 2.2: Predict a vehicle's total stopping distance as a function of its speed. In previous chapters, explicative models were constructed describing the vehicle's behavior. Those models will be reviewed in the next section. For the time being, suppose there is no explicative model but only the data displayed in Table 4.19.

If the modeler is interested in constructing an empirical model by smoothing the data, a divided difference table can be constructed as displayed in Table 4.20.

An examination of the table reveals that the third divided differences are small in magnitude compared to the data and that negative signs have begun to appear. As previously discussed, the negative signs may indicate the presence of measurement error or variations in the data that will not be captured with low-order polynomials. The negative signs will also have a detrimental effect on the differences in the remaining columns. Here, we may decide to use a quadratic model, reasoning that higher-order terms will not reduce the deviations sufficiently to justify their inclusion, but our judgment is qualitative. The cubic term will probably account for some for the deviations not accounted for by the best quadratic model (otherwise, the optimal value of the coefficient of the cubic term would be zero, causing

Table 4.19 Data relating total stopping distance and speed

Speed v (mph)	20	25	30	35	40	45	50	55	60	65	70	75	80
Distance d (ft)	42	56	73.5	91.5	116	142.5	173	209.5	248	292.5	343	401	464

Table 4.20 A divided difference table for the data relating total vehicular stopping distance and speed

Data		Divided differences			
v_i	d_i	Δ	Δ^2	Δ^3	Δ^4
20	42				
		2.2800			
25	56		0.0700		
		3.5000		−0.0040	
30	73.5		0.0100		0.0006
		3.6000		0.0080	
35	91.5		0.1300		−0.0007
		4.9000		−0.0060	
40	116		0.0400		0.0004
		5.3000		0.0027	
45	142.5		0.0800		0.0000
		6.1000		0.0027	
50	173		0.1200		−0.0004
		7.3000		−0.0053	
55	209.5		0.0400		0.0005
		7.7000		0.0053	
60	248		0.1200		−0.0003
		8.9000		0.0000	
65	292.5		0.1200		0.0001
		10.1000		0.0020	
70	343		0.1500		−0.0003
		11.6000		−0.0033	
75	401		0.1000		
		12.6000			
80	464				

the quadratic and cubic polynomials to coincide), but the addition of higher-order terms increases the complexity of the model, its susceptibility to oscillation, and its sensitivity to data errors. These considerations are studied in statistics.

In the following model, v is the speed of the vehicle, $P(v)$ is the stopping distance, and a, b, and c are constants to be determined.

$$P(v) = a + bv + cv^2$$

Our problem is to determine the constants a, b, and c so that the resultant quadratic model best fits the data. Although other criteria might be used, we will find the quadratic that minimizes the sum of the squared deviations. Mathematically, the problem is

$$\text{Minimize } S = \sum_{i=1}^{m} \left[d_i - \left(a + bv_i + cv_i^2 \right) \right]^2$$

The necessary conditions for a minimum to exist ($\partial S / \partial a = \partial S / \partial b = \partial S / \partial c = 0$) yield the following equations:

$$ma + \left(\sum v_i \right) b + \left(\sum v_i^2 \right) c = \sum d$$

$$\left(\sum v_i \right) a + \left(\sum v_i^2 \right) b + \left(\sum v_i^3 \right) c = \sum v_i d_i$$

$$\left(\sum v_i^2 \right) a + \left(\sum v_i^3 \right) b + \left(\sum v_i^4 \right) c = \sum v_i^2 d_i$$

Substitution from the data in Table 4.19 gives the system

$$13a + 650b + 37{,}050c = 2652.5$$
$$650a + 37{,}050b + 2{,}307{,}500c = 163{,}970$$
$$37{,}050a + 2{,}307{,}500b + 152{,}343{,}750c = 10{,}804{,}975$$

which leads to the solution $a = 50.0594$, $b = -1.9701$, and $c = 0.0886$ (rounded to four decimals). Therefore, the empirical quadratic model is given by

$$P(v) = 50.0594 - 1.9701v + 0.0886v^2$$

Finally, the fit of $P(v)$ is analyzed in Table 4.21. This empirical model fits better than the model

$$d = 1.104v + 0.0542v^2$$

determined in Section 3.4 because there is an extra parameter (the constant a in this case) that absorbs some of the error. Note, however, that the empirical model predicts a stopping distance of approximately 50 ft when the velocity is zero. ■ ■ ■

Table 4.21 Smoothing the stopping distance using a quadratic polynomial

v_i	20	25	30	35	40	45	50
d_i	42	56	73.5	91.5	116	142.5	173
$d_i - P(v_i)$	−4.097	−0.182	2.804	1.859	2.985	1.680	−0.054

v_i	55	60	65	70	75	80
d_i	209.5	248	292.5	343	401	464
$d_i - P(v_i)$	−0.719	−2.813	−3.838	−3.292	0.323	4.509

EXAMPLE 4 *Growth of a Yeast Culture*

In this example we consider a collection of data points for which a divided difference table can help in deciding whether a low-order polynomial will provide a satisfactory empirical model. The data represent the population of yeast cells in a culture measured over time (in hours). A divided difference table for the population data is given by Table 4.22.

Note that the first divided differences Δ are increasing until $t = 8$ hr, when they begin to decrease. This characteristic is reflected in the column Δ^2 with the appearance of a consecutive string of negative signs, indicating a change in concavity. Thus, we cannot hope to capture the trend of these data with a quadratic function that has only a single concavity. In the Δ^3 column, additional negative signs appear sporadically, although the magnitude of the numbers is relatively large.

A scatterplot of the data is given in Figure 4.19. Although the divided difference table suggests that a quadratic function would not be a good model, for illustrative purposes suppose we try to fit a quadratic anyway. Using the least-squares criterion and the equations developed previously for the tape recorder example, we determine the following quadratic model for the data in Table 4.22:

$$P = -93.82 + 65.70t - 1.12t^2$$

Table 4.22 A divided difference table for the growth of yeast in a culture

| Data | | Divided differences | | | |
t_i	P_i	Δ	Δ^2	Δ^3	Δ^4
0	9.60				
		8.70			
1	18.30		1.00		
		10.70		0.92	
2	29.00		3.75		−0.31
		18.20		−0.30	
3	47.20		2.85		0.84
		23.90		3.07	
4	71.10		12.05		−1.46
		48.00		−2.77	
5	119.10		3.75		1.51
		55.50		3.28	
6	174.60		13.60		−1.51
		82.70		−2.75	
7	257.30		5.35		0.11
		93.40		−2.30	
8	350.70		−1.55		−0.05
		90.30		−2.48	
9	441.00		−9.00		0.29
		72.30		−1.32	
10	513.30		−12.95		0.94
		46.40		2.43	
11	559.70		−5.65		−0.16
		35.10		1.80	
12	594.80		−0.25		−1.40
		34.60		−3.78	
13	629.40		−11.60		1.87
		11.40		3.68	
14	640.80		−0.55		−1.10
		10.30		−0.73	
15	651.10		−2.75		0.37
		4.80		0.73	
16	655.90		−0.55		−0.20
		3.70		−0.07	
17	659.60		−0.75		
		2.20			
18	661.80				

■ **Figure 4.19**

A scatterplot of the "yeast growth in a culture" data

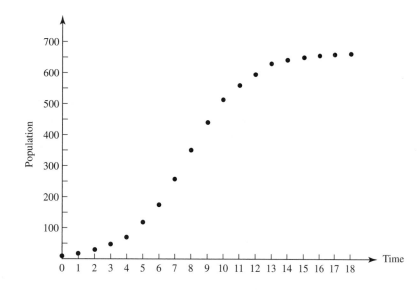

■ Figure 4.20

The best-fitting quadratic model fails to capture the trend of the data; note the magnitude of the deviations $P_i - P(t_i)$.

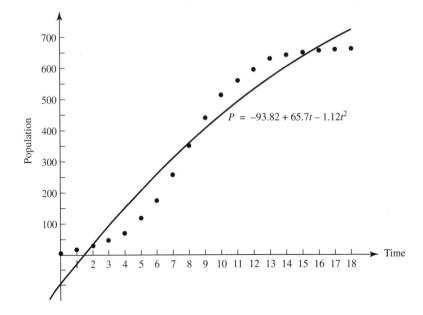

$P = -93.82 + 65.7t - 1.12t^2$

The model and the data points are plotted in Figure 4.20. The model fits very poorly, as we expected, and fails to capture the trend of the data. In the problem set you will be asked to fit the cubic model and check its reasonableness. In the next section you will be asked to construct a cubic spline model that fits the data much better. ■ ■ ■

4.3 PROBLEMS

For the data sets in Problems 1–4, construct a divided difference table. What conclusions can you make about the data? Would you use a low-order polynomial as an empirical model? If so, what order?

1.

x	0	1	2	3	4	5	6	7
y	2	8	24	56	110	192	308	464

2.

x	0	1	2	3	4	5	6	7
y	23	48	73	98	123	148	173	198

3.

x	0	1	2	3	4	5	6	7
y	7	15	33	61	99	147	205	273

4.

x	0	1	2	3	4	5	6	7
y	1	4.5	20	90	403	1808	8103	36,316

5. Construct a scatterplot for the "yeast growth in a culture" data. Do the data seem reasonable? Construct a divided difference table. Try smoothing with a low-order cubic polynomial using an appropriate criterion. Analyze the fit and compare your model to the quadratic we developed in this section. Graph your model, the data points, and the deviations.

In Problems 6–12, construct a scatterplot of the given data. Is there a trend in the data? Are any of the data points outliers? Construct a divided difference table. Is smoothing with a low-order polynomial appropriate? If so, choose an appropriate polynomial and fit using the least-squares criterion of best fit. Analyze the goodness of fit by examining appropriate indicators and graphing the model, the data points, and the deviations.

6. In the following data, X is the Fahrenheit temperature and Y is the number of times a cricket chirps in 1 min (see Problem 3, Section 4.2).

X	46	49	51	52	54	56	57	58	59	60
Y	40	50	55	63	72	70	77	73	90	93

X	61	62	63	64	66	67	68	71	72
Y	96	88	99	110	113	120	127	137	132

7. In the following data, X represents the diameter of a ponderosa pine measured at breast height, and Y is a measure of volume—number of board feet divided by 10 (see Problem 4, Section 4.2).

X	17	19	20	22	23	25	31	32	33	36	37	38	39	41
Y	19	25	32	51	57	71	141	123	187	192	205	252	248	294

8. The following data represent the population of the United States from 1790 to 2000.

Year	Observed population
1790	3,929,000
1800	5,308,000
1810	7,240,000
1820	9,638,000
1830	12,866,000
1840	17,069,000
1850	23,192,000
1860	31,443,000
1870	38,558,000
1880	50,156,000
1890	62,948,000
1900	75,995,000
1910	91,972,000
1920	105,711,000
1930	122,755,000
1940	131,669,000
1950	150,697,000
1960	179,323,000
1970	203,212,000
1980	226,505,000
1990	248,709,873
2000	281,416,000

9. The following data were obtained for the growth of a sheep population introduced into a new environment on the island of Tasmania. (Adapted from J. Davidson, "On the Growth of the Sheep Population in Tasmania," *Trans. Roy. Soc. S. Australia* 62(1938): 342–346.)

t (year)	1814	1824	1834	1844	1854	1864
$P(t)$	125	275	830	1200	1750	1650

10. The following data represent the "pace of life" data (see Problem 1, Section 4.1). P is the population and V is the mean velocity in feet per second over a 50-ft course.

P	365	2500	5491	14000	23700	49375	70700	78200
V	2.76	2.27	3.31	3.70	3.27	4.90	4.31	3.85

P	138000	304500	341948	867023	1092759	1340000	2602000
V	4.39	4.42	4.81	5.21	5.88	5.62	5.05

11. The following data represent the length of a bass fish and its weight.

Length (in.)	12.5	12.625	14.125	14.5	17.25	17.75
Weight (oz)	17	16.5	23	26.5	41	49

12. The following data represent the weight-lifting results from the 1976 Olympics.

Bodyweight class (lb)		Total winning lifts (lb)		
	Max. weight	Snatch	Jerk	Total weight
Flyweight	114.5	231.5	303.1	534.6
Bantamweight	123.5	259.0	319.7	578.7
Featherweight	132.5	275.6	352.7	628.3
Lightweight	149.0	297.6	380.3	677.9
Middleweight	165.5	319.7	418.9	738.5
Light-heavyweight	182.0	358.3	446.4	804.7
Middle-heavyweight	198.5	374.8	468.5	843.3
Heavyweight	242.5	385.8	496.0	881.8

4.4 Cubic Spline Models

The use of polynomials in constructing empirical models that capture the trend of the data is appealing because polynomials are so easy to integrate and differentiate. High-order polynomials, however, tend to oscillate near the endpoints of the data interval, and the coefficients can be quite sensitive to small changes in the data. Unless the data are essentially quadratic or cubic in nature, smoothing with a low-order polynomial may yield a relatively poor fit somewhere over the range of the data. For instance, the quadratic model

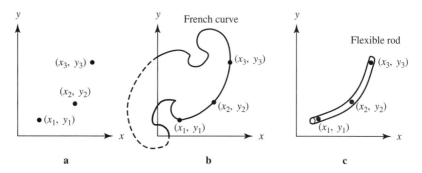

■ Figure 4.21

A draftsperson might attempt to draw a smooth curve through the data points using a French curve or a thin flexible rod called a *spline*.

that was fit to the data collected for the vehicular braking distance problem in Section 4.3 did not fit well at high velocities.

In this section a very popular modern technique called **cubic spline interpolation** is introduced. By using different cubic polynomials between successive pairs of data points, we can capture the trend of the data regardless of the nature of the underlying relationship, while simultaneously reducing the tendency toward oscillation and the sensitivity to changes in the data.

Consider the data in Figure 4.21a. What would a draftsperson do if asked to draw a smooth curve connecting the points? One solution would be to use a drawing instrument called a French curve (Figure 4.21b), which actually contains many different curves. By manipulating the French curve, we can draw a curve that works reasonably well between two data points and transitions smoothly to another curve for the next pair of data points. Another alternative is to take a very thin flexible rod (called a **spline**) and tack it down at each data point. Cubic spline interpolation is essentially the same idea, except that distinct cubic polynomials are used between successive pairs of data points in a smooth way.

Linear Splines

Probably all of us at one time or another have referred to a table of values (e.g., square root, trigonometric table, or logarithmic table) without locating the value we were seeking. We found, instead, two values that bracket the desired value, and we made a proportionality adjustment.

Table 4.23
Linear interpolation

For example, consider Table 4.23. Suppose an estimate of the value of y at $x = 1.67$ is desired. Probably we would compute $y(1.67) \approx 5 + (2/3)(8 - 5) = 7$. That is, we implicitly assume that the variation in y, between $x = 1$ and $x = 2$, occurs linearly. Similarly, $y(2.33) \approx 13\frac{2}{3}$. This procedure is called **linear interpolation**, and for many applications it yields reasonable results, especially where the data are closely spaced.

Figure 4.22 is helpful in interpreting the process of linear interpolation geometrically in a manner that mimics what is done with cubic spline interpolation. When x is in the interval $x_1 \leq x < x_2$, the model that is used is the **linear spline** $S_1(x)$ passing through the

x_i	$y(x_i)$
1	5
2	8
3	25

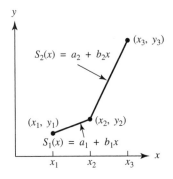

data points (x_1, y_1) and (x_2, y_2):

$$S_1(x) = a_1 + b_1 x \quad \text{for } x \text{ in } [x_1, x_2)$$

Similarly, when $x_2 \leq x < x_3$, the linear spline $S_2(x)$ passing through (x_2, y_2) and (x_3, y_3)
is used:

$$S_2(x) = a_2 + b_2 x \quad \text{for } x \text{ in } [x_2, x_3]$$

Note that both spline segments meet at the point (x_2, y_2).

Let's determine the constants for the respective splines for the data given in Table 4.23.
The spline $S_1(x)$ must pass through the points $(1, 5)$ and $(2, 8)$. Mathematically this implies

$$a_1 + 1b_1 = 5$$
$$a_1 + 2b_1 = 8$$

Similarly, the spline $S_2(x)$ must pass through the points $(2, 8)$ and $(3, 25)$, yielding

$$a_2 + 2b_2 = 8$$
$$a_2 + 3b_2 = 25$$

Solution of these linear equations yields $a_1 = 2$, $b_1 = 3$, $a_2 = -26$, and $b_2 = 17$. The
linear spline model for the data in Table 4.23 is summarized in Table 4.24. To illustrate how
the linear spline model is used, let's predict $y(1.67)$ and $y(2.33)$. Because $1 \leq 1.67 < 2$,
$S_1(x)$ is selected to compute $S_1(1.67) \approx 7.01$. Likewise, $2 \leq 2.33 \leq 3$ gives rise to the
prediction $S_2(2.33) \approx 13.61$.

Although the linear spline method is sufficient for many applications, it fails to capture
the trend of the data. Furthermore, if we examine Figure 4.23, we see that the linear spline

Table 4.24   A linear spline
model for the data of Table 4.23

Interval	Spline model
$1 \leq x < 2$	$S_1(x) = 2 + 3x$
$2 \leq x \leq 3$	$S_2(x) = -26 + 17x$

■ Figure 4.23

The linear spline does not appear smooth because the first derivative is not continuous.

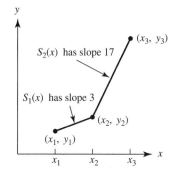

model does not appear smooth. That is, in the interval $[1, 2)$, $S_1(x)$ has a constant slope 3, whereas in the interval $[2, 3]$ $S_2(x)$ has a constant slope of 17. Thus, at $x = 2$ there is an abrupt change in the slope of the model from 3 to 17 so that the first derivatives $S_1'(x)$ and $S_2'(x)$ fail to agree at $x = 2$. As we will discover, the process of cubic splines interpolation incorporates smoothness into the empirical model by requiring that both the first and second derivatives of adjacent splines agree at each data point.

Cubic Splines

Consider now Figure 4.24. In a manner analogous to linear splines, we define a separate spline function for the intervals $x_1 \leq x < x_2$ and $x_2 \leq x < x_3$ as follows:

$$S_1(x) = a_1 + b_1 x + c_1 x^2 + d_1 x^3 \quad \text{for } x \text{ in } [x_1, x_2)$$
$$S_2(x) = a_2 + b_2 x + c_2 x^2 + d_2 x^3 \quad \text{for } x \text{ in } [x_2, x_3]$$

Because we will want to refer to the first and second derivatives, let's define them as well:

$$S_1'(x) = b_1 + 2c_1 x + 3d_1 x^2 \quad \text{for } x \text{ in } [x_1, x_2)$$
$$S_1''(x) = 2c_1 + 6d_1 x \qquad\qquad \text{for } x \text{ in } [x_1, x_2)$$
$$S_2'(x) = b_2 + 2c_2 x + 3d_2 x^2 \quad \text{for } x \text{ in } [x_2, x_3]$$
$$S_2''(x) = 2c_2 + 6d_2 x \qquad\qquad \text{for } x \text{ in } [x_2, x_3]$$

The model is presented geometrically in Figure 4.24.

■ Figure 4.24

A cubic spline model is a continuous function with continuous first and second derivatives consisting of cubic polynomial segments.

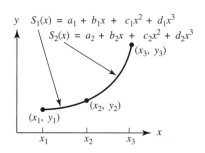

Cubic splines offer the possibility of matching up not only the slopes but also the curvature at each interior data point. To determine the constants defining each cubic spline segment, we appeal to the requirement that each spline pass through two data points specified by the interval over which the spline is defined. For the model depicted in Figure 4.24, this requirement yields the equations

$$y_1 = S_1(x_1) = a_1 + b_1 x_1 + c_1 x_1^2 + d_1 x_1^3$$
$$y_2 = S_1(x_2) = a_1 + b_1 x_2 + c_1 x_2^2 + d_1 x_2^3$$
$$y_2 = S_2(x_2) = a_2 + b_2 x_2 + c_2 x_2^2 + d_2 x_2^3$$
$$y_3 = S_2(x_3) = a_2 + b_2 x_3 + c_2 x_3^2 + d_2 x_3^3$$

Note that there are eight unknowns $(a_1, b_1, c_1, d_1, a_2, b_2, c_2, d_2)$ and only four equations in the preceding system. An additional four independent equations are needed to determine the constants uniquely. Because smoothness of the spline system is also required, adjacent first derivatives must match at the interior data point (in this case, when $x = x_2$). This requirement yields the equation

$$S_1'(x_2) = b_1 + 2c_1 x_2 + 3d_1 x_2^2 = b_2 + 2c_2 x_2 + 3d_2 x_2^2 = S_2'(x_2)$$

It can also be required that adjacent second derivatives match at each interior point as well:

$$S_1''(x_2) = 2c_1 + 6d_1 x_2 = 2c_2 + 6d_2 x_2 = S_2''(x_2)$$

To determine unique constants, we still require two additional independent equations. Although conditions on the derivatives at interior data points have been applied, nothing has been said about the derivatives at the exterior endpoints (x_1 and x_3 in Figure 4.24). Two popular conditions may be specified. One is to require that there be no change in the first derivative at the exterior endpoints. Mathematically, because the first derivative is constant, the second derivative must be zero. Application of this condition at x_1 and x_3 yields

$$S_1''(x_1) = 2c_1 + 6d_1 x_1 = 0$$
$$S_2''(x_3) = 2c_2 + 6d_2 x_3 = 0$$

A cubic spline formed in this manner is called a **natural spline**. If we think again of our analog with the thin flexible rod tacked down at the data points, a natural spline allows the rod to be free at the endpoints to assume whatever direction the data points indicate. The natural spline is interpreted geometrically in Figure 4.25a.

Alternatively, if the values of the first derivative at the exterior endpoints are known, the first derivatives of the exterior splines can be required to match the known values. Suppose the derivatives at the exterior endpoints are known and are given by $f'(x_1)$ and $f'(x_3)$. Mathematically, this matching requirement yields the equations

$$S_1'(x_1) = b_1 + 2c_1 x_1 + 3d_1 x_1^2 = f'(x_1)$$
$$S_2'(x_3) = b_2 + 2c_2 x_3 + 3d_2 x_3^2 = f'(x_3)$$

A cubic spline formed in this manner is called a **clamped spline**. Again referring to our flexible rod analog, this situation corresponds to clamping the flexible rod in a vise at the

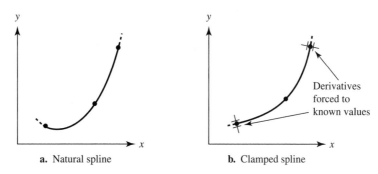

a. Natural spline **b.** Clamped spline

■ **Figure 4.25**

The conditions for the natural and clamped cubic splines result in first derivatives at the two exterior endpoints that are constant; that constant value for the first derivative is specified in the clamped spline, whereas it is free to assume a natural value in the natural spline.

exterior endpoints to ensure that the flexible rod has the proper angle. The clamped cubic spline is interpreted geometrically in Figure 4.25b. Unless *precise* information about the first derivative at the endpoints is known, the natural spline is generally used.

Let's illustrate the construction of a natural cubic spline model using the data displayed in Table 4.23. We illustrate the technique with this simple example because the procedure is readily extended to problems with more data points.

Requiring the spline segment $S_1(x)$ to pass through the two endpoints $(1, 5)$ and $(2, 8)$ of its interval requires that $S_1(1) = 5$ and $S_1(2) = 8$, or

$$a_1 + 1b_1 + 1c_1 + 1d_1 = 5$$
$$a_1 + 2b_1 + 2^2 c_1 + 2^3 d_1 = 8$$

Similarly, $S_2(x)$ must pass through its endpoints of the second interval so that $S_2(2) = 8$ and $S_2(3) = 25$, or

$$a_2 + 2b_2 + 2^2 c_2 + 2^3 d_2 = 8$$
$$a_2 + 3b_2 + 3^2 c_2 + 3^3 d_2 = 25$$

Next, the first derivatives of $S_1(x)$ and $S_2(x)$ are forced to match at the interior data point $x_2 = 2$: $S_1'(2) = S_2'(2)$, or

$$b_1 + 2c_1(2) + 3d_1(2)^2 = b_2 + 2c_2(2) + 3d_2(2)^2$$

Forcing the second derivatives of $S_1(x)$ and $S_2(x)$ to match at $x_2 = 2$ requires $S_1''(2) = S_2''(2)$, or

$$2c_1 + 6d_1(2) = 2c_2 + 6d_2(2)$$

Finally, a natural spline is built by requiring that the second derivatives at the endpoints be zero: $S_1''(1) = S_2''(3) = 0$, or

$$2c_1 + 6d_1(1) = 0$$
$$2c_2 + 6d_2(3) = 0$$

Table 4.25 A natural cubic spline model for the data of Table 4.23

Interval	Model
$1 \le x < 2$	$S_1(x) = 2 + 10x - 10.5x^2 + 3.5x^3$
$2 \le x \le 3$	$S_2(x) = 58 - 74x + 31.5x^2 - 3.5x^3$

■ **Figure 4.26**

The natural cubic spline model for the data in Table 4.23 is a smooth curve that is easily integrated and differentiated.

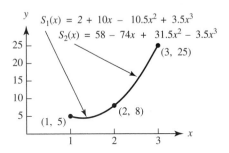

Thus, the procedure has yielded a linear algebraic system of eight equations in eight unknowns that can be solved uniquely. The resulting model is summarized in Table 4.25 and graphed in Figure 4.26.

Let's illustrate the use of the model by again predicting $y(1.67)$ and $y(2.33)$:

$$S_1(1.67) \approx 5.72$$
$$S_2(2.33) \approx 12.32$$

Compare these values with the values predicted by the linear spline. In which values do you have the most confidence? Why?

The construction of cubic splines for more data points proceeds in the same manner. That is, each spline is forced to pass through the endpoints of the interval over which it is defined, the first and second derivatives of adjacent splines are forced to match at the interior data points, and either the clamped or natural conditions are applied at the two exterior data points. For computational reasons it would be necessary to implement the procedure on a computer. The procedure we described here does not give rise to a computationally or numerically efficient computer algorithm. Our approach was designed to facilitate our understanding of the basic concepts underlying the cubic spline interpolation.[3]

It is revealing to view how the graphs of the different cubic splines fit together to form a single composite interpolating curve between the data points. Consider the following data (from Problem 4, Section 3.3.):

x	7	14	21	28	35	42
y	8	41	133	250	280	297

---

[3]For a computationally efficient algorithm, see R. L. Burden and J. D. Faires, *Numerical Analysis*, 7th ed. (Pacific Grove, CA: Brooks/Cole, 2001).

Because there are six data points, five distinct cubic polynomials, S_1–S_5, are calculated to form the composite natural cubic spline. Each of these cubics is graphed and overlaid on the same graph to obtain Figure 4.27. Between any two consecutive data points only one of the five cubic polynomials is active, giving the smooth composite cubic spline shown in Figure 4.28.

You should be concerned with whether the procedure just described results in a unique solution. Also, you may be wondering why we jumped from a linear spline to a cubic spline without discussing quadratic splines. Intuitively, you would think that first derivatives could

■ **Figure 4.27**

Between any two consecutive data points only one cubic spline polynomial is active. (Graphics by Jim McNulty and Bob Hatton)

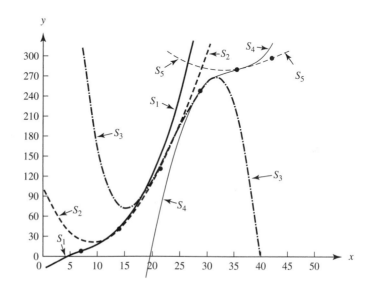

■ **Figure 4.28**

The smooth composite cubic spline from the cubic polynomials in Figure 4.27

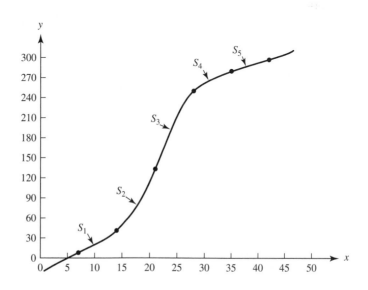

be matched with a quadratic spline. These and related issues are discussed in most numerical analysis texts (for example, see Burden and Faires cited earlier).

EXAMPLE 1   *Vehicular Stopping Distance Revisited*

Consider again the problem posed in Section 2.2: Predict a vehicle's total stopping distance as a function of its speed. In Section 2.2, we reasoned that the model should have the form

$$d = k_1 v + k_2 v^2$$

where d is the total stopping distance, v is the velocity, and k_1 and k_2 are constants of proportionality resulting from the submodels for reaction distance and mechanical braking distance, respectively. We found reasonable agreement between the data furnished for the submodels and the graphically estimated k_1 and k_2 to obtain the model

$$d = 1.1v + 0.054v^2$$

In Section 3.4 we estimated k_1 and k_2 using the least-squares criterion and obtained the model

$$d = 1.104v + 0.0542v^2$$

The fit of the preceding two models was analyzed in Table 3.7. Note in particular that both models break down at high speeds, where they increasingly underestimate the stopping distances. In Section 4.3, we constructed an empirical model by smoothing the data with a quadratic polynomial, and we analyzed the fit.

Now suppose that we are not satisfied with the predictions made by our analytic models or are unable to construct an analytic model, but we find it necessary to make predictions. If we are reasonably satisfied with the collected data, we might consider constructing a cubic spline model for the data presented in Table 4.26.

Using a computer code, we obtained the cubic spline model summarized in Table 4.27. The first three spline segments are plotted in Figure 4.29. Note how each segment passes through the data points at either end of its interval, and note the smoothness of the transition across adjacent segments.

Table 4.26   Data relating total stopping distance and speed

Speed, v (mph)	20	25	30	35	40	45	50
Distance, d (ft)	42	56	73.5	91.5	116	142.5	173

Speed, v (mph)	55	60	65	70	75	80
Distance, d (ft)	209.5	248	292.5	343	401	464

Table 4.27 A cubic spline model for vehicular stopping distance

Interval	Model
$20 \leq v < 25$	$S_1(v) = 42 + 2.596(v - 20) + 0.008(v - 20)^3$
$25 \leq v < 30$	$S_2(v) = 56 + 3.208(v - 25) + 0.122(v - 25)^2 - 0.013(v - 25)^3$
$30 \leq v < 35$	$S_3(v) = 73.5 + 3.472(v - 30) - 0.070(v - 30)^2 + 0.019(v - 30)^3$
$35 \leq v < 40$	$S_4(v) = 91.5 + 4.204(v - 35) + 0.216(v - 35)^2 - 0.015(v - 35)^3$
$40 \leq v < 45$	$S_5(v) = 116 + 5.211(v - 40) - 0.015(v - 40)^2 + 0.006(v - 40)^3$
$45 \leq v < 50$	$S_6(v) = 142.5 + 5.550(v - 45) + 0.082(v - 45)^2 + 0.005(v - 45)^3$
$50 \leq v < 55$	$S_7(v) = 173 + 6.787(v - 50) + 0.165(v - 50)^2 - 0.012(v - 50)^3$
$55 \leq v < 60$	$S_8(v) = 209.5 + 7.503(v - 55) - 0.022(v - 55)^2 + 0.012(v - 55)^3$
$60 \leq v < 65$	$S_9(v) = 248 + 8.202(v - 60) + 0.161(v - 60)^2 - 0.004(v - 60)^3$
$65 \leq v < 70$	$S_{10}(v) = 292.5 + 9.489(v - 65) + 0.096(v - 65)^2 + 0.005(v - 65)^3$
$70 \leq v < 75$	$S_{11}(v) = 343 + 10.841(v - 70) + 0.174(v - 70)^2 - 0.005(v - 70)^3$
$75 \leq v < 80$	$S_{12}(v) = 401 + 12.245(v - 75) + 0.106(v - 75)^2 - 0.007(v - 75)^3$

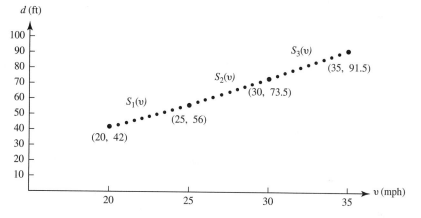

■ **Figure 4.29**

A plot of the cubic spline model for vehicular stopping distance for $20 \leq v \leq 35$

Summary: Constructing Empirical Models

We conclude this chapter by presenting a summary and suggesting a procedure for constructing empirical models using the techniques presented. Begin the procedure by examining the data in search of suspect data points that you need to examine more closely and perhaps discard or obtain anew. Simultaneously, look to see if a trend exists in the data. Normally, these questions are best considered by constructing a scatterplot, possibly with the aid of a computer if a great many data are under consideration. If a trend does appear to exist, first investigate simple one-term models to see if one adequately captures the trend of the data. You can then identify a one-term model using a transformation that converts the data into a straight line (approximately). You can find the transformation among the ladder of transformations or among the transformations discussed in Chapter 3. A graphical plot of the transformed data is often useful to determine whether the one-term model linearizes the data. If you find an adequate fit, the chosen model can be fit graphically or analytically using

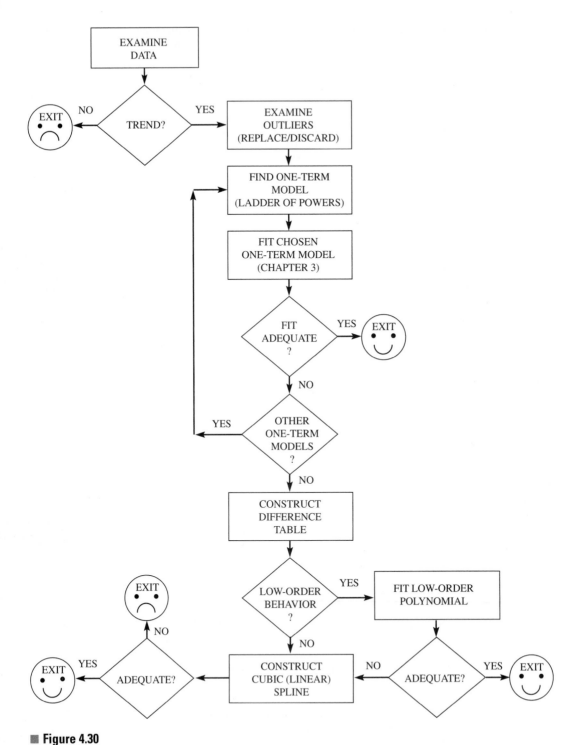

■ **Figure 4.30**

A flowchart for empirical model building

one of the criteria discussed in Chapter 3. Next, conduct a careful analysis to determine how well the model fits the data points by examining indicators such as the sum of the absolute deviations, the largest absolute deviation, and the sum of squared deviations. A plot of the deviations as a function of the independent variable(s) may be useful for determining where the model does not fit well. If the fit proves unsatisfactory, consider other one-term models.

If you determine that one-term models are inadequate, use polynomials. If there is a small number of data points, try an $(n-1)$st-order polynomial through the n data points. Be sure to note any oscillations, especially near the endpoints of the interval. A careful plot of the polynomial will help reveal this feature. If there is a large number of data points, consider a low-order polynomial to smooth the data. A divided difference table is a qualitative aid in determining whether a low-order polynomial is appropriate and in choosing the order of that polynomial. After the order of the polynomial is chosen, you may fit and analyze the polynomial according to the techniques discussed in Chapter 3. If smoothing with a low-order polynomial proves inadequate, use cubic (or linear) splines. See Figure 4.30 for a flowchart summary of this discussion.

4.4 PROBLEMS

1. For each of the following data sets, write a system of equations to determine the coefficients of the natural cubic splines passing through the given points. If a computer program is available, solve the system of equations and graph the splines.

   a.

x	2	4	7
y	2	8	12

   b.

x	3	4	6
y	10	15	35

   c.

x	0	1	2
y	0	10	30

   d.

x	0	2	4
y	5	10	40

For Problems 2 and 3, find the natural cubic splines that pass through the given data points. Use the splines to answer the requirements.

2.

x	3.0	3.1	3.2	3.3	3.4	3.5	3.6	3.7	3.8	3.9
y	20.08	22.20	24.53	27.12	29.96	33.11	36.60	40.45	44.70	49.40

   a. Estimate the derivative evaluated at $x = 3.45$. Compare your estimate with the derivative of e^x evaluated at $x = 3.45$.

   b. Estimate the area under the curve from 3.3 to 3.6. Compare with

$$\int_{3.3}^{3.6} e^x \, dx$$

3.

x	0	$\pi/6$	$\pi/3$	$\pi/2$	$2\pi/3$	$5\pi/6$	π
y	0.00	0.50	0.87	1.00	0.87	0.50	0.00

4. For the data collected in the tape recorder problem (Sections 4.2 and 4.3) relating the elapsed time with the counter reading, construct the natural spline that passes through the data points. Compare this model with previous models that you have constructed. Which model makes the best predictions?

5. *The Cost of a Postage Stamp*—Consider the following data. Use the procedures in this chapter to capture the trend of the data if one exists. Would you eliminate any data points? Why? Would you be willing to use your model to predict the price of a postage stamp on January 1, 2010? What do the various models you construct predict about the price on January 1, 2010? When will the price reach $1? You might enjoy reading the article on which this problem is based: Donald R. Byrkit and Robert E. Lee, "The Cost of a Postage Stamp, or Up, Up, and Away," *Mathematics and Computer Education* 17, no. 3 (Summer 1983): 184–190.

Date	First-class stamp	
1885–1917	$0.02	
1917–1919	0.03	(Wartime increase)
1919	0.02	(Restored by Congress)
July 6, 1932	0.03	
August 1, 1958	0.04	
January 7, 1963	0.05	
January 7, 1968	0.06	
May 16, 1971	0.08	
March 2, 1974	0.10	
December 31, 1975	0.13	(Temporary)
July 18, 1976	0.13	
May 15, 1978	0.15	
March 22, 1981	0.18	
November 1, 1981	0.20	
February 17, 1985	0.22	
April 3, 1988	0.25	
February 3, 1991	0.29	
January 1, 1995	0.32	
January 10, 1999	0.33	
January 7, 2001	0.34	
June 30, 2002	0.37	
January 8, 2006	0.39	
May 14, 2007	0.41	
May 12, 2008	0.42	

4.4 PROJECTS

1. Construct a computer code for determining the coefficients of the natural splines that pass through a given set of data points. See Burden and Faires, cited earlier in this chapter, for an efficient algorithm.

For Projects 2–8, use the software you developed in Project 1 to find the splines that pass through the given data points. Use graphics software, if available, to sketch the resulting splines.

2. The data presented in Table 4.22 on the growth of yeast in a culture (data from R. Pearl, "The Growth of Population," *Quart. Rev. Biol.* 2(1927): 532–548).

3. The following data represent the population of the United States from 1790 to 2000.

Year	Observed population
1790	3,929,000
1800	5,308,000
1810	7,240,000
1820	9,638,000
1830	12,866,000
1840	17,069,000
1850	23,192,000
1860	31,443,000
1870	38,558,000
1880	50,156,000
1890	62,948,000
1900	75,995,000
1910	91,972,000
1920	105,711,000
1930	122,755,000
1940	131,669,000
1950	150,697,000
1960	179,323,000
1970	203,212,000
1980	226,505,000
1990	248,709,873
2000	281,416,000

4. The following data were obtained for the growth of a sheep population introduced into a new environment on the island of Tasmania (adapted from J. Davidson, "On the Growth of the Sheep Population in Tasmania," *Trans. Roy. Soc. S. Australia* 62(1938): 342–346).

t (year)	1814	1824	1834	1844	1854	1864
$P(t)$	125	275	830	1200	1750	1650

5. The following data represent the pace of life (see Problem 1, Section 4.1). P is the population and V is the mean velocity in feet per second over a 50-ft course.

P	365	2500	5491	14000	23700	49375	70700	78200
V	2.76	2.27	3.31	3.70	3.27	4.90	4.31	3.85
P	138000	304500	341948	867023	1092759	1340000	2602000	
V	4.39	4.42	4.81	5.21	5.88	5.62	5.05	

6. The following data represent the length and weight of a fish (bass).

Length (in.)	12.5	12.625	14.125	14.5	17.25	17.75
Weight (oz)	17	16.5	23	26.5	41	49

7. The following data represent the weight-lifting results from the 1976 Olympics.

Bodyweight class (lb)		Total winning lifts (lb)		
	Max. weight	Snatch	Jerk	Total weight
Flyweight	114.5	231.5	303.1	534.6
Bantamweight	123.5	259.0	319.7	578.7
Featherweight	132.5	275.6	352.7	628.3
Lightweight	149.0	297.6	380.3	677.9
Middleweight	165.5	319.7	418.9	738.5
Light-heavyweight	182.0	358.3	446.4	804.7
Middle-heavyweight	198.5	374.8	468.5	843.3
Heavyweight	242.5	385.8	496.0	881.8

8. You can use the cubic spline software you developed, coupled with some graphics, to draw smooth curves to represent a figure you wish to draw on the computer. Overlay a piece of graph paper on a picture or drawing that you wish to produce on a computer. Record enough data to gain smooth curves ultimately. Take more data points where there are abrupt changes (Figure 4.31).

■ Figure 4.31

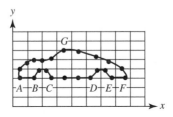

Now take the data points and determine the splines that pass through them. Note that if natural discontinuities in the derivative occur in the data, such as at points A–G in Figure 4.31, you will want to terminate one set of spline functions and begin another. You can then graph the spline functions using graphics software. In essence, you are using the computer to connect the dots with smooth curves. Select a figure of interest to you, such as your school mascot, and draw it on the computer.

5 Simulation Modeling

Introduction

In many situations a modeler is unable to construct an analytic (symbolic) model adequately explaining the behavior being observed because of its complexity or the intractability of the proposed explicative model. Yet if it is necessary to make predictions about the behavior, the modeler may conduct experiments (or gather data) to investigate the relationship between the dependent variable(s) and selected values of the independent variable(s) within some range. We constructed empirical models based on collected data in Chapter 4. To collect the data, the modeler may observe the behavior directly. In other instances, the behavior might be duplicated (possibly in a scaled-down version) under controlled conditions, as we will do when predicting the size of craters in Section 9.4.

In some circumstances, it may not be feasible either to observe the behavior directly or to conduct experiments. For instance, consider the service provided by a system of elevators during morning rush hour. After identifying an appropriate problem and defining what is meant by good service, we might suggest some alternative delivery schemes, such as assigning elevators to even and odd floors or using express elevators. Theoretically, each alternative could be tested for some period of time to determine which one provided the best service for particular arrival and destination patterns of the customers. However, such a procedure would probably be very disruptive because it would be necessary to harass the customers constantly as the required statistics were collected. Moreover, the customers would become very confused because the elevator delivery system would keep changing. Another problem concerns testing alternative schemes for controlling automobile traffic in a large city. It would be impractical to constantly change directions of the one-way streets and the distribution of traffic signals to conduct tests.

In still other situations, the system for which alternative procedures need to be tested *may not even exist yet*. An example is the situation of several proposed communications networks, with the problem of determining which is best for a given office building. Still another example is the problem of determining locations of machines in a new industrial plant. The *cost* of conducting experiments may be prohibitive. This is the case when an agency tries to predict the effects of various alternatives for protecting and evacuating the population in case of failure of a nuclear power plant.

In cases where the behavior cannot be explained analytically or data collected directly, the modeler might *simulate* the behavior indirectly in some manner and then test the various alternatives under consideration to estimate how each affects the behavior. Data can then be collected to determine which alternative is best. An example is to determine the drag force on a proposed submarine. Because it is infeasible to build a prototype, we can build a scaled model to simulate the behavior of the actual submarine. Another example of this type of simulation is using a scaled model of a jet airplane in a wind tunnel to estimate the effects of very high speeds for various designs of the aircraft. There is yet another type of simulation, which we will study in this chapter. This **Monte Carlo simulation** is typically accomplished with the aid of a computer.

Suppose we are investigating the service provided by a system of elevators at morning rush hour. In Monte Carlo simulation, the arrival of customers at the elevators during the hour and the destination floors they select need to be replicated. That is, the distribution of arrival times and the distribution of floors desired on the simulated trial must portray a possible rush hour. Moreover, after we have simulated many trials, the daily distribution of arrivals and destinations that occur must mimic the real-world distributions in proper proportions. When we are satisfied that the behavior is adequately duplicated, we can investigate various alternative strategies for operating the elevators. Using a large number of trials, we can gather appropriate statistics, such as the average total delivery time of a customer or the length of the longest queue. These statistics can help determine the best strategy for operating the elevator system.

This chapter provides a brief introduction to Monte Carlo simulation. Additional studies in probability and statistics are required to delve into the intricacies of computer simulation and understand its appropriate uses. Nevertheless, you will gain some appreciation of this powerful component of mathematical modeling. Keep in mind that there is a danger in placing too much confidence in the predictions resulting from a simulation, especially if the assumptions inherent in the simulation are not clearly stated. Moreover, the appearance of using large amounts of data and huge amounts of computer time, coupled with the fact the lay people can understand a simulation model and computer output with relative ease, often leads to overconfidence in the results.

When any Monte Carlo simulation is performed, random numbers are used. We discuss how to generate random numbers in Section 5.2. Loosely speaking, a "sequence of random numbers uniformly distributed in an interval m to n" is a set of numbers with no apparent pattern, where each number between m and n can appear with equal likelihood. For example, if you toss a six-sided die 100 times and write down the number showing on the die each time, you will have written down a sequence of 100 random integers approximately uniformly distributed over the interval 1 to 6. Now, suppose that random numbers consisting of six digits can be generated. The tossing of a coin can be duplicated by generating a random number and assigning it a head if the random number is even and a tail if the random number is odd. If this trial is replicated a large number of times, you would expect heads to occur about 50% of the time. However, there is an element of chance involved. It is possible that a run of 100 trials could produce 51 heads and that the next 10 trials could produce all heads (although this is not very likely). Thus, the estimate with 110 trials would actually be worse than the estimate with 100 trials. Processes with an element of chance involved are called **probabilistic**, as opposed to **deterministic**, processes. Monte Carlo simulation is therefore a probabilistic model.

The modeled behavior may be either deterministic or probabilistic. For instance, the area under a curve is deterministic (even though it may be impossible to find it precisely). On the other hand, the time between arrivals of customers at the elevator on a particular day is probabilistic behavior. Referring to Figure 5.1, we see that a deterministic model can be used to approximate either a deterministic or a probabilistic behavior, and likewise, a Monte Carlo simulation can be used to approximate a deterministic behavior (as you will see with

■ **Figure 5.1**

The behavior and the model can be either deterministic or probabilistic.

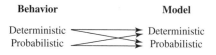

a Monte Carlo approximation to an area under a curve) or a probabilistic one. However, as we would expect, the real power of Monte Carlo simulation lies in modeling a probabilistic behavior.

A principal advantage of Monte Carlo simulation is the relative ease with which it can sometimes be used to approximate very complex probabilistic systems. Additionally, Monte Carlo simulation provides performance estimation over a wide range of conditions rather than a very restricted range as often required by an analytic model. Furthermore, because a particular submodel can be changed rather easily in a Monte Carlo simulation (such as the arrival and destination patterns of customers at the elevators), there is the potential of conducting a sensitivity analysis. Still another advantage is that the modeler has control over the level of detail in a simulation. For example, a very long time frame can be compressed or a small time frame expanded, giving a great advantage over experimental models. Finally, there are very powerful, high-level simulation languages (such as GPSS, GASP, PROLOG, SIMAN, SLAM, and DYNAMO) that eliminate much of the tedious labor in constructing a simulation model.

On the negative side, simulation models are typically expensive to develop and operate. They may require many hours to construct and large amounts of computer time and memory to run. Another disadvantage is that the probabilistic nature of the simulation model limits the conclusions that can be drawn from a particular run unless a sensitivity analysis is conducted. Such an analysis often requires many more runs just to consider a small number of combinations of conditions that can occur in the various submodels. This limitation then forces the modeler to estimate which combination might occur for a particular set of conditions.

5.1 Simulating Deterministic Behavior: Area Under a Curve

In this section we illustrate the use of Monte Carlo simulation to model a deterministic behavior, the area under a curve. We begin by finding an approximate value to the area under a nonnegative curve. Specifically, suppose $y = f(x)$ is some given continuous function satisfying $0 \le f(x) \le M$ over the closed interval $a \le x \le b$. Here, the number M is simply some constant that *bounds* the function. This situation is depicted in Figure 5.2. Notice that the area we seek is wholly contained within the rectangular region of height M and length $b - a$ (the length of the interval over which f is defined).

Now we select a point $P(x, y)$ at random from within the rectangular region. We will do so by generating two random numbers, x and y, satisfying $a \le x \le b$ and $0 \le y \le M$, and interpreting them as a point P with coordinates x and y. Once $P(x, y)$ is selected, we ask whether it lies within the region below the curve. That is, does the y-coordinate satisfy $0 \le y \le f(x)$? If the answer is yes, then count the point P by adding 1 to some counter.

■ **Figure 5.2**

The area under the nonnegative curve $y = f(x)$ over $a \le x \le b$ is contained within the rectangle of height M and base length $b - a$.

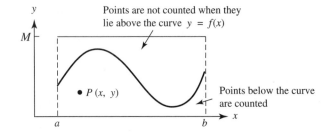

Two counters will be necessary: one to count the total points generated and a second to count those points that lie below the curve (Figure 5.2). You can then calculate an approximate value for the area under the curve by the following formula:

$$\frac{\text{area under curve}}{\text{area of rectangle}} \approx \frac{\text{number of points counted below curve}}{\text{total number of random points}}$$

As discussed in the Introduction, the Monte Carlo technique is probabilistic and typically requires a large number of trials before the deviation between the predicted and true values becomes small. A discussion of the number of trials needed to ensure a predetermined level of confidence in the final estimate requires a background in statistics. However, as a general rule, to double the accuracy of the result (i.e., to cut the expected error in half), about four times as many experiments are necessary.

The following algorithm gives the sequence of calculations needed for a general computer simulation of this Monte Carlo technique for finding the area under a curve.

Monte Carlo Area Algorithm

Input   Total number n of random points to be generated in the simulation.

Output   AREA = approximate area under the specified curve $y = f(x)$ over the given interval $a \le x \le b$, where $0 \le f(x) < M$.

Step 1   Initialize: COUNTER = 0.

Step 2   For $i = 1, 2, \ldots, n$, do Steps 3–5.

Step 3   Calculate random coordinates x_i and y_i that satisfy $a \le x_i \le b$ and $0 \le y_i < M$.

Step 4   Calculate $f(x_i)$ for the random x_i coordinate.

Step 5   If $y_i \le f(x_i)$, then increment the COUNTER by 1. Otherwise, leave COUNTER as is.

Step 6   Calculate AREA = $M(b - a)$ COUNTER$/n$.

Step 7   OUTPUT (AREA)
STOP

Table 5.1 gives the results of several different simulations to obtain the area beneath the curve $y = \cos x$ over the interval $-\pi/2 \le x \le \pi/2$, where $0 \le \cos x < 2$.

The actual area under the curve $y = \cos x$ over the given interval is 2 square units. Note that even with the relatively large number of points generated, the error is significant. For functions of one variable, the Monte Carlo technique is generally not competitive with quadrature techniques that you will learn in numerical analysis. The lack of an error bound and the difficulty in finding an upper bound M are disadvantages as well. Nevertheless, the

Table 5.1 Monte Carlo approximation to the area under the curve $y = \cos x$ over the interval $-\pi/2 \le x \le \pi/2$

Number of points	Approximation to area	Number of points	Approximation to area
100	2.07345	2000	1.94465
200	2.13628	3000	1.97711
300	2.01064	4000	1.99962
400	2.12058	5000	2.01429
500	2.04832	6000	2.02319
600	2.09440	8000	2.00669
700	2.02857	10000	2.00873
800	1.99491	15000	2.00978
900	1.99666	20000	2.01093
1000	1.96664	30000	2.01186

Monte Carlo technique can be extended to functions of several variables and becomes more practical in that situation.

Volume Under a Surface

Let's consider finding part of the volume of the sphere

$$x^2 + y^2 + z^2 \le 1$$

that lies in the first octant, $x > 0$, $y > 0$, $z > 0$ (Figure 5.3).

The methodology to approximate the volume is very similar to that of finding the area under a curve. However, now we will use an approximation for the volume under the surface by the following rule:

$$\frac{\text{volume under surface}}{\text{volume of box}} \approx \frac{\text{number of points counted below surface in 1st octant}}{\text{total number of points}}$$

The following algorithm gives the sequence of calculations required to employ Monte Carlo techniques to find the approximate volume of the region.

■ Figure 5.3

Volume of a sphere $x^2 + y^2 + z^2 \le 1$ that lies in the first octant, $x > 0$, $y > 0$, $z > 0$

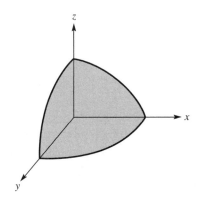

Monte Carlo Volume Algorithm

Input   Total number n of random points to be generated in the simulation.

Output   VOLUME = approximate volume enclosed by the specified function, $z = f(x, y)$ in the first octant, $x > 0$, $y > 0$, $z > 0$.

Step 1   Initialize: COUNTER $= 0$.

Step 2   For $i = 1, 2, \ldots, n$, do Steps 3–5.

   Step 3   Calculate random coordinates x_i, y_i, z_i that satisfy $0 \le x_i \le 1, 0 \le y_i \le 1, 0 \le z_i \le 1$. (In general, $a \le x_i \le b, c \le y_i \le d, 0 \le z_i \le M$.)

   Step 4   Calculate $f(x_i, y_i)$ for the random coordinate (x_i, y_i).

   Step 5   If random $z_i \le f(x_i, y_i)$, then increment the COUNTER by 1. Otherwise, leave COUNTER as is.

Step 6   Calculate VOLUME $= M(d - c)(b - a)$COUNTER$/n$.

Step 7   OUTPUT (VOLUME)

         STOP

Table 5.2 gives the results of several Monte Carlo runs to obtain the approximate volume of

$$x^2 + y^2 + z^2 \le 1$$

that lies in the first octant, $x > 0$, $y > 0$, $z > 0$.

Table 5.2 Monte Carlo approximation to the volume in the first octant under the surface $x^2 + y^2 + z^2 \le 1$

Number of points	Approximate volume
100	0.4700
200	0.5950
300	0.5030
500	0.5140
1,000	0.5180
2,000	0.5120
5,000	0.5180
10,000	0.5234
20,000	0.5242

The actual volume in the first octant is found to be approximately 0.5236 cubic units ($\pi/6$). Generally, though not uniformly, the error becomes smaller as the number of points generated increases.

5.1 PROBLEMS

1. Each ticket in a lottery contains a single "hidden" number according to the following scheme: 55% of the tickets contain a 1, 35% contain a 2, and 10% contain a 3. A participant in the lottery wins a prize by obtaining all three numbers 1, 2, and 3. Describe

an experiment that could be used to determine how many tickets you would expect to buy to win a prize.

2. Two record companies, A and B, produce classical music recordings. Label A is a budget label, and 5% of A's new compact discs exhibit significant degrees of warpage. Label B is manufactured under tighter quality control (and consequently more expensive) than A, so only 2% of its compact discs are warped. You purchase one label A and one label B recording at your local store on a regular basis. Describe an experiment that could be used to determine how many times you would expect to make such a purchase before buying two warped compact discs for a given sale.

3. Using Monte Carlo simulation, write an algorithm to calculate an approximation to π by considering the number of random points selected inside the quarter circle

$$Q : x^2 + y^2 = 1, x \geq 0, y \geq 0$$

where the quarter circle is taken to be inside the square

$$S : 0 \leq x \leq 1 \text{ and } 0 \leq y \leq 1$$

Use the equation $\pi/4 = \text{area } Q/\text{area } S$.

4. Use Monte Carlo simulation to approximate the area under the curve $f(x) = \sqrt{x}$, over the interval $\frac{1}{2} \leq x \leq \frac{3}{2}$.

5. Find the area trapped between the two curves $y = x^2$ and $y = 6 - x$ and the x- and y-axes.

6. Using Monte Carlo simulation, write an algorithm to calculate that part of the volume of an ellipsoid

$$\frac{x^2}{2} + \frac{y^2}{4} + \frac{z^2}{8} \leq 16$$

that lies in the first octant, $x > 0, y > 0, z > 0$.

7. Using Monte Carlo simulation, write an algorithm to calculate the volume trapped between the two paraboloids

$$z = 8 - x^2 - y^2 \quad \text{and} \quad z = x^2 + 3y^2$$

Note that the two paraboloids intersect on the elliptic cylinder

$$x^2 + 2y^2 = 4$$

5.2 Generating Random Numbers

In the previous section, we developed algorithms for Monte Carlo simulations to find areas and volumes. A key ingredient common to these algorithms is the need for random numbers. Random numbers have a variety of applications, including gambling problems, finding an area or volume, and modeling larger complex systems such as large-scale combat operations or air traffic control situations.

In some sense a computer does not really generate random numbers because computers employ deterministic algorithms. However, we can generate sequences of pseudorandom numbers that, for all practical purposes, may be considered random. There is no single best random number generator or best test to ensure randomness.

There are complete courses of study for random numbers and simulations that cover in depth the methods and tests for pseudorandom number generators. Our purpose here is to introduce a few random number methods that can be utilized to generate sequences of numbers that are nearly random.

Many programming languages, such as Pascal and Basic, and other software (e.g., Minitab, MATLAB, and EXCEL) have built-in random number generators for user convenience.

Middle-Square Method

The middle-square method was developed in 1946 by John Von Neuman, S. Ulm, and N. Metropolis at Los Alamos Laboratories to simulate neutron collisions as part of the Manhattan Project. Their middle-square method works as follows:

1. Start with a four-digit number x_0, called the *seed*.

2. Square it to obtain an eight-digit number (add a leading zero if necessary).

3. Take the middle four digits as the next random number.

Continuing in this manner, we obtain a sequence that appears to be random over the integers from 0 to 9999. These integers can then be scaled to any interval a to b. For example, if we wanted numbers from 0 to 1, we would divide the four-digit numbers by 10,000. Let's illustrate the middle-square method.

Pick a seed, say $x_0 = 2041$, and square it (adding a leading zero) to get 04165681. The middle four digits give the next random number, 1656. Generating 9 random numbers in this way yields

n	0	1	2	3	4	5	6	7	8	9	10	11	12
x_n	2041	1656	7423	1009	0180	0324	1049	1004	80	64	40	16	2

We can use more than 4 digits if we wish, but we always take the middle number of digits equal to the number of digits in the seed. For example, if $x_0 = 653217$ (6 digits), its square 426,692,449,089 has 12 digits. Thus, take the middle 6 digits as the random number, namely, 692449.

The middle-square method is reasonable, but it has a major drawback in its tendency to degenerate to zero (where it will stay forever). With the seed 2041, the random sequence does seem to be approaching zero. How many numbers can be generated until we are almost at zero?

Linear Congruence

The linear congruence method was introduced by D. H. Lehmer in 1951, and a majority of pseudorandom numbers used today are based on this method. One advantage it has over other methods is that seeds can be selected that generate patterns that eventually cycle (we illustrate this concept with an example). However, the length of the cycle is so large that the pattern does not repeat itself on large computers for most applications. The method requires the choice of three integers: a, b, and c. Given some initial seed, say x_0, we generate a sequence by the rule

$$x_{n+1} = (a \times x_n + b)\,\text{mod}(c)$$

where c is the modulus, a is the multiplier, and b is the increment. The qualifier mod(c) in the equation means to obtain the remainder after dividing the quantity $(a \times x_n + b)$ by c. For example, with $a = 1, b = 7$, and $c = 10$,

$$x_{n+1} = (1 \times x_n + 7)\,\text{mod}(10)$$

means x_{n+1} is the integer remainder upon dividing $x_n + 7$ by 10. Thus, if $x_n = 115$, then $x_{n+1} = remainder\left(\frac{122}{10}\right) = 2$.

Before investigating the linear congruence methodology, we need to discuss **cycling**, which is a major problem that occurs with random numbers. Cycling means the sequence repeats itself, and, although undesirable, it is unavoidable. At some point, all pseudorandom number generators begin to cycle. Let's illustrate cycling with an example.

If we set our seed at $x_0 = 7$, we find $x_1 = (1 \times 7 + 7)\,\text{mod}(10)$ or 14 mod(10), which is 4. Repeating this same procedure, we obtain the sequence

$$7, 4, 1, 8, 5, 2, 9, 6, 3, 0, 7, 4, \ldots$$

and the original sequence repeats again and again. Note that there is cycling after 10 numbers. The methodology produces a sequence of integers between 0 and $c - 1$ inclusively before cycling (which includes the possible remainders after dividing the integers by c). Cycling is guaranteed with at most c numbers in the random number sequence. Nevertheless, c can be chosen to be very large, and a and b can be chosen in such a way as to obtain a full set of c numbers before cycling begins to occur. Many computers use $c = 2^{31}$ for the large value of c. Again, we can scale the random numbers to obtain a sequence between any limits a and b, as required.

A second problem that can occur with the linear congruence method is lack of statistical independence among the members in the list of random numbers. Any correlations between the nearest neighbors, the next-nearest neighbors, the third-nearest neighbors, and so forth are generally unacceptable. (Because we live in a three-dimensional world, third-nearest neighbor correlations can be particularly damaging in physical applications.) Pseudorandom number sequences can never be completely statistically independent because they are generated by a mathematical formula or algorithm. Nevertheless, the sequence will appear (for practical purposes) independent when it is subjected to certain statistical tests. These concerns are best addressed in a course in statistics.

5.2   PROBLEMS

1. Use the middle-square method to generate
   a. 10 random numbers using $x_0 = 1009$.
   b. 20 random numbers using $x_0 = 653217$.
   c. 15 random numbers using $x_0 = 3043$.
   d. Comment about the results of each sequence. Was there cycling? Did each sequence degenerate rapidly?

2. Use the linear congruence method to generate
   a. 10 random numbers using $a = 5$, $b = 1$, and $c = 8$.
   b. 15 random numbers using $a = 1$, $b = 7$, and $c = 10$.
   c. 20 random numbers using $a = 5$, $b = 3$, and $c = 16$.
   d. Comment about the results of each sequence. Was there cycling? If so, when did it occur?

5.2   PROJECTS

1. Complete the requirement for UMAP module 269, "Monte Carlo: The Use of Random Digits to Simulate Experiments," by Dale T. Hoffman. The Monte Carlo technique is presented, explained, and used to find approximate solutions to several realistic problems. Simple experiments are included for student practice.

2. Refer to "Random Numbers" by Mark D. Myerson, UMAP 590. This module discusses methods for generating random numbers and presents tests for determining the randomness of a string of numbers. Complete this module and prepare a short report on testing for randomness.

3. Write a computer program to generate uniformly distributed random integers in the interval $m < x < n$, where m and n are integers, according to the following algorithm:

Step 1   Let $d = 2^{31}$ and choose N (the number of random numbers to generate).

Step 2   Choose any seed integer Y such that
$$999999 > Y > 100000$$

Step 3   Let $i = 1$.

Step 4   Let $Y = (15625\,Y + 22221)\,\text{mod}(d)$.

Step 5   Let $X_i = m + \text{floor}[(n - m + 1)Y/d]$.

Step 6   Increment i by 1: $i = i + 1$.

Step 7   Go to Step 4 unless $i = N + 1$.

Here, floor (p) means the largest integer not exceeding p.
   For most choices of Y, the numbers X_1, X_2, ... form a sequence of (pseudo)random integers as desired. One possible recommended choice is $Y = 568731$. To generate

random numbers (not just integers) in an interval a to b with $a < b$, use the preceding algorithm, replacing the formula in Step 5 by

$$\text{Let } X_i = a + \frac{Y(b-a)}{d-1}$$

4. Write a program to generate 1000 integers between 1 and 5 in a random fashion so that 1 occurs 22% of the time, 2 occurs 15% of the time, 3 occurs 31% of the time, 4 occurs 26% of the time, and 5 occurs 6% of the time. Over what interval would you generate the random numbers? How do you decide which of the integers from 1 to 5 has been generated according to its specified chance of selection?

5. Write a program or use a spreadsheet to find the approximate area or volumes in Problems 3–7 in Section 5.1.

5.3 Simulating Probabilistic Behavior

One of the keys to good Monte Carlo simulation practices is an understanding of the axioms of probability. The term *probability* refers to the study of both randomness and uncertainty, as well as to quantifying of the likelihoods associated with various outcomes. Probability can be seen as a long-term average. For example, if the probability of an event occurring is 1 out of 5, then in the long run, the chance of the event happening is 1/5. Over the long haul, the probability of an event can be thought of as the ratio

$$\frac{\text{number of favorable events}}{\text{total number of events}}$$

Our goal in this section is to show how to model simple probabilistic behavior to build intuition and understanding before developing submodels of probabilistic processes to incorporate in simulations (Sections 5.4 and 5.5).

We examine three simple probabilistic models:

1. Flip of a fair coin
2. Roll of a fair die or pair of dice
3. Roll of an unfair die or pair of unfair dice

A Fair Coin

Most people realize that the chance of obtaining a head or a tail on a coin is 1/2. What happens if we actually start flipping a coin? Will one out of every two flips be a head? Probably not. Again, probability is a long-term average. Thus, in the long run, the ratio of heads to the number of flips approaches 0.5. Let's define $f(x)$ as follows, where x is a random number between [0, 1].

$$f(x) = \begin{cases} \text{Head, } 0 \le x \le 0.5 \\ \text{Tail, } 0.5 < x \le 1 \end{cases}$$

Note that $f(x)$ assigns the outcome head or tail to a number between $[0, 1]$. We want to take advantage of the cumulative nature of this function as we make random assignments to numbers between $[0, 1]$. In the long run, we expect to find the following percent occurrences:

Random number interval	Cumulative occurrences	Percent occurrence
$x < 0$	0	0.00
$0 < x < 0.5$	0.5	0.50
$0.5 < x < 1.0$	1	0.50

Let's illustrate using the following algorithm:

Monte Carlo Fair Coin Algorithm

Input   Total number n of random flips of a fair coin to be generated in the simulation.

Output   Probability of getting a head when we flip a fair coin.

Step 1   Initialize: COUNTER $= 0$.

Step 2   For $i = 1, 2, \ldots, n$, do Steps 3 and 4.

Step 3   Obtain a random number x_i between 0 and 1.

Step 4   If $0 \leq x_i \leq 0.5$, then COUNTER $=$ COUNTER $+ 1$. Otherwise, leave COUNTER as is.

Step 5   Calculate $P(\text{head}) =$ COUNTER$/n$.

Step 6   OUTPUT Probability of heads, $P(\text{head})$.
STOP

Table 5.3 illustrates our results for various choices n of the number of random x_i generated. Note that as n gets large, the probability of heads occurring is 0.5, or half the time.

Table 5.3 Results from flipping a fair coin

Number of flips	Number of heads	Percent heads
100	49	0.49
200	102	0.51
500	252	0.504
1,000	492	0.492
5,000	2469	0.4930
10,000	4993	0.4993

Roll of a Fair Die

Rolling a fair die adds a new twist to the process. In the flip of a coin, only one event is assigned (with two possible answers, yes or no). Now we must devise a method to assign six events because a die consists of the numbers $\{1, 2, 3, 4, 5, 6\}$. The probability of each event occurring is $1/6$ because each number is equally likely to occur. As before, this probability of a particular number occurring is defined to be

$$\frac{\text{number of occurences of the particular number } \{1, 2, 3, 4, 5, 6\}}{\text{total number of trials}}$$

We can use the following algorithm to generate our experiment for a roll of a die.

Monte Carlo Roll of a Fair Die Algorithm

Input Total number n of random rolls of a die in the simulation.

Output The percentage or probability for rolls $\{1, 2, 3, 4, 5, 6\}$.

Step 1 Initialize COUNTER 1 through COUNTER 6 to zero.

Step 2 For $i = 1, 2, \ldots, n$, do Steps 3 and 4.

Step 3 Obtain a random number satisfying $0 \le x_i \le 1$.

Step 4 If x_i belongs to these intervals, then increment the appropriate COUNTER.

$$
\begin{array}{ll}
0 \le x_i \le \frac{1}{6} & \text{COUNTER 1} = \text{COUNTER 1} + 1 \\
\frac{1}{6} < x_i \le \frac{2}{6} & \text{COUNTER 2} = \text{COUNTER 2} + 1 \\
\frac{2}{6} < x_i \le \frac{3}{6} & \text{COUNTER 3} = \text{COUNTER 3} + 1 \\
\frac{3}{6} < x_i \le \frac{4}{6} & \text{COUNTER 4} = \text{COUNTER 4} + 1 \\
\frac{4}{6} < x_i \le \frac{5}{6} & \text{COUNTER 5} = \text{COUNTER 5} + 1 \\
\frac{5}{6} < x_i \le 1 & \text{COUNTER 6} = \text{COUNTER 6} + 1
\end{array}
$$

Step 5 Calculate probability of each roll $j = \{1, 2, 3, 4, 5, 6\}$ by $\text{COUNTER}(j)/n$.

Step 6 OUTPUT probabilities.

STOP

Table 5.4 illustrates the results for 10, 100, 1000, 10,000, and 100,000 runs. We see that with 100,000 runs we are close (for these trials) to the expected results.

Table 5.4 Results from a roll of a fair die ($n =$ number of trials)

Die value	10	100	1000	10,000	100,000	Expected results
1	0.300	0.190	0.152	0.1703	0.1652	0.1667
2	0.00	0.150	0.152	0.1652	0.1657	0.1667
3	0.100	0.090	0.157	0.1639	0.1685	0.1667
4	0.00	0.160	0.180	0.1653	0.1685	0.1667
5	0.400	0.150	0.174	0.1738	0.1676	0.1667
6	0.200	0.160	0.185	0.1615	0.1652	0.1667

Roll of an Unfair Die

Let's consider a probability model in which the events are not all equally likely. Assume the die is loaded or biased according to the following empirical distribution:

Roll value	$P(\text{roll})$
1	0.1
2	0.1
3	0.2
4	0.3
5	0.2
6	0.1

The cumulative occurrences for the function to be used in our algorithm would be

Value of x_i	Assignment
$[0, 0.1]$	ONE
$(0.1, 0.2]$	TWO
$(0.2, 0.4]$	THREE
$(0.4, 0.7]$	FOUR
$(0.7, 0.9]$	FIVE
$(0.9, 1.0]$	SIX

We model the roll of an unfair die using the following algorithm:

Monte Carlo Roll of an Unfair Die Algorithm

Input   Total number n of random rolls of a die in the simulation.

Output   The percentage or probability for rolls $\{1, 2, 3, 4, 5, 6\}$.

Step 1   Initialize COUNTER 1 through COUNTER 6 to zero.

Step 2   For $i = 1, 2, \ldots, n$, do Steps 3 and 4.

Step 3   Obtain a random number satisfying $0 \le x_i \le 1$.

Step 4   If x_i belongs to these intervals, then increment the appropriate COUNTER.

$$0 \le x_i \le 0.1 \quad \text{COUNTER 1} = \text{COUNTER 1} + 1$$
$$0.1 < x_i \le 0.2 \quad \text{COUNTER 2} = \text{COUNTER 2} + 1$$
$$0.2 < x_i \le 0.4 \quad \text{COUNTER 3} = \text{COUNTER 3} + 1$$
$$0.4 < x_i \le 0.7 \quad \text{COUNTER 4} = \text{COUNTER 4} + 1$$
$$0.7 < x_i \le 0.9 \quad \text{COUNTER 5} = \text{COUNTER 5} + 1$$
$$0.9 < x_i \le 1.0 \quad \text{COUNTER 6} = \text{COUNTER 6} + 1$$

Step 5   Calculate probability of each roll $j = \{1, 2, 3, 4, 5, 6\}$ by COUNTER$(j)/n$.

Step 6   OUTPUT probabilities.

STOP

The results are shown in Table 5.5. Note that a large number of trials are required for the model to approach the long-term probabilities.

In the next section, we will see how to use these ideas to simulate a real-world probabilistic situation.

Table 5.5 Results from a roll of an unfair die

Die value	100	1000	5000	10,000	40,000	Expected results
1	0.080	0.078	0.094	0.0948	0.0948	0.1
2	0.110	0.099	0.099	0.0992	0.0992	0.1
3	0.230	0.199	0.192	0.1962	0.1962	0.2
4	0.360	0.320	0.308	0.3082	0.3081	0.3
5	0.110	0.184	0.201	0.2012	0.2011	0.2
6	0.110	0.120	0.104	0.1044	0.1045	0.1

5.3 PROBLEMS

1. You arrive at the beach for a vacation and are dismayed to learn that the local weather station is predicting a 50% chance of rain every day. Using Monte Carlo simulation, predict the chance that it rains three consecutive days during your vacation.

2. Use Monte Carlo simulation to approximate the probability of three heads occurring when five fair coins are flipped.

3. Use Monte Carlo simulation to simulate the sum of 100 consecutive rolls of a fair die.

4. Given loaded dice according to the following distribution, use Monte Carlo simulation to simulate the sum of 300 rolls of two unfair dice.

Roll	Die 1	Die 2
1	0.1	0.3
2	0.1	0.1
3	0.2	0.2
4	0.3	0.1
5	0.2	0.05
6	0.1	0.25

5. Make up a game that uses a flip of a fair coin, and then use Monte Carlo simulation to predict the results of the game.

5.3 PROJECTS

1. *Blackjack*—Construct and perform a Monte Carlo simulation of blackjack (also called twenty-one). The rules of blackjack are as follows:

> Most casinos use six or eight decks of cards when playing this game to inhibit "card counters." You will use two decks of cards in your simulation (104 cards total). There are only two players, you and the dealer. Each player receives two cards to begin play. The cards are worth their face value for 2–10, 10 for face cards (jack, queen, and king), and either 1 or 11 points for aces. The object of the game is to obtain a total as close to 21 as possible without going over (called "busting") so that your total is more than the dealer's.
>
> If the first two cards total 21 (ace–10 or ace–face card), this is called blackjack and is an automatic winner (unless both you and the dealer have blackjack, in which case it is a tie, or "push," and your bet remains on the table). Winning via blackjack pays you 3 to 2, or 1.5 to 1 (a $1 bet reaps $1.50, and you do not lose the $1 you bet).
>
> If neither you nor the dealer has blackjack, you can take as many cards as you want, one at a time, to try to get as close to 21 as possible. If you go over 21, you lose and the game ends. Once you are satisfied with your score, you "stand." The dealer then draws cards according to the following rules:
>
> The dealer stands on 17, 18, 19, 20, or 21. The dealer must draw a card if the total is 16 or less. The dealer always counts aces as 11 unless it causes him or her to bust, in which case the ace is counted as a 1. For example, an ace–6 combo for the dealer is 17, not 7 (the dealer

has no option), and the dealer must stand on 17. However, if the dealer has an ace–4 (for 15) and draws a king, then the new total is 15 because the ace reverts to its value of 1 (so as not to go over 21). The dealer would then draw another card.

   If the dealer goes over 21, you win (even your bet money; you gain $1 for every $1 you bet). If the dealer's total exceeds your total, you lose all the money you bet. If the dealer's total equals your total, it is a push (no money exchanges hands; you do not lose your bet, but neither do you gain any money).

What makes the game exciting in a casino is that the dealer's original two cards are one up, one down, so you do not know the dealer's total and must play the odds based on the one card showing. You do not need to incorporate this twist into your simulation for this project. Here's what you are required to do:

   Run through 12 sets of two decks playing the game. You have an unlimited bankroll (don't you wish!) and bet $2 on each hand. Each time the two decks run out, the hand in play continues with two fresh decks (104 cards). At that point record your standing (plus or minus X dollars). Then start again at 0 for the next set of decks. Thus your output will be the 12 results from playing each of the 12 sets of decks, which you can then average or total to determine your overall performance.

   What about *your* strategy? That's up to you! But here's the catch—you will assume that you can see neither of the dealer's cards (so you have no idea what cards the dealer has). Choose a strategy to play, and then play it throughout the entire simulation. (Blackjack enthusiasts can consider implementing doubling down and splitting pairs into their simulation, but this is not necessary.)

   Provide your instructor with the simulation algorithm, computer code, and output results from each of the 12 decks.

2. *Darts*—Construct and perform a Monte Carlo simulation of a darts game. The rules are

Dart board area	Points
Bullseye	50
Yellow ring	25
Blue ring	15
Red ring	10
White ring	5

From the origin (the center of the bullseye), the radius of each ring is as follows:

Ring	Thickness (in.)	Distance to outer ring edge from the origin (in.)
Bullseye	1.0	1.0
Yellow	1.5	2.5
Blue	2.5	5.0
Red	3.0	8.0
White	4.0	12.0

The board has a radius of 1 ft (12 in.).

Make an assumption about the distribution of how the darts hit on the board. Write an algorithm, and code it in the computer language of your choice. Run 1000 simulations to determine the mean score for throwing five darts. Also, determine which ring has the highest expected value (point value times the probability of hitting that ring).

3. *Craps*—Construct and perform a Monte Carlo simulation of the popular casino game of craps. The rules are as follows:

> There are two basic bets in craps, pass and don't pass. In the *pass* bet, you wager that the shooter (the person throwing the dice) will win; in the *don't pass* bet, you wager that the shooter will lose. We will play by the rule that on an initial roll of 12 ("boxcars"), both pass and don't pass bets are losers. Both are even-money bets.
>
> Conduct of the game:
>
> Roll a 7 or 11 on the first roll: Shooter wins (pass bets win and don't pass bets lose).
> Roll a 12 on the first roll: Shooter loses (boxcars; pass and don't pass bets lose).
> Roll a 2 or 3 on the first roll: Shooter loses (pass bets lose, don't pass bets win).
> Roll 4, 5, 6, 8, 9, 10 on the first roll: This becomes the point. The object then becomes to roll the point again before rolling a 7.
>
> The shooter continues to roll the dice until the point or a 7 appears. Pass bettors win if the shooter rolls the point again before rolling a 7. Don't pass bettors win if the shooter rolls a 7 before rolling the point again.

Write an algorithm and code it in the computer language of your choice. Run the simulation to estimate the probability of winning a pass bet and the probability of winning a don't pass bet. Which is the better bet? As the number of trials increases, to what do the probabilities converge?

4. *Horse Race*—Construct and perform a Monte Carlo simulation of a horse race. You can be creative and use odds from the newspaper, or simulate the Mathematical Derby with the entries and odds shown in following table.

Mathematical Derby

Entry's name	Odds
Euler's Folly	7–1
Leapin' Leibniz	5–1
Newton Lobell	9–1
Count Cauchy	12–1
Pumped up Poisson	4–1
Loping L'Hôpital	35–1
Steamin' Stokes	15–1
Dancin' Dantzig	4–1

Construct and perform a Monte Carlo simulation of 1000 horse races. Which horse won the most races? Which horse won the fewest races? Do these results surprise you? Provide the tallies of how many races each horse won with your output.

5. *Roulette*—In American roulette, there are 38 spaces on the wheel: 0, 00, and 1–36. Half the spaces numbered 1–36 are red and half are black. The two spaces 0 and 00 are green.

Simulate the playing of 1000 games betting either red or black (which pay even money, 1:1). Bet $1 on each game and keep track of your earnings. What are the earnings per game betting red/black according to your simulation? What was your longest winning streak? Your longest losing streak?

Simulate 1000 games betting green (pays 17:1, so if you win, you add $17 to your kitty, and if you lose, you lose $1). What are your earnings per game betting green according to your simulation? How does it differ from your earnings betting red/black? What was your longest winning streak betting green? Longest losing streak? Which strategy do you recommend using, and why?

6. *The Price Is Right*—On the popular TV game show *The Price Is Right*, at the end of each half hour, the three winning contestants face off in the Showcase Showdown. The game consists of spinning a large wheel with 20 spaces on which the pointer can land, numbered from $0.05 to $1.00 in 5¢ increments. The contestant who has won the least amount of money at this point in the show spins first, followed by the one who has won the next most, followed by the biggest winner for that half hour.

The objective of the game is to obtain as close to $1.00 as possible without going over that amount with a maximum of two spins. Naturally, if the first player does not go over, the other two will use one or both spins in an attempt to overtake the leader.

However, what of the person spinning first? If he or she is an expected value decision maker, how high a value on the first spin does he or she need to not want a second spin? Remember, the person will lose if

a. Either of the other two players surpasses the player's total or

b. The player spins again and goes over $1.

7. *Let's Make a Deal*—You are dressed to kill in your favorite costume and the host picks you out of the audience. You are offered the choice of three wallets. Two wallets contain a single $50 bill, and the third contains a $1000 bill. You choose one of the wallets, 1, 2, or 3. The host, who knows which wallet contains the $1000, then shows you one of the other two wallets, which has $50 inside. The host does this purposely because he has at least one wallet with $50 inside. If he also has the $1000 wallet, he simply shows you the $50 one he holds. Otherwise, he just shows you one of his two $50 wallets. The host then asks you if you want to trade your choice for the one he's still holding. Should you trade?

Develop an algorithm and construct a computer simulation to support your answer.

5.4 Inventory Model: Gasoline and Consumer Demand

In the previous section, we began modeling probabilistic behaviors using Monte Carlo simulation. In this section we will learn a method to approximate more probabilistic processes. Additionally, we will check to determine how well the simulation duplicates the process under examination. We begin by considering an inventory control problem.

You have been hired as a consultant by a chain of gasoline stations to determine how often and how much gasoline should be delivered to the various stations. Each time gasoline is delivered, a cost of d dollars is incurred, which is in addition to the cost of gasoline and is independent of the amount delivered. The gasoline stations are near interstate highways, so demand is fairly constant. Other factors determining the costs include the capital tied up in the inventory, the amortization costs of equipment, insurance, taxes, and security measures. We assume that in the short run, the demand and price of gasoline are constant for each station, yielding a constant total revenue as long as the station does not run out of gasoline. Because total profit is total revenue minus total cost, and total revenue is constant by assumption, total profit can be maximized by minimizing total cost. Thus, we identify the following problem: *Minimize the average daily cost of delivering and storing sufficient gasoline at each station to meet consumer demand.*

After discussing the relative importance of the various factors determining the average daily cost, we develop the following model:

$$\text{average daily cost} = f(\text{storage costs, delivery costs, demand rate})$$

Turning our attention to the various submodels, we argue that although the cost of storage may vary with the amount stored, it is reasonable to assume the cost per unit stored would be constant over the range of values under consideration. Similarly, the delivery cost is assumed constant per delivery, independent of the amount delivered, over the range of values under consideration. Plotting the daily demand for gasoline at a particular station is very likely to give a graph similar to the one shown in Figure 5.4a. If the frequency of each demand level over a fixed time period (e.g., 1 year) is plotted, then a plot similar to that shown in Figure 5.4b might be obtained.

If demands are tightly packed around the most frequently occurring demand, then we would accept the daily demand as being constant. In some cases, it may be reasonable to assume a constant demand. Finally, even though the demands occur in discrete time periods, a continuous submodel for demand can be used for simplicity. A continuous submodel is depicted in Figure 5.4c, where the slope of the line represents the constant daily demand. Notice the importance of each of the preceding assumptions in producing the linear submodel.

From these assumptions we will construct, in Chapter 13, an analytic model for the average daily cost and use it to compute an optimal time between deliveries and an optimal

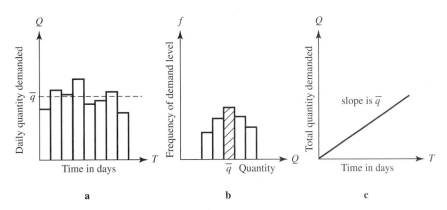

■ **Figure 5.4**

A constant demand rate

delivery quantity:

$$T^* = \sqrt{\frac{2d}{sr}}$$

$$Q^* = rT^*$$

where

$T^* =$ optimal time between deliveries in days

$Q^* =$ optimal delivery quantity of gasoline in gallons

$r =$ demand rate in gallons per day

$d =$ delivery cost in dollars per delivery

$s =$ storage cost per gallon per day

We will see how the analytic model depends heavily on a set of conditions that, although reasonable in some cases, would never be met precisely in the real world. It is difficult to develop analytic models that take into account the probabilistic nature of the submodels.

Suppose we decide to check our submodel for constant demand rate by inspecting the sales for the past 1000 days at a particular station. Thus, the data displayed in Table 5.6 are collected.

For each of the 10 intervals of demand levels given in Table 5.6, compute the relative frequency of occurrence by dividing the number of occurrences by the total number of days, 1000. This computation results in an estimate of the probability of occurrence for each demand level. These probabilities are displayed in Table 5.7 and plotted in histogram form in Figure 5.5.

If we are satisfied with the assumption of a constant demand rate, we might estimate this rate at 1550 gallons per day (from Figure 5.5). Then the analytic model could be used to compute the optimal time between deliveries and the delivery quantities from the delivery and storage costs.

Suppose, however, that we are not satisfied with the assumption of constant daily demand. How could we simulate the submodel for the demand suggested by Figure 5.5? First,

Table 5.6 History of demand at a particular gasoline station

Number of gallons demanded	Number of occurrences (in days)
1000–1099	10
1100–1199	20
1200–1299	50
1300–1399	120
1400–1499	200
1500–1599	270
1600–1699	180
1700–1799	80
1800–1899	40
1900–1999	30
	1000

Table 5.7   Probability of the occurrence of each demand level

Number of gallons demanded	Probability of occurrence
1000–1099	0.01
1100–1199	0.02
1200–1299	0.05
1300–1399	0.12
1400–1499	0.20
1500–1599	0.27
1600–1699	0.18
1700–1799	0.08
1800–1899	0.04
1900–1999	0.03
	1.00

■ **Figure 5.5**

The relative frequency of each of the demand intervals in Table 5.7

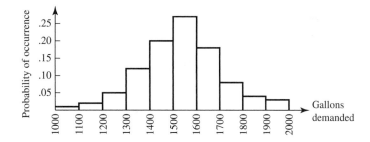

we could build a cumulative histogram by consecutively adding together the probabilities of each individual demand level, as displayed in Figure 5.6. Note in Figure 5.6 that the difference in height between adjacent columns represents the probability of occurrence of the subsequent demand interval. Thus, we can construct a correspondence between the numbers in the interval $0 \leq x \leq 1$ and the relative occurrence of the various demand intervals. This correspondence is displayed in Table 5.8.

■ **Figure 5.6**

A cumulative histogram of the demand submodel from the data in Table 5.7

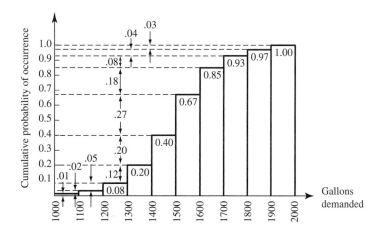

Table 5.8 Using random numbers uniformly distributed over $0 \le x \le 1$ to duplicate the occurrence of the various demand intervals

Random number	Corresponding demand	Percent occurrence
$0 \le x < 0.01$	1000–1099	0.01
$0.01 \le x < 0.03$	1100–1199	0.02
$0.03 \le x < 0.08$	1200–1299	0.05
$0.08 \le x < 0.20$	1300–1399	0.12
$0.20 \le x < 0.40$	1400–1499	0.20
$0.40 \le x < 0.67$	1500–1599	0.27
$0.67 \le x < 0.85$	1600–1699	0.18
$0.85 \le x < 0.93$	1700–1799	0.08
$0.93 \le x < 0.97$	1800–1899	0.04
$0.97 \le x \le 1.00$	1900–1999	0.03

Thus, if the numbers between 0 and 1 are randomly generated so that each number has an equal probability of occurring, the histogram of Figure 5.5 can be approximated. Using a random number generator on a handheld programmable calculator, we generated random numbers between 0 and 1 and then used the assignment procedure suggested by Table 5.7 to determine the demand interval corresponding to each random number. The results for 1000 and 10,000 trials are presented in Table 5.9.

Table 5.9 A Monte Carlo approximation of the demand submodel

Interval	Number of occurrences/expected no. of occurrences	
	1000 trials	10,000 trials
1000–1099	8/10	91/100
1100–1199	16/20	198/200
1200–1299	46/50	487/500
1300–1399	118/120	1205/1200
1400–1499	194/200	2008/2000
1500–1599	275/270	2681/2700
1600–1699	187/180	1812/1800
1700–1799	83/80	857/800
1800–1899	34/40	377/400
1900–1999	39/30	284/300
	1000/1000	10,000/10,000

For the gasoline inventory problem, we ultimately want to be able to determine a specific demand, rather than a demand interval, for each day simulated. How can this be accomplished? There are several alternatives. Consider the plot of the midpoints of each demand interval as displayed in Figure 5.7. Because we want a continuous model capturing the trend of the plotted data, we can use methods discussed in Chapter 4.

In many instances, especially where the subintervals are small and the data fairly approximate, a linear spline model is suitable. A linear spline model for the data displayed in Figure 5.7 is presented in Figure 5.8, and the individual spline functions are given in Table 5.10. The interior spline functions—$S_2(q)$–$S_9(q)$—were computed by passing a line

■ Figure 5.7

A cumulative plot of the demand submodel displaying only the center point of each interval

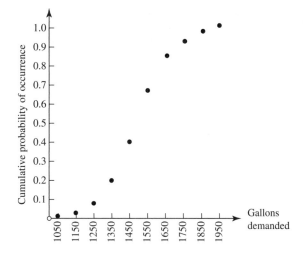

■ Figure 5.8

A linear spline model for the demand submodel

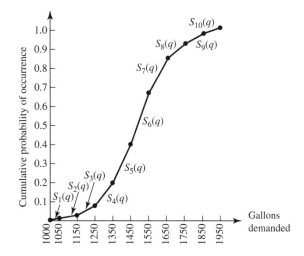

Table 5.10 Linear splines for the empirical demand submodel

Demand interval	Linear spline
$1000 \leq q < 1050$	$S_1(q) = 0.0002q - 0.2$
$1050 \leq q < 1150$	$S_2(q) = 0.0002q - 0.2$
$1150 \leq q < 1250$	$S_3(q) = 0.0005q - 0.545$
$1250 \leq q < 1350$	$S_4(q) = 0.0012q - 1.42$
$1350 \leq q < 1450$	$S_5(q) = 0.002q - 2.5$
$1450 \leq q < 1550$	$S_6(q) = 0.0027q - 3.515$
$1550 \leq q < 1650$	$S_7(q) = 0.0018q - 2.12$
$1650 \leq q < 1750$	$S_8(q) = 0.0008q - 0.47$
$1750 \leq q < 1850$	$S_9(q) = 0.0004q + 0.23$
$1850 \leq q \leq 2000$	$S_{10}(q) = 0.0002q + 0.6$

through the adjacent data points. $S_1(q)$ was computed by passing a line through $(1000, 0)$ and the first data point $(1050, 0.01)$. $S_{10}(q)$ was computed by passing a line through the points $(1850, 0.97)$ and $(2000, 1.00)$. Note that if we use the midpoints of the intervals, we have to make a decision on how to construct the two exterior splines. If the intervals are small, it is usually easy to construct a linear spline function that captures the trend of the data.

Now suppose we wish to simulate a daily demand for a given day. To do this, we generate a random number x between 0 and 1 and compute a corresponding demand, q. That is, x is the independent variable from which a unique corresponding q is calculated. This calculation is possible because the function depicted in Figure 5.8 is strictly increasing. (Think about whether this situation will always be the case.) Thus, the problem is to find the inverse functions for the splines listed in Table 5.10. For instance, given $x = S_1(q) = 0.0002q - 0.2$, we can solve for $q = (x + 0.2)5000$. In the case of linear splines, it is very easy to find the inverse functions summarized in Table 5.11.

Table 5.11   Inverse linear splines provide for the daily demand as a function of a random number in [0, 1]

Random number	Inverse linear spline
$0 \le x < 0.01$	$q = (x + 0.2)5000$
$0.01 \le x < 0.03$	$q = (x + 0.2)5000$
$0.03 \le x < 0.08$	$q = (x + 0.545)2000$
$0.08 \le x < 0.20$	$q = (x + 1.42)833.3\underline{3}$
$0.20 \le x < 0.40$	$q = (x + 2.5)500$
$0.40 \le x < 0.67$	$q = (x + 3.515)370.37$
$0.67 \le x < 0.85$	$q = (x + 2.12)555.5\underline{5}$
$0.85 \le x < 0.93$	$q = (x + 0.47)1250$
$0.93 \le x < 0.97$	$q = (x - 0.23)2500$
$0.97 \le x \le 1.00$	$q = (x - 0.6)5000$

Let's illustrate how Table 5.11 can be used to represent the daily demand submodel. To simulate a demand for a given day, we generate a random number between 0 and 1, say $x = 0.214$. Because $0.20 \le 0.214 < 0.40$, the spline $q = (x + 2.5)500$ is used to compute $q = 1357$. Thus, 1357 gallons is the simulated demand for that day.

Note that the inverse splines presented in Table 5.11 could have been constructed directly from the data in Figure 5.6 by choosing x as the independent variable instead of q. We will follow this procedure later when computing the cubic spline demand submodel. (The preceding development was presented to help you understand the process and also because it mimics what you will do after studying probability.) Figure 5.6 is an example of a cumulative distribution function. Many types of behavior approximate well-known probability distributions, which can be used as the basis for Figure 5.6 rather than experimental data. The inverse function must then be found to use as the demand submodel in the simulation, and this may prove to be difficult. In such cases, the inverse function is approximated with an empirical model, such as a linear spline or cubic spline. For an excellent introduction to some types of behavior that follow well-known probability distributions, see UMAP 340, "The Poisson Random Process," by Carroll Wilde, listed in the projects at the end of this section.

If we want a smooth continuous submodel for demand, we can construct a cubic spline submodel. We will construct the splines directly as a function of the random number x. That is, using a computer program, we calculate the cubic splines for the following data points:

x	0	0.01	0.03	0.08	0.2	0.4	0.67	0.85	0.93	0.97	1.0
q	1000	1050	1150	1250	1350	1450	1550	1650	1750	1850	2000

The splines are presented in Table 5.12. If the random number $x = 0.214$ is generated, the empirical cubic spline model yields the demand $q = 1350 + 715.5(0.014) - 1572.5(0.014)^2 + 2476(0.014)^3 = 1359.7$ gal.

Table 5.12 An empirical cubic spline model for demand

Random number	Cubic spline
$0 \leq x < 0.01$	$S_1(x) = 1000 + 4924.92x + 750788.75x^3$
$0.01 \leq x < 0.03$	$S_2(x) = 1050 + 5150.18(x - 0.01) + 22523.66(x - 0.01)^2 - 1501630.8(x - 0.01)^3$
$0.03 \leq x < 0.08$	$S_3(x) = 1150 + 4249.17(x - 0.03) - 67574.14(x - 0.03)^2 + 451815.88(x - 0.03)^3$
$0.08 \leq x < 0.20$	$S_4(x) = 1250 + 880.37(x - 0.08) + 198.24(x - 0.08)^2 - 4918.74(x - 0.08)^3$
$0.20 \leq x < 0.40$	$S_5(x) = 1350 + 715.46(x - 0.20) - 1572.51(x - 0.20)^2 + 2475.98(x - 0.20)^3$
$0.40 \leq x < 0.67$	$S_6(x) = 1450 + 383.58(x - 0.40) - 86.92(x - 0.40)^2 + 140.80(x - 0.40)^3$
$0.67 \leq x < 0.85$	$S_7(x) = 1550 + 367.43(x - 0.67) + 27.12(x - 0.67)^2 + 5655.69(x - 0.67)^3$
$0.85 \leq x < 0.93$	$S_8(x) = 1650 + 926.92(x - 0.85) + 3081.19(x - 0.85)^2 + 11965.43(x - 0.85)^3$
$0.93 \leq x < 0.97$	$S_9(x) = 1750 + 1649.66(x - 0.93) + 5952.90(x - 0.93)^2 + 382645.25(x - 0.93)^3$
$0.97 \leq x \leq 1.00$	$S_{10}(x) = 1850 + 3962.58(x - 0.97) + 51870.29(x - 0.97)^2 - 576334.88(x - 0.97)^3$

An empirical submodel for demand can be constructed in a variety of other ways. For example, rather than using the intervals for gallons demanded as given in Table 5.6, we can use smaller intervals. If the intervals are small enough, the midpoint of an interval could be a reasonable approximation to the demand for the entire interval. Thus, a cumulative histogram similar to that in Figure 5.6 could serve as a submodel directly. If preferred, a continuous submodel could be constructed readily from the refined data.

The purpose of our discussion has been to demonstrate how a submodel for a probabilistic behavior can be constructed using Monte Carlo simulation and experimental data. Now let's see how the inventory problem can be simulated in general terms.

An inventory strategy consists of specifying a delivery quantity Q and a time T between deliveries, given values for storage cost per gallon per day s and a delivery cost d. If s and d are known, then a specific inventory strategy can be tested using a Monte Carlo simulation algorithm, as follows:

Summary of Monte Carlo Inventory Algorithm Terms

Q Delivery quantity of gasoline in gallons
T Time between deliveries in days
I Current inventory in gallons
d Delivery cost in dollars per delivery
s Storage cost per gallon per day
C Total running cost
c Average daily cost

N Number of days to run the simulation
K Days remaining in the simulation
x_i A random number in the interval $[0, 1]$
q_i A daily demand
Flag An indicator used to terminate the algorithm

Monte Carlo Inventory Algorithm

Input	Q, T, d, s, N
Output	c
Step 1	Initialize:

$$K = N$$
$$I = 0$$
$$C = 0$$
$$\text{Flag} = 0$$

Step 2   Begin the next inventory cycle with a delivery:
$$I = I + Q$$
$$C = C + d$$

Step 3   Determine if the simulation will terminate during this cycle:
If $T \geq K$, then set $T = K$ and Flag $= 1$

Step 4   Simulate each day in the inventory cycle (or portion remaining):
For $i = 1, 2, \ldots, T$, do Steps 5–9

Step 5   Generate the random number x_i.

Step 6   Compute q_i using the demand submodel.

Step 7   Update the current inventory: $I = I - q_i$.

Step 8   Compute the daily storage cost and total running cost, unless the inventory has been depleted:
If $I \leq 0$, then set $I = 0$ and GOTO Step 9.
Else $C = C + I * s$.

Step 9   Decrement the number of days remaining in the simulation:
$$K = K - 1$$

Step 10   If Flag $= 0$, then GOTO Step 2. Else GOTO Step 11.

Step 11   Compute the average daily cost: $c = C/N$.

Step 12   Output c.
STOP

Various strategies can now be tested with the algorithm to determine the average daily costs. You probably want to refine the algorithm to keep track of other measures of effectiveness, such as unsatisfied demands and number of days without gasoline, as suggested in the following problem set.

5.4 PROBLEMS

1. Modify the inventory algorithm to keep track of unfilled demands and the total number of days that the gasoline station is without gasoline for at least part of the day.

2. Most gasoline stations have a storage capacity Q_{max} that cannot be exceeded. Refine the inventory algorithm to take this consideration into account. Because of the probabilistic

nature of the demand submodel at the end of the inventory cycle, there might still be significant amounts of gasoline remaining. If several cycles occur in succession, the excess might build up to Q_{max}. Because there is a financial cost in carrying excess inventory, this situation would be undesirable. What alternatives can you suggest? Modify the inventory algorithm to take your alternatives into account.

3. In many situations, the time T between deliveries and the order quantity Q is not fixed. Instead, an order is placed for a specific amount of gasoline. Depending on how many orders are placed in a given time interval, the time to fill an order varies. You have no reason to believe that the performance of the delivery operation will change. Therefore, you have examined records for the past 100 deliveries and found the following lag times, or extra days, required to fill your order:

Lag time (in days)	Number of occurrences
2	10
3	25
4	30
5	20
6	13
7	2
Total:	100

Construct a Monte Carlo simulation for the lag time submodel. If you have a hand-held calculator or computer available, test your submodel by running 1000 trials and comparing the number of occurrences of the various lag times with the historical data.

4. Problem 3 suggests an alternative inventory strategy. When the inventory reaches a certain level (an order point), an order can be placed for an optimal amount of gasoline. Construct an algorithm that simulates this process and incorporates probabilistic submodels for demand and lag times. How could you use this algorithm to search for the optimal order point and the optimal order quantity?

5. In the case in which a gasoline station runs out of gas, the customer is simply going to go to another station. In many situations (name a few), however, some customers will place a back order or collect a rain check. If the order is not filled within a time period varying from customer to customer in a probabilistic fashion, the customer will cancel his or her order. Suppose we examine historical data for 1000 customers and find the data shown in Table 5.13. That is, 200 customers will not even place an order, and an additional 150 customers will cancel if the order is not filled within 1 day.

a. Construct a Monte Carlo simulation for the back order submodel. If you have a calculator or computer available, test your submodel by running 1000 trials and comparing the number of occurrences of the various cancellations with the historical data.

b. Consider the algorithm you modified in Problem 1. Further modify the algorithm to consider back orders. Do you think back orders should be penalized in some fashion? If so, how would you do it?

Table 5.13 Hypothetical data for a back order submodel

Number of days customer is willing to wait before canceling	Number of occurrences	Cumulative occurrences
0	200	200
1	150	350
2	200	550
3	200	750
4	150	900
5	50	950
6	50	1000
	1000	

5.4 PROJECTS

1. Complete the requirements of UMAP module 340, "The Poisson Random Process," by Carroll O. Wilde. Probability distributions are introduced to obtain practical information on random arrival patterns, interarrival times or gaps between arrivals, waiting line buildup, and service loss rates. The Poisson distribution, the exponential distribution, and Erlang's formulas are used. The module requires an introductory probability course, the ability to use summation notation, and basic concepts of the derivative and the integral from calculus. Prepare a 10-min summary of the module for a classroom presentation.

2. Assume a storage cost of $0.001 per gallon per day and a delivery charge of $500 per delivery. Construct a computer code of the algorithm you constructed in Problem 4, and compare various order points and order quantity strategies.

5.5 Queuing Models

EXAMPLE 1 *A Harbor System*

Consider a small harbor with unloading facilities for ships. Only one ship can be unloaded at any one time. Ships arrive for unloading of cargo at the harbor, and the time between the arrival of successive ships varies from 15 to 145 min. The unloading time required for a ship depends on the type and amount of cargo and varies from 45 to 90 min. We seek answers to the following questions:

1. What are the average and maximum times per ship in the harbor?
2. If the *waiting time* for a ship is the time between its arrival and the start of unloading, what are the average and maximum waiting times per ship?
3. What percentage of the time are the unloading facilities idle?
4. What is the length of the longest queue?

To obtain some reasonable answers, we can simulate the activity in the harbor using a computer or programmable calculator. We assume the arrival times between successive ships and the unloading time per ship are uniformly distributed over their respective time intervals. For instance, the arrival time between ships can be any integer between 15 and 145, and any integer within that interval can appear with equal likelihood. Before giving a general algorithm to simulate the harbor system, let's consider a hypothetical situation with five ships.

We have the following data for each ship:

	Ship 1	Ship 2	Ship 3	Ship 4	Ship 5
Time between successive ships	20	30	15	120	25
Unloading time	55	45	60	75	80

Because Ship 1 arrives 20 min after the clock commences at $t = 0$ min, the harbor facilities are idle for 20 min at the start. Ship 1 immediately begins to unload. The unloading takes 55 min; meanwhile, Ship 2 arrives on the scene at $t = 20 + 30 = 50$ min after the clock begins. Ship 2 cannot start to unload until Ship 1 finishes unloading at $t = 20 + 55 = 75$ min. This means that Ship 2 must wait $75 - 50 = 25$ min before unloading begins. The situation is depicted in the following timeline diagram:

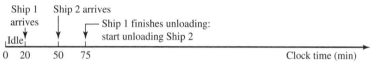

Timeline 1

Now before Ship 2 starts to unload, Ship 3 arrives at time $t = 50 + 15 = 65$ min. Because the unloading of Ship 2 starts at $t = 75$ min and it takes 45 min to unload, unloading Ship 3 cannot start until $t = 75 + 45 = 120$ min, when Ship 2 is finished. Thus, Ship 3 must wait $120 - 65 = 55$ min. The situation is depicted in the next timeline diagram:

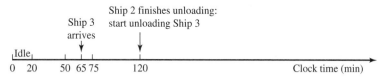

Timeline 2

Ship 4 does not arrive in the harbor until $t = 65 + 120 = 185$ min. Therefore, Ship 3 has already finished unloading at $t = 120 + 60 = 180$ min, and the harbor facilities are idle for $185 - 180 = 5$ min. Moreover, the unloading of Ship 4 commences immediately upon its arrival, as depicted in the next diagram:

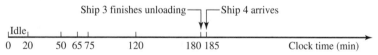

Timeline 3

Finally, Ship 5 arrives at $t = 185 + 25 = 210$ min, before Ship 4 finishes unloading at $t = 185 + 75 = 260$ min. Thus, Ship 5 must wait $260 - 210 = 50$ min before it starts to unload. The simulation is complete when Ship 5 finishes unloading at $t = 260 + 80 = 340$ min. The final situation is shown in the next diagram:

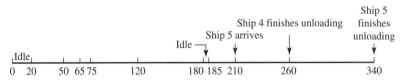

Timeline 4

In Figure 5.9, we summarize the waiting and unloading times for each of the five hypothetical ship arrivals. In Table 5.14, we summarize the results of the entire simulation of the five hypothetical ships. Note that the total waiting time spent by all five ships before unloading is 130 min. This waiting time represents a cost to the shipowners and is a source of customer dissatisfaction with the docking facilities. On the other hand, the docking facility has only 25 min of total idle time. It is in use 315 out of the total 340 min in the simulation, or approximately 93% of the time.

Suppose the owners of the docking facilities are concerned with the quality of service they are providing and want various management alternatives to be evaluated to determine whether improvement in service justifies the added cost. Several statistics can help in evaluating the quality of the service. For example, the maximum time a ship spends in the harbor is 130 min by Ship 5, whereas the average is 89 min (Table 5.14). Generally, customers are very sensitive to the amount of time spent waiting. In this example, the maximum time spent waiting for a facility is 55 min, whereas the average time spent waiting is 26 min. Some customers are apt to take their business elsewhere if queues are too long. In this case, the longest queue is two. The following Monte Carlo simulation algorithm computes such statistics to assess various management alternatives.

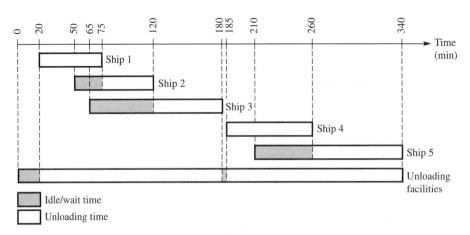

■ **Figure 5.9**

Idle and unloading times for the ships and docking facilities

Table 5.14 Summary of the harbor system simulation

Ship no.	Random time between ship arrivals	Arrival time	Start service	Queue length at arrival	Wait time	Random unload time	Time in harbor	Dock idle time
1	20	20	20	0	0	55	55	20
2	30	50	75	1	25	45	70	0
3	15	65	120	2	55	60	115	0
4	120	185	185	0	0	75	75	5
5	25	210	260	1	50	80	130	0
Total (if appropriate):					130			25
Average (if appropriate):					26	63	89	

Note: All times are given in minutes after the start of the clock at time $t = 0$.

Summary of Harbor System Algorithm Terms

between$_i$ Time between successive arrivals of Ships i and $i - 1$ (a random integer varying between 15 and 145 min)

arrive$_i$ Time from start of clock at $t = 0$ when Ship i arrives at the harbor for unloading

unload$_i$ Time required to unload Ship i at the dock (a random integer varying between 45 and 90 min)

start$_i$ Time from start of clock at which Ship i commences its unloading

idle$_i$ Time for which dock facilities are idle immediately *before* commencement of unloading Ship i

wait$_i$ Time Ship i waits in the harbor after arrival before unloading commences

finish$_i$ Time from start of clock at which service for Ship i is completed at the unloading facilities

harbor$_i$ Total time Ship i spends in the harbor

HARTIME   Average time per ship in the harbor

MAXHAR   Maximum time of a ship in the harbor

WAITIME   Average waiting time per ship before unloading

MAXWAIT   Maximum waiting time of a ship

IDLETIME   Percentage of total simulation time unloading facilities are idle

Harbor System Simulation Algorithm

Input   Total number n of ships for the simulation.

Output   HARTIME, MAXHAR, WAITIME, MAXWAIT, and IDLETIME.

Step 1   Randomly generate between$_1$ and unload$_1$. Then set arrive$_1$ = between$_1$.

Step 2   Initialize all output values:
HARTIME = unload$_1$,   MAXHAR = unload$_1$,
WAITIME = 0,   MAXWAIT = 0,   IDLETIME = arrive$_1$

Step 3   Calculate finish time for unloading of Ship$_1$:
finish$_1$ = arrive$_1$ + unload$_1$

Step 4   For $i = 2, 3, \ldots, n$, do Steps 5–16.

Step 5   Generate the random pair of integers between$_i$ and unload$_i$ over their respective time intervals.

Step 6   Assuming the time clock begins at $t = 0$ min, calculate the time of arrival for Ship$_i$:
arrive$_i$ = arrive$_{i-1}$ + between$_i$

Step 7   Calculate the time difference between the arrival of Ship$_i$ and the finish time for unloading the previous Ship$_{i-1}$:
timediff = arrive$_i$ − finish$_{i-1}$

Step 8 For nonnegative timediff, the unloading facilities are idle:
$$idle_i = timediff \quad and \quad wait_i = 0$$
For negative timediff, $Ship_i$ must wait before it can unload:
$$wait_i = -timediff \quad and \quad idle_i = 0$$

Step 9 Calculate the start time for unloading $Ship_i$:
$$start_i = arrive_i + wait_i$$

Step 10 Calculate the finish time for unloading $Ship_i$:
$$finish_i = start_i + unload_i$$

Step 11 Calculate the time in harbor for $Ship_i$:
$$harbor_i = wait_i + unload_i$$

Step 12 Sum $harbor_i$ into total harbor time HARTIME for averaging.

Step 13 If $harbor_i > $ MAXHAR, then set MAXHAR $= harbor_i$. Otherwise leave MAXHAR as is.

Step 14 Sum $wait_i$ into total waiting time WAITIME for averaging.

Step 15 Sum $idle_i$ into total idle time IDLETIME.

Step 16 If $wait_i > $ MAXWAIT, then set MAXWAIT $= wait_i$. Otherwise leave MAXWAIT as is.

Step 17 Set HARTIME $=$ HARTIME$/n$, WAITIME $=$ WAITIME$/n$, and IDLETIME $=$ IDLETIME$/finish_n$.

Step 18 OUTPUT (HARTIME, MAXHAR, WAITIME, MAXWAIT, IDLETIME)
STOP

Table 5.15 gives the results, according to the preceding algorithm, of six independent simulation runs of 100 ships each.

Now suppose you are a consultant for the owners of the docking facilities. What would be the effect of hiring additional labor or acquiring better equipment for unloading cargo so that the unloading time interval is reduced to between 35 and 75 min per ship? Table 5.16 gives the results based on our simulation algorithm.

Table 5.15 Harbor system simulation results for 100 ships

Average time of a ship in the harbor	106	85	101	116	112	94
Maximum time of a ship in the harbor	287	180	233	280	234	264
Average waiting time of a ship	39	20	35	50	44	27
Maximum waiting time of a ship	213	118	172	203	167	184
Percentage of time dock facilities are idle	0.18	0.17	0.15	0.20	0.14	0.21

Note: All times are given in minutes. Time between successive ships is 15–145 min. Unloading time per ship varies from 45 to 90 min.

Table 5.16 Harbor system simulation results for 100 ships

Average time of a ship in the harbor	74	62	64	67	67	73
Maximum time of a ship in the harbor	161	116	167	178	173	190
Average waiting time of a ship	19	6	10	12	12	16
Maximum waiting time of a ship	102	58	102	110	104	131
Percentage of time dock facilities are idle	0.25	0.33	0.32	0.30	0.31	0.27

Note: All times are given in minutes. Time between successive ships is 15–145 min. Unloading time per ship varies from 35 to 75 min.

You can see from Table 5.16 that a reduction of the unloading time per ship by 10 to 15 min decreases the time ships spend in the harbor, especially the waiting times. However, the percentage of the total time during which the dock facilities are idle nearly doubles. The situation is favorable for shipowners because it increases the availability of each ship for hauling cargo over the long run. Thus, the traffic coming into the harbor is likely to increase. If the traffic increases to the extent that the time between successive ships is reduced to between 10 and 120 min, the simulated results are as shown in Table 5.17. We can see from this table that the ships again spend more time in the harbor with the increased traffic, but now harbor facilities are idle much less of the time. Moreover, both the shipowners and the dock owners are benefiting from the increased business.

Suppose now that we are not satisfied with the assumption that the arrival time between ships (i.e., their interarrival times) and the unloading time per ship are uniformly distributed over the time intervals $15 \leq \text{between}_i \leq 145$ and $45 \leq \text{unload}_i \leq 90$, respectively. We decide to collect experimental data for the harbor system and incorporate the results into our model, as discussed for the demand submodel in the previous section. We observe (hypothetically) 1200 ships using the harbor to unload their cargoes, and we collect the data displayed in Table 5.18.

Table 5.17 Harbor system simulation results for 100 ships

Average time of a ship in the harbor	114	79	96	88	126	115
Maximum time of a ship in the harbor	248	224	205	171	371	223
Average waiting time of a ship	57	24	41	35	71	61
Maximum waiting time of a ship	175	152	155	122	309	173
Percentage of time dock facilities are idle	0.15	0.19	0.12	0.14	0.17	0.06

Note: All times are given in minutes. Time between successive ships is 10–120 min. Unloading time per ship varies from 35 to 75 min.

Table 5.18 Data collected for 1200 ships using the harbor facilities

Time between arrivals	Number of occurrences	Probability of occurrence	Unloading time	Number of occurrences	Probability of occurrence
15–24	11	0.009			
25–34	35	0.029			
35–44	42	0.035	45–49	20	0.017
45–54	61	0.051	50–54	54	0.045
55–64	108	0.090	55–59	114	0.095
65–74	193	0.161	60–64	103	0.086
75–84	240	0.200	65–69	156	0.130
85–94	207	0.172	70–74	223	0.185
95–104	150	0.125	75–79	250	0.208
105–114	85	0.071	80–84	171	0.143
115–124	44	0.037	85–90	109	0.091
125–134	21	0.017		1200	1.000
135–145	3	0.003			
	1200	1.000			

Note: All times are given in minutes.

■ Figure 5.10

Cumulative histograms of the time between ship arrivals and the unloading times, from the data in Table 5.18

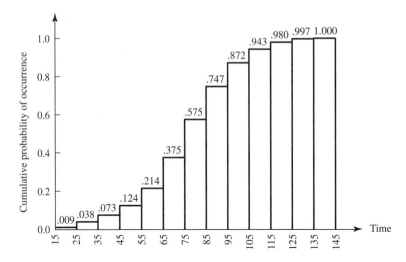

a. Time between arrivals

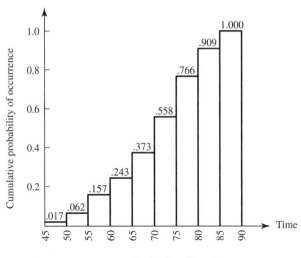

b. Unloading time

Following the procedures outlined in Section 5.4, we consecutively add together the probabilities of each individual time interval between arrivals as well as probabilities of each individual unloading time interval. These computations result in the cumulative histograms depicted in Figure 5.10.

Next we use random numbers uniformly distributed over the interval $0 \le x \le 1$ to duplicate the various interarrival times and unloading times based on the cumulative histograms. We then use the midpoints of each interval and construct linear splines through adjacent data points. (We ask you to complete this construction in Problem 1.) Because it is easy to calculate the inverse splines directly, we do so and summarize the results in Tables 5.19 and 5.20.

Table 5.19 Linear segment submodels provide for the time between arrivals of successive ships as a function of a random number in the interval [0, 1].

Random number interval	Corresponding arrival time	Inverse linear spline
$0 \leq x < 0.009$	$15 \leq b < 20$	$b = 555.6x + 15.0000$
$0.009 \leq x < 0.038$	$20 \leq b < 30$	$b = 344.8x + 16.8966$
$0.038 \leq x < 0.073$	$30 \leq b < 40$	$b = 285.7x + 19.1429$
$0.073 \leq x < 0.124$	$40 \leq b < 50$	$b = 196.1x + 25.6863$
$0.124 \leq x < 0.214$	$50 \leq b < 60$	$b = 111.1x + 36.2222$
$0.214 \leq x < 0.375$	$60 \leq b < 70$	$b = 62.1x + 46.7080$
$0.375 \leq x < 0.575$	$70 \leq b < 80$	$b = 50.0x + 51.2500$
$0.575 \leq x < 0.747$	$80 \leq b < 90$	$b = 58.1x + 46.5698$
$0.747 \leq x < 0.872$	$90 \leq b < 100$	$b = 80.0x + 30.2400$
$0.872 \leq x < 0.943$	$100 \leq b < 110$	$b = 140.8x - 22.8169$
$0.943 \leq x < 0.980$	$110 \leq b < 120$	$b = 270.3x - 144.8649$
$0.980 \leq x < 0.997$	$120 \leq b < 130$	$b = 588.2x - 456.4706$
$0.997 \leq x \leq 1.000$	$130 \leq b \leq 145$	$b = 5000.0x - 4855$

Table 5.20 Linear segment submodels provide for the unloading time of a ship as a function of a random number in the interval [0, 1].

Random number interval	Corresponding unloading time	Inverse linear spline
$0 \leq x < 0.017$	$45 \leq u < 47.5$	$u = 147x + 45.000$
$0.017 \leq x < 0.062$	$47.5 \leq u < 52.5$	$u = 111x + 45.611$
$0.062 \leq x < 0.157$	$52.5 \leq u < 57.5$	$u = 53x + 49.237$
$0.157 \leq x < 0.243$	$57.5 \leq u < 62.5$	$u = 58x + 48.372$
$0.243 \leq x < 0.373$	$62.5 \leq u < 67.5$	$u = 38.46x + 53.154$
$0.373 \leq x < 0.558$	$67.5 \leq u < 72.5$	$u = 27x + 57.419$
$0.558 \leq x < 0.766$	$72.5 \leq u < 77.5$	$u = 24x + 59.087$
$0.766 \leq x < 0.909$	$77.5 \leq u < 82.5$	$u = 35x + 50.717$
$0.909 \leq x \leq 1.000$	$82.5 \leq u \leq 90$	$u = 82.41x + 7.582$

Finally, we incorporate our linear spline submodels into the simulation model for the harbor system by generating between$_i$ and unload$_i$ for $i = 1, 2, \ldots, n$ in Steps 1 and 5 of our algorithm, according to the rules displayed in Tables 5.19 and 5.20. Employing these submodels, Table 5.21 gives the results of six independent simulation runs of 100 ships each.   ▪ ▪ ▪

EXAMPLE 2 *Morning Rush Hour*

In the previous example, we initially considered a harbor system with a single facility for unloading ships. Such problems are often called *single-server queues*. In this example, we

Table 5.21 Harbor system simulation results for 100 ships

Average time of a ship in the harbor	108	95	125	78	123	101
Maximum time of a ship in the harbor	237	188	218	133	250	191
Average waiting time of a ship	38	25	54	9	53	31
Maximum waiting time of a ship	156	118	137	65	167	124
Percentage of time dock facilities are idle	0.09	0.09	0.08	0.12	0.06	0.10

Note: Based on the data exhibited in Table 5.18. All times are given in minutes.

consider a system with four elevators, illustrating *multiple-server queues*. We discuss the problem and present the algorithm in Appendix B.

Consider an office building with 12 floors in a metropolitan area of some city. During the morning rush hour, from 7:50 to 9:10 A.M., workers enter the lobby of the building and take an elevator to their floor. There are four elevators servicing the building. The time between arrivals of the customers at the building varies in a probabilistic manner every 0–30 sec, and upon arrival each customer selects the first available elevator (numbered 1–4). When a person enters an elevator and selects the floor of destination, the elevator waits 15 sec before closing its doors. If another person arrives within the 15-sec interval, the waiting cycle is repeated. If no person arrives within the 15-sec interval, the elevator departs to deliver all of its passengers. We assume no other passengers are picked up along the way. After delivering its last passenger, the elevator returns to the main floor, picking up no passengers on the way down. The maximum occupancy of an elevator is 12 passengers. When a person arrives in the lobby and no elevator is available (because all four elevators are transporting their load of passengers), a queue begins to form in the lobby.

The management of the building wants to provide good elevator service to its customers and is interested in exactly what service it is now giving. Some customers claim that they have to wait too long in the lobby before an elevator returns. Others complain that they spend too much time riding the elevator, and still others say that there is considerable congestion in the lobby during the morning rush hour. What is the real situation? Can the management resolve these complaints by a more effective means of scheduling or utilizing the elevators?

We wish to simulate the elevator system using an algorithm for computer implementation that will give answers to the following questions:

1. How many customers are actually being serviced in a typical morning rush hour?

2. If the *waiting time* of a person is the time the person stands in a queue—the time from arrival at the lobby until entry into an available elevator—what are the average and maximum times a person waits in a queue?

3. What is the length of the longest queue? (The answer to this question will provide the management with information about congestion in the lobby.)

4. If the *delivery time* is the time it takes a customer to reach his or her floor after arrival in the lobby, including any waiting time for an available elevator, what are the average and maximum delivery times?

5. What are the average and maximum times a customer actually spends in the elevator?

6. How many stops are made by each elevator? What percentage of the total morning rush hour time is each elevator actually in use?

An algorithm is presented in Appendix B.

5.5 PROBLEMS

1. Using the data from Table 5.18 and the cumulative histograms of Figure 5.10, construct cumulative plots of the time between arrivals and unloading time submodels (as in Figure 5.7). Calculate equations for the linear splines over each random number interval. Compare your results with the inverse splines given in Tables 5.19 and 5.20.

2. Use a smooth polynomial to fit the data in Table 5.18 to obtain arrivals and unloading times. Compare results to those in Tables 5.19 and 5.20.

3. Modify the ship harbor system algorithm to keep track of the number of ships waiting in the queue.

4. Most small harbors have a maximum number of ships N_{max} that can be accommodated in the harbor area while they wait to be unloaded. If a ship cannot get into the harbor, assume it goes elsewhere to unload its cargo. Refine the ship harbor algorithm to take these considerations into account.

5. Suppose the owners of the docking facilities decide to construct a second facility to accommodate the unloading of more ships. When a ship enters the harbor, it goes to the next available facility, which is facility 1 if both facilities are available. Using the same assumption for interarrival times between successive ships and unloading times as in the initial text example, modify the algorithm for a system with two facilities.

6. Construct a Monte Carlo simulation of a baseball game. Use individual batting statistics to simulate the probability of a single, double, triple, home run, or out. In a more refined model, how would you handle walks, hit batsman, steals, and double plays?

5.5 PROJECTS

1. Write a computer simulation to implement the ship harbor algorithm.

2. Write a computer simulation to implement a baseball game between your two favorite teams (see Problem 6).

3. Pick a traffic intersection with a traffic light. Collect data on vehicle arrival times and clearing times. Build a Monte Carlo simulation to model traffic flow at this intersection.

4. In the Los Angeles County School District, substitute teachers are placed in a pool and paid whether they teach or not. It is assumed that if the need for substitutes exceeds the size of the pool, classes can be covered by regular teachers, but at a higher pay rate. Letting x represent the number of substitutes needed on a given day, S the pool size, p the amount of pay for pool members, and r the daily overtime rate, we have for the cost

$$C(x, S) = \begin{cases} pS & \text{if } x < S \\ pS + (x - S)r & \text{if } x \geq S \end{cases}$$

Here, we assume $p < r$.

   a. Use the data provided for the number of substitutes needed on Mondays to simulate the situation in an attempt to optimize the pool size. The optimized pool will be the one with the lowest expected cost to the school district. Use, for pay rates, $p = \$45$ and $r = \$81$. Assume that the data are distributed uniformly.

i. Make 500 simulations at each value of S from $S = 100$ to $S = 900$ in steps of 100, using the averages of the 500 runs to estimate the cost for each value of S.

ii. Narrow the search for the best value of S to an interval of length 200 and make runs of 1000 simulations for each of ten equally spaced values of S in this interval.

iii. Continue the process narrowing the search for the optimal size of the pool, each time stepping the values of S a smaller amount and increasing the number of iterations for better accuracy. When you have determined the optimal value for the pool size, S, submit your choice with substantiating evidence.

Demand for substitute teachers on Mondays

Number of teachers	Relative percentage	Cumulative percentage
201–275	2.7	2.7
276–350	2.7	5.4
351–425	2.7	8.1
426–500	2.7	10.8
501–575	16.2	27
576–650	10.8	37.8
651–725	48.6	86.4
726–800	8.1	94.5
801–875	2.7	97.2
876–950	2.7	99.9

b. Redo part (a) with $p = \$36$ and $r = \$81$.

c. Redo part (a) using the data provided for Tuesdays. Assume $p = \$45$ and $r = \$81$.

d. Redo part (c) using the data for Tuesdays assuming $p = \$36$ and $r = \$81$.

Demand for substitute teachers on Tuesdays

Number of teachers	Relative percentage	Cumulative percentage
201–275	2.5	2.5
276–350	2.5	5.0
351–425	5.0	10.0
426–500	7.5	17.5
501–575	12.5	30.0
576–650	17.5	47.5
651–725	42.5	90.0
726–800	5.0	95.0
801–875	2.5	97.5
876–950	2.5	100.0

6 Discrete Probabilistic Modeling

Introduction

We have been developing models using proportionality and determining the constants of proportionality. However, what if variation actually takes place, as in the demand for gasoline at the gasoline station? In this chapter, we will allow the constants of proportionality to vary in a *random* fashion rather than to be fixed. We begin by revisiting discrete dynamical systems from Chapter 1 and introduce scenarios that have probabilistic parameters.

6.1 Probabilistic Modeling with Discrete Systems

In this section we revisit the systems of difference equations studied in Section 1.4, but now we allow the coefficients of the systems to vary in a probabilistic manner. A special case, called a Markov chain, is a process in which there are the same finite number of states or outcomes that can be occupied at any given time. The states do not overlap and cover all possible outcomes. In a Markov process, the system may move from one state to another, one for each time step, and there is a probability associated with this transition for each possible outcome. The sum of the probabilities for transitioning from the *present state* to the *next state* is equal to 1 for each state at each time step. A Markov process with two states is illustrated in Figure 6.1.

■ **Figure 6.1**

A Markov chain with two states; the sum of the probabilities for the transition from a present state is 1 for each state (e.g., $p + (1 - p) = 1$ for state 1).

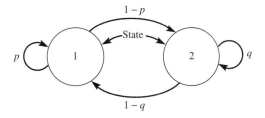

EXAMPLE 1 *Rental Car Company Revisited*

Consider a rental car company with branches in Orlando and Tampa. Each rents cars to Florida tourists. The company specializes in catering to travel agencies that want to arrange tourist activities in both Orlando and Tampa. Consequently, a traveler will rent a car in one city and drop off the car in either city. Travelers begin their itinerary in either city. Cars can be returned to either location, which can cause an imbalance in available cars to rent. The following historical data on the percentages of cars rented and returned to these locations have been collected for the previous few years.

		Next state	
		Orlando	Tampa
Present state	Orlando	0.6	0.4
	Tampa	0.3	0.7

This array of data is called a **transition matrix** and shows that the probability for returning a car to Orlando that was also rented in Orlando is 0.6, whereas the probability that it will be returned in Tampa is 0.4. Likewise, a car rented in Tampa has a 0.3 likelihood of being returned to Orlando and a 0.7 probability of being returned to Tampa. This represents a Markov process with two states: Orlando and Tampa. Notice that the sum of the probabilities for transitioning from a present state to the next state, which is the sum of the probabilities in each row, equals 1 because all possible outcomes are taken into account. The process is illustrated in Figure 6.2.

■ Figure 6.2

Two-state Markov chain
for the rental car example

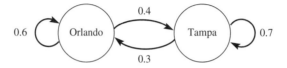

Model Formulation   Let's define the following variables:

p_n = the percentage of cars available to rent in Orlando at the end of period n

q_n = the percentage of cars available to rent in Tampa at the end of period n

Using the previous data and discrete modeling ideas from Section 1.5, we construct the following probabilistic model:

$$p_{n+1} = 0.6p_n + 0.3q_n$$
$$q_{n+1} = 0.4p_n + 0.7q_n$$

(6.1)

Model Solution   Assuming that all of the cars are originally in Orlando, numerical solutions to System (6.1) give the long-term behavior of the percentages of cars available at each location. The sum of these long-term percentages or probabilities also equals 1. Table 6.1 and Figure 6.3 show the results in table form and graphically:
Notice that

$$p_k \rightarrow 3/7 = 0.428571$$
$$q_k \rightarrow 4/7 = 0.571429$$

Model Interpretation   If the two branches begin the year with a total of n cars, then after 14 time periods, or days, approximately 57% of the cars will be in Tampa and 43% will be

Table 6.1 Iterated solution to the rental car example

n	Orlando	Tampa
0	1	0
1	0.6	0.4
2	0.48	0.52
3	0.444	0.556
4	0.4332	0.5668
5	0.42996	0.57004
6	0.428988	0.571012
7	0.428696	0.571304
8	0.428609	0.571391
9	0.428583	0.571417
10	0.428575	0.571425
11	0.428572	0.571428
12	0.428572	0.571428
13	0.428572	0.571428
14	0.428571	0.571429

■ **Figure 6.3**

Graphical solution to the rental car example

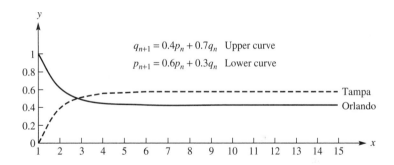

$$q_{n+1} = 0.4p_n + 0.7q_n \quad \text{Upper curve}$$
$$p_{n+1} = 0.6p_n + 0.3q_n \quad \text{Lower curve}$$

in Orlando. Thus, starting with 100 cars in each location, about 114 cars will be based out of Tampa and 86 will be based out of Orlando in the steady state (and it would take only approximately 5 days to reach this state). ■ ■ ■

EXAMPLE 2 *Voting Tendencies*

Presidential voting tendencies are of interest every 4 years. During the past decade, the Independent party has emerged as a viable alternative for voters in the presidential race. Let's consider a three-party system with Republicans, Democrats, and Independents.

Problem Identification   *Can we find the long-term behavior of voters in a presidential election?*

Assumptions   During the past decade, the trends have been to vote less strictly along party lines. We provide hypothetical historical data for voting trends in the past 10 years of statewide voting. The data are presented in the following hypothetical transition matrix and shown in Figure 6.4.

	Next state		
	Republicans	Democrats	Independents
Present state Republicans	0.75	0.05	0.20
Democrats	0.20	0.60	0.20
Independents	0.40	0.20	0.40

◼ Figure 6.4

Three-state Markov chain for presidential voting tendencies

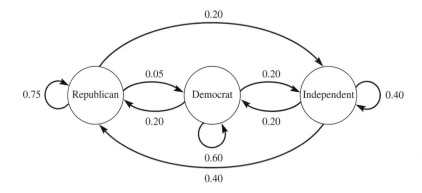

Model Formulation   Let's define the following variables:

$$R_n = \text{the percentage of voters to vote Republican in period } n$$
$$D_n = \text{the percentage of voters to vote Democratic in period } n$$
$$I_n = \text{the percentage of voters to vote Independent in period } n$$

Using the previous data and the ideas on discrete dynamical systems from Chapter 1, we can formulate the following system of equations giving the percentage of voters who vote Republican, Democratic, or Independent at each time period.

$$R_{n+1} = 0.75R_n + 0.20D_n + 0.40I_n$$
$$D_{n+1} = 0.05R_n + 0.60D_n + 0.20I_n \qquad (6.2)$$
$$I_{n+1} = 0.20R_n + 0.20D_n + 0.40I_n$$

Model Solution   Assume that initially 1/3 of the voters are Republican, 1/3 are Democrats, and 1/3 are Independent. We then obtain the numerical results shown in Table 6.2 for the percentage of voters in each group at each period n. The table shows that in the long run (and after approximately 10 time periods), approximately 56% of the voters cast their ballots for the Republican candidate, 19% vote Democrat, and 25% vote Independent. Figure 6.5 shows these results graphically. ▪ ▪ ▪

Let's summarize the ideas of a Markov chain. A **Markov chain** is a process consisting of a sequence of events with the following properties:

1. An event has a finite number of outcomes, called **states**. The process is always in one of these states.

Table 6.2 Iterated solution to the presidential voting problem

n	Republican	Democrat	Independent
0	0.33333	0.33333	0.33333
1	0.449996	0.283331	0.266664
2	0.500828	0.245831	0.253331
3	0.52612	0.223206	0.250664
4	0.539497	0.210362	0.250131
5	0.546747	0.203218	0.250024
6	0.550714	0.199273	0.250003
7	0.552891	0.1971	0.249999
8	0.554088	0.195904	0.249998
9	0.554746	0.195247	0.249998
10	0.555108	0.194885	0.249998
11	0.555307	0.194686	0.249998
12	0.555416	0.194576	0.249998
13	0.555476	0.194516	0.249998
14	0.55551	0.194483	0.249998

■ **Figure 6.5**

Graphical solution to the presidential voting tendencies example

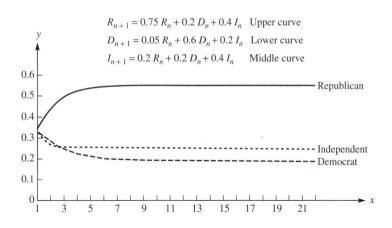

$R_{n+1} = 0.75 \, R_n + 0.2 \, D_n + 0.4 \, I_n$ Upper curve

$D_{n+1} = 0.05 \, R_n + 0.6 \, D_n + 0.2 \, I_n$ Lower curve

$I_{n+1} = 0.2 \, R_n + 0.2 \, D_n + 0.4 \, I_n$ Middle curve

2. At each stage or period of the process, a particular outcome can transition from its present state to any other state or remain in the same state.

3. The probability of going from one state to another in a single stage is represented by a **transition matrix** for which the entries in each row are between 0 and 1; each row sums to 1. These probabilities depend only on the present state and not on past states.

6.1 PROBLEMS

1. Consider a model for the long-term dining behavior of the students at College USA. It is found that 25% of the students who eat at the college's Grease Dining Hall return to eat there again, whereas those who eat at Sweet Dining Hall have a 93% return rate. These are the only two dining halls available on campus, and assume that all students eat at one of these halls. Formulate a model to solve for the long-term percentage of students eating at each hall.

2. Consider adding a pizza delivery service as an alternative to the dining halls. Table 6.3 gives the transition percentages based on a student survey. Determine the long-term percentages eating at each place.

Table 6.3 Survey of dining at College USA

		Next state		
		Grease Dining Hall	Sweet Dining Hall	Pizza delivery
Present state	Grease Dining Hall	0.25	0.25	0.50
	Sweet Dining Hall	0.10	0.30	0.60
	Pizza delivery	0.05	0.15	0.80

3. In Example 1, it was assumed that all the cars were initially in Orlando. Try several different starting values. Is equilibrium achieved in each case? If so, what is the final distribution of cars in each case?

4. In Example 2, it was assumed that initially the voters were equally divided among the three parties. Try several different starting values. Is equilibrium achieved in each case? If so, what is the final distribution of voters in each case?

6.1 PROJECT

1. Consider the pollution in two adjoining lakes in which the only flow is between the lakes as illustrated in Figure 6.6. To simplify matters, assume that 100% of the water from Lake A comes from Lake B. Let a_n and b_n be the total amounts of pollution in Lake A and Lake B, respectively, after n years. It has also been determined that we can measure the amount of pollutants given the lake in which they originated. Figure 6.7 shows a diagram reflecting the situation. Formulate and solve the model of the pollution flow as a Markov chain using dynamical systems.

Figure 6.6

Pollution of the Great Lakes

■ Figure 6.7

Two-state Markov chain for the Great Lakes pollution project

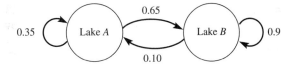

6.2 Modeling Component and System Reliability

Do your personal computer and your automobile perform well for a reasonably long period of time? If they do, we say that these systems are *reliable*. The **reliability** of a component or system is the probability that it will not fail over a specific time period n. Let's define $f(t)$ to be the failure rate of an item, component, or system over time t, so $f(t)$ is a probability distribution. Let $F(t)$ be the cumulative distribution function corresponding to $f(t)$ as we discussed in Section 5.3. We define the reliability of the item, component, or system by

$$R(t) = 1 - F(t) \tag{6.3}$$

Thus, the reliability at any time t is 1 minus the expected cumulative failure rate at time t.

Human–machine systems, whether electronic or mechanical, consist of components, some of which may be combined to form subsystems. (Consider systems such as your personal computer, your stereo, or your automobile.) We want to build simple models to examine the reliability of complex systems. We now consider design relationships in series, parallel, or combinations of these. Although individual item failure rates can follow a wide variety of distributions, we consider only a few elementary examples.

EXAMPLE 1 *Series Systems*

A **series system** is one that performs well as long as *all* of the components are fully functional. Consider a NASA space shuttle's rocket propulsion system as displayed in Figure 6.8. This is an example of a series system because failure for any one of the *independent* booster rockets will result in a failed mission. If the reliabilities of the three components are given by $R_1(t) = 0.90$, $R_2(t) = 0.95$, and $R_3(t) = 0.96$, respectively, then the **system reliability** is defined to be the product

$$R_s(t) = R_1(t)R_2(t)R_3(t) = (0.90)(0.95)(0.96) = 0.8208$$

Note that in a series relationship the reliability of the whole system is less than any single component's reliability because each component has a reliability that is less than 1.

■ ■ ■

■ Figure 6.8

A NASA rocket propulsion system for a space shuttle showing the booster rockets in series (for three stages)

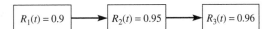

EXAMPLE 2 *Parallel Systems*

A **parallel system** is one that performs as long as a single one of its components remains operational. Consider the communication system of a NASA space shuttle, as displayed in Figure 6.9. Note that there are two separate and *independent* communication systems, either of which can operate to provide satisfactory communications with NASA control. If the independent component reliabilities for these communication systems are $R_1(t) = 0.95$ and $R_2(t) = 0.96$, then we define the **system reliability** to be

$$R_s(t) = R_1(t) + R_2(t) - R_1(t)R_2(t) = 0.95 + 0.96 - (0.95)(0.96) = 0.998$$

■ **Figure 6.9**

Two NASA space shuttle communication systems operating in parallel

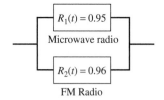

Note that in parallel relationships, the system reliability is higher than any of the individual component reliabilities. ■ ■ ■

EXAMPLE 3 *Series and Parallel Combinations*

Let's now consider a system combining series and parallel relationships as we join together the previous two subsystems for a controlled propulsion ignition system (Figure 6.10). We examine each subsystem. Subsystem 1 (the communication system) is in a parallel relationship, and we found its reliability to be 0.998. Subsystem 2 (the propulsion system) is in a series relationship, and we found its system reliability to be 0.8208. These two systems are in a series relationship, so the reliability for the entire system is the product of the two subsystem reliabilities:

$$R_s(t) = R_{s_1}(t) \cdot R_{s_2}(t) = (0.998)(0.8208) = 0.8192$$ ■ ■ ■

■ **Figure 6.10**

A NASA-controlled space shuttle propulsion ignition system

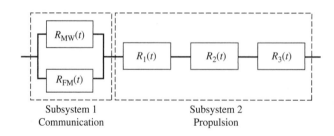

6.2 PROBLEMS

1. Consider a stereo with CD player, FM–AM radio tuner, speakers (dual), and power amplifier components, as displayed with the reliabilities shown in Figure 6.11. Determine the system's reliability. What assumptions are required in your model?

■ **Figure 6.11**

Reliability of stereo components

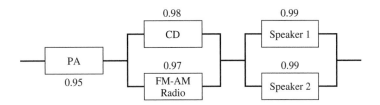

2. Consider a personal computer with each item's reliability as shown in Figure 6.12. Determine the reliability of the computer system. What assumptions are required?

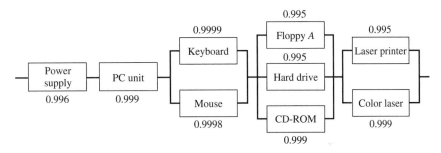

■ **Figure 6.12**

Personal computer reliabilities

3. Consider a more advanced stereo system with component reliabilities as displayed in Figure 6.13. Determine the system's reliability. What assumptions are required?

■ **Figure 6.13**

Advanced stereo system

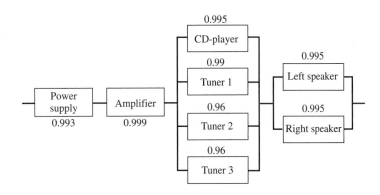

6.2   PROJECT

1. Two alternative designs are submitted for a landing module to enable the transport of astronauts to the surface of Mars. The mission is to land safely on Mars, collect several hundred pounds of samples from the planet's surface, and then return to the shuttle in its orbit around Mars. The alternative designs are displayed together with their reliabilities in Figure 6.14. Which design would you recommend to NASA? What assumptions are required? Are the assumptions reasonable?

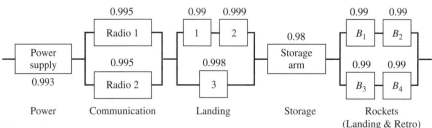

Alternative 1

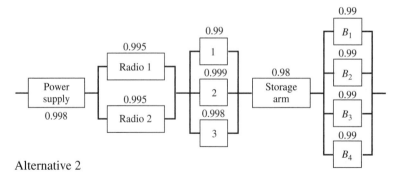

Alternative 2

■ **Figure 6.14**

Alternative designs for the Mars module

6.3   Linear Regression

In Chapter 3 we discussed various criteria for fitting models to collected data. In particular, the least-squares criterion that minimizes the sum of the squared deviations was presented. We showed that the formulation of minimizing the sum of the squared deviations is an optimization problem. Until now, we have considered only a single observation y_i for each value of the independent variable x_i. However, what if we have multiple observations? In this section we explore a statistical methodology for minimizing the sum of the squared deviations, called **linear regression**. Our objectives are

1. To illustrate the basic linear regression model and its assumptions.

2. To define and interpret the statistic R^2.

3. To illustrate a graphical interpretation for the fit of the linear regression model by examining and interpreting the residual scatterplots.

We introduce only basic concepts and interpretations here. More in-depth studies of the subject of linear regression are given in advanced statistics courses.

The Linear Regression Model

The basic linear regression model is defined by

$$y_i = ax_i + b \quad \text{for } i = 1, 2, \ldots, m \text{ data points} \tag{6.4}$$

In Section 3.3 we derived the *normal equations*

$$a \sum_{i=1}^{m} x_i^2 + b \sum_{i=1}^{m} x_i = \sum_{i=1}^{m} x_i y_i$$
$$a \sum_{i=1}^{m} x_i + mb = \sum_{i=1}^{m} y_i \tag{6.5}$$

and solved them to obtain the slope a and y-intercept b for the least-squares best-fitting line:

$$a = \frac{m \sum x_i y_i - \sum x_i \sum y_i}{m \sum x_i^2 - \left(\sum x_i \right)^2}, \text{ the slope} \tag{6.6}$$

and

$$b = \frac{\sum x_i^2 \sum y_i - \sum x_i y_i \sum x_i}{m \sum x_i^2 - \left(\sum x_i \right)^2}, \text{ the intercept} \tag{6.7}$$

We now add several additional equations to aid in the statistical analysis of the basic model (6.4).

The first of these is the **error sum of squares** given by

$$\text{SSE} = \sum_{i=1}^{m} [y_i - (ax_i + b)]^2 \tag{6.8}$$

which reflects variation about the regression line. The second concept is the **total corrected sum of squares** of y defined by

$$\text{SST} = \sum_{i=1}^{m} (y_i - \bar{y})^2 \tag{6.9}$$

where $\bar{y}$ is the average of the y values for the data points (x_i, y_i), $i = 1, \ldots, m$. (The number $\bar{y}$ is also the average value of the linear regression line $y = ax + b$ over the range of data.) Equations (6.8) and (6.9) then produce the **regression sum of squares** given by the equation

$$\text{SSR} = \text{SST} - \text{SSE} \tag{6.10}$$

The quantity SSR reflects the amount of variation in the y values explained by the linear regression line $y = ax + b$ when compared with the variation in the y values about the line $y = \bar{y}$.

From Equation (6.10), SST is always at least as large as SSE. This fact prompts the following definition of the coefficient of determination R^2, which is a measure of fit for the regression line.

$$R^2 = 1 - \frac{\text{SSE}}{\text{SST}} \tag{6.11}$$

The number R^2 expresses the proportion of the total variation in the y variable of the actual data (when compared with the line $y = \bar{y}$) that can be accounted for by the straight-line model, whose values are given by $ax + b$ (the predicted values) and calculated in terms of the x variable. If $R^2 = 0.81$, for instance, then 81% of the total variation of the y values (from the line $y = \bar{y}$) is accounted for by a linear relationship with the values of x. Thus, the closer the value of R^2 is to 1, the better the fit of the regression line model to the actual data. If $R^2 = 1$, then the data lie perfectly along the regression line. (Note that $R^2 \leq 1$ always holds.) The following are additional properties of R^2:

1. The value of R^2 does not depend on which of the two variables is labeled x and which is labeled y.

2. The value of R^2 is independent of the units of x and y.

Another indicator of the reasonableness of the fit is a plot of the **residuals** versus the independent variable. Recall that the residuals are the *errors* between the actual and predicted values:

$$r_i = y_i - f(x_i) = y_i - (ax_i - b) \tag{6.12}$$

If we plot the residuals versus the independent variable, we obtain some valuable information about them:

1. The residuals should be randomly distributed and contained in a reasonably small band that is commensurate with the accuracy of the data.

2. An extremely large residual warrants further investigation of the associated data point to discover the cause of the large residual.

3. A pattern or trend in the residuals indicates that a forecastable effect remains to be modeled. The nature of the pattern often provides clues to how to refine the model, if a refinement is needed. These ideas are illustrated in Figure 6.15.

EXAMPLE 1   *Ponderosa Pines*

Recall the ponderosa pine data from Chapter 2, provided in Table 6.4. Figure 6.16 on page 228 shows a scatterplot of the data and suggests a trend. The plot is concave up and increasing, which could suggest a power function model (or an exponential model).

Problem Identification   *Predict the number of board-feet as a function of the diameter of the ponderosa pine.*

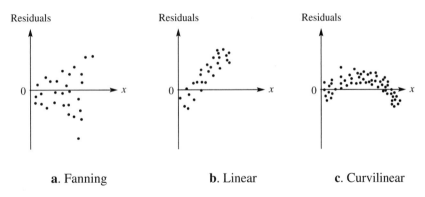

a. Fanning   **b.** Linear   **c.** Curvilinear

■ **Figure 6.15**

Possible patterns of residuals plots

Table 6.4 Ponderosa pine data

Diameter (in.)	Board feet
36	192
28	113
28	88
41	294
19	28
32	123
22	51
38	252
25	56
17	16
31	141
20	32
25	86
19	21
39	231
33	187
17	22
37	205
23	57
39	265

Assumptions   We assume that the ponderosa pines are geometrically similar and have the shape of a right circular cylinder. This allows us to use the diameter as a characteristic dimension to predict the volume. We can reasonably assume that the height of a tree is proportional to its diameter.

Model Formulation   Geometric similarity gives the proportionality

$$V \propto d^3 \qquad\qquad (6.13)$$

■ Figure 6.16

Scatterplot of the
ponderosa pine data

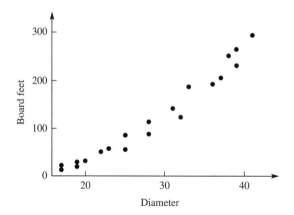

where d is the diameter of a tree (measured at ground level). If we further assume that the
ponderosa pines have constant height (rather than assuming height is proportional to the
diameter), then we obtain

$$V \propto d^2 \qquad (6.14)$$

Assuming there is a constant volume associated with the underground root system would
then suggest the following refinements of these proportionality models:

$$V = ad^3 + b \qquad (6.15)$$

and

$$V = \alpha d^2 + \beta \qquad (6.16)$$

Let's use linear regression to find the constant parameters in the four model types and then
compare the results.

Model Solution   The following solutions were obtained using a computer and applying
linear regression on the four model types:

$$V = 0.00431d^3$$
$$V = 0.00426d^3 + 2.08$$
$$V = 0.152d^2$$
$$V = 0.194d^2 - 45.7$$

Table 6.5 displays the results of these regression models.

Notice that the R^2 values are all quite large (close to 1), which indicates a strong linear
relationship. The residuals are calculated using Equation (6.12), and their plots are shown in
Figure 6.17. (Recall that we are searching for a random distribution of the residuals having
no apparent pattern.) Note that there is an apparent trend in the errors corresponding to the
model $V = 0.152d^2$. We would probably *reject* (or *refine*) this model based on this plot,
while accepting the other models that appear reasonable.

Table 6.5 Key information from regression models using the ponderosa pine data

Model	SSE	SSR	SST	R^2
$V = 0.00431d^3$	3,742	458,536	462,278	0.9919
$V = 0.00426d^3 + 2.08$	3,712	155,986	159,698	0.977
$V = 0.152d^2$	12,895	449,383	462,278	0.9721
$V = 0.194d^2 - 45.7$	3,910	155,788	159,698	0.976

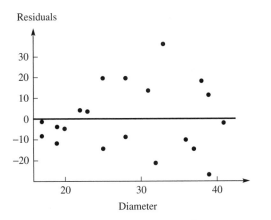

a. Residual plot for board-feet $= 0.00431d^3$; the model appears to be adequate because no apparent trend appears in the residual plot

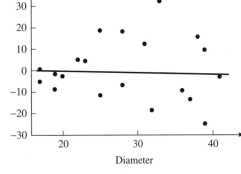

b. Residual plot for board-feet $= 0.00426d^3 + 2.08$; the model appears to be adequate because no trend appears in the plot

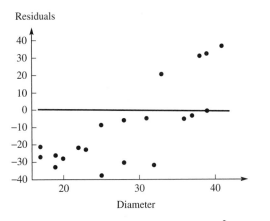

c. Residual plot for the model board-feet $= 0.152d^2$; the model does not appear to be adequate because there is a linear trend in the residual plot

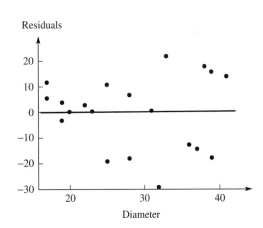

d. Residual plot for the model board-feet $= 0.194d^2 - 45.7$; the model appears to be adequate because there is no apparent trend in the residual plot

■ **Figure 6.17**

Residual plots for various models for board-feet $= f(\text{diameter})$ for the ponderosa pine data

EXAMPLE 2   *The Bass Fishing Derby Revisited*

Let's revisit the bass fishing problem from Section 2.3. We have collected much more data and now have 100 data points to use for fitting the model. These data are given in Table 6.6 and plotted in Figure 6.18. Based on the analysis in Section 2.3, we assume the following model type:

$$W = al^3 + b \tag{6.17}$$

where W is the weight of the fish and l is its length.

Table 6.6   Data for bass fish with weight (W) in oz and length (l) in in.

W	13	13	13	13	13	14	14	15	15	15
l	12	12.25	12	12.25	14.875	12	12	12.125	12.125	12.25
W	15	15	15	16	16	16	16	16	16	16
l	12	12.5	12.25	12.675	12.5	12.75	12.75	12	12.75	12.25
W	16	16	16	16	16	17	17	17	17	17
l	12	13	12.5	12.5	12.25	12.675	12.25	12.75	12.75	13.125
W	17	17	17	18	18	18	18	18	18	18
l	15.25	12.5	13.5	12.5	13	13.125	13	13.375	16.675	13
W	18	19	19	19	19	19	19	20	20	20
l	13.375	13.25	13.25	13.5	13.5	13.5	13	13.75	13.125	13.75
W	20	20	20	20	20	20	21	21	21	22
l	13.5	13.75	13.5	13.75	17	14.5	13.75	13.5	13.25	13.765
W	22	22	23	23	23	24	24	24	24	24
l	14	14	14.25	14.375	14	14.75	13.5	13.5	14.5	14
W	24	25	25	26	26	27	27	28	28	28
l	17	14.25	14.25	14.375	14.675	16.75	14.25	14.75	13	14.75
W	28	29	29	30	35	36	40	41	41	44
l	14.875	14.5	13.125	14.5	12.5	15.75	16.25	17.375	14.5	13.25
W	45	46	47	47	48	49	53	56	62	78
l	17.25	17	18	16.5	18	17.5	18	18.375	19.25	20

The linear regression solution is

$$W = 0.008l^3 + 0.95 \tag{6.18}$$

and the analysis of variance is illustrated in Table 6.7.

    The R^2 value for our model is 0.735, which is reasonably close to 1, considering the nature of the behavior being modeled. The residuals are found from Equation (6.12) and plotted in Figure 6.19. We see no apparent trend in the plot, so there is no suggestion on how to refine the model for possible improvement.   ■ ■ ■

■ **Figure 6.18**

Scatterplot of the bass fish data

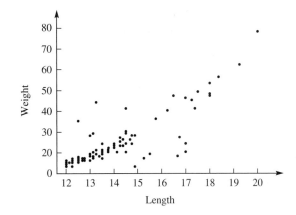

Table 6.7   Key information from a regression model for the bass fish example with 100 data points

Model	SSE	SSR	SST	R^2
$W = 0.008l^3 + 0.95$	3758	10,401	14,159	0.735

■ **Figure 6.19**

Residual plot for $W = 0.008l^3 + 0.95$; the model appears to be adequate because no trends appear in the residual plot.

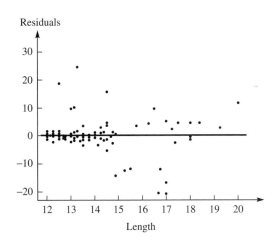

6.3 PROBLEMS

Use the basic linear model $y = ax + b$ to fit the following data sets. Provide the model, provide the values of SSE, SSR, SST, and R^2, and provide a residual plot.

1. For Table 2.7, predict weight as a function of height.

2. For Table 2.7, predict weight as a function of the cube of the height.

6.3 PROJECT

1. Use linear regression to formulate and analyze Projects 1–5 in Section 2.3.

6.3 Further Reading

Mendenhall, William, & Terry Sincich. *A Second Course in Statistics: Regression Analysis*, 6th ed. Upper Saddle River, NJ: Prentice Hall, 2003.

Neter, John, M. Kutner, C. Nachstheim, & W. Wasserman. *Applied Statistical Models*, 4th ed. Boston: McGraw–Hill, 1996.

7 Optimization of Discrete Models

Introduction

In Chapter 3 we considered three criteria for fitting a selected model to a collection of data:

1. Minimize the sum of the absolute deviations.
2. Minimize the largest of the absolute deviations (Chebyshev criterion).
3. Minimize the sum of the squared deviations (least-squares criterion).

Also in Chapter 3 we used calculus to solve the optimization problem resulting from the application of the least-squares criterion. Although we formulated several optimization problems resulting from the first criterion to minimize the sum of the absolute deviations, we were unable to solve the resulting mathematical problem. In Section 7.6 we study several search techniques that allow us to find good solutions to that curve-fitting criterion, and we examine many other optimization problems as well.

In Chapter 3 we interpreted the Chebyshev criterion for several models. For example, given a collection of m data points (x_i, y_i), $i = 1, 2, \ldots, m$, fit the collection to that line $y = ax + b$ (determined by the parameters a and b) that minimizes the greatest distance r_{max} between any data point (x_i, y_i) and its corresponding point $(x_i, ax_i + b)$ on the line. That is, the largest absolute deviation, $r = \text{Maximum } \{|y_i - y(x_i)|\}$, is minimized over the entire collection of data points. This criterion defines the optimization problem

$$\text{Minimize } r$$

subject to

$$\left. \begin{array}{l} r - r_i \geq 0 \\ r + r_i \geq 0 \end{array} \right\} \quad \text{for } i = 1, 2, \ldots, m$$

which is a *linear program* for many applications. You will learn how to solve linear programs geometrically and algebraically in Sections 7.2–7.4. You will learn how to determine the sensitivity of the optimal solution to the coefficients appearing in the linear program in Section 7.5. We begin by providing a general classification of discrete optimization problems. Our emphasis is on *model formulation*, which will allow you additional practice on the first several steps of the modeling process, while simultaneously providing a preview of the kinds of problems you will learn to solve in advanced mathematics courses.

7.1 An Overview of Optimization Modeling

We offer a basic model providing a framework for discussing a class of optimization problems. The problems are classified according to various characteristics of the basic model intrinsic to the particular problem. We also discuss variations from the basic model. The basic model is

$$\text{Optimize } f_j(\mathbf{X}) \text{ for } j \text{ in } J \tag{7.1}$$

subject to

$$g_i(\mathbf{X}) \left\{ \begin{array}{c} \geq \\ = \\ \leq \end{array} \right\} b_i \text{ for all } i \text{ in } I$$

Now let's explain the notation. To optimize means to maximize or minimize. The subscript j indicates that there may be one or more functions to optimize. The functions are distinguished by the integer subscripts that belong to the finite set J. We seek the vector $\mathbf{X}_0$ giving the optimal value for the set of functions $f_j(\mathbf{X})$. The various components of the vector $\mathbf{X}$ are called the **decision variables** of the model, whereas the functions $f_j(\mathbf{X})$ are called the **objective functions**. By adding *subject to*, we indicate that certain side conditions must be met. For example, if the objective is to minimize costs of producing a particular product, it might be specified that all contractual obligations for the product must be met. Side conditions are typically called **constraints**. The integer subscript i indicates that there may be one or more constraint relationships that must be satisfied. A constraint may be an equality (such as precisely meeting the demand for a product) or an inequality (such as not exceeding budgetary limitations or providing the minimal nutritional requirements in a diet problem). Finally, each constant b_i represents the level that the associated constraint function $g_i(\mathbf{X})$ must achieve and, because of the way optimization problems are typically written, is often called the **right-hand side** in the model. Thus, the solution vector $\mathbf{X}_0$ must optimize each of the objective functions $f_j(\mathbf{X})$ and simultaneously satisfy each constraint relationship. We now consider one simple problem illustrating the basic ideas.

EXAMPLE 1 *Determining a Production Schedule*

A carpenter makes tables and bookcases. He is trying to determine how many of each type of furniture he should make each week. The carpenter wishes to determine a weekly production schedule for tables and bookcases that maximizes his profits. It costs $5 and $7 to produce tables and bookcases, respectively. The revenues are estimated by the expressions

$$50x_1 - 0.2x_1^2, \text{ where } x_1 \text{ is the number of tables produced per week}$$

and

$$65x_2 - 0.3x_2^2, \text{ where } x_2 \text{ is the number of bookcases produced per week}$$

In this example, the problem is to decide how many tables and bookcases to make every week. Consequently, the decision variables are the quantities of tables and bookcases to be made per week. We assume this is a schedule, so noninteger values of tables and bookcases make sense. That is, we will permit x_1 and x_2 to take on any real values within their respective ranges. The objective function is a nonlinear expression representing the net weekly profit to be realized from selling the tables and bookcases. Profit is revenue minus costs. The profit function is

$$f(x_1, x_2) = 50x_1 - 0.2x_1^2 + 65x_2 - 0.3x_2^2 - 5x_1 - 7x_2$$

There are no constraints in this problem.

Let's consider a variation to the previous scenario. The carpenter realizes a net unit profit of $25 per table and $30 per bookcase. He is trying to determine how many of each piece of furniture he should make each week. He has up to 690 board-feet of lumber to devote weekly to the project and up to 120 hr of labor. He can use lumber and labor productively elsewhere if they are not used in the production of tables and bookcases. He estimates that it requires 20 board-feet of lumber and 5 hr of labor to complete a table and 30 board-feet of lumber and 4 hr of labor for a bookcase. Moreover, he has signed contracts to deliver four tables and two bookcases every week. The carpenter wishes to determine a weekly production schedule for tables and bookcases that maximizes his profits. The formulation yields

$$\text{Maximize } 25x_1 + 30x_2$$

subject to

$$
\begin{aligned}
20x_1 + 30x_2 &\leq 690 \quad \text{(lumber)} \\
5x_1 + 4x_2 &\leq 120 \quad \text{(labor)} \\
x_1 &\geq 4 \quad\;\; \text{(contract)} \\
x_2 &\geq 2 \quad\;\; \text{(contract)}
\end{aligned}
$$

Classifying Some Optimization Problems

There are various ways of classifying optimization problems. These classifications are not meant to be mutually exclusive but just to describe certain mathematical characteristics possessed by the problem under investigation. We now describe several of these classifications.

An optimization problem is said to be **unconstrained** if there are no constraints and **constrained** if one or more side conditions are present. The first production schedule problem described in Example 1 illustrates an unconstrained problem.

An optimization problem is said to be a **linear program** if it satisfies the following properties:

1. There is a unique objective function.

2. Whenever a decision variable appears in either the objective function or one of the constraint functions, it must appear only as a power term with an exponent of 1, possibly multiplied by a constant.

3. No term in the objective function or in any of the constraints can contain products of the decision variables.

4. The coefficients of the decision variables in the objective function and each constraint are constant.

5. The decision variables are permitted to assume fractional as well as integer values.

These properties ensure, among other things, that the effect of any decision variable is *proportional* to its value. Let's examine each property more closely.

Property 1 limits the problem to a single objective function. Problems with more than one objective function are called **multiobjective** or **goal programs**. Properties 2 and 3 are self-explanatory, and any optimization problem that fails to satisfy either one of them is said to be **nonlinear**. The first production schedule objective function had both decision variables as squared terms and thus violated Property 2. Property 4 is quite restrictive for many scenarios you might wish to model. Consider examining the amount of board-feet and labor required to make tables and bookcases. It might be possible to know exactly the number of board-feet and labor required to produce each item and incorporate these into constraints. Often, however, it is impossible to predict the required values precisely (consider trying to predict the market price of corn), or the coefficients represent average values with rather large deviations from the actual values occurring in practice. The coefficients may be time dependent as well. Time-dependent problems in a certain class are called **dynamic programs**. If the coefficients are not constant but instead are probabilistic in nature, the problem is classified as a **stochastic program**. Finally, if one or more of the decision variables are restricted to integer values (hence violating Property 5), the resulting problem is called an **integer program** (or a **mixed-integer program** if the integer restriction applies only to a subset of the decision variables). In the variation of the production scheduling problem, it makes sense to allow fractional numbers of tables and bookcases in determining a weekly schedule, because they can be completed during the following week. Classifying optimization problems is important because different solution techniques apply to distinct categories. For example, linear programming problems can be solved efficiently by the Simplex Method presented in Section 7.4.

Unconstrained Optimization Problems

A criterion considered for fitting a model to data points is minimizing the sum of absolute deviations. For the model $y = f(x)$, if $y(x_i)$ represents the function evaluated at $x = x_i$ and (x_i, y_i) denotes the corresponding data point for $i = 1, 2, \ldots, m$ points, then this criterion can be formulated as follows: Find the parameters of the model $y = f(x)$ to

$$\text{Minimize} \sum_{i=1}^{m} |y_i - y(x_i)|$$

This last condition illustrates an unconstrained optimization problem. Because the derivative of the function being minimized fails to be continuous (because of the presence of the absolute value), it is impossible to solve this problem with a straightforward application of the elementary calculus. A numerical solution based on pattern search is presented in Section 7.6.

Integer Optimization Programs

The requirements specified for a problem may restrict one or more of its decision variables to integer values. For example, in seeking the right mix of various-sized cars, vans, and trucks for a company's transportation fleet to minimize cost under some set of conditions, it would not make sense to determine a fractional part of a vehicle. Integer optimization problems also arise in coded problems, where binary (0 and 1) variables represent specific states, such as yes/no or on/off.

EXAMPLE 2 *Space Shuttle Cargo*

There are various items to be taken on a space shuttle. Unfortunately, there are restrictions on the allowable weight and volume capacities. Suppose there are m different items, each given some numerical value c_j and having weight w_j and volume v_j. (How might you determine c_j in an actual problem?) Suppose the goal is to maximize the value of the items that are to be taken without exceeding the weight limitation W or the volume limitation V. We can formulate this model

$$\text{Let } y_j = \begin{cases} 1, & \text{if item } j \text{ is taken (yes)} \\ 0, & \text{if item } j \text{ is not taken (no)} \end{cases}$$

Then the problem is

$$\text{Maximize } \sum_{j=1}^{m} c_j y_j$$

subject to

$$\sum_{j=1}^{m} v_j y_j \leq V$$

$$\sum_{j=1}^{m} w_j y_j \leq W$$

The ability to use binary variables such as y_j permits great flexibility in modeling. They can be used to represent yes/no decisions, such as whether to finance a given project in a capital budgeting problem, or to restrict variables to being "on" or "off." For example, the variable x can be restricted to the values 0 and a by using the binary variable y as a multiplier:

$$x = ay, \quad \text{where } y = 0 \text{ or } 1$$

Another illustration restricts x either to be in the interval (a, b) or to be zero by using the binary variable y:

$$ay < x < yb, \quad \text{where } y = 0 \text{ or } 1$$

EXAMPLE 3   *Approximation by a Piecewise Linear Function*

The power of using binary variables to represent intervals can be more fully appreciated when the object is to approximate a nonlinear function by a piecewise linear one. Specifically, suppose the nonlinear function in Figure 7.1a represents a cost function and we want to find its minimum value over the interval $0 \leq x \leq a_3$. If the function is particularly complicated, it could be approximated by a piecewise linear function such as that shown in Figure 7.1b. (The piecewise linear function might occur naturally in a problem, such as when different rates are charged for electrical use based on the amount of consumption.)

When we use the approximation suggested in Figure 7.1, our problem is to find the minimum of the function:

$$c(x) = \begin{cases} b_1 + k_1(x - 0) & \text{if } 0 \leq x \leq a_1 \\ b_2 + k_2(x - a_1) & \text{if } a_1 \leq x \leq a_2 \\ b_3 + k_3(x - a_2) & \text{if } a_2 \leq x \leq a_3 \end{cases}$$

Define the three new variables $x_1 = (x - 0)$, $x_2 = (x - a_1)$, and $x_3 = (x - a_2)$ for each of the three intervals, and use the binary variables y_1, y_2, and y_3 to restrict the x_i to the appropriate interval:

$$0 \leq x_1 \leq y_1 a_1$$
$$0 \leq x_2 \leq y_2(a_2 - a_1)$$
$$0 \leq x_3 \leq y_3(a_3 - a_2)$$

where y_1, y_2, and y_3 equal 0 or 1. Because we want exactly one x_i to be active at any time, we impose the following constraint:

$$y_1 + y_2 + y_3 = 1$$

Now that only one of the x_i is active at any one time, our objective function becomes

$$c(x) = y_1(b_1 + k_1 x_1) + y_2(b_2 + k_2 x_2) + y_3(b_3 + k_3 x_3)$$

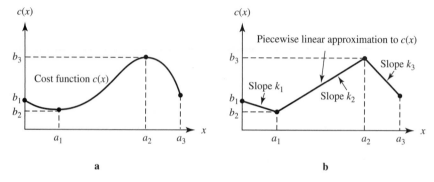

a   b

■ **Figure 7.1**
Using a piecewise linear function to approximate a nonlinear function

Observe that whenever $y_i = 0$, the variable $x_i = 0$ as well. Thus, products of the form $x_i y_i$ are redundant, and the objective function can be simplified to give the following model:

$$\text{Minimize } k_1 x_1 + k_2 x_2 + k_3 x_3 + y_1 b_1 + y_2 b_2 + y_3 b_3$$

subject to

$$0 \leq x_1 \leq y_1 a_1$$
$$0 \leq x_2 \leq y_2 (a_2 - a_1)$$
$$0 \leq x_3 \leq y_3 (a_3 - a_2)$$
$$y_1 + y_2 + y_3 = 1$$

where y_1, y_2, and y_3 equal 0 or 1.

The model in Example 3 is classified as a **mixed-integer programming** problem because only some decision variables are restricted to integer values. The Simplex Method does not solve integer or mixed-integer problems directly. There are many difficulties associated with solving such problems, which are addressed in more advanced courses. One methodology that has proved successful is to devise rules to find good feasible solutions quickly and then to use tests to show which of the remaining feasible solutions can be discarded. Unfortunately, some tests work well for certain classes of problems but fail miserably for others.

Multiobjective Programming: An Investment Problem

Consider the following problem: An investor has $40,000 to invest. She is considering investments in savings at 7%, municipal bonds at 9%, and stocks that have been consistently averaging 14%. Because there are varying degrees of risk involved in the various investments, the investor has listed the following goals for her portfolio:

1. A yearly return of at least $5000.
2. An investment of at least $10,000 in stocks.
3. The investment in stocks should not exceed the combined total in bonds and savings.
4. A liquid savings account between $5000 and $15,000.
5. The total investment must not exceed $40,000.

We can see from the portfolio that the investor has more than one objective. Unfortunately, as is often the case with real-world problems, not all goals can be achieved simultaneously. If the investment returning the lowest yield is set as low as possible (in this example, $5000 into savings), the best return possible without violating Goals 2–5 is obtained by investing $15,000 in bonds and $20,000 in stocks. However, this portfolio falls short of the desired yearly return of $5000. How are problems with more than one objective reconciled?

Let's begin by formulating each objective mathematically. Let x denote the investment in savings, y the investment in bonds, and z the investment in stocks. Then the goals are as follows:

$$\text{Goal 1.} \quad 0.07x + 0.09y + 0.14z \geq 5{,}000$$
$$\text{Goal 2.} \quad z \geq 10{,}000$$
$$\text{Goal 3.} \quad z \leq x + y$$
$$\text{Goal 4.} \quad 5{,}000 \leq x \leq 15{,}000$$
$$\text{Goal 5.} \quad x + y + z \leq 40{,}000$$

We have seen that the investor will have to compromise on one or more of her goals to find a feasible solution. Suppose the investor feels she must have a return of $5000, must invest at least $10,000 in stocks, and cannot spend more than $40,000 but is willing to compromise on Goals 3 and 4. However, she wants to find a solution that minimizes the amounts by which these two goals fail to be met. Let's formulate these new requirements mathematically and illustrate a method applicable to similar problems. Thus, let G_3 denote the amount by which Goal 3 fails to be satisfied, and G_4 denote the amount by which Goal 4 fails to be met. Then the model is

$$\text{Minimize } G_3 + G_4$$

subject to

$$0.07x + 0.09y + 0.14z \geq 5{,}000$$
$$z \geq 10{,}000$$
$$z - G_3 \leq x + y$$
$$5000 - G_4 \leq x \leq 15{,}000$$
$$x + y + z \leq 40{,}000$$

where x, y, and z are positive.

This last condition is included to ensure that negative investments are not allowed. This problem is now a linear program that can be solved by the Simplex Method (Section 7.4). If the investor believes that some goals are more important than others, the objective function can be weighted to emphasize those goals. Furthermore, a sensitivity analysis of the weights in the objective function will identify the breakpoints for the range over which various solutions are optimal. This process generates a number of solutions to be carefully considered by the investor before making her investments. Normally, this is the best that can be accomplished when qualitative decisions are to be made.

Dynamic Programming Problems

Often, the optimization model requires that decisions be made at various time intervals rather than all at once. In the 1950s, the American mathematician Richard Bellman developed a technique for optimizing such models in *stages* rather than simultaneously and referred to such problems as **dynamic programs**. Here's a problem scenario that lends itself to solution by dynamic programming techniques:

A rancher enters the cattle business with an initial herd of k cattle. He intends to retire and sell any remaining cattle after N years. Each year the rancher is faced with the decision of how many cattle to sell and how many to keep. If he sells, he estimates his profit to be p_i in year i. Also, the number of cattle kept in year i is expected to double in year $i + 1$.

Although this scenario neglects many factors that an analyst might consider in the real world (name several), we can see that the cattle rancher is faced with a trade-off decision each year: take a profit or build for the future.

In the next several sections we focus our attention on solving linear programming problems, first geometrically and then by the Simplex Method.

7.1   PROBLEMS

Use the model-building process described in Chapter 2 to analyze the following scenarios. After identifying the problem to be solved using the process, you may find it helpful to answer the following questions in words before formulating the optimization model.

a. Identify the decision variables: What decision is to be made?

b. Formulate the objective function: How do these decisions affect the objective?

c. Formulate the constraint set: What constraints must be satisfied? Be sure to consider whether negative values of the decision variables are allowed by the problem, and ensure they are so constrained if required.

After constructing the model, check the assumptions for a linear program and compare the form of the model to the examples in this section. Try to determine which method of optimization may be applied to obtain a solution.

1. *Resource Allocation*—You have just become the manager of a plant producing plastic products. Although the plant operation involves many products and supplies, you are interested in only three of the products: (1) a vinyl–asbestos floor covering, the output of which is measured in boxed lots, each covering a certain area; (2) a pure vinyl counter top, measured in linear yards; and (3) a vinyl–asbestos wall tile, measured in squares, each covering 100 ft^2.

   Of the many resources needed to produce these plastic products, you have identified four: vinyl, asbestos, labor, and time on a trimming machine. A recent inventory shows that on any given day you have 1500 lb of vinyl and 200 lb of asbestos available for use. Additionally, after talking to your shop foreman and to various labor leaders, you realize that you have 3 person-days of labor available for use per day and that your trimming machine is available for 1 machine-day on any given day. The following table indicates the amount of each of the four resources required to produce a unit of the three desired products, where the units are 1 box of floor cover, 1 yard of counter top, and 1 square of wall tiles. Available resources are also tabulated.

	Vinyl (lb)	Asbestos (lb)	Labor (person-days)	Machine (machine-days)	Profit
Floor cover (per box)	30	3	0.02	0.01	$0.8
Countertop (per yard)	20	0	0.1	0.05	5.0
Wall tile (per square)	50	5	0.2	0.05	5.5
Available (per day)	1500	200	3	1	—

Formulate a mathematical model to help determine how to allocate resources to maximize profits.

2. *Nutritional Requirements*—A rancher has determined that the minimum weekly nutritional requirements for an average-sized horse include 40 lb of protein, 20 lb of carbohydrates, and 45 lb of roughage. These are obtained from the following sources in varying amounts at the prices indicated:

	Protein (lb)	Carbohydrates (lb)	Roughage (lb)	Cost
Hay (per bale)	0.5	2.0	5.0	$1.80
Oats (per sack)	1.0	4.0	2.0	3.50
Feeding blocks (per block)	2.0	0.5	1.0	0.40
High-protein concentrate (per sack)	6.0	1.0	2.5	1.00
Requirements per horse (per week)	40.0	20.0	45.0	

Formulate a mathematical model to determine how to meet the minimum nutritional requirements at minimum cost.

3. *Scheduling Production*—A manufacturer of an industrial product has to meet the following shipping schedule:

Month	Required shipment (units)
January	10,000
February	40,000
March	20,000

The monthly production capacity is 30,000 units and the production cost per unit is $10. Because the company does not warehouse, the service of a storage company is utilized whenever needed. The storage company determines its monthly bill by multiplying the number of units in storage on the last day of the month by $3. On the first day of January the company does not have any beginning inventory, and it does not want to have any ending inventory at the end of March. Formulate a mathematical model to assist in minimizing the sum of the production and storage costs for the 3-month period.

How does the formulation change if the production cost is $10x + 10$ dollars, where x is the number of items produced?

4. *Mixing Nuts*—A candy store sells three different assortments of mixed nuts, each assortment containing varying amounts of almonds, pecans, cashews, and walnuts. To preserve the store's reputation for quality, certain maximum and minimum percentages of the various nuts are required for each type of assortment, as shown in the following table:

Nut assortment	Requirements	Selling price per pound
Regular	Not more than 20% cashews Not less than 40% walnuts Not more than 25% pecans No restriction on almonds	$0.89
Deluxe	Not more than 35% cashews Not less than 25% almonds No restriction on walnuts and pecans	1.10
Blue Ribbon	Between 30% and 50% cashews Not less than 30% almonds No restriction on walnuts and pecans	1.80

The following table gives the cost per pound and the maximum quantity of each type of nut available from the store's supplier each week.

Nut type	Cost per pound	Maximum quantity available per week (lb)
Almonds	$0.45	2000
Pecans	0.55	4000
Cashews	0.70	5000
Walnuts	0.50	3000

The store would like to determine the exact amounts of almonds, pecans, cashews, and walnuts that should go into each weekly assortment to maximize its weekly profit. Formulate a mathematical model that will assist the store management in solving the mixing problem. *Hint:* How many decisions need to be made? For example, do you need to distinguish between the cashews in the regular mix and the cashews in the deluxe mix?

5. *Producing Electronic Equipment*—An electronics firm is producing three lines of products for sale to the government: transistors, micromodules, and circuit assemblies. The firm has four physical processing areas designated as follows: transistor production, circuit printing and assembly, transistor and module quality control, and circuit assembly test and packing.

The various production requirements are as follows: Production of one transistor requires 0.1 standard hour of transistor production area capacity, 0.5 standard hour of transistor quality control area capacity, and $0.70 in direct costs. Production of micro-modules requires 0.4 standard hour of the quality control area capacity, three transistors, and $0.50 in direct costs. Production of one circuit assembly requires 0.1 standard hour of the capacity of the circuit printing area, 0.5 standard hour of the capacity of the test and packing area, one transistor, three micromodules, and $2.00 in direct costs.

Suppose that the three products (transistors, micromodules, and circuit assemblies) may be sold in unlimited quantities at prices of $2, $8, and $25 each, respectively. There are 200 hours of production time open in each of the four process areas in the coming month. Formulate a mathematical model to help determine the production that will produce the highest revenue for the firm.

6. *Purchasing Various Trucks*—A truck company has allocated $800,000 for the purchase of new vehicles and is considering three types. Vehicle A has a 10-ton payload capacity and is expected to average 45 mph; it costs $26,000. Vehicle B has a 20-ton payload capacity and is expected to average 40 mph; it costs $36,000. Vehicle C is a modified form of B and carries sleeping quarters for one driver. This modification reduces the capacity to an 18-ton payload and raises the cost to $42,000, but its operating speed is still expected to average 40 mph.

Vehicle A requires a crew of one driver and, if driven on three shifts per day, could be operated for an average of 18 hr per day. Vehicles B and C must have crews of two drivers each to meet local legal requirements. Vehicle B could be driven an average of 18 hr per day with three shifts, and vehicle C could average 21 hr per day with three shifts. The company has 150 drivers available each day to make up crews and will not be able to hire additional trained crews in the near future. The local labor union prohibits any driver from working more than one shift per day. Also, maintenance facilities are such that the total number of vehicles must not exceed 30. Formulate a mathematical model to help determine the number of each type of vehicle the company should purchase to maximize its shipping capacity in ton-miles per day.

7. *A Farming Problem*—A farm family owns 100 acres of land and has $25,000 in funds available for investment. Its members can produce a total of 3500 work-hours worth of labor during the winter months (mid-September to mid-May) and 4000 work-hours during the summer. If any of these work-hours are not needed, younger members of the family will use them to work on a neighboring farm for $4.80 per hour during the winter and $5.10 per hour during the summer.

Cash income may be obtained from three crops (soybeans, corn, and oats) and two types of livestock (dairy cows and laying hens). No investment funds are needed for the crops. However, each cow requires an initial investment outlay of $400, and each hen requires $3. Each cow requires 1.5 acres of land and 100 work-hours of work during the winter months and another 50 work-hours during the summer. Each cow produces a net annual cash income of $450 for the family. The corresponding figures for the hen are as follows: no acreage, 0.6 work-hour during the winter, 0.3 more work-hour in the summer, and an annual net cash income of $3.50. The chicken house accommodates a maximum of 3000 hens, and the size of the barn limits the cow herd to a maximum of 32 head.

Estimated work-hours and income per acre planted in each of the three crops are as shown in the following table:

Crop	Winter work-hours	Summer work-hours	Net annual cash income (per acre)
Soybean	20	30	$175.00
Corn	35	75	300.00
Oats	10	40	120.00

Formulate a mathematical model to assist in determining how much acreage should be planted in each of the crops and how many cows and hens should be kept to *maximize net cash income.*

7.1 PROJECTS

For Projects 1–5, complete the requirements in the referenced UMAP module or monograph. (See enclosed CD for UMAP modules.)

1. "Unconstrained Optimization," by Joan R. Hundhausen and Robert A. Walsh, UMAP 522. This unit introduces gradient search procedures with examples and applications. Acquaintance with elementary partial differentiation, chain rules, Taylor series, gradients, and vector dot products is required.

2. "Calculus of Variations with Applications in Mechanics," by Carroll O. Wilde, UMAP 468. This module provides a brief introduction to finding functions that yield the maximum or minimum value of certain definite integral forms, with applications in mechanics. Students learn Euler's equations for some definite integral forms and learn Hamilton's principle and its application to conservative dynamical systems. The basic physics of kinetic and potential energy, the multivariate chain rules, and ordinary differential equations are required.

3. *The High Cost of Clean Water: Models for Water Quality Management*, by Edward Beltrami, UMAP Expository Monograph. To cope with the severe wastewater disposal problems caused by increases in the nation's population and industrial activity, the U.S. Environmental Protection Agency (EPA) has fostered the development of regional wastewater management plans. This monograph discusses the EPA plan developed for Long Island and formulates a model that allows for the articulation of the trade-offs between cost and water quality. The mathematics involves partial differential equations and mixed-integer linear programming.

4. "Geometric Programming," by Robert E. D. Woolsey, UMAP 737. This unit provides some alternative optimization formulations, including geometric programming. Familiarity with basic differential calculus is required.

5. "Municipal Recycling: Location and Optimality," by Jannett Highfill and Michael McAsey, *UMAP Journal* Vol. 15(1), 1994. This article considers optimization in municipal recycling. Read the article and prepare a 10-min classroom presentation.

7.2   Linear Programming 1: Geometric Solutions

Consider using the Chebyshev criterion to fit the model $y = cx$ to the following data set:

x	1	2	3
y	2	5	8

The optimization problem that determines the parameter c to minimize the largest absolute deviation $r_i = |y_i - y(x_i)|$ (residual or error) is the linear program

$$\text{Minimize } r$$

subject to

$$
\begin{array}{ll}
r - (2 - c) \geq 0 & \text{(constraint 1)} \\
r + (2 - c) \geq 0 & \text{(constraint 2)} \\
r - (5 - 2c) \geq 0 & \text{(constraint 3)} \\
r + (5 - 2c) \geq 0 & \text{(constraint 4)} \\
r - (8 - 3c) \geq 0 & \text{(constraint 5)} \\
r + (8 - 3c) \geq 0 & \text{(constraint 6)}
\end{array}
\right\} \tag{7.2}
$$

In this section we solve this problem geometrically.

Interpreting a Linear Program Geometrically

Linear programs can include a set of constraints that are linear equations or linear inequalities. Of course, in the case of two decision variables, an equality requires that solutions to the linear program lie precisely on the line representing the equality. What about inequalities? To gain some insight, consider the constraints

$$x_1 + 2x_2 \leq 4 \tag{7.3}$$
$$x_1, x_2 \geq 0$$

The **nonnegativity** constraints $x_1, x_2 \geq 0$ mean that possible solutions lie in the first quadrant. The inequality $x_1 + 2x_2 \leq 4$ divides the first quadrant into two regions. The *feasible region* is the half-space in which the constraint is satisfied. The feasible region can be found by graphing the equation $x_1 + 2x_2 = 4$ and determining which half-plane is feasible, as shown in Figure 7.2.

If the feasible half-plane fails to be obvious, choose a convenient point (such as the origin) and substitute it into the constraint to determine whether it is satisfied. If it is, then all points on the same side of the line as this point will also satisfy the constraint.

A linear program has the important property that the points satisfying the constraints form a *convex set*. A set is *convex* if for every pair of points in the set, the line segment joining them lies wholly in the set. The set depicted in Figure 7.3a fails to be convex, whereas the set in Figure 7.3b is convex.

An extreme point (corner point) of a convex set is any boundary point in the convex set that is the unique intersection point of two of the (straight-line) boundary segments.

■ Figure 7.2

The feasible region for the constraints $x_1 + 2x_2 \leq 4$, $x_1, x_2 \geq 0$

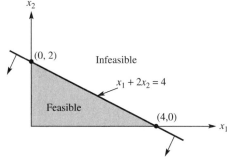

■ Figure 7.3

The set shown in **a** is not convex, whereas the set shown in **b** is convex.

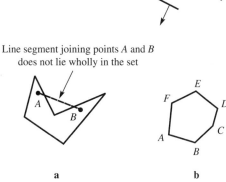

In Figure 7.3b, points A–F are extreme points. Let's now find the feasible region and the optimal solution for the carpenter's problem formulated in Example 1 in Section 7.1.

EXAMPLE 1 *The Carpenter's Problem*

Consider a version of the carpenter's problem from Section 7.1. The carpenter realizes a net unit profit of $25 per table and $30 per bookcase. He is trying to determine how many tables (x_1) and how many bookcases (x_2) he should make each week. He has up to 690 board-feet of lumber and up to 120 hours of labor to devote weekly to the project. The lumber and labor can be used productively elsewhere if not used in the production of tables and bookcases. He estimates that it requires 20 board-feet of lumber and 5 hr of labor to complete a table and 30 board-feet of lumber and 4 hr of labor for a bookcase. This formulations yields

$$\text{Maximize } 25x_1 + 30x_2$$

subject to

$$
\begin{aligned}
20x_1 + 30x_2 &\leq 690 &&\text{(lumber)} \\
5x_1 + 4x_2 &\leq 120 &&\text{(labor)} \\
x_1, x_2 &\geq 0 &&\text{(nonnegativity)}
\end{aligned}
$$

The convex set for the constraints in the carpenter's problem is graphed and given by the polygon region $ABCD$ in Figure 7.4. Note that there are six intersection points of the constraints, but only four of these points (namely, A–D) satisfy all of the constraints and hence belong to the convex set. The points A–D are the **extreme points** of the polygon.

■ **Figure 7.4**

The set of points satisfying the constraints of the carpenter's problem form a convex set.

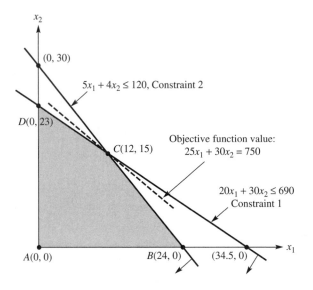

If an optimal solution to a linear program exists, it must occur among the extreme points of the convex set formed by the set of constraints. The values of the objective function (profit for the carpenter's problem) at the extreme points are

Extreme point	Objective function value
A (0, 0)	\$0
B (24, 0)	600
C (12, 15)	750
D (0, 23)	690

Thus, the carpenter should make 12 tables and 15 bookcases each week to earn a maximum weekly profit of \$750. We provide further geometric evidence later in this section that extreme point C is optimal.   ■ ■ ■

Before considering a second example, let's summarize the ideas presented thus far. The constraint set to a linear program is a convex set, which generally contains an infinite number of feasible points to the linear program. If an optimal solution to the linear program exists, it must be taken on at one or more of the extreme points. Thus, to find an optimal solution, we choose from among all the extreme points the one with the best value for the objective function.

EXAMPLE 2   *A Data-Fitting Problem*

Let's now solve the linear program represented by Equation (7.2). Given the model $y = cx$ and the data set

x	1	2	3
y	2	5	8

■ Figure 7.5

The feasible region for fitting $y = cx$ to a collection of data

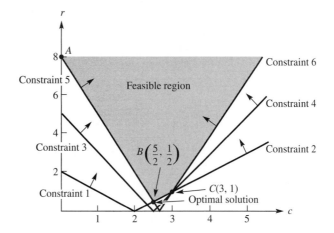

we wish to find a value for c such that the resulting largest absolute deviation r is as small as possible. In Figure 7.5 we graph the set of six constraints

$$r - (2 - c) \geq 0 \qquad \text{(constraint 1)}$$
$$r + (2 - c) \geq 0 \qquad \text{(constraint 2)}$$
$$r - (5 - 2c) \geq 0 \qquad \text{(constraint 3)}$$
$$r + (5 - 2c) \geq 0 \qquad \text{(constraint 4)}$$
$$r - (8 - 3c) \geq 0 \qquad \text{(constraint 5)}$$
$$r + (8 - 3c) \geq 0 \qquad \text{(constraint 6)}$$

by first graphing the equations

$$r - (2 - c) = 0 \qquad \text{(constraint 1 boundary)}$$
$$r + (2 - c) = 0 \qquad \text{(constraint 2 boundary)}$$
$$r - (5 - 2c) = 0 \qquad \text{(constraint 3 boundary)}$$
$$r + (5 - 2c) = 0 \qquad \text{(constraint 4 boundary)}$$
$$r - (8 - 3c) = 0 \qquad \text{(constraint 5 boundary)}$$
$$r + (8 - 3c) = 0 \qquad \text{(constraint 6 boundary)}$$

We note that constraints 1, 3, and 5 are satisfied above and to the right of the graph of their boundary equations. Similarly, constraints 2, 4, and 6 are satisfied above and to the left of their boundary equations.

The intersection of all the feasible regions for constraints 1–6 forms a convex set in the c, r plane, with extreme points labeled A–C in Figure 7.5. The point A is the intersection of constraint 5 and the r-axis: $r - (8 - 3c) = 0$ and $c = 0$, or $A = (0, 8)$. Similarly, B is the intersection of constraints 5 and 2:

$$r - (8 - 3c) = 0 \quad \text{or} \quad r + 3c = 8$$
$$r + (2 - c) = 0 \quad \text{or} \quad r - c = -2$$

yielding $c = \frac{5}{2}$ and $r = \frac{1}{2}$, or $B = (\frac{5}{2}, \frac{1}{2})$. Finally, C is the intersection of constraints 2 and 4 yielding $C = (3, 1)$. Note that the set is *unbounded*. (We discuss unbounded convex sets later.) If an optimal solution to the problem exists, at least one extreme point must take on the optimal solution. We now evaluate the objective function $f(r) = r$ at each of the three extreme points.

Extreme point	Objective function value
(c, r)	$f(r) = r$
A	8
B	$\frac{1}{2}$
C	1

The extreme point with the smallest value of r is the extreme point B with coordinates $(\frac{5}{2}, \frac{1}{2})$. Thus, $c = \frac{5}{2}$ is the optimal value of c. No other value of c will result in a largest absolute deviation as small as $|r_{max}| = \frac{1}{2}$. ■ ■ ■

Model Interpretation

Let's interpret the optimal solution for the data-fitting problem in Example 2. Resolving the linear program, we obtained a value of $c = \frac{5}{2}$ corresponding to the model $y = \frac{5}{2}x$. Furthermore, the objective function value $r = \frac{1}{2}$ should correspond to the largest deviation resulting from the fit. Let's check to see if that is true.

The data points and the model $y = \frac{5}{2}x$ are plotted in Figure 7.6. Note that a largest deviation of $r_i = \frac{1}{2}$ occurs for both the first and third data points. Fix one end of a ruler at the origin. Now rotate the ruler to convince yourself geometrically that no other line passing through the origin can yield a smaller largest absolute deviation. Thus, the model $y = \frac{5}{2}x$ is optimal by the Chebyshev criterion.

■ **Figure 7.6**

The line $y = (5/2)x$ results in a largest absolute deviation $r_{max} = \frac{1}{2}$, the smallest possible r_{max}.

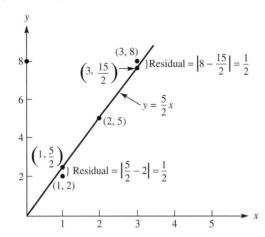

Empty and Unbounded Feasible Regions

We have been careful to say that if an optimal solution to the linear program exists, at least one of the extreme points must take on the optimal value for the objective function. When does an optimal solution fail to exist? Moreover, when does more than one optimal solution exist?

If the feasible region is empty, no feasible solution can exist. For example, given the constraints

$$x_1 \leq 3$$

and

$$x_1 \geq 5$$

there is no value of x_1 that satisfies both of them. We say that such constraint sets are *inconsistent*.

There is another reason an optimal solution may fail to exist. Consider Figure 7.5 and the constraint set for the data-fitting problem in which we noted that the feasible region is *unbounded* (in the sense that either x_1 or x_2 can become arbitrarily large). Then it would be impossible to

$$\text{Maximize } x_1 + x_2$$

over the feasible region because x_1 and x_2 can take on arbitrarily large values. Note, however, that even though the feasible region is unbounded, an optimal solution *does* exist for the objective function we considered in Example 2, so it is not *necessary* for the feasible region to be bounded for an optimal solution to exist.

Level Curves of the Objective Function

Consider again the carpenter's problem. The objective function is $25x_1 + 30x_2$ and in Figure 7.7 we plot the lines

$$25x_1 + 30x_2 = 650$$
$$25x_1 + 30x_2 = 750$$
$$25x_1 + 30x_2 = 850$$

in the first quadrant.

■ **Figure 7.7**

The level curves of the objective function f are parallel line segments in the first quadrant; the objective function either increases or decreases as we move in a direction perpendicular to the level curves.

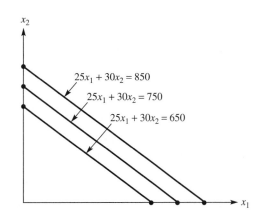

Note that the objective function has constant values along these line segments. The line segments are called **level curves** of the objective function. As we move in a direction perpendicular to these line segments, the objective function either increases or decreases. Now superimpose the constraint set from the carpenter's problem

$$20x_1 + 30x_2 \leq 690 \quad \text{(lumber)}$$
$$5x_1 + 4x_2 \leq 120 \quad \text{(labor)}$$
$$x_1, x_2 \geq 0 \quad \text{(nonnegativity)}$$

onto these level curves (Figure 7.8). Notice that the level curve with value 750 is the one that intersects the feasible region exactly once at the extreme point $C(12, 15)$.

■ **Figure 7.8**

The level curve $25x_1 + 30x_2 = 750$ is tangent to the feasible region at extreme point C.

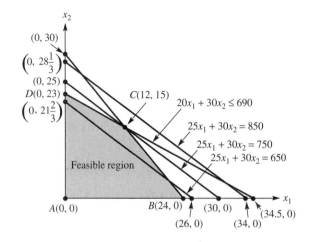

Can there be more than one optimal solution? Consider the following slight variation of the carpenter's problem in which the labor constraint has been changed:

$$\text{Maximize } 25x_1 + 30x_2$$

subject to

$$20x_1 + 30x_2 \leq 690 \quad \text{(lumber)}$$
$$5x_1 + 6x_2 \leq 150 \quad \text{(labor)}$$
$$x_1, x_2 \geq 0 \quad \text{(nonnegativity)}$$

The constraint set and the level curve $25x_1 + 30x_2 = 750$ are graphed in Figure 7.9. Notice that the level curve and boundary line for the labor constraint coincide. Thus, both extreme points B and C have the same objective function value of 750, which is optimal. In fact, the entire line segment BC coincides with the level curve $25x_1 + 30x_2 = 750$. Thus, there are infinitely many optimal solutions to the linear program, all along line segment BC.

In Figure 7.10 we summarize the general two-dimensional case for optimizing a linear function on a convex set. The figure shows a typical convex set together with the level curves of a linear objective function. Figure 7.10 provides geometric intuition for the following fundamental theorem of linear programming.

■ Figure 7.9

The line segment BC coincides with the level curve $25x_1 + 30x_2 = 750$; every point between extreme points C and B, as well as extreme points C and B, is an optimal solution.

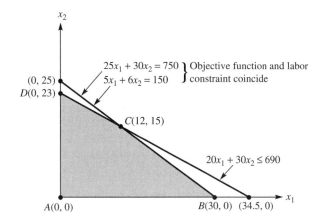

■ Figure 7.10

A linear function assumes its maximum and minimum values on a nonempty and bounded convex set at an extreme point.

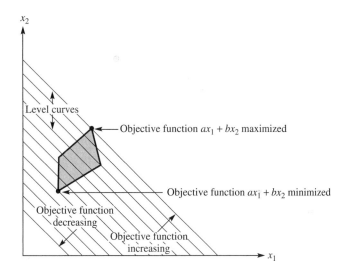

Theorem 1

Suppose the feasible region of a linear program is a nonempty and bounded convex set. Then the objective function must attain both a maximum and a minimum value occurring at extreme points of the region. If the feasible region is unbounded, the objective function need not assume its optimal values. If either a maximum or a minimum does exist, it must occur at one of the extreme points.

The power of this theorem is that it guarantees an optimal solution to a linear program from among the extreme points of a bounded nonempty convex set.

7.2   PROBLEMS

1. Consider a company that carves wooden soldiers. The company specializes in two main types: Confederate and Union soldiers. The profit for each is $28 and $30, respectively. It requires 2 units of lumber, 4 hr of carpentry, and 2 hr of finishing to complete a Confederate soldier. It requires 3 units of lumber, 3.5 hr of carpentry, and 3 hr of finishing to complete a Union soldier. Each week the company has 100 units of lumber delivered. There are 120 hr of carpenter machine time available and 90 hr of finishing time available. Determine the number of each wooden soldier to produce to maximize weekly profits.

2. A local company restores cars and trucks for resale. Each vehicle must be processed in the refinishing/paint shop and the machine/body shop. Each car (on average) con-tributes $3000 to profit, and each truck contributes (on average) $2000 to profit. The refinishing/paint shop has 2400 work-hours available and the machine/body shop has 2500 work-hours available. A car requires 50 work-hours in the machine/body shop and 40 work-hours in the refinishing/paint shop, whereas a truck requires 50 work-hours in the machine/body shop and 60 work-hours in the refinishing/paint shop. Use graphical linear programming to determine a daily production schedule that will maximize the company's profits.

3. A Montana farmer owns 45 acres of land. She is planning to plant each acre with wheat or corn. Each acre of wheat yields $200 in profits, whereas each acre of corn yields $300 in profits. The labor and fertilizer requirements for each are provided here. The farmer has 100 workers and 120 tons of fertilizer available. Determine how many acres of wheat and corn need to be planted to maximize profits.

	Wheat	Corn
Labor (workers)	3	2
Fertilizer (tons)	2	4

Solve Problems 4–7 using graphical analysis.

4. Maximize $x + y$
   subject to

$$x + y \leq 6$$
$$3x - y \leq 9$$
$$x, y \geq 0$$

5. Minimize $x + y$
   subject to

$$x + y \geq 6$$
$$3x - y \geq 9$$
$$x, y \geq 0$$

6. Maximize $10x + 35y$
   subject to

$$8x + 6y \leq 48 \quad \text{(board-feet of lumber)}$$
$$4x + y \leq 20 \quad \text{(hours of carpentry)}$$
$$y \geq 5 \quad \text{(demand)}$$
$$x, y \geq 0 \quad \text{(nonnegativity)}$$

7. Minimize $5x + 7y$
   subject to

$$2x + 3y \geq 6$$
$$3x - y \leq 15$$
$$-x + y \leq 4$$
$$2x + 5y \leq 27$$
$$x \geq 0$$
$$y \geq 0$$

For Problems 8–12, find both the maximum solution and the minimum solution using graphical analysis. Assume $x \geq 0$ and $y \geq 0$ for each problem.

8. Optimize $2x + 3y$
   subject to

$$2x + 3y \geq 6$$
$$3x - y \leq 15$$
$$-x + y \leq 4$$
$$2x + 5y \leq 27$$

9. Optimize $6x + 4y$
   subject to

$$-x + y \leq 12$$
$$x + y \leq 24$$
$$2x + 5y \leq 80$$

10. Optimize $6x + 5y$
   subject to

$$x + y \geq 6$$
$$2x + y \geq 9$$

11. Optimize $x - y$
   subject to

$$x + y \geq 6$$
$$2x + y \geq 9$$

12. Optimize $5x + 3y$
      subject to

$$1.2x + 0.6y \le 24$$
$$2x + 1.5y \le 80$$

13. Fit the model to the data using Chebyshev's criterion to minimize the largest deviation.

a. $y = cx$

y	11	25	54	90
x	5	10	20	30

b. $y = cx^2$

y	10	90	250	495
x	1	3	5	7

7.3 Linear Programming II: Algebraic Solutions

The graphical solution to the carpenter's problem suggests a rudimentary procedure for finding an optimal solution to a linear program with a nonempty and bounded feasible region:

1. Find all intersection points of the constraints.

2. Determine which intersection points, if any, are feasible to obtain the extreme points.

3. Evaluate the objective function at each extreme point.

4. Choose the extreme point(s) with the largest (or smallest) value for the objective function.

To implement this procedure algebraically, we must characterize the intersection points and the extreme points.

The convex set depicted in Figure 7.11 consists of three linear constraints (plus the two nonnegativity constraints). The nonnegative variables y_1, y_2, and y_3 indicated in the figure measure the degree by which a point satisfies the constraints 1, 2, and 3, respectively. The variable y_i is added to the left side of inequality constraint i to convert it to an equality. Thus, $y_2 = 0$ characterizes those points that lie precisely on constraint 2, and a negative value for y_2 indicates the violation of constraint 2. Likewise, the decision variables x_1 and x_2 are constrained to nonnegative values. Thus, the values of the decision variables x_1 and x_2 measure the degree of satisfaction of the nonnegativity constraints, $x_1 \ge 0$ and $x_2 \ge 0$. Note that along the x_1-axis, the decision variable x_2 is 0. Now consider the values for the entire set of variables $\{x_1, x_2, y_1, y_2, y_3\}$. If two of the variables simultaneously have the value 0, then we have characterized an **intersection point** in the x_1x_2-plane. All (possible) intersection points can be determined systematically by setting all possible distinguishable pairs of the five variables to zero and solving for the remaining three dependent variables. If a solution

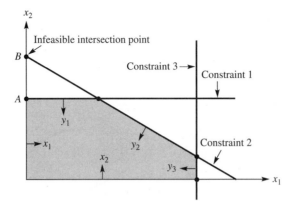

■ Figure 7.11

The variables x_1, x_2, y_1, y_2, and y_3 measure the satisfaction of each of the constraints; intersection point *A* is characterized by $y_1 = x_1 = 0$; intersection point *B* is not feasible because y_1 is negative; the intersection points surrounding the shaded region are all feasible because none of the five variables is negative there.

to the resulting system of equations exists, then it must be an intersection point, which may or may not be a **feasible solution**. A negative value for any of the five variables indicates that a constraint is not satisfied. Such an intersection point would be **infeasible**. For example, the intersection point *B*, where $y_2 = 0$ and $x_1 = 0$, gives a negative value for y_1 and hence is not feasible. Other pairs of variables such as x_1 and y_3, cannot simultaneously be zero because they represent constraints that are parallel lines. Let's illustrate the procedure by solving the carpenter's problem alg

EXAMPLE 1   *Solving the Carpenter's Pr*

The carpenter's model is

subject to

We convert each of "slack" variables y_1 and the problem becomes

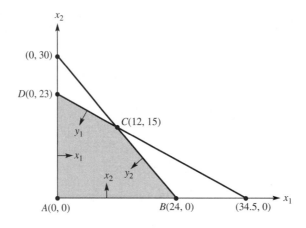

■ **Figure 7.12**

The variables $\{x_1, x_2, y_1, y_2\}$ measure the satisfaction of each constraint; an intersection point is characterized by setting two of the variables to zero.

subject to

$$20x_1 + 30x_2 + y_1 = 690$$
$$5x_1 + 4x_2 + y_2 = 120$$
$$x_1, x_2, y_1, y_2 \geq 0$$

We now consider the entire set of four variables $\{x_1, x_2, y_1, y_2\}$, which are interpreted geometrically in Figure 7.12. To determine a possible intersection point in the $x_1 x_2$-plane, assign two of the four variables the value zero. There are $\frac{4!}{2!\,2!} = 6$ possible intersection points to consider in this way (four variables taken two at a time). Let's begin by assigning the variables x_1 and x_2 the value zero, resulting in the following set of equations:

$$y_1 = 690$$
$$y_2 = 120$$

which is a feasible intersection point $A(0, 0)$ because all four variables are nonnegative.

For the second intersection point we choose the variables x_1 and y_1 and set them to zero, resulting in the system

$$30x_2 = 690$$
$$4x_2 + y_2 = 120$$

that has solution $x_2 = 23$ and $y_2 = 28$, which is also a feasible intersection point $D(0, 23)$.

For the third intersection point we choose x_1 and y_2 and set them to zero, yielding the

$$30x_2 + y_1 = 690$$
$$4x_2 = 120$$

nd $y_1 = -210$. Thus, the first constraint is violated by 210 units, on point $(0, 30)$ is infeasible.

ing y_1 and y_2 and setting them to zero gives $x_1 = 12$ and rsection point $C(12, 15)$, which is feasible. Our fifth choice is to choose the variables x_2 and y_1 and set them to zero, giving values of $x_1 = 34.5$ and $y_2 = -52.5$, so the second constraint is not satisfied. Thus, the intersection point $(34.5, 0)$ is infeasible.

Finally we determine the sixth intersection point by setting the variables x_2 and y_2 to zero to determine $x_1 = 24$ and $y_1 = 210$; therefore, the intersection point $B(24, 0)$ is feasible.

In summary, of the six possible intersection points in the x_1x_2-plane, four were found to be feasible. For the four we find the value of the objective function by substitution:

Extreme point	Value of objective function
$A(0, 0)$	$0
$D(0, 23)$	690
$C(12, 15)$	750
$B(24, 0)$	600

Our procedure determines that the optimal solution to maximize the profit is $x_1 = 12$ and $x_2 = 15$. That is, the carpenter should make 12 tables and 15 bookcases for a maximum profit of $750. ■ ■ ■

Computational Complexity: Intersection Point Enumeration

We now generalize the procedure presented in the carpenter example. Suppose we have a linear program with m nonnegative decision variables and n constraints, where each constraint is an inequality of the form $\leq$. First, convert each inequality to an equation by adding a nonnegative "slack" variable y_i to the ith constraint. We now have a total of $m + n$ nonnegative variables. To determine an intersection point, choose m of the variables (because we have m decision variables) and set them to zero. There are $\frac{(m+n)!}{m!\,n!}$ possible choices to consider. Obviously, as the size of the linear program increases (in terms of the numbers of decision variables and constraints), this technique of enumerating all possible intersection points becomes unwieldy, even for powerful computers. How can we improve the procedure?

Note that we enumerated some intersection points in the carpenter example that turned out to be infeasible. Is there a way to quickly identify that a possible intersection point is infeasible? Moreover, if we have found an extreme point (i.e., a feasible intersection point) and know the corresponding value of the objective function, can we quickly determine if another proposed extreme point will improve the value of the objective function? In conclusion, we desire a procedure that does not enumerate infeasible intersection points and that enumerates only those extreme points that improve the value of the objective function for the best solution found so far in the search. We study one such procedure in the next section.

7.3 PROBLEMS

1–7. Using the method of this section, resolve Problems 1–6 and 13 in Section 7.2.

8. How many possible intersection points are there in the following cases?

   a. 2 decision variables and $5 \leq$ inequalities

b. 2 decision variables and $10 \leq$ inequalities

c. 5 decision variables and $12 \leq$ inequalities

d. 25 decision variables and $50 \leq$ inequalities

e. 2000 decision variables and $5000 \leq$ inequalities

7.4 Linear Programming III: The Simplex Method

So far we have learned to find an optimal extreme point by searching among all possible intersection points associated with the decision and slack variables. Can we reduce the number of intersection points we actually consider in our search? Certainly, once we find an initial feasible intersection point, we need not consider a potential intersection point that fails to improve the value of the objective function. Can we test the optimality of our current solution against other possible intersection points? Even if an intersection point promises to be more optimal than the current extreme point, it is of no interest if it violates one or more of the constraints. Is there a test to determine whether a proposed intersection point is feasible? The **Simplex Method**, developed by George Dantzig, incorporates both *optimality* and *feasibility* tests to find the optimal solution(s) to a linear program (if one exists).

An **optimality test** shows whether or not an intersection point corresponds to a value of the objective function better than the best value found so far.

A **feasibility test** determines whether the proposed intersection point is feasible.

To implement the Simplex Method we first separate the decision and slack variables into two nonoverlapping sets that we call the **independent** and **dependent** sets. For the particular linear programs we consider, the original independent set will consist of the decision variables, and the slack variables will belong to the dependent set.

Steps of the Simplex Method

1. **Tableau Format:** Place the linear program in Tableau Format, as explained later.

2. **Initial Extreme Point:** The Simplex Method begins with a known extreme point, usually the origin $(0, 0)$.

3. **Optimality Test:** Determine whether an adjacent intersection point improves the value of the objective function. If not, the current extreme point is optimal. If an improvement is possible, the optimality test determines which variable currently in the independent set (having value zero) should *enter* the dependent set and become nonzero.

4. **Feasibility Test:** To find a new intersection point, one of the variables in the dependent set must *exit* to allow the entering variable from Step 3 to become dependent. The feasibility test determines which current dependent variable to choose for exiting, ensuring feasibility.

5. **Pivot:** Form a new, equivalent system of equations by eliminating the new dependent variable from the equations that do not contain the variable that exited in Step 4. Then set the new independent variables to zero in the new system to find the values of the new dependent variables, thereby determining an intersection point.

6. **Repeat Steps 3–5** until an optimal extreme point is found.

■ Figure 7.13

The set of points satisfying the constraints of a linear program (the shaded region) form a convex set.

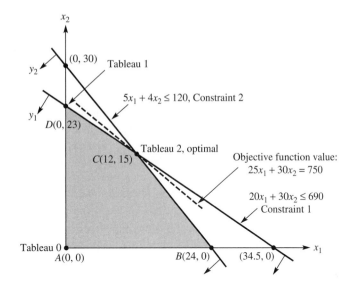

Before detailing each of the preceding steps, let's examine the carpenter's problem (Figure 7.13). The origin is an extreme point, so we choose it as our starting point. Thus, x_1 and x_2 are the current arbitrary independent variables and are assigned the value zero, whereas y_1 and y_2 are the current dependent variables with values of 690 and 120, respectively. The optimality test determines whether a current independent variable assigned the value zero could improve the value of the objective function if it is made dependent and positive. For example, either x_1 or x_2, if made positive, would improve the objective function value. (They have positive coefficients in the objective function we are trying to maximize.) Thus, the optimality test determines a promising variable to enter the dependent set. Later, we give a rule of thumb for choosing which independent variable to enter when more than one candidate exists. In the carpenter's problem at hand, we select x_2 as the new dependent variable.

The variable chosen for entry into the dependent set by the optimality condition replaces one of the current dependent variables. The feasibility condition determines which exiting variable this entering variable replaces. Basically, the entering variable replaces whichever current dependent variable can assume a zero value while maintaining nonnegative values for all the remaining dependent variables. That is, the feasibility condition ensures that the new intersection point will be feasible and hence an extreme point. In Figure 7.13, the feasibility test would lead us to the intersection point $(0, 23)$, which is feasible, and not to $(0, 30)$, which is infeasible. Thus, x_2 replaces y_1 as a dependent or nonzero variable. Therefore, x_2 enters and y_1 exits the set of dependent variables.

Computational Efficiency

The feasibility test does not require actual computation of the values of the dependent variables when selecting an exiting variable for replacement. Instead, we will see that an appropriate exiting variable is selected by quickly determining whether any variable becomes negative if the dependent variable being considered for replacement is assigned the value zero (a ratio test that will be explained later). If any variable would become negative, then the dependent variable under consideration cannot be replaced by the entering variable if feasibility is to be maintained. Once a set of dependent variables corresponding to a more optimal extreme point is found from the optimality and feasibility tests, the values of the new dependent variables are determined by pivoting. The pivoting process essentially solves an equivalent system of equations for the new dependent variables after the exchange of the entering and exiting variables in the dependent set. The values of the new dependent variables are obtained by assigning the independent variables the value zero. Note that only one dependent variable is replaced at each stage. *Geometrically, the Simplex Method proceeds from an initial extreme point to an adjacent extreme point until no adjacent extreme point is more optimal.* At that time, the current extreme point is an optimal solution. We now detail the steps of the Simplex Method.

STEP 1   **TABLEAU FORMAT**   Many formats exist for implementing the Simplex Method. The format we use assumes that the objective function is to be maximized and that the constraints are less than or equal to inequalities. (If the problem is not expressed initially in this format, it can easily be changed to this format.) For the carpenter's example, the problem is

$$\text{Maximize } 25x_1 + 30x_2$$

subject to

$$20x_1 + 30x_2 \le 690$$
$$5x_1 + 4x_2 \le 120$$
$$x_1, x_2 \ge 0$$

Next we adjoin a new constraint to ensure that any solution improves the best value of the objective function found so far. Take the initial extreme point as the origin, where the value of the objective function is zero. We want to constrain the objective function to be better than its current value, so we require

$$25x_1 + 30x_2 \ge 0$$

Because all the constraints must be $\le$ inequalities, multiply the new constraint by -1 and adjoin it to the original constraint set:

$$20x_1 + 30x_2 \le 690 \quad \text{(constraint 1, lumber)}$$
$$5x_1 + 4x_2 \le 120 \quad \text{(constraint 2, labor)}$$
$$-25x_1 - 30x_2 \le 0 \quad \text{(objective function constraint)}$$

The Simplex Method implicitly assumes that all variables are nonnegative, so we do not repeat the nonnegativity constraints in the remainder of the presentation.

Next, we convert each inequality to an equality by adding a *nonnegative* new variable y_i (or z), called a *slack variable* because it measures the slack or degree of satisfaction of the constraint. A negative value for y_i indicates the constraint is not satisfied. (We use the variable z for the objective function constraint to avoid confusion with the other constraints.)

This process gives the *augmented constraint set*

$$20x_1 + 30x_2 + y_1 = 690$$
$$5x_1 + 4x_2 + y_2 = 120$$
$$-25x_1 - 30x_2 + z = 0$$

where the variables x_1, x_2, y_1, y_2 are nonnegative. The value of the variable z represents the value of the objective function, as we shall see later. (Note from the last equation that $z = 25x_1 + 30x_2$ is the value of the objective function.)

STEP 2   **INITIAL EXTREME POINT** Because there are two decision variables, all possible intersection points lie in the x_1x_2-plane and can be determined by setting two of the variables $\{x_1, x_2, y_1, y_2\}$ to zero. (The variable z is *always* a dependent variable and represents the value of the objective function at the extreme point in question.) The origin is feasible and corresponds to the extreme point characterized by $x_1 = x_2 = 0$, $y_1 = 690$, and $y_2 = 120$. Thus, x_1 and x_2 are independent variables assigned the value 0; y_1, y_2, and z are dependent variables whose values are then determined. As we will see, z conveniently records the current value of the objective function at the extreme points of the convex set in the x_1x_2-plane as we compute them by elimination.

STEP 3   **THE OPTIMALITY TEST FOR CHOOSING AN ENTERING VARIABLE** In the preceding format, a negative coefficient in the last (or objective function) equation indicates that the corresponding variable could improve the current objective function value. Thus, the coefficients -25 and -30 indicate that either x_1 or x_2 could enter and improve the current objective function value of $z = 0$. (The current constraint corresponds to $z = 25x_1 + 30x_2 \geq 0$, with x_1 and x_2 currently independent and 0.) When more than one candidate exists for the entering variable, a rule of thumb for selecting the variable to enter the dependent set is to select that variable with the largest (in absolute value) negative coefficient in the objective function row. If no negative coefficients exist, the current solution is optimal. In the case at hand, we choose x_2 as the new entering variable. That is, x_2 will increase from its current value of zero. The next step determines how great an increase is possible.

STEP 4   **THE FEASIBILITY CONDITION FOR CHOOSING AN EXITING VARIABLE** The entering variable x_2 (in our example) must replace either y_1 or y_2 as a dependent variable (because z *always* remains the third dependent variable). To determine which of these variables is to exit the dependent set, first divide the right-hand-side values 690 and 120 (associated with the original constraint inequalities) by the components for the entering variable in each inequality (30 and 4, respectively, in our example) to obtain the ratios $\frac{690}{30} = 23$ and $\frac{120}{4} = 30$. From the subset of ratios that are positive (both in this case), the variable corresponding to the minimum ratio is chosen for replacement (y_1, which corresponds to 23 in this case). *The ratios represent the value the entering variable would obtain if the corresponding exiting variable were assigned the value 0.* Thus, only positive values are considered and the smallest positive value is chosen so as not to drive any variable negative. For instance, if y_2 were chosen as the exiting variable and assigned the value 0, then x_2 would assume a value 30 as the new dependent variable. However, then y_1 would be negative, indicating that the intersection point $(0, 30)$ does not satisfy the first constraint. Note that the intersection point $(0, 30)$ is not feasible in Figure 7.13. The **minimum positive ratio rule** illustrated previously obviates enumeration of any infeasible intersection points. In the case at hand, the dependent variable corresponding to the smallest ratio 23 is y_1, so it becomes the exiting variable. Thus, x_2, y_2, and z form the new set of dependent variables, and x_1 and y_1 form the new set of independent variables.

STEP 5 **PIVOTING TO SOLVE FOR THE NEW DEPENDENT VARIABLE VALUES** Next we derive a new (equivalent) system of equations by eliminating the entering variable x_2 in all the equations of the previous system that do not contain the exiting variable y_1. There are numerous ways to execute this step, such as the method of elimination used in Section 7.3. Then we find the values of the dependent variables x_2, y_2, and z when the independent variables x_1 and y_1 are assigned the value 0 in the new system of equations. This is called the **pivoting procedure**. The values of x_1 and x_2 give the new extreme point (x_1, x_2), and z is the (improved) value of the objective function at that point.

After performing the pivot, apply the optimality test again to determine whether another candidate entering variable exists. If so, choose an appropriate one and apply the feasibility test to choose an exiting variable. Then the pivoting procedure is performed again. The process is repeated until no variable has a negative coefficient in the objective function row. We now summarize the procedure and use it to solve the carpenter's problem.

Summary of the Simplex Method

STEP 1 **PLACE THE PROBLEM IN TABLEAU FORMAT.** Adjoin slack variables as needed to convert inequality constraints to equalities. Remember that all variables are nonnegative. Include the objective function constraint as the last constraint, including its slack variable z.

STEP 2 **FIND ONE INITIAL EXTREME POINT.** (For the problems we consider, the origin will be an extreme point.)

STEP 3 **APPLY THE OPTIMALITY TEST.** Examine the last equation (which corresponds to the objective function). If all its coefficients are nonnegative, then stop: The current extreme point is optimal. Otherwise, some variables have negative coefficients, so choose the variable with the largest (in absolute value) negative coefficient as the new entering variable.

STEP 4 **APPLY THE FEASIBILITY TEST.** Divide the current right-hand-side values by the corresponding coefficient values of the entering variable in each equation. Choose the exiting variable to be the one corresponding to the smallest positive ratio after this division.

STEP 5 **PIVOT.** Eliminate the entering variable from all the equations that do not contain the exiting variable. (For example, you can use the elimination procedure presented in Section 7.2.) Then assign the value 0 to the variables in the new independent set (consisting of the exited variable and the variables remaining after the entering variable has left to become dependent). The resulting values give the new extreme point (x_1, x_2) and the objective function value z for that point.

STEP 6 **REPEAT STEPS 3–5** until an optimal extreme point is found.

EXAMPLE 1   *The Carpenter's Problem Revisited*

STEP 1   The Tableau Format gives

$$20x_1 + 30x_2 + y_1 = 690$$
$$5x_1 + 4x_2 + y_2 = 120$$
$$-25x_1 - 30x_2 + z = 0$$

STEP 2 The origin $(0, 0)$ is an initial extreme point for which the independent variables are $x_1 = x_2 = 0$ and the dependent variables are $y_1 = 690$, $y_2 = 120$, and $z = 0$.

STEP 3 We apply the optimality test to choose x_2 as the variable entering the dependent set because it corresponds to the negative coefficient with the largest absolute value.

STEP 4 Applying the feasibility test, we divide the right-hand-side values 690 and 120 by the components for the entering variable x_2 in each equation (30 and 4, respectively), yielding the ratios $\frac{690}{30} = 23$ and $\frac{120}{4} = 30$. The smallest positive ratio is 23, corresponding to the first equation that has the slack variable y_1. Thus, we choose y_1 as the exiting dependent variable.

STEP 5 We pivot to find the values of the new dependent variables x_2, y_2, and z when the independent variables x_1 and y_1 are set to the value 0. After eliminating the new dependent variable x_2 from each previous equation that does not contain the exiting variable y_1, we obtain the equivalent system

$$\frac{2}{3}x_1 \quad + x_2 + \frac{1}{30}y_1 \qquad\qquad = 23$$
$$\frac{7}{3}x_1 \qquad\quad - \frac{2}{15}y_1 + y_2 \qquad = 28$$
$$-5x_1 \qquad\qquad + y_1 \qquad\quad + z = 690$$

Setting $x_1 = y_1 = 0$, we determine $x_2 = 23$, $y_2 = 28$, and $z = 690$. These results give the extreme point $(0, 23)$ where the value of the objective function is $z = 690$.

Applying the optimality test again, we see that the current extreme point $(0, 23)$ is not optimal (because there is a negative coefficient -5 in the last equation corresponding to the variable x_1). Before continuing, observe that we really do not need to write out the entire symbolism of the equations in each step. We merely need to know the coefficient values associated with the variables in each of the equations together with the right-hand side. A table format, or *tableau*, is commonly used to record these numbers. We illustrate the completion of the carpenter's problem using this format, where the headers of each column designate the variables; the abbreviation RHS heads the column where the values of the right-hand side appear. We begin with Tableau 0, corresponding to the initial extreme point at the origin.

Tableau 0 (Original Tableau)

x_1	x_2	y_1	y_2	z	RHS
20	30	1	0	0	$690\,(= y_1)$
5	4	0	1	0	$120\,(= y_2)$
-25	$\boxed{-30}$	0	0	1	$0\,(= z)$

Dependent variables: $\{y_1, y_2, z\}$
Independent variables: $x_1 = x_2 = 0$
Extreme point: $(x_1, x_2) = (0, 0)$
Value of objective function: $z = 0$

Optimality Test   The entering variable is x_2 (corresponding to -30 in the last row).

Feasibility Test   Compute the ratios for the RHS divided by the coefficients in the column labeled x_2 to determine the minimum positive ratio.

—Entering variable

x_1	x_2	y_1	y_2	z	RHS	Ratio
20	30	1	0	0	690	㉓$(= 690/30)$ ← Exiting variable
5	4	0	1	0	120	30 $(= 120/4)$
−25	⓪−30	0	0	1	0	*

Choose y_1 corresponding to the minimum positive ratio 23 as the exiting variable.

Pivot   Divide the row containing the exiting variable (the first row in this case) by the coefficient of the entering variable in that row (the coefficient of x_2 in this case), giving a coefficient of 1 for the entering variable in this row. Then eliminate the entering variable x_2 from the remaining rows (which do not contain the exiting variable y_1 and have a zero coefficient for it). The results are summarized in the next tableau, where we use five-place decimal approximations for the numerical values.

Tableau 1

x_1	x_2	y_1	y_2	z	RHS
0.66667	1	0.03333	0	0	23 $(= x_2)$
2.33333	0	−0.13333	1	0	28 $(= y_2)$
−5.00000	0	1.00000	0	1	690 $(= z)$

Dependent variables: $\{x_2, y_2, z\}$
Independent variables: $x_1 = y_1 = 0$
Extreme point: $(x_1, x_2) = (0, 23)$
Value of objective function: $z = 690$

The pivot determines that the new dependent variables have the values $x_2 = 23$, $y_2 = 28$, and $z = 690$.

Optimality Test   The entering variable is x_1 (corresponding to the coefficient −5 in the last row).

Feasibility Test   Compute the ratios for the RHS.

—Entering variable

x_1	x_2	y_1	y_2	z	RHS	Ratio
0.66667	1	0.03333	0	0	23	34.5 $(= 23/0.66667)$
2.33333	0	−0.13333	1	0	28	⑫12.0 $(= 28/2.33333)$ ← Exiting variable
−5.00000	0	1.00000	0	1	690	*

Choose y_2 as the exiting variable because it corresponds to the minimum positive ratio 12.

Pivot   Divide the row containing the exiting variable (the second row in this case) by the coefficient of the entering variable in that row (the coefficient of x_1 in this case), giving a coefficient of 1 for the entering variable in this row. Then eliminate the entering variable x_1 from the remaining rows (which do not contain the exiting variable y_2 and have a zero coefficient for it). The results are summarized in the next tableau.

Tableau 2

x_1	x_2	y_1	y_2	z	RHS
0	1	0.071429	−0.28571	0	$15\,(=x_2)$
1	0	−0.057143	0.42857	0	$12\,(=x_1)$
0	0	0.714286	2.14286	1	$750\,(=z)$

Dependent variables: $\{x_2, x_1, z\}$
Independent variables: $y_1 = y_2 = 0$
Extreme point: $(x_1, x_2) = (12, 15)$
Value of objective function: $z = 750$

Optimality Test   Because there are no negative coefficients in the bottom row, $x_1 = 12$ and $x_2 = 15$ gives the optimal solution $z = \$750$ for the objective function. Note that starting with an initial extreme point, we had to enumerate only two of the possible six intersection points. The power of the Simplex Method is its reduction of the computations required to find an optimal extreme point.   ▪ ▪ ▪

EXAMPLE 2   *Using the Tableau Format*

Solve the problem

$$\text{Maximize } 3x_1 + x_2$$

subject to

$$2x_1 + x_2 \leq 6$$
$$x_1 + 3x_2 \leq 9$$
$$x_1, x_2 \geq 0.$$

The problem in Tableau Format is

$$2x_1 + x_2 + y_1 = 6$$
$$x_1 + 3x_2 + y_2 = 9$$
$$-3x_1 - x_2 + z = 0$$

where $x_1, x_2, y_1, y_2,$ and $z \geq 0$.

Tableau 0 (Original Tableau)

x_1	x_2	y_1	y_2	z	RHS
2	1	1	0	0	$6\,(= y_1)$
1	3	0	1	0	$9\,(= y_2)$
$\;\ominus3\;$	-1	0	0	1	$0\,(= z)$

Dependent variables: $\{y_1, y_2, z\}$
Independent variables: $x_1 = x_2 = 0$
Extreme point: $(x_1, x_2) = (0, 0)$
Value of objective function: $z = 0$

Optimality Test The entering variable is x_1 (corresponding to -3 in the bottom row).

Feasibility Test Compute the ratios of the RHS divided by the column labeled x_1 to determine the minimum positive ratio.

x_1	x_2	y_1	y_2	z	RHS	Ratio
2	1	1	0	0	6	$\,③\,(= 6/2)\;\leftarrow$ Exiting variable
1	3	0	1	0	9	$9\;\,(= 9/1)$
$\ominus3$	-1	0	0	1	0	$*$

$\uparrow$
└─Entering variable

Choose y_1 corresponding to the minimum positive ratio 3 as the exiting variable.

Pivot Divide the row containing the exiting variable (the first row in this case) by the coefficient of the entering variable in that row (the coefficient of x_1 in this case), giving a coefficient of 1 for the entering variable in this row. Then eliminate the entering variable x_1 from the remaining rows (which do not contain the exiting variable y_1 and have a zero coefficient for it). The results are summarized in the next tableau.

Tableau 1

x_1	x_2	y_1	y_2	z	RHS
1	$\frac{1}{2}$	$\frac{1}{2}$	0	0	$3\,(= x_1)$
0	$\frac{5}{2}$	$-\frac{1}{2}$	1	0	$6\,(= y_2)$
0	$\frac{1}{2}$	$\frac{3}{2}$	0	1	$9\,(= z)$

Dependent variables: $\{x_1, y_2, z\}$
Independent variables: $x_2 = y_1 = 0$
Extreme point: $(x_1, x_2) = (3, 0)$
Value of objective function: $z = 9$

The pivot determines that the dependent variables have the values $x_1 = 3$, $y_2 = 6$, and $z = 9$.

Optimality Test There are no negative coefficients in the bottom row. Thus, $x_1 = 3$ and $x_2 = 0$ is an extreme point giving the optimal objective function value $z = 9$. ▪ ▪ ▪

Remarks   We have assumed that the origin is a feasible extreme point. If it is not, then some extreme point must be found before the Simplex Method can be used as presented. We have also assumed that the linear program is not degenerate in the sense that no more than two constraints intersect at the same point. These restrictions and other topics are studied in more advanced treatments of linear programming.

7.4   PROBLEMS

1–7. Use the Simplex Method to resolve Problems 1–6 and 13 in Section 7.2.

Use the Simplex Method to find both the maximum solution and the minimum solution to Problems 8–12. Assume $x \geq 0$ and $y \geq 0$ for each problem.

8. Optimize $2x + 3y$
   subject to

$$2x + 3y \geq 6$$
$$3x - y \leq 15$$
$$-x + y \leq 4$$
$$2x + 5y \leq 27$$

9. Optimize $6x + 4y$
   subject to

$$-x + y \leq 12$$
$$x + y \leq 24$$
$$2x + 5y \leq 80$$

10. Optimize $6x + 5y$
   subject to

$$x + y \geq 6$$
$$2x + y \geq 9$$

11. Optimize $x - y$
   subject to

$$x + y \geq 6$$
$$2x + y \geq 9$$

12. Optimize $5x + 3y$
   subject to

$$1.2x + 0.6y \leq 24$$
$$2x + 1.5y \leq 80$$

7.4   PROJECT

1. Write a computer code to perform the basic simplex algorithm. Solve Problem 3 using your code.

7.5   Linear Programming IV: Sensitivity Analysis

A mathematical model typically approximates a problem under study. For example, the coefficients in the objective function of a linear program may only be estimates. Or the amount of the resources constraining production made available by management may vary, depending on the profit returned per unit of resource invested. (Management may be willing to procure additional resources if the additional profit is high enough.) Thus, management would like to know whether the potential additional profit justifies the cost of another unit of resource. If so, over what range of values for the resources is the analysis valid? Hence, in addition to solving a linear program, we would like to know how sensitive the optimal solution is to changes in the various constants used to formulate the program. In this section we analyze graphically the effect on the optimal solution of changes in the coefficients of the objective function and the amount of resource available. Using the carpenter's problem as an example, we answer the following questions:

1. Over what range of values for the profit per table does the current solution remain optimal?

2. What is the value of another unit of the second resource (labor)? That is, how much will the profit increase if another unit of labor is obtained? Over what range of labor values is the analysis valid? What is required to increase profit beyond this limit?

Sensitivity of the Optimal Solution to Changes in the Coefficients of the Objective Function

The objective function in the carpenter's problem is to maximize profits where each table nets $25 profit and each bookcase $30. If z represents the amount of profit, then we wish to

$$\text{Maximize } z = 25x_1 + 30x_2$$

Note that z is a function of two variables and we can draw the level curves of z in the x_1x_2-plane. In Figure 7.14, we graph the level curves $z = 650$, $z = 750$, and $z = 850$ for illustrative purposes.

Note that every level curve is a line with slope $-\frac{5}{6}$. In Figure 7.15, we superimpose on the previous graph the constraint set for the carpenter's problem and see that the optimal solution (12, 15) gives an optimal objective function value of $z = 750$.

Now we ask the following question: What is the effect of changing the value of the profit for each table? Intuitively, if we increase the profit sufficiently, we eventually make only tables (giving the extreme point of 24 tables and 0 bookcases), instead of the current mix of 12 tables and 15 bookcases. Similarly, if we decrease the profit per table sufficiently,

▇ Figure 7.14

Some level curves of $z = 25x_1 + 30x_2$ in the $x_1 x_2$-plane have a slope of $-5/6$.

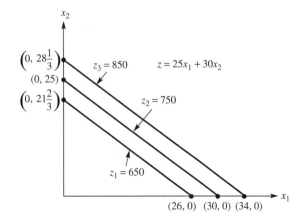

▇ Figure 7.15

The level curve $z = 750$ is tangent to the convex set of feasible solutions at extreme point $C(12, 15)$.

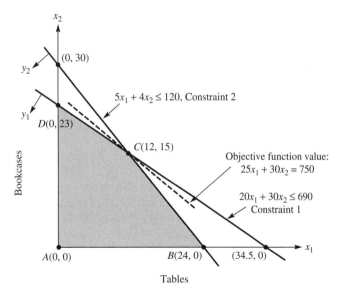

we should make only bookcases at the extreme point $(0, 23)$. Note again that the slope of the level curves of the objective function is $-\frac{5}{6}$. If we let c_1 represent the profit per table, then the objective function becomes

$$\text{Maximize } z = c_1 x_1 + 30 x_2$$

with slope $-c_1/30$ in the $x_1 x_2$-plane. As we vary c_1, the slope of the level curves of the objective function changes. Examine Figure 7.16 to convince yourself that the current extreme point $(12, 15)$ remains optimal as long as the slope of the objective function is between the slopes of the two binding constraints. In this case, the extreme point $(12, 15)$ remains optimal as long as the slope of the objective function is less than $-\frac{2}{3}$ but greater than $-\frac{5}{4}$, the slopes of the lumber and labor constraints, respectively. If we start with the slope for the objective function as $-\frac{2}{3}$, as we increase c_1, we rotate the level curve of the objective function clockwise. If we rotate clockwise, the optimal extreme point changes to

■ **Figure 7.16**

The extreme point (12, 15) remains optimal for objective functions with a slope between −5/4 and −2/3.

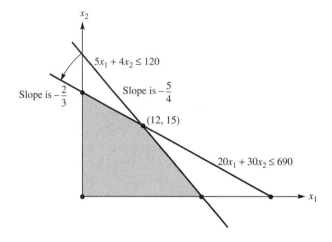

(24, 0) if the slope of the objective function is less than $-\frac{5}{4}$. Thus, the *range* of values for which the current extreme point remains optimal is given by the inequality

$$-\frac{5}{4} \leq -\frac{c_1}{30} \leq -\frac{2}{3}$$

or

$$20 \leq c_1 \leq 37.5$$

Interpreting this result, if the profit per table exceeds 37.5, the carpenter should produce only tables (i.e., 24 tables). If the profit per table is reduced below 20, the carpenter should produce only bookcases (i.e., 23 bookcases). If c_1 is between 20 and 37.5, he should produce the mix of 12 bookcases and 15 tables. Of course, as we change c_1 over the range [20, 37.5], the value of the objective function changes even though the location of the extreme point does not. Because he is making 12 tables, the objective function changes by a factor of 12 times the change in c_1. Note that at the limit $c_1 = 20$ there are *two* extreme points C and B, which produce the same value for the objective function. Likewise, if $c_1 = 37.5$, the extreme points D and C produce the same value for the objective function. In such cases, we say that there are *alternative optimal solutions.*

Changes in the Amount of Resource Available

Currently, there are 120 units of labor available, all of which are used to produce the 12 tables and 15 bookcases represented by the optimal solution. What is the effect of increasing the amount of labor? If b_2 represents the units of available labor (the second resource constraint), the constraint can be rewritten as

$$5x_1 + 4x_2 \leq b_2$$

What happens geometrically as we vary b_2? To answer this question, graph the constraint set for the carpenter's problem with the original value of $b_2 = 120$ and a second value, such

■ Figure 7.17

As the amount of labor resource b_2 increases from 120 to 150 units, the optimal solution moves from A to A' along the lumber constraint, increasing x_1 and decreasing x_2.

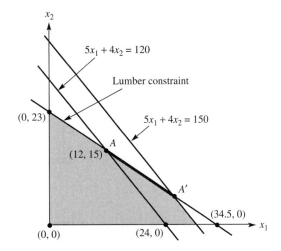

as $b_2 = 150$ (Figure 7.17). Note that the effect of increasing b_2 is to translate the constraint upward and to the right. As this happens, the optimal value of the objective function moves along the line segment AA', which lies on the lumber constraint. As the optimal solution moves along the line segment from A to A', the value for x_1 increases and the value of x_2 decreases. The net effect of increasing b_2 is to increase the value of the objective function, but by how much? One goal is to determine how much the objective function value changes as b_2 *increases by 1 unit*.

Note, however, that if b_2 increases beyond $5 \times 34.5 = 172.5$, the optimal solution remains at the extreme point $(34.5, 0)$. That is, at $(34.5, 0)$ the lumber constraint must also be increased if the objective function is to be increased further. Thus, increasing the labor constraint to 200 units results in some excess labor that cannot be used unless the amount of lumber is increased beyond its present value of 690 (Figure 7.18). Following a similar analysis, if b_2 is decreased, the value of the objective function moves along the lumber constraint until the extreme point $(0, 23)$ is reached. Further reductions in b_2 would cause the optimal solution to move from $(0, 23)$ down the y-axis to the origin.

■ Figure 7.18

As resource b_2 increases from 120 to 172.5, the optimal solution moves from A to A' along the line segment AA'; increasing b_2 beyond $b_2 = 172.5$ does not increase the value of the objective function unless the lumber constraint is also increased (moving it upward to the right).

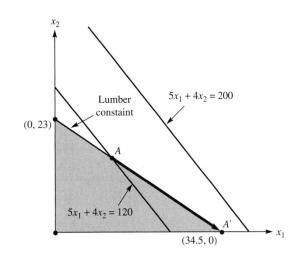

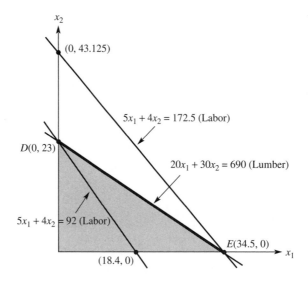

■ **Figure 7.19**

As b_2 increases from 92 to 172.5, the optimal solution moves from point $D(0, 23)$ to point $E(34.5, 0)$ along the line segment DE, the lumber constraint.

Now let's find the range of values of b_2 for which the optimal solution moves along the lumber constraint as the amount of labor varies. Refer to Figure 7.19 and convince yourself that we wish to find the value for b_2 for which the labor constraint would intersect the lumber constraint on the x_1-axis, or point E (34.5, 0). At point E (34.5, 0), the amount of labor is $5 \times 34.5 + 4 \times 0 = 172.5$. Similarly, we wish to find the value for b_2 at point D (0, 23), which is $5 \times 0 + 4 \times 23 = 92$. Summarizing, as b_2 changes, the optimal solution moves along the lumber constraint as long as

$$92 \leq b_2 \leq 172.5$$

However, by how much does the objective function change as b_2 increases by 1 unit within the range $92 \leq b_2 \leq 172.5$? We will analyze this in two ways. First, suppose $b_2 = 172.5$. The optimal solution is then the "new" extreme point E (34.5, 0) and the value of the objective function is $34.5 \times 25 = 862.5$ at E. Thus the objective function increased by $862.5 - 750 = 112.5$ units when b_2 increased by $172.5 - 120 = 52.5$ units. Hence, the change in the objective function for 1 unit of change in labor is

$$\frac{862.5 - 750}{172.5 - 120} = 2.14$$

Now let's analyze the value of a unit change of labor in another way. If b_2 increases by 1 unit from 120 to 121, then the new extreme point A' represented by the intersection of the constraints

$$20x_1 + 30x_2 = 690$$
$$5x_1 + 4x_2 = 121$$

is the point $A'(12.429, 14.714)$, which has an objective function value of 752.14. Thus, the net effect as b_2 increases by 1 unit is to increase the objective function by 2.14 units.

Economic Interpretation of a Unit Change in a Resource

In the foregoing analysis, we saw that as 1 more unit of labor is added, the objective function increases by 2.14 as long as the total amount of labor does not exceed 172.5 units. Thus, in terms of the objective function, an *additional* unit of labor is worth 2.14 units. If management can procure a unit of labor for less than 2.14 it would be profitable to do so. Conversely, if management can sell labor for more than 2.14 (which is valid until labor is reduced to 92 units) it should also consider doing that. Note that our analysis gives the value of a unit of resource in terms of the value of the objective function at the optimal extreme point, which is a *marginal value*.

Sensitivity analysis is a powerful methodology for interpreting linear programs. The information embodied in a carefully accomplished sensitivity analysis is often at least as valuable to the decision maker as the optimal solution to the linear program. In advanced courses in optimization, you can learn how to perform a sensitivity analysis algebraically. Moreover, the coefficients in the constraint set, as well as the right-hand side of the constraints, can be analyzed for their sensitivity.

Sections 7.2–7.5 discussed a solution method for a certain class of optimization problem: those with linear constraints and a linear objective function. Unfortunately, many optimization problems encountered in practice do not fall into this category. For example, rather than allowing the variables to take on any real value, we sometimes restrict them to a discrete set of values. Examples would include allowing variables to take on any integer values and allowing only binary values (either 0 or 1). We will consider some discrete optimization problems in the next chapter.

Another class of problem arises when we allow the objective function and/or constraints to be nonlinear. Recall from Section 7.1 that an optimization problem is nonlinear if it fails to satisfy Properties 2 and 3 on pages 235 and 236. In the next section, we will briefly turn our attention to solving unconstrained *nonlinear optimization problems*.

7.5 PROBLEMS

1. For the example problem in this section, determine the sensitivity of the optimal solution to a change in c_2 using the objective function $25x_1 + c_2x_2$.

2. Perform a complete sensitivity analysis (objective function coefficients and right-hand-side values) of the wooden toy soldier problem in Section 7.2 (Problem 1).

3. Why is sensitivity analysis important in linear programming?

7.5 PROJECTS

1. With the rising cost of gasoline and increasing prices to consumers, the use of additives to enhance performance of gasoline may be considered. Suppose there are two additives, Additive 1 and Additive 2, and several restrictions must hold for their use: First, the quantity of Additive 2 plus twice the quantity of Additive 1 must be at least 1/2 lb per car. Second, 1 lb of Additive 1 will add 10 octane units per tank, and 1 lb of Additive 2 will add 20 octane units per tank. The total number of octane units added must not be

less than 6. Third, additives are expensive and cost $1.53 per pound for Additive 1 and $4.00 per pound for Additive 2.

   a. Build a linear programming model and determine the quantity of each additive that meets the restrictions and minimizes their cost.

   b. Perform a sensitivity analysis on the cost coefficients and the resource values. Prepare a letter discussing your conculsions from your sensitivity analysis.

2. A farmer has 30 acres on which to grow tomatoes and corn. Each 100 bushels of tomatoes require 1000 gallons of water and 5 acres of land. Each 100 bushels of corn require 6000 gallons of water and 2.5 acres of land. Labor costs are $1 per bushel for both corn and tomatoes. The farmer has available 30,000 gallons of water and $750 in capital. He knows that he cannot sell more than 500 bushels of tomatoes or 475 bushels of corn. He estimates a profit of $2 on each bushel of tomatoes and $3 on each bushel of corn.

   a. How many bushels of each should he raise to maximize profits?

   b. Next, assume that the farmer has the oppportunity to sign a nice contract with a grocery store to grow and deliver at least 300 bushels of tomatoes and at least 500 bushels of corn. Should the farmer sign the contract? Support your recommendation.

   c. Now assume that the farmer can obtain an additional 10,000 gallons of water for a total cost of $50. Should he obtain the additional water? Support your recommendation.

3. Firestone, headquartered in Akron, Ohio, has a plant in Florence, South Carolina, that manufactures two types of tires: SUV 225 radials and SUV 205 radials. Demand is high because of the recent recall of tires. Each batch of 100 SUV 225 radial tires requires 100 gal of synthetic plastic and 5 lb of rubber. Each batch of 100 SUV 205 radial tires requires 60 gal of synthetic plastic and 2.5 lb of rubber. Labor costs are $1 per tire for each type of tire. The manufacturer has weekly quantities available of 660 gal of synthetic plastic, $750 in capital, and 300 lb of rubber. The company estimates a profit of $3 on each SUV 225 radial and $2 on each SUV 205 radial.

   a. How many of each type of tire should the company manufacture in order to maximize its profits?

   b. Assume now that the manufacturer has the opportunity to sign a nice contract with a tire outlet store to deliver at least 500 SUV 225 radial tires and at least 300 SUV 205 radial tires. Should the manufacturer sign the contract? Support your recommendation.

   c. If the manufacturer can obtain an additional 1000 gal of synthetic plastic for a total cost of $50, should he? Support your recommendation.

7.6   Numerical Search Methods

Consider the problem of maximizing a differentiable function $f(x)$ over some interval (a, b). Students of calculus will recall that if we compute the first derivative $f(x)$ and solve $f(x) = 0$ for x, we obtain the *critical points* of $f(x)$. The second derivative test may then be employed to characterize the nature of these critical points. We also know that we may have to check the endpoints and points where the first derivative fails to exist. However, it may be impossible to solve algebraically the equation resulting from setting the first derivative equal to zero. In such cases, we can use a search procedure to approximate the optimal solution.

Various search methods permit us to approximate solutions to nonlinear optimization problems with a single independent variable. Two search methods commonly used are the Dichotomous and Golden Section methods. Both share several features common to most search methods.

A *unimodal function* on an interval has exactly one point where a maximum or minimum occurs in the interval. If the function is known (or assumed to be) multimodal, then it must be subdivided into separate unimodal functions. (In most practical problems, the optimal solution is known to lie in some restricted range of the independent variable.) More precisely, $f(x)$ is a **unimodal function** with an interior local maximum on an interval $[a, b]$ if for some point x^* on $[a, b]$, the function is strictly increasing on $[a, x^*]$ and strictly decreasing on $[x^*, b]$. A similar statement holds for $f(x)$ being unimodal with an interior local minimum. These concepts are illustrated in Figure 7.20.

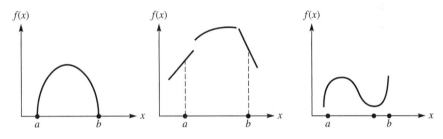

a. Unimodal functions on the interval $[a, b]$

b. A function that is not unimodal on the interval $[a, b]$

■ **Figure 7.20**

Examples of unimodal functions

The unimodal assumption is important to find the subset of the interval $[a, b]$ that contains the optimal point x to maximize (or minimize) $f(x)$.

The Search Method Paradigm

With most search methods, we divide the region $[a, b]$ into two overlapping intervals $[a, x_2]$ and $[x_1, b]$ after placing two test points x_1 and x_2 in the original interval $[a, b]$ according to some criterion of our chosen search method, as illustrated in Figure 7.21. We next determine the subinterval where the optimal solution lies and then use that subinterval to continue the search based on the function evaluations $f(x_1)$ and $f(x_2)$. There are three cases (illustrated in Figure 7.22) in the maximization problem (the minimization problem is analogous) with experiments x_1 and x_2 placed between $[a, b]$ according to the chosen search method (fully discussed later):

Case 1: $f(x_1) < f(x_2)$. Because $f(x)$ is unimodal, the solution cannot occur in the interval $[a, x_1]$. The solution must lie in the interval $[x_2, b]$.

Figure 7.21

Location of test points for
search methods
(overlapping intervals)

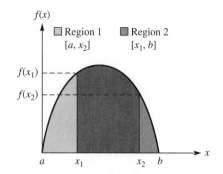

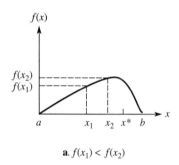

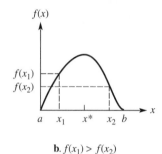

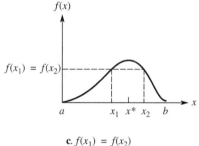

a. $f(x_1) < f(x_2)$ **b.** $f(x_1) > f(x_2)$ **c.** $f(x_1) = f(x_2)$

Figure 7.22

Cases where $f(x_1) < f(x_2)$, $f(x_1) > f(x_2)$, and $f(x_1) = f(x_2)$

Case 2: $f(x_1) > f(x_2)$. Because $f(x)$ is unimodal, the solution cannot occur in the
interval $[x_2, b]$. The solution must lie in the interval $[a, x_2]$.

Case 3: $f(x_1) = f(x_2)$. The solution must lie somewhere in the interval (x_1, x_2).

Dichotomous Search Method

Assume we have a function $f(x)$ to maximize over a specified interval $[a, b]$. The dichotomous method computes the midpoint $\frac{a+b}{2}$ and then moves slightly to either side of the midpoint to compute two test points: $\frac{a+b}{2} \pm \varepsilon$, where ε is some very small real number. In
practice, the number ε is chosen as small as the accuracy of the computational device will
permit, the objective being to place the two experimental points as close together as possible.
Figure 7.23 illustrates this procedure for a maximization problem. The procedure continues
until it gets within some small interval containing the optimal solution. Table 7.1 lists the
steps in this algorithm.

In this presentation, the number of iterations to perform is determined by the reduction
in the length of uncertainty desired. Alternatively, one may wish to continue to iterate until
the change in the dependent variable is less than some predetermined amount, such as Δ.
That is, continue to iterate until $f(a) - f(b) \le \Delta$. For example, in an application in which
$f(x)$ represents the profit realized by producing x items, it might make more sense to stop
when the change in profit is less than some acceptable amount. To minimize a function
$y = f(x)$, either maximize $-y$ or switch the directions of the signs in Steps 4a and 4b.

■ Figure 7.23

Dichotomous search computes the two test points from the midpoint of the interval.

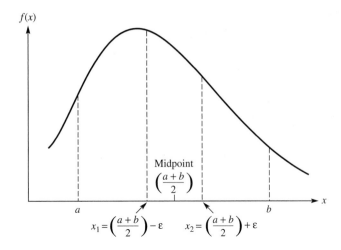

$$x_1 = \left(\frac{a+b}{2}\right) - \varepsilon \qquad x_2 = \left(\frac{a+b}{2}\right) + \varepsilon$$

Table 7.1 Dichotomous Search algorithm to maximize $f(x)$ over the interval $a \le x \le b$

STEP 1 Initialize: Choose a small number $\varepsilon > 0$, such as 0.01. Select a small $t > 0$, between $[a, b]$, called the *length of uncertainty* for the search. Calculate the number of iterations n using the formula

$$(0.5)^n = t/[b - a]$$

STEP 2 For $k = 1$ to n, do Steps 3 and 4.

STEP 3

$$x_1 = \left(\frac{a+b}{2}\right) - \varepsilon \qquad \text{and} \qquad x_2 = \left(\frac{a+b}{2}\right) + \varepsilon$$

STEP 4 (For a maximization problem)
a. If $f(x_1) \ge f(x_2)$, then let

$$a = a$$
$$b = x_2$$
$$k = k + 1$$

Return to Step 3.

b. If $f(x_1) < f(x_2)$, then let

$$b = b$$
$$a = x_1$$
$$k = k + 1$$

Return to Step 3.

STEP 5 Let $x^* = \frac{a+b}{2}$ and MAX $= f(x^*)$.
STOP

EXAMPLE 1   *Using the Dichotomous Search Method*

Suppose we want to maximize $f(x) = -x^2 - 2x$ over the interval $-3 \le x \le 6$. Assume we want the optimal tolerance to be less than 0.2. We arbitrarily choose ε (the distinguishability constant) to be 0.01. Next we determine the number n of iterations using the relationship $(0.5)^n = 0.2/(6 - (-3))$, or $n \ln(0.5) = \ln(0.2/9)$, which implies that $n = 5.49$ (and we round to the next higher integer, $n = 6$). Table 7.2 gives the results for implementing the search in the algorithm.

Table 7.2   Results of a Dichotomous Search for Example 1*

a	b	x_1	x_2	$f(x_1)$	$f(x_2)$
-3	6	1.49	1.51	-5.2001	-5.3001
-3	1.51	-0.755	-0.735	0.9400	0.9298
-3	-0.735	-1.8775	-1.8575	0.2230	0.2647
-1.8775	-0.735	-1.3163	-1.2963	0.9000	0.9122
-1.3163	-0.735	-1.0356	-1.0156	0.9987	0.9998
-1.0356	-0.735	-0.8953	-0.8753	0.9890	0.9845
-1.0356	-0.8753				

*The numerical results in this section were computed carrying the 13-place accuracy of the computational device being used. The results were then rounded to 4 places for presentation.

The length of the final interval of uncertainty is less than the 0.2 tolerance initially specified. From Step 5 we estimate the location of a maximum at

$$x^* = \frac{-1.0356 - 0.8753}{2} = -0.9555$$

with $f(-0.9555) = 0.9980$. (Examining our table we see that $f(-1.0156) = 0.9998$, a better estimate.) We note that the number of evaluations $n = 6$ refers to the number of intervals searched. (For this example, we can use calculus to find the optimal solution $f(-1) = 1$ at $x = -1$.) ■ ■ ■

Golden Section Search Method

The Golden Section Search Method is a procedure that uses the **golden ratio**. To better understand the golden ratio, divide the interval $[0, 1]$ into two separate subintervals of lengths r and $1 - r$, as shown in Figure 7.24. These subintervals are said to be divided into the *golden ratio* if the length of the whole interval is to the length of the longer segment as the length of the longer segment is to the length of the smaller segment. Symbolically, this can be written as $1/r = r/(1 - r)$ or $r^2 + r - 1 = 0$, because $r > 1 - r$ in the figure.

■ **Figure 7.24**
Golden ratio using a line segment

Solving this last equation gives the two roots

$$r_1 = (\sqrt{5} - 1)/2 \quad \text{and} \quad r_2 = (-\sqrt{5} - 1)/2$$

Only the positive root r_1 lies in the given interval $[0, 1]$. The numerical value of r_1 is approximately 0.618 and is known as the golden ratio.

The Golden Section Method incorporates the following assumptions:

1. The function $f(x)$ must be unimodal over the specified interval $[a, b]$.

2. The function must have a maximum (or minimum) value over a known interval of uncertainty.

3. The method gives an approximation to the maximum rather than the exact maximum.

The method will determine a final interval containing the optimal solution. The length of the final interval can be controlled and made arbitrarily small by the selection of a tolerance value. The length of the final interval will be less than our specified tolerance level.

The search procedure to find an approximation to the maximum value is iterative. It requires evaluations of $f(x)$ at the test points $x_1 = a + (1 - r)(b - a)$ and $x_2 = a + r(b - a)$ and then determines the new interval of search (Figure 7.25). If $f(x_1) < f(x_2)$, then the new interval is (x_1, b); if $f(x_1) > f(x_2)$, then the new interval is (a, x_2), as in the Dichotomous Search Method. The iterations continue until the final interval length is less than the tolerance imposed, and the final interval contains the optimal solution point. The length of this final interval determines the accuracy in finding the approximate optimal solution point. The number of iterations required to achieve the tolerance length can be found as the integer greater than k, where $k = \ln[(\text{tolerance})/(b-a)]/\ln[0.618]$. Alternatively, the method can be stopped when an interval $[a, b]$ is less than the required tolerance. Table 7.3 summarizes the steps of the Golden Section Search Method.

To minimize a function $y = f(x)$, either maximize $-y$ or switch the directions of the signs in Steps 4a and 4b. Note that the advantage of the Golden Section Search Method is that only one new test point (and one evaluation of the function at the test point) must be computed at each successive iteration, compared with two new test points (and two evaluations of the function at those test points) for the Dichotomous Search Method. Using

■ **Figure 7.25**

Location of x_1 and x_2 for the Golden Section Search

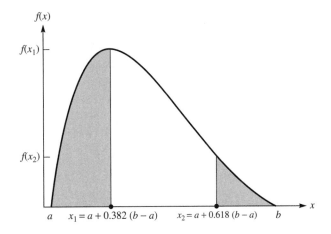

Table 7.3 Golden Section Search Method to maximize $f(x)$ over the interval $a \leq x \leq b$

STEP 1 Initialize: Choose a tolerance $t > 0$.

STEP 2 Set $r = 0.618$ and define the test points:

$$x_1 = a + (1 - r)(b - a)$$
$$x_2 = a + r(b - a)$$

STEP 3 Calculate $f(x_1)$ and $f(x_2)$.

STEP 4 (For a maximization problem) Compare $f(x_1)$ with $f(x_2)$:
  (a) If $f(x_1) \leq f(x_2)$, then the new interval is (x_1, b):
  a becomes the previous x_1.
  b does not change.
  x_1 becomes the previous x_2.
  Find the new x_2 using the formula in **Step 2**.
  (b) If $f(x_1) > f(x_2)$, then the new interval is (a, x_2):
  a remains unchanged.
  b becomes the previous x_2.
  x_2 becomes the previous x_1.
  Find the new x_1 using the formula in **Step 2**.

STEP 5 If the new interval [either (x_1, b) or (a, x_2)] is less than the tolerance t specified, then stop. Otherwise go back to **Step 3**.

STEP 6 Estimate x* as the midpoint of the final interval $x^* = \frac{a+b}{2}$ and compute MAX $= f(x^*)$.
STOP

the Golden Section Search Method, the length of the interval of uncertainty is 61.8% of the length of the previous interval of uncertainty. Thus, for large n, the interval of uncertainty is reduced by approximately $(0.618)^n$ after n test points are computed. (Compare with $(0.5)^{n/2}$ for the Dichotomous Search Method.)

EXAMPLE 2 *Using the Golden Section Search Method*

Suppose we want to maximize $f(x) = -3x^2 + 21.6x + 1$ over $0 \leq x \leq 25$, with a tolerance of $t = 0.25$. Determine the first two test points and evaluate $f(x)$ at each test point:

$$x_1 = a + 0.382(b - a) \rightarrow x_1 = 0 + 0.382(25 - 0) = 9.55$$

and

$$x_2 = a + 0.618(b - a) \rightarrow x_2 = 0 + 0.618(25 - 0) = 15.45$$

Then

$$f(x_1) = -66.3275 \quad \text{and} \quad f(x_2) = -381.3875$$

Because $f(x_1) > f(x_2)$, we discard all values in $[x_2, b]$ and select the new interval $[a, b] = [0, 15.45]$. Then $x_2 = 9.55$, which is the previous x_1, and $f(x_2) = -66.2972$. We must now find the position of the new test point x_1 and evaluate $f(x_1)$:

$$x_1 = 0 + 0.382(15.45 - 0) = 5.9017$$
$$f(x_1) = 23.9865$$

Again, $f(x_1) > f(x_2)$, so the new interval $[a, b]$ is $[0, 9.5592]$. Then the new $x_2 = 5.9017$ and $f(x_1) = 23.9865$. We find a new x_1 and $f(x_1)$:

$$x_1 = 0 + 0.382(9.55 - 0) = 3.6475$$
$$f(x_1) = 39.8732$$

Because $f(x_1) > f(x_2)$, we discard $[x_2, b]$ and our new search interval is $[a, b] = [0, 5.9017]$. Then $x_2 = 3.6475$, with $f(x_2) = 39.8732$. We find a new x_1 and $f(x_1)$:

$$x_1 = 0 + 0.382(5.9017 - 0) = 2.2542$$
$$f(x_1) = 34.4469$$

Because $f(x_2) > f(x_1)$, we discard $[a, x_1]$ and the new interval is $[a, b] = [2.2542, 5.9017]$. The new $x_1 = 3.6475$, with $f(x_1) = 39.8732$. We find a new x_2 and $f(x_2)$:

$$x_2 = 2.2545 + (0.618)(5.9017 - 2.2542) = 4.5085$$
$$f(x_2) = 37.4039$$

This process continues until the length of the interval of uncertainty, $b - a$, is less than the tolerance, $t = 0.25$. This requires 10 iterations. The results of the Golden Section Search Method for Example 2 are summarized in Table 7.4.

The final interval $[a, b] = [3.4442, 3.6475]$ is the first interval of our $[a, b]$ intervals that is less than our 0.25 tolerance. The value of x that maximizes the given function over

Table 7.4 Golden Section Search Method results for Example 2

k	a	b	x_1	x_2	$f(x_1)$	$f(x_2)$
0	0	25	9.5491	15.4509	−66.2972	−381.4479
1	0	15.4506	5.9017	9.5491	23.9865	−66.2972
2	0	9.5592	3.6475	5.9017	39.8732	23.9865
3	0	5.9017	2.2542	3.6475	34.4469	39.8732
4	2.2542	5.9017	3.6475	4.5085	39.8732	37.4039
5	2.2542	4.5085	3.1153	3.6475	39.1752	39.8732
6	3.1153	4.5085	3.6475	3.9763	39.8732	39.4551
7	3.1153	3.9763	3.4442	3.6475	39.8072	39.8732
8	3.4442	3.9763	3.6475	3.7731	39.8732	39.7901
9	3.4442	3.7731	3.5698	3.6475	39.8773	39.8732
10	3.4442	3.6475				

the interval must lie within this final interval of uncertainty [3.4442, 3.6475]. We estimate $x^* = (3.4442 + 3.6475)/2 = 3.5459$ and $f(x^*) = 39.8712$. The actual maximum, which in this case can be found by calculus, occurs at $x^* = 3.60$, where $f(3.60) = 39.88$.

■ ■ ■

As illustrated, we stopped when the interval of uncertainty was less than 0.25. Alternatively, we can compute the number of iterations required to attain the accuracy specified by the tolerance. Because the interval of uncertainty is 61.8% of the length of the interval of uncertainty at each stage, we have

$$\frac{\text{length of final interval (tolerance } t)}{\text{length of initial interval}} = 0.618^k$$

$$\frac{0.25}{25} = 0.618^k$$

$$k = \frac{\ln 0.01}{\ln 0.618} = 9.57, \text{ or 10 iterations}$$

In general, the number of iterations k required is given by

$$k = \frac{\ln(\text{tolerance}/(b - a))}{\ln 0.618}$$

EXAMPLE 3   *Model-Fitting Criterion Revisited*

Recall the curve-fitting procedure from Chapter 3 using the criterion

$$\text{Minimize} \sum |y_i - y(x_i)|$$

Let's use the Golden Section Search Method to fit the model $y = cx^2$ to the following data for this criterion:

x	1	2	3
y	2	5	8

The function to be minimized is

$$f(c) = |2 - c| + |5 - 4c| + |8 - 9c|$$

and we will search for an optimal value of c in the closed interval [0, 3]. We choose a tolerance $t = 0.2$. We apply the Golden Section Search Method until an interval of uncertainty is less than 0.2. The results are summarized in Table 7.5.

The length of the final interval is less than 0.2. We can estimate $c^* = (0.8115 + 0.9787)/2 = 0.8951$, with $f(0.8951) = 2.5804$. In the problem set, we ask you to show analytically that the optimal value for c is $c = \frac{8}{9} \approx 0.8889$.

■ ■ ■

Table 7.5 Find the best c to minimize the sum of the absolute deviations for the model $y = cx^2$.

Iteration k	a	b	c_1	c_2	$f(c_1)$	$f(c_2)$
1	0	3	1.1459	1.8541	3.5836	11.2492
2	0	1.8541	0.7082	1.1459	5.0851	3.5836
3	0.7082	1.8541	1.1459	1.4164	3.5836	5.9969
4	0.7082	1.4164	0.9787	1.1459	2.9149	3.5836
5	0.7082	1.1459	0.8754	0.9787	2.7446	2.9149
6	0.7082	0.9787	0.8115	0.8754	3.6386	2.7446
	0.8115	0.9787				

EXAMPLE 4 *Optimizing Industrial Flow*

Figure 7.26 represents a physical system engineers might need to consider for an industrial flow process. As shown, let x represent the flow rate of dye into the coloring process of cotton fabric. Based on this rate, the reaction differs with the other substances in the process as evidenced by the step function shown in Figure 7.26. The step function is defined as

$$f(x) = \begin{cases} 2 + 2x - x^2 & \text{for } 0 < x \le \dfrac{3}{2} \\ -x + \dfrac{17}{4} & \text{for } \dfrac{3}{2} < x \le 4 \end{cases}$$

The function defining the process is unimodal. The company wants to find the flow rate x that maximizes the reaction of the other substances $f(x)$. Through experimentation the engineers have found that the process is sensitive to within about 0.020 of the actual value of x. They have also found that the flow is either *off* ($x = 0$) or *on* ($x > 0$). The process will not allow for turbulent flow that occurs above $x = 4$ for this process. Thus, $x \le 4$ and we use a tolerance of 0.20 to maximize $f(x)$ over $[0, 4]$. Using the Golden Section Search

■ **Figure 7.26**

Industrial flow process function

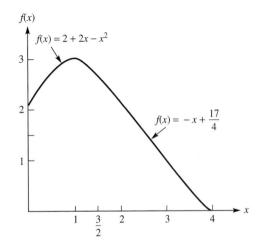

Method, we locate the first two test points,

$$x_1 = 0 + 4(0.382) = 1.5279$$
$$x_2 = 0 + 4(0.618) = 2.4721$$

and evaluate the function at the test points,

$$f(x_1) = 2.7221$$
$$f(x_2) = 1.7779$$

The results of the search are given in Table 7.6.

Table 7.6 Results of Golden Section Search Method for Example 4

a	b	x_1	x_2	$f(x_1)$	$f(x_2)$
0	4	1.5279	2.4721	2.7221	1.7779
0	2.4721	0.9443	1.5279	2.9969	2.7221
0	1.5279	0.5836	0.9443	2.8266	2.9969
0.5836	1.5279	0.9443	1.1672	2.9969	2.9720
0.5836	1.1672	0.8065	0.9443	2.9626	2.9969
0.8065	1.1672	0.9443	1.0294	2.9969	2.9991
0.9443	1.1672	1.0294	1.0820	2.9991	2.9933
0.9443	1.0820				

We stop because $(1.0820 - 0.9443) < 0.20$. The midpoint of the interval is 1.0132, where the value of the function is 2.9998. In the problem set, you are asked to show analytically that a maximum value of $f(x^*) = 3$ occurs at $x = 1$. ■ ■ ■

7.6 PROBLEMS

1. Use the Dichotomous Search Method with a tolerance of $t = 0.2$ and $\varepsilon = 0.01$.
   a. Minimize $f(x) = x^2 + 2x, -3 \le x \le 6$
   b. Maximize $f(x) = -4x^2 + 3.2x + 3, [-2 \le x \le 2]$

2. Use the Golden Section Search Method with a tolerance of $t = 0.2$.
   a. Minimize $f(x) = x^2 + 2x, -3 \le x \le 6$
   b. Maximize $f(x) = -4x^2 + 3.2x + 3, [-2 \le x \le 2]$

3. Use the curve-fitting criterion to minimize the sum of the absolute deviations for the following models and data set:
   a. $y = ax$
   b. $y = ax^2$
   c. $y = ax^3$

x	7	14	21	28	35	42
y	8	41	133	250	280	297

4. For Example 2, show that the optimal value of c is $c^* = \frac{8}{9}$. *Hint:* Apply the definition of the absolute value to obtain a piecewise continuous function. Then find the minimum value of the function over the interval $[0, 3]$.

5. For Example 3, show that the optimal value of x is $x^* = 1$.

7.6 PROJECTS

1. *Fibonacci Search*—One of the more interesting search techniques uses the Fibonacci sequence. This search method can be employed even if the function is not continuous. The method uses the Fibonacci numbers to place test points for the search. These Fibonacci numbers are defined as follows: $F_0 = F_1 = 1$ and $F_n = F_{n-1} + F_{n-2}$, for $n = 2, 3, 4, \ldots$, yielding the sequence 1, 1, 2, 3, 5, 8, 13, 21, 34, 55, 89, 144, 233, 377, 510, 887, 1397, and so forth.

   a. Find the ratio between successive Fibonacci numbers using the preceding sequences. Then find the numerical limit as n gets large. How is this limiting ratio related to the Golden Section Search Method?

   b. Research and make a short presentation on the Fibonacci Search Method. Present your results to the class.

2. *Methods Using Derivatives: Newton's Method*—One of the best-known interpolation methods is Newton's Method, which exploits a quadratic approximation to the function $f(x)$ at a given point x_1. The quadratic approximation q is given by

$$q(x) = f(x_1) + f'(x_1)(x - x_1) + \frac{1}{2}f''(x_1)(x - x_1)^2$$

The point x_2 will be the point where q' equals zero. Continuing this procedure,

$$x_{k+1} = x_k - [f'(x_k)/f''(x_k)]$$

for $k = 1, 2, 3, \ldots$. This procedure is terminated when either $|x_{k+1} - x_k| < \varepsilon$ or $|f'(x_k)| < \varepsilon$, where ε is some small number. This procedure can be applied to twice-differentiable functions only if $f''(x)$ never equals zero.

   a. Starting with $x = 4$ and a tolerance of $\varepsilon = 0.01$, use Newton's Method to minimize $f(x) = x^2 + 2x$, over $-3 \leq x \leq 6$.

   b. Use Newton's Method to minimize

$$f(x) = \begin{cases} 4x^3 - 3x^4 & \text{for } x > 0 \\ 4x^3 + 3x^4 & \text{for } x < 0 \end{cases}$$

   Let the tolerance be $\varepsilon = 0.01$ and start with $x = 0.4$.

   c. Repeat part (b), starting at $x = 0.6$. Discuss what happens when you apply the method.

7.6 Further Reading

Bazarra, M., Hanif D. Sherali, & C. M. Shetty. *Nonlinear Programming: Theory and Algorithms*, 2nd ed. New York: Wiley, 1993.

Rao, S. S. *Engineering Optimization: Theory and Practice*, 3rd ed. New York: Wiley, 1996.

Winston, Wayne. *Operations Research: Applications and Algorithms*, 3rd ed. Belmont, CA: Duxbury Press, 1994.

Winston, Wayne. *Introduction to Mathematical Programming: Applications and Algorithms (for Windows)*, 2nd ed. Belmont, CA: Duxbury Press, 1997.

Winston, Wayne. *Mathematical Programming: Applications and Algorithms*, 4th ed. Belmont, CA: Duxbury Press, 2002.

Djanogly LRC - Issue Receipt

Customer name: Hewakandamby, Buddhika

Title: A first course in mathematical modeling / Frank R. Giordano ... [et al.].
ID: 1006388256
Due: 30/06/2011 23:59

Total items: 1
15/04/2011 15:18

All items must be returned before the due date and time.

The Loan period may be shortened if the item is requested.

WWW.nottingham.ac.uk/is

8 Modeling Using Graph Theory

Introduction

The manager of a recreational softball team has 15 players on her roster: Al, Bo, Che, Doug, Ella, Fay, Gene, Hal, Ian, John, Kit, Leo, Moe, Ned, and Paul. She has to pick a starting team, which consists of 11 players to fill 11 positions: pitcher (1), catcher (2), first base (3), second base (4), third base (5), shortstop (6), left field (7), left center (8), right center (9), right field (10) and additional hitter (11). Table 8.1 summarizes the positions each player can play.

Table 8.1 Positions players can play

Al	Bo	Che	Doug	Ella	Fay	Gene	Hal	Ian	John	Kit	Leo	Moe	Ned	Paul
2, 8	1, 5, 7	2, 3	1, 4, 5, 6, 7	3, 8	10, 11	3, 8, 11	2, 4, 9	8, 9, 10	1, 5, 6, 7	8, 9	3, 9, 11	1, 4, 6, 7		9, 10

Can you find an assignment where all of the 11 starters are in a position they can play? If so, is it the only possible assignment? Can you determine the "best" assignment? Suppose players' talents were as summarized in Table 8.2 instead of Table 8.1. Table 8.2 is the same as Table 8.1, except that now Hal can't play second base (position 4). Now can you find a feasible assignment?

Table 8.2 Positions players can play (updated)

Al	Bo	Che	Doug	Ella	Fay	Gene	Hal	Ian	John	Kit	Leo	Moe	Ned	Paul
2, 8	1, 5, 7	2, 3	1, 4, 5, 6, 7	3, 8	10, 11	3, 8, 11	2, 9	8, 9, 10	1, 5, 6, 7	8, 9	3, 9, 11	1, 4, 6, 7		9, 10

This is just one of an almost unlimited number of real-world situations that can be modeled using a graph. A graph is a mathematical object that we will learn how to leverage in this chapter in order to solve relevant problems.

We won't attempt to be comprehensive in our coverage of graphs. In fact, we're going to consider only a few ideas from the branch of mathematics known as *graph theory*.

8.1   Graphs as Models

So far in this book we have seen a variety of mathematical models. Graphs are mathematical models too. In this section, we will look at two examples.

The Seven Bridges of Königsberg

In his 1736 paper *Solutio problematic ad geometriam situs pertinenis* ("The solution to a problem pertinent to the geometry of places"), the famous Swiss mathematician Leonhard Euler (pronounced "Oiler") addressed a problem of interest to the citizens of Königsberg, Prussia. At the time, there were seven bridges crossing various branches of the river Pregel, which runs through the city. Figure 8.1 is based on the diagram that appeared in Euler's paper. The citizens of Königsberg enjoyed walking through the city—and specifically across the bridges. They wanted to start at some location in the city, cross each bridge *exactly* once, and end at the same (starting) location. Do you think this can be done? Euler developed a representation, or mathematical model, of the problem by using a graph. With the help of this model, he was able to answer the walkers' question.

■ **Figure 8.1**

The seven bridges of Königsberg

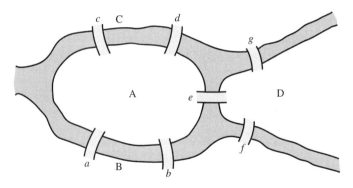

Euler's Problem   *Can the seven bridges be traversed exactly once starting and ending at the same place?* Try to answer this question yourself before reading on.

There are really two problems to solve here. One is transforming the map in Figure 8.1 into a graph. In this sense of the word, a **graph** is a mathematical way of describing relationships between things. In this case, we have a set of bridges and a set of land masses. We also have a relationship between these things; that is, each bridge joins exactly two specific land masses. The graph in Figure 8.2 is a mathematical model of the situation in Königsberg in 1736 with respect to what we can call the bridge-walking problem.

But Figure 8.2 doesn't answer our original question directly. That brings us to the second problem. Given a graph that *models* the bridges and land masses, how can you tell whether it is possible to start on some land mass, cross every bridge exactly once, and end up where you started? Does it matter where you started? Euler answered these questions in 1736, and if you think about it for a while, you might be able to answer them too.

Euler actually showed that it was *impossible* to walk through Königsberg in the stated manner, regardless of where the walker started. Euler's solution answered the bridge-walking question not only for Königsberg but also for every other city on Earth! In fact, in a

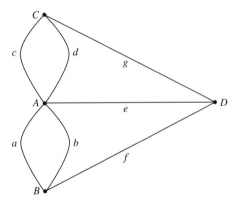

■ Figure 8.2

Graph model of Königsberg

sense he answered it for every possible city that might ever be built that has land masses and bridges that connect them. That's because he chose to solve the following problem instead.

Euler's Problem (restated)   *Given a graph, under what conditions is it possible to find a closed walk that traverses every edge exactly once?* We will call graphs for which this kind of walk is possible **Eulerian**.

What graphs are Eulerian? If you think for a moment, it is pretty clear that the graph has to be *connected*—that is, there must be a path between every pair of vertices. Another observation you might make is that whenever you walk over bridges between land masses and return to the starting point, the number of times you enter a land mass is the same as the number of times you leave it. If you add the number of times you enter a specific land mass to the number of times you leave it, you therefore get an even number. This means, in terms of the graph that represents the bridges and land masses, that an even number of edges *incident* with each vertex is needed. In the language of graph theory, we say that every vertex has *even degree* or that the graph has *even degree*.

Thus we have reasoned that Eulerian graphs must be connected and must have even degree. In other words, for a graph to be Eulerian, it is *necessary* that it both be connected and have even degree. But it is also true that for a graph to be Eulerian, it is *sufficient* that it be connected with even degree. Establishing necessary and sufficient conditions between two concepts—in this case, "all Eulerian graphs" and "all connected graphs with even degree"—is an important idea in mathematics with practical consequences. Once we establish that being connected with even degree is necessary and sufficient for a graph to be Eulerian, we need only model a situation with a graph, and then check to see whether the graph is connected and each vertex of the graph has even degree. Almost any textbook on graph theory will contain a proof of Euler's result; see 8.2 Further Reading.

Graph Coloring

Our second example is also easy to describe.

Four-Color Problem   *Given a geographic map, is it possible to color it with four colors so that any two regions that share a common border (of length greater than 0) are assigned different colors?* Figure 8.3 illustrates this problem.

■ **Figure 8.3**
United States map

The four-color problem can be modeled with a graph. Figure 8.4 shows a map with a vertex representing each state in the continental United States and an edge between every pair of vertices corresponding to states that share a common (land) border. Note that Utah and New Mexico, for example, are not considered adjacent because their common border is only a point.

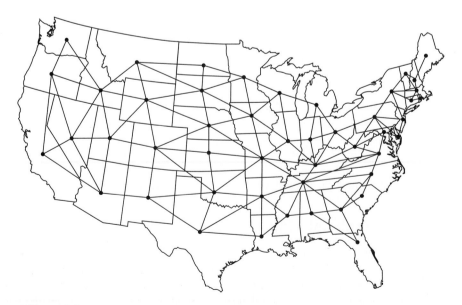

■ **Figure 8.4**
United States map with graph superimposed

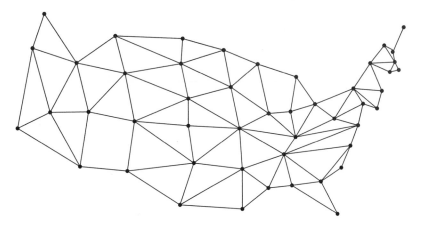

■ **Figure 8.5**

Graph-coloring problem

Next, Figure 8.5 shows the graph only. Now the original question we posed about the map of the United States has been transformed into a question about a graph.

Four-Color Problem (restated) *Using only four colors, can you color the vertices of a graph so that no vertex gets the same color as an adjacent vertex?* Try it on the graph in Figure 8.5. We'll call a coloring **proper** if no two adjacent vertices share the same color. Assuming you can find a proper four coloring of the graph, it is very easy to color the map accordingly.

Unfortunately, properly coloring graphs with a minimum number of colors is in general a hard problem. But for graphs that can be drawn in the plane without edges crossing, there is a fascinating and fairly recent result that at least puts an upper bound on the number of colors needed. Clearly, the graph in Figure 8.5 can be drawn in the plane without edges crossing; it *is* drawn that way! The good news is that every graph that arises from any possible political map can be colored with only four colors. The result that allows us to say this is the celebrated Four-Color Theorem that was first proved by Appel, Haken, and Koch in 1977.

This result was remarkable for several reasons. The question of how many colors are needed to color maps had been around for a long time. This means that lots of mathematicians and other people had devoted considerable effort to proving the result, over a period of at least a century. When this many people think for this long about a problem that's so simple to state and understand, it usually gets solved! But nobody could prove that four colors were enough to color any map drawn in the plane. Thus for decades, the Four-Color Theorem was known as the Four-Color Conjecture; it was one of the most famous conjectures in all of mathematics.

A proof of the Four-Color Conjecture had to wait for the dawn of the information age. This was no coincidence; the proof actually uses a computer. The proof that turned the Four-Color Conjecture into the Four-Color Theorem relied on computer analysis of a very large number of possible cases called configurations. This made the result remarkable too—it was one of the first widely circulated mathematical proofs that used a computer. It also caused quite a stir in the international mathematics community. Some mathematicians didn't believe that a proof that used a computer was a proof at all. Much of the early skepticism has since died down, however. Other researchers, notably Robertson, Sanders, Seymour, and Thomas, have in some sense improved on Appel and Haken's work. A wonderful summary of the Four-Color Problem can be found at www.math.gatech.edu/~thomas/FC/fourcolor.html.

Figure 8.6 shows a solution to the four-color problem on the graph in Figure 8.5.

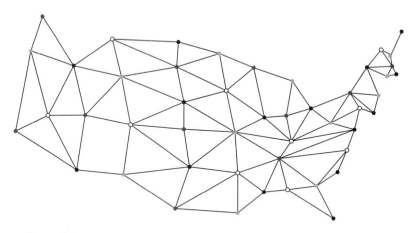

■ **Figure 8.6**

Graph-coloring solution

The four-color problem is really a special case of a more general graph-coloring problem: Given a graph, find the smallest number of colors needed to color the vertices properly. Graph coloring turns out to be a good model for a large variety of real-world problems. We explore a few of these below.

Applications of Graph Coloring   One problem typically modeled with graph coloring is final exam scheduling. Suppose that a university has n courses in which a final exam will be given and that it desires to minimize the number of exam periods used, while avoiding "conflicts." A conflict happens when a student is scheduled for two exams at the same time. We can model this problem using a graph as follows. We start by creating a vertex for each course. Then we draw an edge between two vertices whenever there is a student enrolled in both courses corresponding to those vertices. Now we solve the graph-coloring problem; that is, we properly color the vertices of the resulting graph in a way that minimizes the number of colors used. The color classes are the time periods. If some vertices are colored blue in a proper coloring, then there can be no edges between any pair of them. This means that no student is enrolled in more than one of the classes corresponding to these vertices. Then each color class can be given its own time slot.

8.1 PROBLEMS

1. Solve the softball manager's problem (both versions) from the Introduction to this chapter.

2. The bridges and land masses of a certain city can be modeled with graph G in Figure 8.7.

■ **Figure 8.7**

Graph G

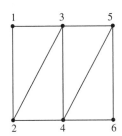

    a. Is G Eulerian? Why or why not?

    b. Suppose we relax the requirement of the walk so that the walker need not start and end at the same land mass but still must traverse every bridge exactly once. Is this type of walk possible in a city modeled by the graph in Figure 8.7? If so, how? If not, why not?

3. Find a political map of Australia. Create a graph model where there is a vertex for each of the six mainland states (Victoria, South Australia, Western Australia, Northern Territory, Queensland, and New South Wales) and an edge between two vertices if the corresponding states have a common border. Is the resulting graph Eulerian? Now suppose you add a seventh state (Tasmania) that is deemed to be adjacent (by boat) to South Australia, Northern Territory, Queensland, and New South Wales. Is the new graph Eulerian? If so, find a "walkabout" (a list of states) that shows this.

4. Can you think of other real-world problems that can be solved using techniques from the section about the bridges of Königsberg?

5. Consider the two political maps of Australia described in Problem 3. What is the smallest number of colors needed to color these maps?

6. Consider the graph of Figure 8.8.

■ **Figure 8.8**

Graph for Problem 6

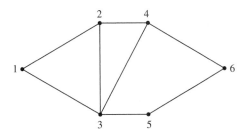

    a. Color the graph with three colors.

    b. Now suppose that vertices 1 and 6 must be colored red. Can you still color the graph with three colors (including red)?

7. The Mathematics Department at a small college plans to schedule final exams. The class rosters for all the upper-class math courses are listed in Table 8.3. Find an exam schedule that minimizes the number of time periods used.

Table 8.3   Mathematics course rosters at Sunnyvale State

Course	Students			
math 350	Jimi	B. B.	Eric	
math 365	Ry	Jimmy P.	Carlos	
math 385	Jimi	Chrissie	Bonnie	Brian
math 420	Bonnie	Robin	Carlos	
math 430	Ry	B. B.	Buddy	Robin
math 445	Brian	Buddy		
math 460	Jimi	Ry	Brian	Mark

8.1   Further Reading

Appel, K., W. Haken, & J. Koch. "Every planar map is four-colorable." *Illinois J. Math.,* 21(1977); 429–567.

Robertson, N., D. Sanders, P. Seymour, & R. Thomas. "The four colour theorem." *J. Combin. Theory Ser. B.,* 70 (1997): 2–44.

8.2   Describing Graphs

Before we proceed, we need to develop some basic notation and terminolgy that we can use to describe graphs. Our intent is to present just enough information to begin our discussion about modeling with graphs.

As we have noted, a **graph** is a mathematical way of describing relationships between things. A graph G consists of two sets: a **vertex set** $V(G)$ and an **edge set** $E(G)$. Each element of $E(G)$ is a pair of elements of $V(G)$. Figure 8.9 shows an example. When we refer to the vertices of a graph, we often write them using set notation. In our example we would write $V(G) = \{a, b, c, d, e, f, g, h, i\}$. The edge set is often described as a set of pairs of vertices; for our example we could write $E(G) = \{ac, ad, af, bd, bg, ch, di, ef, ei, fg, gh, hi\}$.

■ Figure 8.9

Example graph

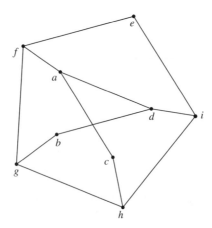

Vertices don't have to be labeled with letters. They can instead be labeled with numbers, or the names of things they represent, and so forth. Note that when we draw a graph, we might decide to draw it so that edges cross (or we might be forced to do so by the nature of the graph). Places on the paper where edges cross are not necessarily vertices; the example of Figure 8.9 shows a crossing point where there is no vertex.

When an edge ij has a vertex j as one of its endpoints, we say edge ij is **incident** with vertex j. For example, in the graph of Figure 8.9, edge bd is incident with vertex b but not with vertex a. When there is an edge ij between two vertices, we say vertices i and j are **adjacent**. In our example, vertices c and h are adjacent but a and b are not. The **degree** of a vertex j, $\deg(j)$, is the number of incidences between j and an edge. In our example, $\deg(b) = 2$ and $\deg(a) = 3$. As we saw in the previous section, a vertex v is said to have **even degree** if $\deg(v)$ is an even number (a number divisible by 2). Similarly, a graph is said to have even degree if every vertex in the graph has even degree.

Because graphs are often described using sets, we need to introduce some set notation. If S is a set, then $|S|$ denotes the number of elements in S. In our example, $|V(G)| = 9$ and $|E(G)| = 12$. We use the symbol $\in$ as shorthand for "is an element of" and $\notin$ for "is not an element of." In our example, $c \in V(G)$ and $bd \in E(G)$, but $m \notin V(G)$ and $b \notin E(G)$ (because b is a vertex and not an edge).

We will also use *summation* notation. The Greek letter $\sum$ (that's a capital letter sigma) is used to represent the idea of adding things up. For example, suppose we have a set $Q = \{q_1, q_2, q_3, q_4\} = \{1, 3, 5, 7\}$. We can succinctly express the idea of adding up the elements of Q using the summation symbol like this:

$$\sum_{q_i \in Q} q_i = 1 + 3 + 5 + 7 = 16$$

If we were reading the previous line aloud, we would say, "The sum of q sub i, for all q sub i in the set Q, equals 1 plus 3 plus 5 plus 7, which equals 16." Another way to express the same idea is

$$\sum_{i=1}^{4} q_i = 1 + 3 + 5 + 7 = 16$$

In this case, we say, "The sum, for i equals 1 to 4, of q sub i, equals 1 plus 3 plus 5 plus 7, which equals 16." We can also perform the summation operation on other functions of q. For example,

$$\sum_{i=1}^{4} (q_i^2 + 4) = (1^2 + 4) + (3^2 + 4) + (5^2 + 4) + (7^2 + 4) = 100$$

8.2 | PROBLEMS

1. Consider the graph in Figure 8.10.
   a. Write down the set of edges $E(G)$.
   b. Which edges are incident with vertex b?

    c. Which vertices are adjacent to vertex c?

    d. Compute $\deg(a)$.

    e. Compute $|E(G)|$.

■ **Figure 8.10**

Graph for Problem 1

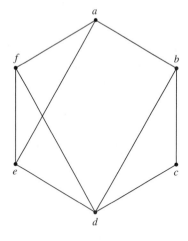

2. Suppose $r_1 = 4$, $r_2 = 3$, and $r_3 = 7$.

    a. Compute $\sum_{i=1}^{3} r_i$.

    b. Compute $\sum_{i=1}^{3} r_i^2$.

    c. Compute $\sum_{i=1}^{3} i r_i$.

3. At a large meeting of business executives, lots of people shake hands. Everyone at the meeting is asked to keep track of the number of times she or he shook hands, and as the meeting ends, these data are collected. Explain why you will obtain an even number if you add up all the individual handshake numbers collected. What does this have to do with graphs? Express this idea using the notation introduced in this section.

8.2 Further Reading

Buckley, F., & M. Lewinter. *A Friendly Introduction to Graph Theory.* Upper Saddle River, NJ: Pearson Education, 2003.

Chartrand, G., & P. Zhang. *Introduction to Graph Theory.* New York: McGraw Hill, 2005.

Chung, F., & R. Graham. *Erdös on Graphs.* Wellesley, MA: A. K. Peters, 1998.

Diestel, R. *Graph Theory.* Springer Verlag, 1992.

Goodaire, E., & M. Parmenter. *Discrete Mathematics with Graph Theory*, 2nd ed. Upper Saddle River, NJ: Prentice Hall, 2002.

West, D. *Introduction to Graph Theory*, 2nd ed. Upper Saddle River, NJ: Prentice Hall, 2001.

8.3 Graph Models

Bacon Numbers

You may have heard of the popular trivia game "Six Degrees of Kevin Bacon." In this game, players attempt to connect an actor to Kevin Bacon quickly by using a small number of connections. A *connection* in this context is a movie in which two actors you are connecting both appeared. The *Bacon number* of an actor is the smallest number of connections needed to connect him or her with Kevin Bacon. The *Bacon number problem* is the problem of computing Bacon numbers. An *instance* of the Bacon number problem is computing the Bacon number for a specific actor. For example, consider Elvis Presley. Elvis was in the movie *Change of Habit* with Ed Asner in 1969, and Ed Asner was in *JFK* with Kevin Bacon in 1991. This means that Elvis Presley's Bacon number is at most 2. Because Elvis never appeared in a movie with Kevin Bacon, his Bacon number can't be 1, so we know Elvis Presley's Bacon number *is* 2.

For another example, consider Babe Ruth, the New York Yankees baseball star of the 1930s. "The Babe" actually had a few acting parts, and he has a Bacon number of 3. He was in *The Pride of the Yankees* in 1942 with Teresa Wright; Teresa Wright was in *Somewhere in Time* (1980) with JoBe Cerny; and JoBe Cerny was in *Novocaine* (2001) with Kevin Bacon. How can we determine the Bacon numbers of other actors? By now, you won't be surprised to hear that the problem can be modeled using a graph.

To model the problem of determining an actor's Bacon number, let $G = (V(G), E(G))$ be a graph with one vertex for each actor who ever played in a movie (here we use the word *actor* in the modern sense to include both sexes). There is an edge between vertices representing two actors if they appeared together in a movie.

We should pause for a moment to consider some practical aspects of the resulting graph. For one thing, it is enormous! The Internet Movie Database (www.imdb.com) lists over a million actors—meaning there are over a million vertices in the graph—and over 300,000 movies. There is a potential new edge for every distinct pair of actors in a movie. For example, consider a small movie with 10 actors. Assuming these actors have not worked together on another movie, the first actor's vertex has 9 new edges going to the other 9 actors' vertices. The second actor's vertex will have 8 new edges, since we don't want to double-count the connection between the first and second actors. The pattern continues, so this single movie creates $9 + 8 + 7 + \cdots + 1 = 45$ new edges. In general, a movie with n actors has the potential to add $\frac{n(n-1)}{2}$ edges to the graph. We say "potential" to acknowledge the possibility that some of the n actors have appeared together in another movie. Of course, the total number of vertices in the graph grows too, because the list of actors who have appeared in a movie grows over time.

Despite the size of the associated graph, solving the Bacon number problem for a given actor is simple in concept. You just need to find the length of a shortest path in the graph from the vertex corresponding to your actor to the vertex corresponding to Kevin Bacon. Note that we said *a* shortest path, not *the* shortest path. The graph might have several paths that are "tied" for being the shortest. We consider the question of finding a shortest path between two vertices in Section 8.4.

The Bacon number graph is just one of a broader class of models called social networks. A **social network** consists of a set of individuals, groups, or organizations and certain social relationships between them. These networks can be modeled with a graph. For example, a **friendship network** is a graph where the vertices are people and there is an edge between two people if they are friends. Once a social network model is constructed, a variety of mathematical techniques can be applied to gain insight into the situation under study. For example, a recent Harvard Medical School study used friendship networks to investigate the prevelance of obesity in people. The investigators found that having fat friends is a strong predictor of obesity. That is, people with obese friends are more likely to be obese themselves than are people without obese friends. There are many other social relationships that could be modeled and then analyzed in this way. Powerful computational tools have recently been developed to keep track of and analyze huge social networks. This sort of analysis is a part of an emerging field called **network science** that may shed new light on many old problems.

Fitting a Piecewise Linear Function to Data

Suppose you have a collection of data $p_1, p_2, \ldots, p_n$, where each p_i is an ordered pair (x_i, y_i). Further suppose that the data are ordered so that $x_1 \leq x_2 \leq \cdots \leq x_n$. You could plot these data by thinking of each p_i as a point in the xy-plane. A small example with the data set $S = \{(0, 0), (2, 5), (3, 1), (6, 4), (8, 10), (10, 13), (13, 11)\}$ is given in Figure 8.11.

■ **Figure 8.11**

Data example

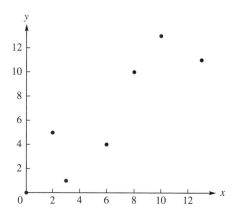

There are many applications where a piecewise linear function that goes through *some* of the points is desired. On the one hand, we could build our model to go through just the first and last points. To do this, we simply draw a line from the first point $p_1 = (x_1, y_1)$ to the last point $p_n = (x_n, y_n)$, obtaining the model

$$y = \frac{y_n - y_1}{x_n - x_1}(x - x_1) + y_1 \tag{8.1}$$

Figure 8.12 shows Model (8.1) displayed on the data.

On the other hand, we could draw a line segment from (x_1, y_1) to (x_2, y_2), another from (x_2, y_2) to (x_3, y_3), and so forth, all the way to (x_n, y_n). There are $n - 1$ line segments in all, and they can be described as follows:

$$y = \frac{y_{i+1} - y_i}{x_{i+1} - x_i}(x - x_i) + y_i \quad \text{for } x_i \leq x \leq x_{i+1} \tag{8.2}$$

where $i = 1, 2, \ldots, n - 1$. Figure 8.13 shows the data with Model (8.2).

■ Figure 8.12

Data with Model (8.1)

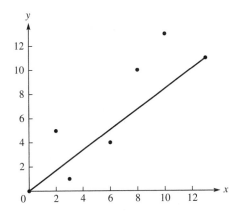

■ Figure 8.13

Data with Model (8.2)

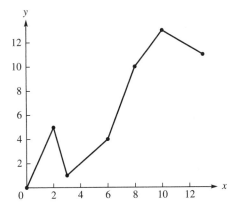

There is a trade-off between these two extreme choices. Model (8.1) is simple to obtain and simple to use, but it might miss some of the points by a great amount. Model (8.2) doesn't miss any points, but it is complicated, especially if the data set contains many points—that is, when n is large.

One option is to build a model that uses more than the one line segment present in Model (8.1), but fewer than the $n - 1$ line segments used in Model (8.2). Suppose we assume that the model at least goes through the first point and the last point (in this case p_1 and p_7, respectively). One possible model for the specific data set given is to choose to go through points $p_1 = (0, 0)$, $p_4 = (6, 4)$, $p_5 = (8, 10)$, and $p_7 = (13, 11)$. Here is an algebraic description of this idea:

$$y = \begin{cases} \frac{2}{3}x & \text{for } 0 \le x < 6 \\ 3(x - 6) + 4 & \text{for } 6 \le x < 8 \\ \frac{1}{5}(x - 8) + 10 & \text{for } 8 \le x \le 13 \end{cases} \tag{8.3}$$

A graphical interpretation of the same idea appears in Figure 8.14.

Figure 8.14

Data with Model (8.3)

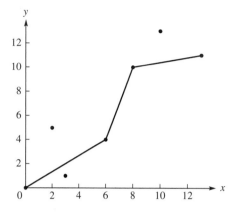

Now which model is *best*? There isn't a universal answer to that question; the situation and context must dictate which choice is best. We can, however, develop a framework for analyzing how good a given model is. Suppose there is a cost associated with having the model not go through a given point. In this case, we say the model "misses" that point. There are many ways to formalize this idea, but one natural way is to use the least-squares criteria discussed in Chapter 3. To capture the idea of a trade-off, there also needs to be a cost associated with each line segment that we use in our model. Thus Model (8.1) has a relatively high cost for missing points, but a minimum cost for using line segments. Model (8.2) has a minimum cost for missing points (in fact that cost is 0!) but has the largest reasonable cost for line segments.

Let's be a little more specific and see how we can find the best possible model. First recall that we always go through p_1 and p_n. So we must decide which of the other points $p_2, p_3, \ldots, p_{n-1}$ we will visit. Suppose we are considering going directly from p_1 to p_4, as we did in Model (8.3). What is the cost of this decision? We have to pay the fixed cost of adding a line segment to our model. Let's use α to represent that cost parameter. We also have to pay for all the points we miss along the way. We will use the notation $f_{i,j}$ to represent the line segment from p_i to p_j. In Model (8.3), $f_{1,4}(x) = \frac{2}{3}x$ and $f_{4,5}(x) = 3(x-6)+4$, because those are the algebraic descriptions of the line segments going from p_1 to p_4 and from p_4 to p_5, respectively. Accordingly, we can use f to compute the y value associated with a specific part of the model. For example, $f_{1,4}(x_4) = \frac{2}{3}x_4 = \frac{2}{3}(6) = 4$. This isn't surprising. Recall that $p_4 = (x_4, y_4) = (6, 4)$, so we expect $f_{1,4}(x_4)$ to equal 4 because we have chosen a model that goes through p_4. On the other hand, $f_{1,4}(x_3) = \frac{2}{3}x_3 = \frac{2}{3}(3) = 2$, but $y_3 = 1$, so our line drawn from p_1 to p_4 misses p_3. In fact, the amount it misses by, or the **error**, is the absolute difference between $f_{1,4}(x_3)$ and y_3, which is $|2-1| = 1$. To avoid some of the complexities of dealing with the absolute value function, we will square these errors, as we did in Chapter 3. When we chose to skip points p_2 and p_3 in Model (8.3), we did so believing that it was reasonable to pay the price of missing these points in exchange for using only one line segment on the interval from x_1 to x_4. The standard least-squares measure of the extent to which we missed points on this interval is

$$\sum_{k=1}^{4}(f_{1,4}(x_k) - y_k)^2 \tag{8.4}$$

The expression 8.4 simply adds up the squared vertical distances, which are the squared errors, between the line segment $f_{1,4}$ and the original data points. It does this for each point p_1, p_2, p_3, and p_4. Table 8.4 shows how the sum of squared errors is used to compute the value of (8.4).

Table 8.4 Computing the sum of squared errors for (8.4)

k	x_k	$f_{1,4}(x_k)$	y_k	$(f_{1,4}(x_k) - y_k)^2$
1	0	0	0	0
2	2	$\frac{4}{3}$	5	$\frac{121}{9}$
3	3	2	1	1
4	6	4	4	0

Thus

$$\sum_{k=1}^{4}(f_{1,4}(x_k) - y_k)^2 = 0 + \frac{121}{9} + 1 + 0 = \frac{130}{9} \approx 14.4444$$

Verbally, it means that the line that goes from $p_1 = (0, 0)$ to $p_4 = (6, 4)$ (we call that line $f_{1,4}$) doesn't miss p_1 at all, misses p_2 by $\frac{11}{3}$ units (note that $\left(\frac{11}{3}\right)^2 = \frac{121}{9}$), misses p_3 by 1 unit (as we saw in the previous paragraph), and doesn't miss p_4. The total of $\frac{130}{9} \approx 14.4444$ is a measure of the extent to which $f_{1,4}$ fails to accurately model the behavior of the data we have between the first and fourth points.

Although the sum of the squared errors may not describe the "cost" of missing the points along the way, we can scale the sum so that it does. We will use β to represent the cost per unit of summed squared errors. The modeler can choose a value of β that reflects his or her aversion to missing points, also considering the extent to which they are missed. Similarly, we can let α be the fixed cost associated with each line segment. A model with only one line segment will pay the cost α once, and a model with, say, five line segments will pay α five times. Then the total cost of the portion of the model going from p_1 to p_4 is

$$\alpha + \beta \sum_{k=1}^{4}(f_{1,4}(x_k) - y_k)^2 \tag{8.5}$$

If we choose parameter values $\alpha = 10$ and $\beta = 1$, then the total cost of the model over the interval from p_1 to p_4 is

$$\alpha + \beta \sum_{k=1}^{4}(f_{1,4}(x_k) - y_k)^2 = 10 + 1\frac{130}{9} \approx 24.4444 \tag{8.6}$$

It is easy, though somewhat tedious, to employ this same procedure for the other possible choices for our model. Building on what we have already done, we could easily compute the cost associated with the portion of our model that goes from p_4 to p_5. It is 10 because all we are paying for is the line segment cost (no points are missed in this case, on this interval). Finally, we could compute the cost of the remaining portion of the model.

So we just compute

$$\alpha + \beta \sum_{k=5}^{7} (f_{5,7}(x_k) - y_k)^2 = 10 + 1\frac{169}{25} = 16.76 \qquad (8.7)$$

The details of this computation are left to the reader. Thus the total cost of Model (8.3) is about $24.4444 + 10 + 16.76 = 51.2044$.

We could also compute the cost of choices we did *not* make in Model (8.3). In fact, if we compute the cost of *each possible choice* we can make in creating a similar model, we can compare options to see which model is best, given the data and our selected values of α and β. Table 8.5 shows the computed costs of all possible segments of a piecewise linear model for our data and parameter values.

Table 8.5 Cost for each line segment

	2	3	4	5	6	7
1	10	28.7778	24.4444	36.0625	38.77	55.503
2		10	24.0625	52.1389	61	56.9917
3			10	15.76	14.7755	51
4				10	12.25	51
5					10	16.76
6						10

Now we can use Table 8.5 to compute the total cost of any peicewise linear model we want to consider in the following way. Let $c_{i,j}$ be the cost of a line segment from point p_i to point p_j. In other words, $c_{i,j}$ is the entry in row i and column j in Table 8.5. Recall our first model, (8.1). Because this model had one line segment going from p_1 to p_7, we can compute its cost by looking at the entry in row 1 and column 7 of Table 8.5, which is $c_{1,7} = 55.503$. Our next model, (8.2), included six line segments; from p_1 to p_2, from p_2 to p_3, and so on, up to p_6 to p_7. These correspond to the lowest entry in each column of Table 8.5, so the total cost of this model is $c_{1,2} + c_{2,3} + \cdots + c_{6,7} = 10 + 10 + 10 + 10 + 10 + 10 = 60$. Other models can be considered as well. For example, suppose we decided to go from p_1 to p_3 to p_6 to p_7. This model costs $c_{1,3} + c_{3,6} + c_{6,7} = 28.7778 + 14.7755 + 10 = 53.5533$.

It is natural at this point to ask which model is best of all the possible models. We are looking for a piecewise linear model that starts at p_1 and ends at p_7 and has the smallest possible cost. We just have to decide which points we will visit. We can visit or not visit each of p_2, p_3, p_4, p_5, and p_6, because the conditions of the problem force us to visit p_1 and p_7. You can think of this as looking for a **path** that goes from p_1 to p_7, visiting any number of the five intermediate points. Of all those paths we want to find the best (cheapest) one.

One way to find the best model is to look at every possible model and pick the best one. For our example, this isn't a bad idea. Because each of the five points p_2, p_3, p_4, p_5, and p_6 is either visited or not visited, there are $2^5 = 32$ possible paths from p_1 to p_7. It would not be too hard to check all 32 of these and pick the best one. But think about what we would be up against if the original data set contained, say, 100 points instead of 7. In this case there would be $2^{98} = 316,912,650,057,350,374,175,801,244$ possible models. Even if we could test a million models every *second*, it would take more than 10,000,000,000,000,000 *years* to check them all. This is almost a million times longer than the time since the big bang! Fortunately, there is a better way to solve this kind of problem.

It turns out that we can also find the best model by solving a simple graph problem. Consider Figure 8.15. This represents the paths from p_1 to p_7 as a directed graph. Each path described in the paragraphs above is represented by a path from vertex 1 to vertex 7 in Figure 8.15. Look again at Table 8.5. If we use the data from this table as edge weights for the graph in Figure 8.15, we can solve our data-fitting problem of finding the best-fitting piecewise linear function by finding a shortest path in the graph.

■ Figure 8.15

Graph model for data fitting

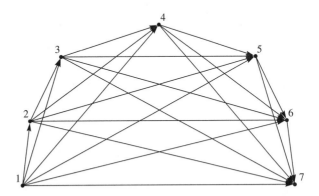

We will learn how to solve shortest-path problems later in this chapter. Once the shortest-path problem on the graph in Figure 8.15 using the cost data in Table 8.5 is solved, it will be clear that the best solution for this particular instance is the path from p_1 to p_2 to p_3 to p_6 to p_7. The model has a total cost of $c_{1,2} + c_{2,3} + c_{3,6} + c_{6,7} = 10 + 10 + 14.7755 + 10 = 44.7755$. This *optimal* model appears in Figure 8.16.

■ Figure 8.16

Optimal piecewise linear model

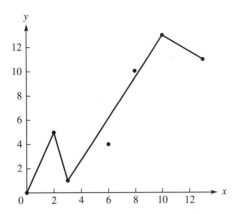

Recall that in Section 8.3 we found that the Bacon number problem can be solved by finding the distance between two vertices in a graph. Finding the distance between two vertices is really a great deal like finding the shortest path between two vertices—we'll state this fact with more precision at the beginning of Section 8.4. One remarkable aspect of mathematics is that technical procedures you can learn in the classroom (such as solving shortest-path problems) can be directly applied in many settings (such as finding Bacon numbers and fitting linear models to data). It is also interesting to observe that mathematics can link two situations that seem vastly different. The data-fitting problem and the Bacon number problem are very similar; they both reduce to a shortest-path problem. In Section 8.4 we will learn about a procedure to solve shortest-path problems on graphs.

The Softball Example Revisited

Recall the softball example from the Introduction to this chapter. Each player can play only in the positions specified in Table 8.6.

Table 8.6   Positions players can play

Al	Bo	Che	Doug	Ella	Fay	Gene	Hal	Ian	John	Kit	Leo	Moe	Ned	Paul
2,8	1,5,7	2,3	1,4,5,6,7	3,8	10,11	3,8,11	2,4,9	8,9,10	1,5,6,7	8,9	3,9,11	1,4,6,7		9,10

This problem can be modeled using a graph. We'll construct a graph from Table 8.6 as follows. For each player, we'll create a vertex. We'll call these vertices *player vertices*. We will also create a vertex for each position and call these *position vertices*. If we let A represent the player vertices and B represent the position vertices, we can write $V(G) = \langle A, B \rangle$ to represent the idea that the vertex set for our graph has two distinct types of vertices. Now we can create an edge in our graph for every entry in Table 8.6. That is, there is an edge between a specific player vertex and a specific position vertex whenever that player can play that position according to our table. For example, our graph has an edge from the vertex representing Che to the vertex representing position 3 (first base), but there is no edge from Che to, say, position 6.

The graph for the softball problem is shown in Figure 8.17. Note that the A vertices are on the left and the B vertices are on the right. Also note that all of the edges in the

Figure 8.17

Graph for the softball problem

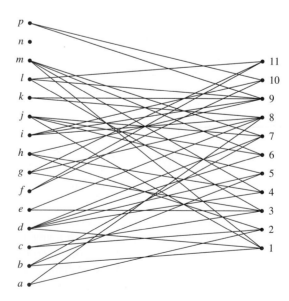

graph have one end in A and the other end in B. Graphs with this special property are called **bipartite**. The name comes from the fact that the vertex set can be partitioned (*-partite*) into two (*bi-*) sets (A and B) so that all of the edges have one end in A and one end in B. As we will see in Section 8.4, we can solve some kinds of problems particlarly easily on bipartite graphs.

Let's return to the problem of solving the softball manager's problem. We'd like an assignment of players to positions (or positions to players) that covers every position. Think about this in terms of Figure 8.17. You can think of the edge from, say, Che to position 3 as having the option of assigning Che to position 3. Note that once you decide to put Che at position 3, Che cannot be assigned a different position, and no other player can play in position 3. In terms of the graph, we can think of choosing some of the edges so that no two of the chosen edges are incident with the same vertex. If we can choose enough of the edges (in this case, 11 of them), we have solved the softball manager's problem.

Given any graph $G = (V(G), E(G))$, a subset $M \subseteq E(G)$ is called a **matching** if no two members of M are incident with the same vertex. A set of edges with this property are said to be **independent**. A **maximum matching** is a matching of the largest possible size. When G is bipartite with bipartition $\langle A, B \rangle$, it is clear that no matching can be bigger than $|A|$, and no matching can be bigger than $|B|$. Therefore, if we can find a matching in the graph for the softball problem that has 11 edges, then there is a feasible solution to the softball manager's problem. Furthermore, if the largest matching in the graph has fewer than 11 edges, *no feasible solution to the softball manager's problem is possible*. That is, the graph has a matching of size eleven **if and only if** there is a feasible solution to the softball manager's problem.

We have shown the relationship between the softball manager's problem and a matching problem in a bipartite graph. In Section 8.4, we will see how to find maximum matchings in bipartite graphs.

A 0–1 Matrix Problem

Informally, a **matrix** is an array or table with rows and columns. A matrix usually has an entry for each possible row and column pair. The entries themselves are ordinarily numbers. An "m by n 0–1 matrix" is a matrix that has m rows and n columns, where each entry is either 0 or 1. Using 0's and 1's in a mathematical model is a common way of expressing ideas such as "yes" or "no," "true" or "false," "on" or "off," and so forth. In this section, we analyze a problem related to 0–1 matrices.

Consider an m by n 0–1 matrix. Let r_i for $i \in \{1, 2, \ldots, m\}$ be the row sum of the ith row, or, equivalently, the number of 1's in the ith row. Similarly, let s_i for $i \in \{1, 2, \ldots, n\}$ be the column sum of the ith column, or, equivalently, the number of 1's in the ith column. Below is an example with $m = 4$, $n = 6$, $r_1 = 3$, $r_2 = 2$, $r_3 = 3$, $r_4 = 4$, $s_1 = 3$, $s_2 = 2$, $s_3 = 2$, $s_4 = 3$, $s_5 = 1$, $s_6 = 1$.

$$\begin{pmatrix} 1 & 0 & 0 & 1 & 1 & 0 \\ 1 & 1 & 0 & 0 & 0 & 0 \\ 1 & 0 & 1 & 1 & 0 & 0 \\ 0 & 1 & 1 & 1 & 0 & 1 \end{pmatrix}$$

Now consider the following problem.

The 0–1 Matrix Problem   *Given values for m and n along with* $r_1, r_2, \ldots, r_m$ *and* $s_1, s_2, \ldots, s_n$, *does there exist a 0–1 matrix with those properties?* If you were given

$$m = 4, n = 6$$
$$r_1 = 3, r_2 = 2, r_3 = 3, r_4 = 4 \tag{8.8}$$
$$s_1 = 3, s_2 = 2, s_3 = 2, s_4 = 3, s_5 = 1, s_6 = 1$$

the answer is clearly "yes," as the matrix above demonstrates. It's easy to say "no" if $r_1 + r_2 + \cdots + r_m \neq s_1 + s_2 + \cdots + s_n$ because the sum of the row sum values must equal the sum of the column sum values (why?). But if you were given values of m and n and $r_1, r_2, \ldots, r_m$ and $s_1, s_2, \ldots, s_n$, where $r_1 + r_2 + \cdots + r_m = s_1 + s_2 + \cdots + s_n$, it might not be so easy to decide whether such a matrix is possible.

Let's pause for a moment to consider why we might *care* about this matrix problem. Some real-world problems can be modeled this way. For example, suppose a supply network has suppliers who supply widgets and has demanders who demand them. Each supplier can send at most one widget to each demander.

The 0–1 Matrix Problem (restated)   *Is there a way to determine which demander each supplier sends widgets to in a way that satisfies constraints on how many widgets each supplier produces and on how many widgets each demander demands?*

We can model this situation with some kind of graph. First draw m "row" vertices and n "column" vertices, along with two additional vertices: one called s and one called t. From s, draw an arc with capacity r_i to the ith row vertex. From the ith column vertex, draw an arc with capacity s_i to t. Then, from every row vertex, draw an arc with capacity 1 to every column vertex.

To decide whether there is a 0–1 matrix with the specified properties, we find the maximum flow from s to t in the directed graph we created above. If there's a flow of magnitude $r_1 + r_2 + \cdots + r_m$, then the answer to the original matrix question is "yes." Furthermore, the arcs that go from row vertices to column vertices that have a flow of 1 on them in the solution to the maximum-flow problem correspond to the places to put the 1's in the matrix to achieve the desired property. If the maximum flow you find is less than $r_1 + r_2 + \cdots + r_m$, we can say "no" to the original matrix question. Figure 8.18 shows the graph constructed in accordance with this transformation with the data specified in (8.8).

■ **Figure 8.18**

Directed graph with arc capacities

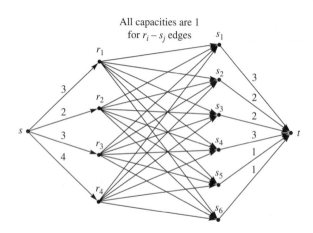

Policing a City

Suppose you are the commander of a company of military police soldiers and you are charged with policing all the roads in a section of a city. The section of the city in question can be modeled with the graph in Figure 8.19, where the road intersections in the city are vertices and the roads between the intersections are edges. Suppose the roads in the city are straight enough and short enough so a soldier stationed at an intersection can effectively police all roads immediately incident with that intersection. For example, a soldier stationed at vertex (intersection) 7 can police the road between 4 and 7, the road between 6 and 7, and the road between 7 and 8.

■ **Figure 8.19**

Vertex cover example graph

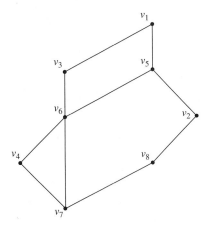

Vertex Cover Problem *What is the smallest number of military police soldiers needed to accomplish the mission of policing all the roads?*

It is easy to see that all roads can be policed by placing a soldier at every vertex. This requires 8 soldiers but may not be the smallest number of soldiers required. A better solution places soldiers at intersections 3, 4, 5, 6, and 8—using only 5 soldiers. Note that you can't improve on this solution simply by removing a soldier (and not moving another one). An even better solution is to locate soldiers at intersections 1, 2, 6, and 7. It turns out that using only 4 soldiers is the best you can do for this graph.

Let's describe this problem in mathematical terms.

Vertex Cover Problem (restated) *Given a graph $G = (V(G), E(G))$, the vertex cover problem seeks a subset $S \subseteq V(G)$ of smallest number of elements such that every edge in the graph is incident with at least one element in S.*

An edge e is said to be **incident** with a vertex v if one of the endpoints of e is v. If G is a graph, we denote the minimum size of a vertex cover as $\beta(G)$. Another way of expressing this is $\beta(G) = \min\{|S| : S \text{ is a vertex cover}\}$.

8.3 PROBLEMS

1. Suppose you are an actor and your Bacon number is 3. Can future events ever cause your Bacon number to rise above 3? What, in general, can you say about an actor's Bacon number in terms of how it can change over time?

2. Just as actors have their Bacon numbers, there is a relation defined between authors of scholarly papers and the prolific Hungarian mathematician Paul Erdös. Use the Internet to find out what you can about Paul Erdös and Erdös numbers. Consider your favorite mathematics professor (no jokes here, please!). Can you determine his or her Erdös number?

3. Can you think of other relations that one could consider?

4. For the example in the text above, explain why, in Table 8.5, the entries in row 3, column 7, and row 4, column 7 are the same. *Hint:* Plot the data and draw a line segment from point 3 to point 7, and another from point 4 to point 7.

5. Using the same data set from the example in the text,

$$S = \{(0, 0), (2, 5), (3, 1), (6, 4), (8, 10), (10, 13), (13, 11)\}$$

recompute Table 8.5 with $\alpha = 2$ and $\beta = 1$.

6. Consider the data set $S = \{(0, 0), (2, 9), (4, 7), (6, 10), (8, 20)\}$. Using $\alpha = 5$ and $\beta = 1$, determine the best piecewise linear function for S.

7. Write computer software that finds the best piecewise linear function given a data set S along with α and β.

8. In the text for this section, there is the sentence "When G is bipartite with bipartition $\langle A, B \rangle$, it is clear that no matching can be bigger than $|A|$, and no matching can be bigger than $|B|$." Explain why this is true.

9. Find a maximum matching in the graph in Figure 8.20. How many edges are in the maximum matching? Now suppose we add the edge bh to the graph. Can you find a larger matching?

■ **Figure 8.20**

Graph for Problem 9

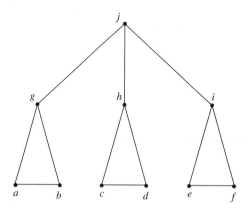

10. A basketball coach needs to find a starting lineup for her team. There are five positions that must be filled: point guard (1), shooting guard (2), swing (3), power forward (4), and center (5). Given the data in Table 8.7, create a graph model and use it to find a feasible starting lineup. What changes if the coach decides she can't play Hermione in position 3?

Table 8.7   Positions players can play

Alice	Bonnie	Courtney	Deb	Ellen	Fay	Gladys	Hermione
1, 2	1	1, 2	3, 4, 5	2	1	3, 4	2, 3

11. Will graphs formed with the procedure used to make the one in Figure 8.17 always be bipartite, regardless of the data? Why or why not?

12. Considering the values below, determine whether there is a 0–1 matrix with m rows and n columns, with row sums r_i and column sums s_j. If there is such a matrix, write it down.

$$m = 3, n = 6$$
$$r_1 = 4, r_2 = 2, r_3 = 3$$
$$s_1 = 2, s_2 = 2, s_3 = 1, s_4 = 0, s_5 = 3, s_6 = 1$$

(8.9)

13. Considering the following values, determine whether there is a 0–1 matrix with m rows and n columns, with row sums r_i and column sums s_j. If there is such a matrix, write it down.

$$m = 3, n = 5$$
$$r_1 = 4, r_2 = 2, r_3 = 3$$
$$s_1 = 3, s_2 = 0, s_3 = 3, s_4 = 0, s_5 = 3$$

(8.10)

14. Explain, in your own words, why a maximum-flow algorithm can solve the matrix problem from this section.

15. A path on n vertices, P_n, is a graph with vertices that can be labeled $v_1, v_2, v_3, \ldots, v_n$ so that there is an edge between v_1 and v_2, between v_2 and v_3, between v_3 and $v_4, \ldots$, and between v_{n-1} and v_n. For example, the graph P_5 appears in Figure 8.21. Compute $\beta(P_5)$. Compute $\beta(P_6)$. Compute $\beta(P_n)$ (your answer should be a function of n).

1   2   3   4   5

■ **Figure 8.21**

P_5

16. Here we consider the *weighted* vertex cover problem. Suppose the graph in Figure 8.22 represents an instance of vertex cover in which the cost of having vertex i in S is $w(i) = (i - 2)^2 + 1$ for $i = 1, 2, 3, 4, 5$. For example, if v_4 is in S, we must use $w(4) = (4 - 2)^2 + 1 = 5$ units of our resource. Now, rather than minimizing the *number* of vertices in S, we seek a solution that minimizes the total amount of resource used $\sum_{i \in S} w(i)$. Using our analogy of soldiers guarding intersections, you can think of $w(i)$ as a description of the number of soldiers needed to guard intersection i. Given the graph in Figure 8.22 and the weighting function $w(i) = (i - 2)^2 + 1$, find a minimum-cost weighted vertex cover.

■ **Figure 8.22**

Graph for Problem 16

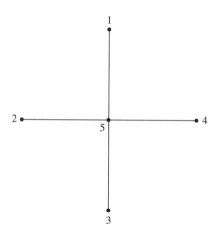

8.3   PROJECTS

1. Investigate a social network that is of interest to you. Carefully define what the vertices represent and what the edges represent. Are there any new modeling techniques that you had to employ?

2. Write a computer program that takes integers m, n, r_i for $1 \le i \le m$ and s_j for $1 \le j \le n$ as input and that either outputs a 0–1 matrix with m rows and n columns with row sums r_i and column sums s_j, or says that no such matrix can exist (some programming experience required).

3. Given a graph $G = (V(G), E(G))$, consider the following strategy for finding a minimum vertex cover in a graph.

   Step 0:   Start with $S = \emptyset$.

   Step 1:   Find a vertex v of maximum degree (one that has a greatest number of incident edges). Add this vertex to S.

   Step 2:   Delete v and all its incident edges from G. If G is now edgeless, stop. Otherwise, repeat Steps 1 and 2.

   Either show that this strategy always finds a minimum vertex cover for G or find a graph for which it fails to do so.

4. Consider a modification of the softball manager's problem where we are interested in the *best* starting lineup. How can our mathematical model be modified to solve this problem? What new techniques are needed to solve models of this type?

8.3   Further Reading

Ahuja, R., T. Magnanti, & J. Orlin. *Network Flows: Theory, Algorithms, and Applications.* Englewood Cliffs, NJ: Prentice Hall, 1993.

Huber, M., & S. Horton. "How Ken Griffey, Jr. is like Kevin Bacon or, degrees of separation in baseball." *Journal of Recreational Mathematics*, 33 (2006): 194–203.

Wilf, H. *Algorithms and Complexity*, 2nd ed. Natick, MA: A. K. Peters, 2002.

Using Graph Models to Solve Problems

In Section 8.3 we saw several real-world problems that can be expressed as problems on graphs. For example, we can solve instances of the Bacon number problem by finding the distance between two vertices in a graph. We also saw that the data-fitting problem could be transformed into a shortest-path problem. We learned that the softball manager's problem and a 0–1 matrix problem could be solved using maximum flows. Finally, we established a relationship between a problem about stationing police on street corners and finding vertex covers in graphs.

In this section we will learn about some simple ways to solve instances of some of these graph problems. This is part of an approach to problem solving similar to our approach in Chapter 1. We start with a real problem, often expressed in nonmathematical terms. Using the insight that comes from practice, experience, and creativity, we recognize that the problem can be expressed in mathematical terms—in these cases, using graphs. Often the resulting graph problem is familiar to us. If so, we solve the problem and then translate the solution back to the language and setting of our original problem.

Solving Shortest-Path Problems

Given a graph $G = (V(G), E(G))$ and a pair of vertices u and v in $V(G)$, we can define the *distance* from u to v as the smallest number of edges in a path from u to v. We use the notation $d(u, v)$ to denote this distance. The *distance problem*, then, is to compute $d(u, v)$ given a graph G and two specific vertices u and v.

Rather than considering methods to solve the distance problem, we will consider a generalization of the problem. Let c_{ij} denote the *length* of the edge ij. Now we can define the length of a shortest path from u to v in a graph to be the minimum (over all possible paths) of the sum of the lengths of the edges in a path from u to v. Given a graph G, edge lengths c_{ij} for each edge $ij \in E(G)$, and two specific vertices u and v, the *shortest-path problem* is to compute the length of a shortest path from u to v in G. It is easy to observe that the distance problem from the previous paragraph is a special case of the shortest-path problem—namely, the case where each $c_{ij} = 1$. Accordingly, we will focus on a technique to solve shortest-path problems, because such a technique will also solve distance problems.

For example, consider the graph in Figure 8.23. Ignoring the edge lengths printed on the figure, it is easy to see that $d(u, y) = 2$ because there is a path of two edges from vertex u to vertex y. However, when the edge lengths are considered, the shortest path from u to y in G goes through vertices v and w and has total length $1 + 2 + 3 = 6$. Thus $d(i, j)$

■ **Figure 8.23**

Example graph for shortest path

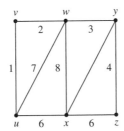

considers only the number of edges between i and j, whereas shortest paths consider edge lengths.

Shortest-path problems have a simple structure that enables us to solve them with very intuitive methods. Consider the following physical analogy. We are given a graph with two specific vertices u and v, as well as edge lengths for each edge. We seek a shortest path from vertex u to vertex v. We'll make our model entirely out of string. We tie the string together with knots that don't slip. The knots correspond to the vertices, and the strings are the edges. Each string is equal in length to the weight (distance) on the edge it represents (plus enough extra length to tie the knots). Of course, if the distances represented in the graph are in miles, we might build our string model scaling, say, miles to inches. Now, to find the length of a shortest path, we hold the two knots representing u and v and pull them apart until some of the strings are tight. Now the length of a shortest path in our original graph corresponds to the distance between the two knots. Furthermore, this model shows which path(s) are shortest (those whose strings are tight). The loose strings are said to have "slack," which means that even if they were slightly shorter, the length of a shortest path would not change.

This analogy underlies the algorithm we will use to solve shortest-path problems. Note that the procedure we will present here works to find a shortest path in any graph satisfying $c_{ij} \geq 0$ for all $ij \in E(G)$. The following procedure will find the length of a shortest path in a graph G from vertex s to vertex t, given nonnegative edge lengths c_{ij}.

Dijkstra's Shortest-Path Algorithm

Input A graph $G = (V(G), E(G))$ with a source vertex s and a sink vertex t and nonnegative edge lengths c_{ij} for each edge $ij \in E(G)$.

Output The length of a shortest path from s to t in G.

Step 0 Start with temporary labels L on each vertex as follows: $L(s) = 0$ and $L(i) = \infty$ for all vertices except s.

Step 1 Find the vertex with the smallest temporary label (if there's a tie, pick one at random). Make that label *permanent* meaning it will never change.

Step 2 For every vertex j without a permanent label that is adjacent to a vertex with a permanent label, compute a new temporary label as follows: $L(j) = \min\{L(i) + c_{ij}\}$, where we minimize over all vertices i with a *permanent* label. Repeat Steps 1 and 2 until all vertices have permanent labels.

This algorithm was formulated by the Dutch computer scientist Edsger Dijkstra (1930–2002). When Dijkstra's Algorithm stops, all vertices have permament labels, and each label $L(j)$ is the length of a shortest path from s to j. We now demonstrate Dijkstra's Algorithm on the graph in Figure 8.23.

Let's start with some notation. Let $L(V) = (L(u), L(v), L(w), L(x), L(y), L(z))$ be the current labels on the vertices of the graph in Figure 8.23, listed in the specified order. We will also add an asterisk to any label we have made permanent. Thus $L(V) = (0^*, 1^*, 3^*, 6, 6, \infty)$ would mean the labels on vertices u, v, w, x, y, and z are 0, 1, 3, 6, 6, and ∞, respectively, and that the labels on u, v, and w are permanent.

Now we are ready to execute Dijkstra's Algorithm to find the shortest distances in G from u to every other vertex. First, in Step 0 we initialize the labels $L(V) = (0, \infty, \infty, \infty, \infty, \infty)$. Next, in Step 1 we make the smallest label permanent, so

$L(V) = (0^*, \infty, \infty, \infty, \infty, \infty)$. In Step 2 we compute a new temporary label for each vertex without a permanent label that is adjacent to one with a permanent label. The only vertex with a permanent label is u, and it is adjacent to v, w, and x. Now we can compute $L(v) = \min\{L(i) + c_{iv}\} = L(u) + c_{uv} = 0 + 1 = 1$. Similarly, $L(w) = \min\{L(i) + c_{iw}\} = L(u) + c_{uw} = 0 + 7 = 7$ and $L(x) = \min\{L(i) + c_{ix}\} = L(u) + c_{ux} = 0 + 6 = 6$. Accordingly, our new label list is $L(V) = (0^*, 1, 7, 6, \infty, \infty)$.

Now we repeat Steps 1 and 2. First we examine the current $L(V)$ and note that the smallest temporary label is $L(v) = 1$. Therefore, we make label $L(v)$ permanent and update our label list accordingly: $L(V) = (0^*, 1^*, 7, 6, \infty, \infty)$. In Step 2, there are two vertices with permanent labels to consider: u and v. The situation at x will not change; $L(x)$ will remain at 6. However, there are now two cases to consider as we recompute the label at w: $L(w) = \min\{L(i) + c_{iw}\} = \min\{L(u) + c_{uw}, L(v) + c_{vw}\} = \min\{0 + 7, 1 + 2\} = \min\{7, 3\} = 3$. Our new label list is $L(V) = (0^*, 1^*, 3, 6, \infty, \infty)$.

We continue repeating Steps 1 and 2. A computer scientist would say that we are now on "iteration three" of the main step of the algorithm. First, we add a permanent label to our list, obtaining $L(V) = (0^*, 1^*, 3^*, 6, \infty, \infty)$. After Step 2 we arrive at $L(V) = (0^*, 1^*, 3^*, 6, 6, \infty)$. In iteration four, we pick one of the vertices with $L(i) = 6$; our algorithm says to break ties at random, so say we pick y to get the permanent label. This results in the label list $L(V) = (0^*, 1^*, 3^*, 6, 6^*, \infty)$. After the next step, we obtain $L(V) = (0^*, 1^*, 3^*, 6, 6^*, 10)$, followed (in iteration five) by $L(V) = (0^*, 1^*, 3^*, 6^*, 6^*, 10)$ and then $L(V) = (0^*, 1^*, 3^*, 6^*, 6^*, 10)$. Note that Step 2 of iteration five resulted in no change to $L(V)$ because the path uxz doesn't improve on the other path $uvwyz$ from u to z that we identified in the previous iteration. Finally, in iteration six, we make the last temporary label permanent. After that, there's nothing left to do, and the algorithm stops with $L(V) = (0^*, 1^*, 3^*, 6^*, 6^*, 10^*)$. This list gives the lengths of shortest paths from u to every vertex in the graph.

Solving Maximum-Flow Problems

We saw in Section 8.3 that both the data-fitting problem and the 0–1 matrix problem can be solved by finding a maximum flow in a related graph. We will see in Section 8.4 that the softball manager's problem can also be solved this way. In fact, many practical problems can be modified as maximum flows in graphs. In this section we present a simple technique for finding a maximum flow in a graph.

So far, all of the graphs we have considered have vertices and edges, but the edges don't have any notion of direction. That is, an edge goes between some pair of vertices rather than specifically from one vertex *to* another. In this section, we will consider directed graphs instead. Directed graphs are just like their undirected siblings, except that each edge has a specific direction associated with it. We will use the term **arc** to refer to a directed edge. More formally, we can define a **directed graph** $G = (V(G), A(G))$ as two sets: a **vertex set** $V(G)$ and an **arc set** $A(G)$. Each element of $A(G)$ is an **ordered pair** of elements of $V(G)$. We will use the notation (i, j) to represent an arc oriented from i to j. Note that undirected graphs can be transformed into directed graphs in many ways. Given an undirected graph, an orientation for each edge in $E(G)$ can be selected. Alternatively, each edge $ij \in E(G)$ can be transformed into two arcs; one from i to j and one from j to i.

Now we return to the problem of solving maximum-flow problems in (directed) graphs. Just as in shortest-path problems, there are many algorithms for finding maximum flows in graphs. These procedures go from very simple and intuitive algorithms to much more detailed methods that can have theoretical or computational advantages over other approaches. Here we will focus on the former. This chapter concludes with references where the interested reader can explore more sophisticated algorithms.

Given a directed graph $G = (V(G), A(G))$, two specific vertices s and t, and a finite flow capacity u_{ij} for each arc (i, j), we can use a very simple technique for finding a maximum flow from s to t in G. We'll demonstrate the technique with an example first, and then we'll describe in more general terms what we did.

Before we start, we need to define the concept of a directed path. A **directed path** is a path that respects the orientation of arcs in the path. For example, consider the graph in Figure 8.24. This directed graph has vertex set $V(G) = \{s, a, b, c, d, t\}$ and arc set $A(G) = \{sa, sb, ab, ac, bc, bd, ct, dc, dt\}$. Note that $ab \in A(G)$ but $ba \notin A(G)$. Accordingly, $s - a - b - d - t$ is a directed path from s to t, but $s - b - a - c - t$ is not (because $ba \notin A(G)$). The graph also has arc flow capacities annotated by each arc.

■ **Figure 8.24**

Example graph for
maximum flow

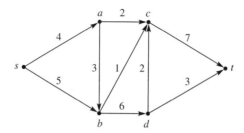

Now we are ready to begin our example. Consider again the graph in Figure 8.24. We start by finding a directed path from s to t. There are several choices, but let's say we pick the path $s - a - c - t$. We observe that the *lowest-capacity* arc on this path is ac, which has capacity $u_{ac} = 2$. Accordingly, we create a new graph that represents the remaining capacity, or *residual capacity* of the network after 2 units of flow are pushed along the path $s - a - c - t$. All we do is account for the flow by reducing the remaining capacity by 2 for each arc in the path. Note that this reduces the residual flow on arc ac to zero, so we delete that arc from the next graph. The result of these changes appears in Figure 8.25. At this point we have managed to get 2 units of flow from s to t. This completes the first iteration of our maximum-flow algorithm.

■ **Figure 8.25**

Example graph for
maximum flow, after
Iteration 1

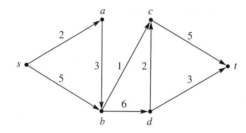

Now we are ready to start the second iteration. Again, we look for a directed path from s to t in the residual graph from the previous iteration (Figure 8.25). The directed path

$s - b - d - t$ is one possiblility. The smallest arc capacity in the residual graph is $u_{dt} = 3$, so we create another residual graph by reducing all of the capacities along the path. In this iteration we pushed 3 more units of flow from s to t, which brings our total so far to 5 units. The result appears in Figure 8.26, and we have finished iteration 2.

■ **Figure 8.26**

Example graph for maximum flow, after Iteration 2

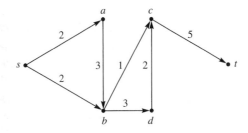

Iteration 3 proceeds like the others. The residual graph in Figure 8.26 has the directed path $s - b - d - c - t$ with a minimum capacity of 2 units of flow. We have now delivered 7 units of flow from s to t. The appropriate modification leads us to Figure 8.27.

■ **Figure 8.27**

Example graph for maximum flow, after Iteration 3

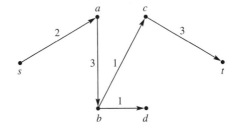

At this point there is only one directed path from s to t; it is $s - a - b - c - t$. Because $u_{bc} = 1$ in Figure 8.27, we add the 1 unit of flow to our previous total, bringing us to 8. After reducing the capacity on each arc in the directed path by 1, we obtain the graph in Figure 8.28.

■ **Figure 8.28**

Example graph for maximum flow, final residual graph

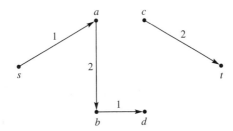

We continue by looking for a directed path from s to t in Figure 8.28. It is clear that there is no such path. Accordingly, we stop and claim that the maximum flow from s to t in our original graph of Figure 8.24 is 8. Note that although we could have made different choices for the paths we selected, the resulting maximum flow would be 8 regardless of which choice we made.

Finally, we can generalize our maximum-flow algorithm. But first, we need to be a little more precise in describing our problem.

Maximum-Flow Problem *Given a directed graph $G = (V(G), A(G))$ with a source vertex s and a sink vertex t and a finite flow capacity u_{ij} for each arc $ij \in A(G)$, find the maximum flow in the graph from vertex s to vertex t.*

The algorithm we used above is more formally and more generally described below.

Maximum-Flow Algorithm

Input A directed graph $G = (V(G), A(G))$ with a source vertex s and a sink vertex t and a finite flow capacity u_{ij} for each arc $ij \in A(G)$.

Output The maximum flow from s to t in G.

Step 0 Set the current flow to zero: $f_c \leftarrow 0$.

Step 1 Find a directed path from s to t in the current graph. If there is no such path, stop. The maximum flow from s to t in G is f_c.

Step 2 Compute $u_{\min}$, the minimum capacity in the current graph of all the arcs in the directed path.

Step 3 For each arc ij in the directed path, update the residual capacity in the current graph: $u_{ij} \leftarrow u_{ij} - u_{\min}$.

Step 4 Set $f_c \leftarrow f_c + u_{\min}$ and return to Step 1.

Now we turn our attention to using the Maximum-Flow Algorithm to solve other problems.

Solving Bipartite Matching Problems Using Maximum Flows In Section 8.3 we learned that the softball manager's problem can be solved by finding a maximum matching in a bipartite graph derived from the problem instance. In this section, we will find that the Maximum-Flow Algorithm from the previous section can be used to find maximum matchings in bipartite graphs. By combining these two, we will be able to solve the softball manager's problem using our maximum-flow algorithm. This is an exciting aspect of mathematical modeling: By looking at problems the right way, we can often solve them using techniques that at first glance seem unrelated.

Recall that a *matching* in a graph G is a subset $S \subseteq E(G)$ such that no two edges in S end at a common vertex. In other words, the edges in S are *independent*. A maximum matching is a matching of largest size. If the graph G is *bipartite*, the maximum-matching problem can be solved using a maximum-flow algorithm. Recall that a graph is bipartite if the vertex set can be partitioned into two sets X and Y such that all of the edges have one end in X and the other end in Y. Figure 8.29 is an example of a bipartite graph. Which edges should we put in S to make $|S|$ as large as possible?

We might start by including $x_1 y_1$ in S. This choice eliminates all other edges incident with x_1 or y_1 from further consideration. Accordingly, we can also include $x_2 y_3$ in S. Now we have two edges in S. Unfortunately, we cannot add more edges to S, at least not without removing some edges that are already in. Can you find a matching larger than size 2 in G? It seems that the initial choices we made were poor in the sense that they blocked too many other options. For this small problem, it was probably easy for you to see how to get a larger matching. In fact you can probably find a maximum matching just by looking at

■ **Figure 8.29**

Maximum-matching
problem

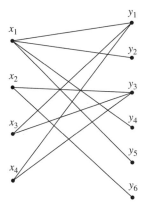

the problem; recall from Problem 8 in Section 8.3 that no matching in this graph can have
more than $\min\{|X|, |Y|\} = 4$ edges.

A larger instance of this problem (that is, a larger graph) might be more difficult to
solve by inspection. As the title of this section suggests, there is a way to solve bipartite
matching problems using a maximum-flow algorithm. Here's how. From the (undirected)
instance bipartite graph with bipartition $V(G) = \langle X, Y \rangle$, we orient each edge so that it goes
from X to Y. We allow each of these arcs to have an unlimited capacity. Then we create
two new vertices s and t. Finally, we create arcs from s to every vertex in X, and arcs from
every vertex in Y to t. Each of these arcs has a capacity of 1. Figure 8.30 demonstrates the
idea. Now we find the maximum flow in the resulting directed graph from s to t, and that
maximum flow will equal the size of a maximum matching in the original graph.

■ **Figure 8.30**

Matching problem
expressed as a
maximum-flow problem

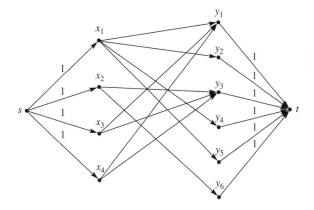

One important feature that shortest paths and maximum flows have in common is that
they can be found *efficiently* by a computer. Unfortunately, not all graph problems have this
property. We now consider some problems that can be described just as easily but seem
to be much harder to solve by algorithmic means. In Section 8.3, we learned about one
such problem: vertex cover. Here we introduce one more, and then we make a surprising
observation about these two problems.

Consider a department store that delivers goods to customers. Every day, a delivery
crew loads a truck at the company's warehouse with the day's orders. Then the truck is

driven to each customer and returned to the warehouse at the end of the day. The delivery crew supervisor needs to decide on a route that visits each customer once and then returns to the starting point. The set of customers who require a delivery changes each day, so the supervisor has a new instance to solve every day. We can state this problem as follows.

Traveling Salesman Problem   *A salesman must visit every location in a list and return to the starting location. In what order should the salesman visit the locations in the list to minimize the distance traveled?*

It is easy to see that this problem can be modeled with a graph. First we note that this problem is usually modeled with a graph that is *complete*. A complete graph has an edge between every pair of vertices. We also need a cost c_{ij} for every edge. This number represents the cost of traveling on the edge between vertex i and vertex j. You can think of this cost as distance, or as time, or as an actual monetary cost. A list of all of the locations, in any order, defines a **tour**. The **cost** of a tour is the sum of the edge costs on that tour. For example, consider the graph in Figure 8.31. Let's say we start at vertex a (it turns out not to matter where we start). It seems reasonable to go first to vertex b, because the cost of doing so is small; only 1 unit. From there, the edge to c looks attractive. So far we have paid $1 + 2 = 3$ units to get this far. From c, we can go to d, and then to e (the only vertex we have not yet visited). Now our cumulative cost is $1 + 2 + 4 + 8 = 15$. Finally, we have to return to the starting point, which adds 3 to our cost, bringing the total to 18.

■ Figure 8.31

Example graph for traveling salesman problem

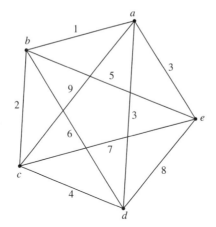

It turns out that this is not the optimal tour. The tour $a - d - c - b - e - a$ has a cost of 17 (the reader should verify this). Note that this improved tour does not use the edge from a to b that seemed so attractive. Now we can restate the problem using graph theory terms.

Traveling Salesman Problem (restated)   *Given a complete graph $G = (V(G), E(G))$ and a cost c_{ij} for each edge in $E(G)$, find a tour of minimum cost.*

Finally, it is important to understand that this problem is different from the shortest-path problems that we have considered. It may seem as though some of the same techniques can be used, but it turns out that this isn't the case. Instances of the shortest-path problem can always be solved efficiently, whereas the Traveling Salesman Problem (and the vertex cover problem) are not so easy to deal with. In fact, this section is about describing *algorithms* to solve various graph problems. It might come as a suprise to learn that there is *no* known efficient algorithm for solving either the Traveling Salesman Problem or the vertex cover problem. For smaller instances, it is often possible to enumerate—that is, to check all of the possible solutions. Unfortunately, that strategy is doomed to take too long for large instances. We will elaborate on this idea in the next section.

8.4 PROBLEMS

1. Find a shortest path from node a to node j in the graph in Figure 8.32 with edge weights shown on the graph.

■ Figure 8.32

Graph for Problem 1

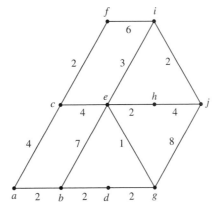

2. A small suburban city is experimenting with a new way to keep its main park clean. From May through October, a crew is needed every day to pick up and remove trash. Rather than contracting with one company for the entire period, the city manager takes bids from firms on its website. Firms submit bids with dates of work and the fee for which the firm is willing to do the clean-up work. On some published date, the city manager reviews all of the bids submitted and decides which ones to accept. For example, a firm might make a bid to perform the work June 7–20 for $1200. Another firm might bid to work June 5–15 for $1000. Because the bids overlap, they cannot both be accepted. How can the city manager use Dijkstra's Algorithm to choose what bids to accept to minimize costs? What assumptions do you need to make?

3. Use our maximum-flow algorithm to find the maximum flow from s to t in the graph of Figure 8.30.

4. Explain why the procedure for using a maximum-flow algorithm to find the size of a maximum matching in a bipartite graph works. Can it be used to find a maximum matching (as opposed to the size of one)? Can it be used for graphs that are not bipartite? Why or why not?

5. In the sport of orienteering, contestants ("orienteers") are given a list of locations on a map ("points") that they need to visit. Orienteering courses are typically set up in natural areas such as forests or parks. The goal is to visit each point and return to the starting

location as quickly as possible. Suppose that an orienteer's estimates of the time it will take her between each pair of points are given in Table 8.8. Find the tour that minimizes the orienteer's time to negotiate the course.

Table 8.8   Orienteer's time estimates between points (in minutes)

	start	a	b	c	d
start	—	15	17	21	31
a	14	—	19	14	17
b	27	22	—	18	19
c	16	19	26	—	15
d	18	22	23	29	—

6. Does Dijkstra's Algorithm work when there might be arcs with *negative* weights?

8.4 Further Reading

Ahuja, R., T. Magnanti, & J. Orlin. *Network Flows: Theory, Algorithms, and Applications.* Englewood Cliffs, NJ: Prentice Hall, 1993.

Buckley, F., & F. Harary. *Distance in Graphs.* NY: Addison-Wesley, 1990.

Lawler, E., J. Lenstra, A. R. Kan, & D. Shmoys, eds. *The Traveling Salesman Problem.* NY: Wiley, 1985.

Skiena, S., & S. Pemmaraju. *Computational Discrete Mathematics: Combinatorics and Graph Theory with Mathematica.* Cambridge: Cambridge University Press, 2003.

Winston, W., & M. Venkataramanan. *Introduction to Mathematical Programming: Applications and Algorithms, Volume 1,* 4th ed. Belmont, CA: Brooks-Cole, 2003.

8.5 Connections to Mathematical Programming

In the previous chapter, we learned about modeling decision problems with linear programs and then using the Simplex Method to solve the associated linear program. In this section, we will consider how linear programming and integer programming can be used to model some of the problems presented in the previous section.

Vertex Cover

Recall the vertex cover problem discussed in Section 8.3. In words, we are looking for a set S that is a subset of $V(G)$ such that every edge in the graph is incident with a member of S, and we want $|S|$ to be as small as possible. We have learned that this can be a hard problem to solve, but sometimes integer programming can help.

Consider the graph in Figure 8.33. Let's try to find a minimum vertex cover. Each vertex we put in S in a sense reduces the number of edges that are still uncovered. Perhaps we should start by being *greedy*—that is, by looking for a vertex of highest degree (the degree of a vertex is the number of edges that are incident with it). This is a greedy choice

■ **Figure 8.33**

Example graph

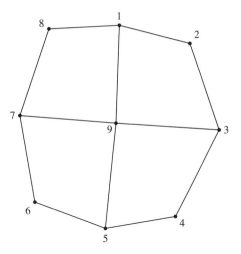

because it gives us the biggest possible return for our "investment" of adding one vertex to S. Vertex 9 is the only vertex that allows us to cover four edges with one vertex, so this seems like a great start. Unfortunately, however, it turns out that this choice isn't optimal; the only minimum vertex cover for this graph is $S = \{1, 3, 5, 7\}$, and our initial choice, vertex 9, is conspicuously absent.

Our small example could certainly be solved by trial and error, but we will see shortly that trial and error can quickly become a completely useless solution strategy for larger graphs: There are simply too many possible solutions to check. Integer programming can sometimes provide an alternative.

An Integer Programming Model of the Vertex Cover Problem   The key to modeling with integer programs is the same as the key to other kinds of modeling. One of the most important steps in any modeling problem is to define your variables. Let's do that now. We can let

$$x_i = \begin{cases} 1 & \text{if } i \in S \\ 0 & \text{otherwise} \end{cases} \tag{8.11}$$

The variable x_i is sometimes called a **decision variable** because each feasible value you choose to assign to x_i represents a decision you might make regarding the vertices of the graph in Figure 8.33 *vis-à-vis* their membership in S. For example, setting $x_5 = 1$ represents including vertex 5 in S, and $x_7 = 0$ represents excluding vertex 7 from S. If we define the vector $x = \langle x_1, x_2, \ldots, x_9 \rangle$, we see that any binary string of length 9 can be thought of as a choice of vertices to include in a vertex cover S.

As an aside, you might think of how many possible decisions there are at this point. Each x_i must take on one of two possible values. Thus there are $2^9 = 512$ possibilities. This isn't too bad, but if we try this problem on a somewhat larger graph of, say, 100 verticies, we would have to consider $2^{100} = 1,267,650,600,228,229,401,496,703,205,376$ different possible choices for which vertices go in S! Clearly, any solution strategy that checks all of these possibilities individually, *even with a very fast computer*, is doomed to take too much time to be useful. (You might enjoy doing a little computation to check this out. See Problem 3 at the end of this section.)

Of course, some choices are better than others! In fact, there are two big issues to consider when we make our choice. First, our choice must be *feasible*. That is, it must satisfy the requirement that every edge be incident with a vertex that is in S. We really have to check all of the edges to make sure this requirement, which is also called a *constraint*, is satisfied. Just saying it this way helps us understand how to write the constraint in the language of integer programming. Consider the edge incident with vertex 4 and vertex 5. The presence of this edge requires that either x_4 or x_5 (or both) equal 1. One way to write this algebraically is

$$(1 - x_4)(1 - x_5) = 0 \qquad (8.12)$$

You should take a moment to convince yourself that this equality is exactly equivalent to the previous sentence. In other words, verify that (8.12) is satisfied if and only if $x_4 = 1$ or $x_5 = 1$ (or both).

Unfortunately, it turns out that (8.12) isn't the best way to write this idea down. In general, software that solves integer programs will perform better if constraints are written as *linear* inequalities (equalities are okay, too). The expression (8.12) isn't linear because we end up multiplying two decision variables. A much better formulation is to express (8.12) this way:

$$x_4 + x_5 \geq 1 \qquad (8.13)$$

Again you can verify that (8.13) is satisfied exactly when $x_4 = 1$ or $x_5 = 1$ (or both).

Certainly the same line of reasoning applies to all 12 edges in the graph. The following is a complete list of these constraints.

$$
\begin{aligned}
x_1 + x_2 &\geq 1 \\
x_2 + x_3 &\geq 1 \\
x_3 + x_4 &\geq 1 \\
x_4 + x_5 &\geq 1 \\
x_5 + x_6 &\geq 1 \\
x_6 + x_7 &\geq 1 \qquad (8.14)\\
x_7 + x_8 &\geq 1 \\
x_8 + x_1 &\geq 1 \\
x_1 + x_9 &\geq 1 \\
x_3 + x_9 &\geq 1 \\
x_5 + x_9 &\geq 1 \\
x_7 + x_9 &\geq 1
\end{aligned}
$$

Rather than write down this long list, we can write a single, more general expression that means the same thing:

$$x_i + x_j \geq 1 \quad \forall ij \in E(G) \qquad (8.15)$$

The inequality (8.15) captures exactly the 12 inequalities in (8.14).

Because we defined the decision variables so that they always take values of 0 or 1, we need a constraint in the integer program to account for that. We can write it this way:

$$x_i \in \{0, 1\} \quad \forall i \in V(G) \tag{8.16}$$

This forces each x_i to be either 0 or 1, as specified by our model.

Now we have taken care of all of the constraints. A choice of $x = \langle x_1, x_2, \ldots, x_9 \rangle$ that satisfies all of the constraints in (8.15) and (8.16) is called feasible or a **feasible point**. Now we simply seek, from all of the feasible points, the *best* one (or perhaps *a* best one). We need some measure of "goodness" to decide which choice for x is best. In this case, as we already said, we want the solution that minimizes $|S|$. In terms of our integer programming formulation, we want to minimize $\sum_{i \in V(G)} x_i$. This is called the **objective function**.

Finally, we can collect our objective function and our constraints (8.15) and (8.16) in one place to obtain a general way to write down *any* vertex cover problem as an integer program.

$$\text{Minimize } z = \sum_{i \in V(G)} x_i$$

subject to (8.17)

$$x_i + x_j \geq 1 \quad \forall ij \in E(G)$$
$$x_i \in \{0, 1\} \quad \forall i \in V(G)$$

Note that (8.17), without the last line $x_i \in \{0, 1\} \ \forall i \in V(G)$, is a *linear* program instead of an *integer* program. It turns out that solving integer programs often involves first solving the linear program obtained by dropping the integrality requirement. There is an extensive body of knowledge about how to formulate and solve linear and integer programming problems, and about how to interpret solution results. The list of readings at the end of the chapter can provide the interested reader with a few places to start.

There is one more point to consider about using integer programming to solve large instances of the vertex cover problem. The formulation 8.17 shows that we can at least formulate any vertex cover problem as an integer program. Unfortunately, it is not known whether there is a fast procedure for solving all integer programs. *Computational complexity* is the branch of theoretical computer science that considers issues such as this. The references at the end of the chapter provide additional information.

Maximum Flows

Now we reconsider the maximum-flow problem on directed graphs. This problem was defined in the previous section. We are given a directed graph $G = (V(G), A(G))$, a source vertex s, a sink vertex t, and a flow capacity u_{ij} for every arc $ij \in A(G)$. We start by defining variables to represent flow. We let x_{ij} represent the flow from vertex i to vertex j.

There are several types of constraints to consider. First, we will allow only nonnegative flows, so

$$x_{ij} \geq 0 \quad \forall ij \in A(G) \tag{8.18}$$

Recall that the symbol $\forall$ means "for all"; thus, the constraint $x_{ij} \geq 0$ applies for every arc ij in the graph's arc set $A(G)$.

We also know that flow on each arc is limited to the capacity, or upper bound, on that arc. The following constraint captures that idea.

$$x_{ij} \leq u_{ij} \quad \forall ij \in A(G) \tag{8.19}$$

Before we consider the last type of constraint, we need to make a key observation. At every vertex in the graph except for s and t, flow is *conserved*. That is, the flow *in* is equal to the flow *out* at every vertex (except s and t). Now we are ready to write down the flow balance constraint.

$$\sum_i x_{ij} = \sum_k x_{jk} \quad \forall j \in V(G) - \{s, t\} \tag{8.20}$$

The set $V(G) - \{s, t\}$ is just $V(G)$ with s and t removed; in this case, $V(G) - \{s, t\} = \{a, b, c, d\}$.

Now we turn our attention to the objective function. We seek to maximize flow from s to t. Because flow is conserved everywhere except at s and t, we observe that any flow going out of s must eventually find its way to t. That is, the quantity we want to maximize is the sum of all the flow out of s (we could instead maximize the sum of all the flow to t—the result would be the same). Therefore, our objective function is

$$\text{Maximize} \quad \sum_j x_{sj} \tag{8.21}$$

Combining (8.18), (8.19), (8.20), and (8.21), we can write any maximum-flow problem as a linear program as follows:

$$\text{Maximize } z = \sum_j x_{sj}$$

subject to

$$\sum_i x_{ij} = \sum_k x_{jk} \quad \forall j \in V(G) - \{s, t\} \tag{8.22}$$
$$x_{ij} \leq u_{ij} \quad \forall ij \in A(G)$$
$$x_{ij} \geq 0 \quad \forall ij \in A(G)$$

When the linear program (8.22) is solved, the resulting flow x is a maximum flow in G, and the largest amount of flow that can go from s to t is $\sum_j x_{sj}$.

Recall our maximum-flow example problem from the previous section (repeated in Figure 8.34 below for convenience).

■ **Figure 8.34**

Example graph for maximum flow

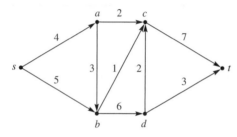

First note that constraints of type (8.18) require flow to be nonnegative on each arc. Thus, $x_{ac} \geq 0$ is one (of nine) such constraints for this instance. Constraints of type (8.19)

require that flow satisfy the upper bound given. The nine constraints of this type for this example include $x_{ac} \leq 2$ and $x_{ct} \leq 7$. Constraints of type (8.20) are for flow balance. There are four such constraints because there are four vertices, not including s and t. At vertex c, the flow balance constraint is $x_{ac} + x_{bc} + x_{dc} = x_{ct}$. The objective function for this instance is to maximize $x_{sa} + x_{sb}$.

8.5 PROBLEMS

1. Consider the graph shown in Figure 8.35.

Figure 8.35

Graph G

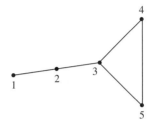

a. Find a minimum vertex cover in the graph in Figure 8.35.

b. Formulate an integer program to find a minimum vertex cover in the graph in Figure 8.35.

c. Solve the integer program in part (b) with computer software.

2. Consider again the graph in Figure 8.35. Now suppose that the cost of placing a vertex in S varies. Suppose the cost of placing vertex i in S is $g(i) = (-i^2 + 6i - 5)^3$ for $i \in \{1, 2, 3, 4, 5\}$. Repeat parts (a), (b), and (c) of the previous problem for this new version of the problem. This is an instance of *weighted* vertex cover.

3. Suppose a computer procedure needs to check all of 2^{100} possibilities to solve a problem. Assume the computer can check 1,000,000 possibilities each second.

a. How long will it take the computer to solve this problem this way?

b. Suppose that the computer company comes out with a new computer that operates 1000 times faster than the old model. How does that change the answer to question a? What is the *practical* impact of the new computer on solving the problem this way?

4. Write down the linear program associated with solving maximum flow from s to t in the graph in Figure 8.36.

Figure 8.36

Graph for Problem 1

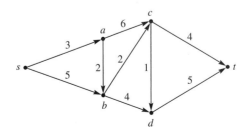

8.5 PROJECT

1. Express the softball manager's problem as a linear or integer program, and solve it with computer software.

8.5 Further Reading

Garey, M., & D. Johnson. *Computers and Intractability.* W. H. Freeman, 1978.

Nemhauser, G., & L. Wolsey. *Interger and Combinatorial Optimization.* New York: Wiley, 1988.

9

Dimensional Analysis and Similitude

Introduction

In the process of constructing a mathematical model, we have seen that the variables influencing the behavior must be identified and classified. We must then determine appropriate relationships among those variables retained for consideration. In the case of a single dependent variable this procedure gives rise to some unknown function:

$$y = f(x_1, x_2, \ldots, x_n)$$

where the x_i measure the various factors influencing the phenomenon under investigation. In some situations the discovery of the nature of the function f for the chosen factors comes about by making some reasonable assumption based on a law of nature or previous experience and construction of a mathematical model. We were able to use this methodology in constructing our model on vehicular stopping distance (see Section 2.2). On the other hand, especially for those models designed to predict some physical phenomenon, we may find it difficult or impossible to construct a solvable or tractable explicative model because of the inherent complexity of the problem. In certain instances we might conduct a series of experiments to determine how the dependent variable y is related to various values of the independent variable(s). In such cases we usually prepare a figure or table and apply an appropriate curve-fitting or interpolation method that can be used to predict the value of y for suitable ranges of the independent variable(s). We employed this technique in modeling the elapsed time of a tape recorder in Sections 4.2 and 4.3.

Dimensional analysis is a method for helping determine how the selected variables are related and for reducing significantly the amount of experimental data that must be collected. It is based on the premise that physical quantities have dimensions and that physical laws are not altered by changing the units measuring dimensions. Thus, the phenomenon under investigation can be described by a dimensionally correct equation among the variables. A dimensional analysis provides qualitative information about the model. It is especially important when it is necessary to conduct experiments in the modeling process because the method is helpful in testing the validity of including or neglecting a particular factor, in reducing the number of experiments to be conducted to make predictions, and in improving the usefulness of the results by providing alternatives for the parameters employed to present them. Dimensional analysis has proved useful in physics and engineering for many years and even now plays a role in the study of the life sciences, economics, and operations research. Let's consider an example illustrating how dimensional analysis can be used in the modeling process to increase the efficiency of an experimental design.

Introductory Example: A Simple Pendulum

Consider the situation of a simple pendulum as suggested in Figure 9.1. Let r denote the length of the pendulum, m its mass, and θ the initial angle of displacement from the vertical. One characteristic that is vital in understanding the behavior of the pendulum is the **period**, which is the time required for the pendulum bob to swing through one complete cycle and return to its original position (as at the beginning of the cycle). We represent the period of the pendulum by the dependent variable t.

■ **Figure 9.1**

A simple pendulum

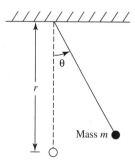

Problem Identification   *For a given pendulum system, determine its period.*

Assumptions   First, we list the factors that influence the period. Some of these factors are the length r, the mass m, the initial angle of displacement θ, the acceleration due to gravity g, and frictional forces such as the friction at the hinge and the drag on the pendulum. Assume initially that the hinge is frictionless, that the mass of the pendulum is concentrated at one end of the pendulum, and that the drag force is negligible. Other assumptions about the frictional forces will be examined in Section 9.3. Thus, the problem is to determine or approximate the function

$$t = f(r, m, \theta, g)$$

and test its worthiness as a predictor.

Experimental Determination of the Model   Because gravity is essentially constant under the assumptions, the period t is a function of the three variables length r, mass m, and initial angle of displacement θ. At this point we could systematically conduct experiments to determine how t varies with these three variables. We would want to choose enough values of the independent variables to feel confident in predicting the period t over that range. How many experiments will be necessary?

For the sake of illustration, consider a function of one independent variable $y = f(x)$, and assume that four points have been deemed necessary to predict y over a suitable domain for x. The situation is depicted in Figure 9.2. An appropriate curve-fitting or interpolation method could be used to predict y within the domain for x.

Next consider what happens when a second independent variable affects the situation under investigation. We then have a function

$$y = f(x, z)$$

Figure 9.2

Four points have been deemed necessary to predict y for this function of one variable x.

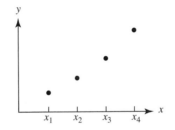

For each data value of x in Figure 9.2, experiments must be conducted to obtain y for four values of z. Thus, 16 (that is, 4^2) experiments are required. These observations are illustrated in Figure 9.3. Likewise, a function of three variables requires 64 (that is, 4^3) experiments. In general, 4^n experiments are required to predict y when n is the number of arguments of the function, assuming four points for the domain of each argument. Thus, a procedure that reduces the number of arguments of the function f will dramatically reduce the total number of required experiments. Dimensional analysis is one such procedure.

Figure 9.3

Sixteen points are necessary to predict y for this function of the two variables x and z.

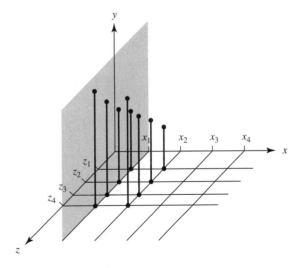

The power of dimensional analysis is also apparent when we examine the interpolation curves that would be determined after collecting the data represented in Figures 9.2 and 9.3. Let's assume it is decided to pass a cubic polynomial through the four points shown in Figure 9.2. That is, the four points are used to determine the four constants C_1–C_4 in the interpolating curve:

$$y = C_1 x^3 + C_2 x^2 + C_3 x + C_4$$

Now consider interpolating from Figure 9.3. If for a fixed value of x, say $x = x_1$, we decide to connect our points using a cubic polynomial in z, the equation of the interpolating surface is

$$y = D_1 x^3 + D_2 x^2 + D_3 x + D_4 + (D_5 x^3 + D_6 x^2 + D_7 x + D_8)z$$
$$+ (D_9 x^3 + D_{10} x^2 + D_{11} x + D_{12})z^2 + (D_{13} x^3 + D_{14} x^2 + D_{15} x + D_{16})z^3$$

Note from the equation that there are 16 constants—$D_1, D_2, \ldots, D_{16}$—to determine, rather than 4 as in the two-dimensional case. This procedure again illustrates the dramatic reduction in effort required when we reduce the number of arguments of the function we will finally investigate.

At this point we make the important observation that the experimental effort required depends more heavily on the number of arguments of the function to be investigated than on the true number of independent variables the modeler originally selected. For example, consider a function of two arguments, say $y = f(x, z)$. The discussion concerning the number of experiments necessary would not be altered if x were some particular combination of several variables. That is, x could be uv/w, where u, v, and w are the variables originally selected in the model.

Consider now the following preview of dimensional analysis, which describes how it reduces our experimental effort. Beginning with a function of n variables (hence, n arguments), the number of arguments is reduced (ordinarily by three) by combining the original variables into products. These resulting $(n - 3)$ products are called **dimensionless products** of the original variables. After applying dimensional analysis, we still need to conduct experiments to make our predictions, but the amount of experimental effort that is required will have been reduced exponentially.

In Chapter 2 we discussed the trade-offs of considering additional variables for increased precision versus neglecting variables for simplification. In constructing models based on experimental data, the preceding discussion suggests that the cost of each additional variable is an exponential increase in the number of experimental trials that must be conducted. In the next two sections, we present the main ideas underlying the dimensional analysis process. You may find that some of these ideas are slightly more difficult than some that we have already investigated, but the methodology is powerful when modeling physical behavior.

9.1 Dimensions as Products

The study of physics is based on abstract concepts such as mass, length, time, velocity, acceleration, force, energy, work, and pressure. To each such concept a unit of measurement is assigned. A physical law such as $F = ma$ is true, provided that the units of measurement are consistent. Thus, if mass is measured in kilograms and acceleration in meters per second squared, then the force must be measured in newtons. These units of measurement belong to the MKS (meter–kilogram–second) mass system. It would be inconsistent with the equation $F = ma$ to measure mass in slugs, acceleration in feet per second squared, and force in newtons. In this illustration, force must be measured in pounds, giving the American Engineering System of measurement. There are other systems of measurement, but all are prescribed by international standards so as to be consistent with the laws of physics.

The three primary physical quantities we consider in this chapter are mass, length, and time. We associate with these quantities the dimensions M, L, and T, respectively. The dimensions are symbols that reveal how the numerical value of a quantity changes when the units of measurement change in certain ways. The dimensions of other quantities follow from definitions or from physical laws and are expressed in terms of M, L, and

T. For example, velocity v is defined as the ratio of distance s (dimension L) traveled to time t (dimension T) of travel—that is, $v = st^{-1}$, so the dimension of velocity is LT^{-1}. Similarly, because area is fundamentally a product of two lengths, its dimension is L^2. These dimension expressions hold true regardless of the particular system of measurement, and they show, for example, that velocity may be expressed in meters per second, feet per second, miles per hour, and so forth. Likewise, area can be measured in terms of square meters, square feet, square miles, and so on.

There are still other entities in physics that are more complex in the sense that they are not usually defined directly in terms of mass, length, and time alone; instead, their definitions include other quantities, such as velocity. We associate dimensions with these more complex quantities in accordance with the algebraic operations involved in the definitions. For example, because momentum is the product of mass with velocity, its dimension is $M(LT^{-1})$ or simply MLT^{-1}.

The basic definition of a quantity may also involve dimensionless constants; these are ignored in finding dimensions. Thus, the dimension of kinetic energy, which is one-half (a dimensionless constant) the product of mass with velocity squared, is $M(LT^{-1})^2$ or simply ML^2T^{-2}. As you will see in Example 2, some constants (dimensional constants), such as gravity g, do have an associated dimension, and these must be considered in a dimensional analysis.

These examples illustrate the following important concepts regarding dimensions of physical quantities.

1. We have based the concept of dimension on three physical quantities: mass m, length s, and time t. These quantities are measured in some appropriate system of units whose choice does not affect the assignment of dimensions. (This underlying system must be linear. A dimensional analysis will not work if the scale is logarithmic, for example.)

2. There are other physical quantities, such as area and velocity, that are defined as simple products involving only mass, length, or time. Here we use the term **product** to include any quotient because we may indicate division by negative exponents.

3. There are still other, more complex physical entities, such as momentum and kinetic energy, whose definitions involve quantities other than mass, length, and time. Because the simpler quantities from (1) and (2) are products, these more complex quantities can also be expressed as products involving mass, length, and time by algebraic simplification. We use the term *product* to refer to any physical quantity from item (1), (2), or (3); a product from (1) is trivial because it has only one factor.

4. To each product, there is assigned a **dimension**—that is, an expression of the form

$$M^n L^p T^q \tag{9.1}$$

where n, p, and q are real numbers that may be positive, negative, or zero.

When a basic dimension is missing from a product, the corresponding exponent is understood to be zero. Thus, the dimension $M^2 L^0 T^{-1}$ may also appear as $M^2 T^{-1}$. When n, p, and q are all zero in an expression of the form (9.1), so that the dimension reduces to

$$M^0 L^0 T^0 \tag{9.2}$$

the quantity, or product, is said to be *dimensionless*.

Special care must be taken in forming sums of products because just as we cannot add apples and oranges, in an equation we cannot add products that have unlike dimensions. For example, if F denotes force, m mass, and v velocity, we know immediately that the equation

$$F = mv + v^2$$

cannot be correct because mv has dimension MLT^{-1}, whereas v^2 has dimension L^2T^{-2}. These dimensions are unlike; hence, the products mv and v^2 cannot be added. An equation such as this—that is, one that contains among its terms two products having unlike dimensions—is said to be *dimensionally incompatible*. Equations that involve only sums of products having the same dimension are *dimensionally compatible*.

The concept of dimensional compatibility is related to another important concept called **dimensional homogeneity**. In general, an equation that is true regardless of the system of units in which the variables are measured is said to be dimensionally homogeneous. For example, $t = \sqrt{2s/g}$ giving the time a body falls a distance s under gravity (neglecting air resistance) is dimensionally homogeneous (true in all systems), whereas the equation $t = \sqrt{s/16.1}$ is not dimensionally homogeneous (because it depends on a particular system). In particular, if an equation involves only sums of dimensionless products (i.e., it is a dimensionless equation), then the equation is dimensionally homogeneous. Because the products are dimensionless, the factors used for conversion from one system of units to another would simply cancel.

The application of dimensional analysis to a real-world problem is based on the assumption that the solution to the problem is given by a dimensionally homogeneous equation in terms of the appropriate variables. Thus, the task is to determine the form of the desired equation by finding an appropriate dimensionless equation and then solving for the dependent variable. To accomplish this task, we must decide which variables enter into the physical problem under investigation and determine all the dimensionless products among them. In general, there may be infinitely many such products, so they will have to be described rather than actually written out. Certain subsets of these dimensionless products are then used to construct dimensionally homogeneous equations. In Section 9.2 we investigate how the dimensionless products are used to find all dimensionally homogeneous equations. The following example illustrates how the dimensionless products may be found.

EXAMPLE 1 *A Simple Pendulum Revisited*

Consider again the simple pendulum discussed in the introduction. Analyzing the dimensions of the variables for the pendulum problem, we have

Variable	m	g	t	r	θ
Dimension	M	LT^{-2}	T	L	$M^0L^0T^0$

Next we find all the dimensionless products among the variables. Any product of these variables must be of the form

$$m^a g^b t^c r^d \theta^e \qquad (9.3)$$

and hence must have dimension

$$(M)^a(LT^{-2})^b(T)^c(L)^d(M^0L^0T^0)^e$$

Therefore, a product of the form (9.3) is dimensionless if and only if

$$M^aL^{b+d}T^{c-2b} = M^0L^0T^0 \tag{9.4}$$

Equating the exponents on both sides of this last equation leads to the system of linear equations

$$\left. \begin{array}{l} a \qquad\qquad\qquad +0e = 0 \\ \quad b \qquad +d +0e = 0 \\ \quad -2b +c \qquad +0e = 0 \end{array} \right\} \tag{9.5}$$

Solution of the system (9.5) gives $a = 0$, $c = 2b$, $d = -b$, where b is arbitrary. Thus, there are infinitely many solutions. Here are some general rules for selecting arbitrary variables: (1) Choose the dependent variable so it will appear only once, (2) select any variable that expedites the solution of the other equations (i.e., a variable that appears in all equations), and (3) choose a variable that always has a zero coefficient, if possible. Notice that the exponent e does not really appear in (9.4) (because it has a zero coefficient in each equation) so it is also arbitrary. One dimensionless product is obtained by setting $b = 0$ and $e = 1$, yielding $a = c = d = 0$. A second, independent dimensionless product is obtained when $b = 1$ and $e = 0$, yielding $a = 0$, $c = 2$, and $d = -1$. These solutions give the dimensionless products

$$\prod_1 = m^0g^0t^0r^0\theta^1 = \theta$$

$$\prod_2 = m^0g^1t^2r^{-1}\theta^0 = \frac{gt^2}{r}$$

In Section 9.2, we will learn a methodology for relating these products to carry the modeling process to completion. For now, we will develop a relationship in an intuitive manner.

Assuming $t = f(r, m, g, \theta)$, to determine more about the function f, we observe that if the units in which we measure mass are made smaller by some factor (e.g., 10), then the measure of the period t will not change because it is measured in units (T) of time. Because m is the only factor whose dimension contains M, it cannot appear in the model. Similarly, if the scale of the units (L) for measuring length is altered, it cannot change the measure of the period. For this to happen, the factors r and g must appear in the model as r/g, g/r, or, more generally, $(g/r)^k$. This ensures that any linear change in the way length is measured will be canceled. Finally, if we make the units (T) that measure time smaller by a factor of 10, for example, the measure of the period will directly increase by this same factor 10. Thus, to have the dimension of T on the right side of the equation $t = f(r, m, g, \theta)$, g and r must appear as $\sqrt{r/g}$ because T appears to the power -2 in the dimension of g. Note that none of the preceding conditions places any restrictions on the angle θ. Thus, the equation of the period should be of the form

$$t = \sqrt{\frac{r}{g}}\, h(\theta)$$

where the function h must be determined or approximated by experimentation.

We note two things in this analysis that are characteristic of a dimensional analysis. First, in the MLT system, three conditions are placed on the model, so we should generally expect to reduce the number of arguments of the function relating the variables by three. In the pendulum problem we reduced the number of arguments from four to one. Second, all arguments of the function present at the end of a dimensional analysis (in this case, θ) are dimensionless products.

In the problem of the undamped pendulum we assumed that friction and drag were negligible. Before proceeding with experiments (which might be costly), we would like to know whether that assumption is reasonable. Consider the model obtained so far:

$$t = \sqrt{\frac{r}{g}}\, h(\theta)$$

Keeping θ constant while allowing r to vary, form the ratio

$$\frac{t_1}{t_2} = \frac{\sqrt{r_1/g}\, h(\theta_0)}{\sqrt{r_2/g}\, h(\theta_0)} = \sqrt{\frac{r_1}{r_2}}$$

Hence the model predicts that t will vary as $\sqrt{r}$ for constant θ. Thus, if we plot t versus r with fixed θ for some observations, we will expect to get a straight line (Figure 9.4). If we do not obtain a reasonably straight line, then we need to reexamine the assumptions. Note that our judgment here is qualitative. The final measure of the adequacy of any model is always how well it predicts or explains the phenomenon under investigation. Nevertheless, this initial test is useful for eliminating obviously bad assumptions and for choosing among competing sets of assumptions.

■ Figure 9.4

Testing the assumptions of the simple pendulum model by plotting the period t versus the square root of the length r for constant displacement θ

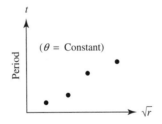

Dimensional analysis has helped construct a model $t = f(r, m, g, \theta)$ for the undamped pendulum as $t = \sqrt{r/g}\, h(\theta)$. If we are interested in predicting the behavior of the pendulum, we could isolate the effect of h by holding r constant and varying θ. This provides the ratio

$$\frac{t_1}{t_2} = \frac{\sqrt{r_0/g}\, h(\theta_1)}{\sqrt{r_0/g}\, h(\theta_2)} = \frac{h(\theta_1)}{h(\theta_2)}$$

Hence a plot of t versus θ for several observations would reveal the nature of h. This plot is illustrated in Figure 9.5. We may never discover the true function h relating the variables. In such cases, an empirical model might be constructed from the experimental data, as discussed in Chapter 4. When we are interested in using our model to predict t, based on

Figure 9.5

Determining the unknown function h

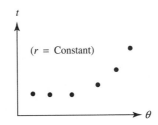

$(r = \text{Constant})$

experimental results, it is convenient to use the equation $t\sqrt{g/r} = h(\theta)$ and to plot $t\sqrt{g/r}$ versus θ, as in Figure 9.6. Then, for a given value of θ, we would determine $t\sqrt{g/r}$, multiply it by $\sqrt{r/g}$ for a specific r, and finally determine t. ■ ■ ■

Figure 9.6

Presenting the results for the simple pendulum

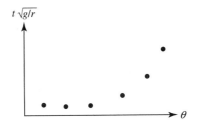

EXAMPLE 2 *Wind Force on a Van*

Suppose you are driving a van down a highway with gusty winds. How does the speed of your vehicle affect the wind force you are experiencing?

The force F of the wind on the van is certainly affected by the speed v of the van and the surface area A of the van directly exposed to the wind's direction. Thus, we might hypothesize that the force is proportional to some power of the speed times some power of the surface area; that is,

$$F = kv^a A^b \qquad (9.6)$$

for some (dimensionless) constant k. Analyzing the dimensions of the variables gives

Variable	F	k	v	A
Dimension	MLT^{-2}	$M^0L^0T^0$	LT^{-1}	L^2

Hence, dimensionally, Equation (9.6) becomes

$$MLT^{-2} = (M^0L^0T^0)(LT^{-1})^a(L^2)^b$$

This last equation cannot be correct because the dimension M for mass does not enter into the right-hand side with nonzero exponent.

So consider again Equation (9.6). What is missing in our assumption concerning the wind force? Wouldn't the strength of the wind be affected by its density? After some reflection we would probably agree that density does have an effect. If we include the density ρ as a factor, then our refined model becomes

$$F = kv^a A^b \rho^c \tag{9.7}$$

Because density is mass per unit volume, the dimension of density is ML^{-3}. Therefore, dimensionally, Equation (9.7) becomes

$$MLT^{-2} = (M^0 L^0 T^0)(LT^{-1})^a (L^2)^b (ML^{-3})^c$$

Equating the exponents on both sides of this last equation leads to the system of linear equations:

$$\left.\begin{array}{rcr} c = & 1 \\ a + 2b - 3c = & 1 \\ -a & = -2 \end{array}\right\} \tag{9.8}$$

Solution of the system (9.8) gives $a = 2$, $b = 1$, and $c = 1$. When substituted into Equation (9.7), these values give the model

$$F = kv^2 A\rho \qquad \blacksquare\;\blacksquare\;\blacksquare$$

At this point we make an important observation. When it was assumed that $F = kv^a A^b$, the constant was assumed to be dimensionless. Subsequently, our analysis revealed that for a particular medium (so ρ is constant)

$$F \propto Av^2$$

giving $F = k_1 Av^2$. However, k_1 does have a dimension associated with it and is called a **dimensional constant**. In particular, the dimension of k_1 is

$$\frac{MLT^{-2}}{L^2(L^2 T^{-2})} = ML^{-3}$$

Dimensional constants contain important information and must be considered when performing a dimensional analysis. We consider dimensional constants again in Section 9.3 when we investigate a damped pendulum.

If we assume the density ρ is constant, our model shows that the force of the wind is proportional to the square of the speed of the van times its surface area directly exposed to the wind. We can test the model by collecting data and plotting the wind force F versus $v^2 A$ to determine whether the graph approximates a straight line through the origin. This example illustrates one of the ways in which dimensional analysis can be used to test our assumptions and check whether we have a faulty list of variables identifying the problem. Table 9.1 gives a summary of the dimensions of some common physical entities.

Table 9.1 Dimensions of physical entities in the MLT system

Mass	M	Momentum	MLT^{-1}
Length	L	Work	ML^2T^{-2}
Time	T	Density	ML^{-3}
Velocity	LT^{-1}	Viscosity	$ML^{-1}T^{-1}$
Acceleration	LT^{-2}	Pressure	$ML^{-1}T^{-2}$
Specific weight	$ML^{-2}T^{-2}$	Surface tension	MT^{-2}
Force	MLT^{-2}	Power	ML^2T^{-3}
Frequency	T^{-1}	Rotational inertia	ML^2
Angular velocity	T^{-1}	Torque	ML^2T^{-2}
Angular acceleration	T^{-2}	Entropy	ML^2T^{-2}
Angular momentum	ML^2T^{-1}	Heat	ML^2T^{-2}
Energy	ML^2T^{-2}		

9.1 PROBLEMS

1. Determine whether the equation

$$s = s_0 + v_0 t - 0.5gt^2$$

is dimensionally compatible, if s is the position (measured vertically from a fixed reference point) of a body at time t, s_0 is the position at $t = 0$, v_0 is the initial velocity, and g is the acceleration caused by gravity.

2. Find a dimensionless product relating the torque τ (ML^2T^{-2}) produced by an automobile engine, the engine's rotation rate ψ (T^{-1}), the volume V of air displaced by the engine, and the air density ρ.

3. The various constants of physics often have physical dimensions (dimensional constants) because their values depend on the system in which they are expressed. For example, Newton's law of gravitation asserts that the attractive force between two bodies is proportional to the product of their masses divided by the square of the distance between them, or, symbolically,

$$F = \frac{Gm_1m_2}{r^2}$$

where G is the gravitational constant. Find the dimension of G so that Newton's law is dimensionally compatible.

4. Certain stars, whose light and radial velocities undergo periodic vibrations, are thought to be pulsating. It is hypothesized that the period t of pulsation depends on the star's radius r, its mass m, and the gravitational constant G. (See Problem 3 for the dimension of G.) Express t as a product of m, r, and G, so the equation

$$t = m^a r^b G^c$$

is dimensionally compatible.

5. In checking the dimensions of an equation, you should note that derivatives also possess dimensions. For example, the dimension of ds/dt is LT^{-1} and the dimension of d^2s/dt^2 is LT^{-2}, where s denotes distance and t denotes time. Determine whether the equation

$$\frac{dE}{dt} = \left[mr^2 \left(\frac{d^2\theta}{dt^2} \right) mgr \sin \theta \right] \frac{d\theta}{dt}$$

for the time rate of change of total energy E in a pendulum system with damping force is dimensionally compatible.

6. For a body moving along a straight-line path, if the mass of the body is changing over time, then an equation governing its motion is given by

$$m\frac{dv}{dt} = F + u\frac{dm}{dt}$$

where m is the mass of the body, v is the velocity of the body, F is the total force acting on the body, dm is the mass joining or leaving the body in the time interval dt, and u is the velocity of dm at the moment it joins or leaves the body (relative to an observer stationed on the body). Show that the preceding equation is dimensionally compatible.

7. In humans, the hydrostatic pressure of blood contributes to the total blood pressure. The hydrostatic pressure P is a product of blood density ρ, height h of the blood column between the heart and some lower point in the body, and gravity g. Determine

$$P = k\rho^a h^b g^c$$

where k is a dimensionless constant.

8. Assume the force F opposing the fall of a raindrop through air is a product of viscosity μ, velocity v, and the diameter r of the drop. Assume that density is neglected. Find

$$F = k\mu^a v^b r^c$$

where k is a dimensionless constant.

9.1 PROJECT

1. Complete the requirements of "Keeping Dimension Straight," by George E. Strecker, UMAP 564. This module is a very basic introduction to the distinction between dimensions and units. It also provides the student with some practice in using dimensional arguments to properly set up solutions to elementary problems and to recognize errors.

9.2 The Process of Dimensional Analysis

In the preceding section we learned how to determine all dimensionless products among the variables selected in the problem under investigation. Now we investigate how to use the dimensionless products to find all possible dimensionally homogeneous equations among the variables. The key result is Buckingham's theorem, which summarizes the entire theory of dimensional analysis.

Example 1 in the preceding section shows that in general, many dimensionless products may be formed from the variables of a given system. In that example we determined every dimensionless product to be of the form

$$g^b t^{2b} r^{-b} \theta^e \tag{9.9}$$

where b and e are arbitrary real numbers. Each one of these products corresponds to a solution of the homogeneous system of linear algebraic equations given by Equation (9.5). The two products

$$\prod_1 = \theta \quad \text{and} \quad \prod_2 = \frac{gt^2}{r}$$

obtained when $b = 0$, $e = 1$, and $b = 1$, $e = 0$, respectively, are special in the sense that any of the dimensionless products (9.9) can be given as a product of some power of $\prod_1$ times some power of $\prod_2$. Thus, for instance,

$$g^3 t^6 r^{-3} \theta^{1/2} = \prod_1^{1/2} \prod_2^3$$

This observation follows from the fact that $b = 0$, $e = 1$ and $b = 1$, $e = 0$ represent, in some sense, independent solutions of the system (9.5). Let's explore these ideas further.

Consider the following system of m linear algebraic equations in the n unknowns $x_1, x_2, \ldots, x_n$:

$$
\begin{aligned}
a_{11}x_1 + a_{12}x_2 + \cdots + a_{1n}x_n &= b_1 \\
a_{21}x_1 + a_{22}x_2 + \cdots + a_{2n}x_n &= b_2 \\
\vdots \qquad \vdots \qquad \vdots \qquad \vdots \qquad \vdots \\
a_{m1}x_1 + a_{m2}x_2 + \cdots + a_{mn}x_n &= b_m
\end{aligned}
\tag{9.10}
$$

The symbols a_{ij} and b_i denote real numbers for each $i = 1, 2, \ldots, m$ and $j = 1, 2, \ldots, n$. The numbers a_{ij} are called the **coefficients** of the system, and the b_i are referred to as the **constants**. The subscript i in the symbol a_{ij} refers to the ith equation of the system (9.10) and the subscript j refers to the jth unknown x_j to which a_{ij} belongs. Thus, the subscripts serve to locate a_{ij}. It is customary to read a_{13} as "a, one, three" and a_{42} as "a, four, two," for example, rather than "a, thirteen" and "a, forty-two."

A **solution** to the system (9.10) is a sequence of numbers $s_1, s_2, \ldots, s_n$ for which $x_1 = s_1, x_2 = s_2, \ldots, x_n = s_n$ solves each equation in the system. If $b_1 = b_2 = \cdots = b_m = 0$, the system (9.10) is said to be **homogeneous**. The solution $s_1 = s_2 = \cdots = s_n = 0$ always solves the homogeneous system and is called the **trivial solution**. For a homogeneous system there are two solution possibilities: Either the trivial solution is the only solution or there are infinitely many solutions.

Whenever $s_1, s_2, \ldots, s_n$ and $s'_1, s'_2, \ldots, s'_n$ are solutions to the homogeneous system, the sequences $s_1 + s'_1, s_2 + s'_2, \ldots, s_n + s'_n$, and $cs_1, cs_2, \ldots, cs_n$ are also solutions for any constant c. These solutions are called the **sum** and **scalar multiple** of the original solutions, respectively. If S and S' refer to the original solutions, then we use the notations $S + S'$ to refer to their sum and cS to refer to a scalar multiple of the first solution. If $S_1, S_2, \ldots, S_k$ is a collection of k solutions to the homogeneous system, then the solution

$$c_1 S_1 + c_2 S_2 + \cdots + c_k S_k$$

is called a **linear combination** of the k solutions, where $c_1, c_2, \ldots, c_k$ are arbitrary real numbers. It is an easy exercise to show that any linear combination of solutions to the homogeneous system is still another solution to the system.

A set of solutions to a homogeneous system is said to be **independent** if no solution in the set is a linear combination of the remaining solutions in the set. A set of solutions is **complete** if it is independent and every solution is expressible as a linear combination of solutions in the set. For a specific homogeneous system, we seek some complete set of solutions because all other solutions are produced from them using linear combinations. For example, the two solutions corresponding to the two choices $b = 0, e = 1$ and $b = 1$, $e = 0$ form a complete set of solutions to the homogeneous system (9.5).

It is not our intent to present the theory of linear algebraic equations. Such a study is appropriate for a course in linear algebra. We do point out that there is an elementary algorithm known as Gaussian elimination for producing a complete set of solutions to a given system of linear equations. Moreover, Gaussian elimination is readily implemented on computers and handheld programmable calculators. The systems of equations we will encounter in this book are simple enough to be solved by the elimination methods learned in intermediate algebra.

How does our discussion relate to dimensional analysis? Our basic goal thus far has been to find all possible dimensionless products among the variables that influence the physical phenomenon under investigation. We developed a homogeneous system of linear algebraic equations to help us determine these dimensionless products. This system of equations usually has infinitely many solutions. Each solution gives the values of the exponents that result in a dimensionless product among the variables. If we sum two solutions, we produce another solution that yields the same dimensionless product as does multiplication of the dimensionless products corresponding to the original two solutions. For example, the sum of the solutions corresponding to $b = 0, e = 1$ and $b = 1, e = 0$ for Equation (9.5) yields the solution corresponding to $b = 1, e = 1$ with the corresponding dimensionless product from Equation (9.9) given by

$$gt^2 r^{-1}\theta = \prod{}_1 \prod{}_2$$

The reason for this result is that the unknowns in the system of equations are the exponents in the dimensionless products, and addition of exponents algebraically corresponds to multiplication of numbers having the same base: $x^{m+n} = x^m x^n$. Moreover, multiplication of a solution by a constant produces a solution that yields the same dimensionless product as does raising the product corresponding to the original solution to the power of the constant. For example, -1 times the solution corresponding to $b = 1, e = 0$ yields the solution corresponding to $b = -1, e = 0$ with the corresponding dimensionless product

$$g^{-1}t^{-2}r = \prod{}_2^{-1}$$

The reason for this last result is that algebraic multiplication of an exponent by a constant corresponds to raising a power to a power, $x^{mn} = (x^m)^n$.

In summary, addition of solutions to the homogeneous system of equations results in multiplication of their corresponding dimensionless products, and multiplication of a solution by a constant results in raising the corresponding product to the power given by that constant. Thus, if S_1 and S_2 are two solutions corresponding to the dimensionless products $\prod_1$ and $\prod_2$, respectively, then the linear combination $aS_1 + bS_2$ corresponds to the dimensionless product

$$\prod_1^a \prod_2^b$$

It follows from our preceding discussion that a complete set of solutions to the homogeneous system of equations produces all possible solutions through linear combination. The dimensionless products corresponding to a complete set of solutions are therefore called a *complete set of dimensionless products*. All dimensionless products can be obtained by forming powers and products of the members of a complete set.

Next, let's investigate how these dimensionless products can be used to produce all possible dimensionally homogeneous equations among the variables. In Section 9.1 we defined an equation to be dimensionally homogeneous if it remains true regardless of the system of units in which the variables are measured. The fundamental result in dimensional analysis that provides for the construction of all dimensionally homogeneous equations from complete sets of dimensionless products is the following theorem.

Theorem 1

Buckingham's Theorem An equation is dimensionally homogeneous if and only if it can be put into the form

$$f\left(\prod_1, \prod_2, \cdots, \prod_n\right) = 0 \qquad (9.11)$$

where f is some function of n arguments and $\{\prod_1, \prod_2, \ldots, \prod_n\}$ is a complete set of dimensionless products.

Let's apply Buckingham's theorem to the simple pendulum discussed in the preceding sections. The two dimensionless products

$$\prod_1 = \theta \quad \text{and} \quad \prod_2 = \frac{gt^2}{r}$$

form a complete set for the pendulum problem. Thus, according to Buckingham's theorem, there is a function f such that

$$f\left(\theta, \frac{gt^2}{r}\right) = 0$$

Assuming we can solve this last equation for gt^2/r as a function of θ, it follows that

$$t = \sqrt{\frac{r}{g}} h(\theta) \qquad (9.12)$$

where h is some function of the single variable θ. Notice that this last result agrees with our intuitive formulation for the simple pendulum presented in Section 9.1. Observe that Equation (9.12) represents only a general form for the relationship among the variables m, g, t, r, and θ. However, it can be concluded from this expression that t does not depend on the mass m and is related to $r^{1/2}$ and $g^{-1/2}$ by some function of the initial angle of displacement θ. Knowing this much, we can determine the nature of the function h experimentally or approximate it, as discussed in Section 9.1.

Consider Equation (9.11) in Buckingham's theorem. For the case in which a complete set consists of a single dimensionless product, for example, $\prod_1$, the equation reduces to the form

$$f\left(\prod_1\right) = 0$$

In this case we assume that the function f has one real root at k (to assume otherwise has little physical meaning). Hence, the solution $\prod_1 = k$ is obtained.

Using Buckingham's theorem, let's reconsider the example from Section 9.1 of the wind force on a van driving down a highway. Because the four variables F, v, A, and ρ were selected and all three equations in (9.8) are independent, a complete set of dimensionless products consists of a single product:

$$\prod_1 = \frac{F}{v^2 A \rho}$$

Application of Buckingham's theorem gives

$$f\left(\prod_1\right) = 0$$

which implies from the preceding discussion that $\prod_1 = k$, or

$$F = k v^2 A \rho$$

where k is a dimensionless constant as before. Thus, when a complete set consists of a *single dimensionless product*, as is generally the case when we begin with four variables, the application of Buckingham's theorem yields the desired relationship *up to a constant of proportionality*. Of course, the predicted proportionality must be tested to determine the adequacy of our list of variables. If the list does prove to be adequate, then the constant of proportionality can be determined by experimentation, thereby completely defining the relationship.

For the case $n = 2$, Equation (9.11) in Buckingham's theorem takes the form

$$f\left(\prod_1, \prod_2\right) = 0 \tag{9.13}$$

If we choose the products in the complete set $\{\prod_1, \prod_2\}$ so that the dependent variable appears in only one of them, for example, $\prod_2$, we can proceed under the assumption that Equation (9.13) can be solved for that chosen product $\prod_2$ in terms of the remaining product $\prod_1$. Such a solution takes the form

$$\prod_2 = H\left(\prod_1\right)$$

and then this latter equation can be solved for the dependent variable. Note that when a complete set consists of more than one dimensionless product, the application of Buckingham's theorem determines the desired relationship *up to an arbitrary function*. After verifying the adequacy of the list of variables, we may be lucky enough to recognize the underlying functional relationship. However, in general we can expect to construct an empirical model, although the task has been eased considerably.

For the general case of n dimensionless products in the complete set for Buckingham's theorem, we again choose the products in the complete set $\{\prod_1, \prod_2, \ldots, \prod_n\}$ so that the dependent variable appears in only one of them, say $\prod_n$ for definiteness. Assuming we can solve Equation (9.11) for that product $\prod_n$ in terms of the remaining ones, we have the form

$$\prod_n = H\left(\prod_1, \prod_2, \ldots, \prod_{n-1}\right)$$

We then solve this last equation for the dependent variable.

Summary of Dimensional Analysis Methodology

STEP 1 Decide which variables enter the problem under investigation.

STEP 2 Determine a complete set of dimensionless products $\{\prod_1, \prod_2, \ldots, \prod_n\}$ among the variables. Make sure the dependent variable of the problem appears in only one of the dimensionless products.

STEP 3 Check to ensure that the products found in the previous step are dimensionless and independent. Otherwise you have an algebra error.

STEP 4 Apply Buckingham's theorem to produce all possible dimensionally homogeneous equations among the variables. This procedure yields an equation of the form (9.11).

STEP 5 Solve the equation in Step 4 for the dependent variable.

STEP 6 Test to ensure that the assumptions made in Step 1 are reasonable. Otherwise the list of variables is faulty.

STEP 7 Conduct the necessary experiments and present the results in a useful format.

Let's illustrate the first five steps of this procedure.

EXAMPLE 1 *Terminal Velocity of a Raindrop*

Consider the problem of determining the terminal velocity v of a raindrop falling from a motionless cloud. We examined this problem from a very simplistic point of view in Chapter 2, but let's take another look using dimensional analysis.

What are the variables influencing the behavior of the raindrop? Certainly the terminal velocity will depend on the size of the raindrop given by, say, its radius r. The density ρ of the air and the viscosity μ of the air will also affect the behavior. (Viscosity measures resistance to motion—a sort of internal molecular friction. In gases this resistance is caused by collisions between fast-moving molecules.) The acceleration due to gravity g is another variable to consider. Although the surface tension of the raindrop is a factor that does influence the behavior of the fall, we will ignore this factor. If necessary, surface tension can be taken into account in a later, refined model. These considerations give the following table relating the selected variables to their dimensions:

Variable	v	r	g	ρ	μ
Dimension	LT^{-1}	L	LT^{-2}	ML^{-3}	$ML^{-1}T^{-1}$

Next we find all the dimensionless products among the variables. Any such product must be of the form

$$v^a r^b g^c \rho^d \mu^e \tag{9.14}$$

and hence must have dimension

$$(LT^{-1})^a (L)^b (LT^{-2})^c (ML^{-3})^d (ML^{-1}T^{-1})^e$$

Therefore, a product of the form (9.14) is dimensionless if and only if the following system of equations in the exponents is satisfied:

$$\left.\begin{array}{r} d + e = 0 \\ a + b + c - 3d - e = 0 \\ - a - 2c - e = 0 \end{array}\right\} \tag{9.15}$$

Solution of the system (9.15) gives $b = (3/2)d - (1/2)a$, $c = (1/2)d - (1/2)a$, and $e = -d$, where a and d are arbitrary. One dimensionless product $\prod_1$ is obtained by setting $a = 1$, $d = 0$; another, independent dimensionless product $\prod_2$ is obtained when $a = 0$, $d = 1$. These solutions give

$$\prod_1 = v r^{-1/2} g^{-1/2} \quad \text{and} \quad \prod_2 = r^{3/2} g^{1/2} \rho \mu^{-1}$$

Next, we check the results to ensure that the products are indeed dimensionless:

$$\frac{LT^{-1}}{L^{1/2}(LT^{-2})^{1/2}} = M^0 L^0 T^0$$

and

$$\frac{L^{3/2}(LT^{-2})^{1/2}(ML^{-3})}{ML^{-1}T^{-1}} = M^0 L^0 T^0$$

Thus, according to Buckingham's theorem, there is a function f such that

$$f\left(v r^{-1/2} g^{-1/2}, \frac{r^{3/2} g^{1/2} \rho}{\mu} \right) = 0$$

Assuming we can solve this last equation for $v r^{-1/2} g^{-1/2}$ as a function of the second product $\prod_2$, it follows that

$$v = \sqrt{rg}\, h\left(\frac{r^{3/2} g^{1/2} \rho}{\mu} \right)$$

where h is some function of the single product $\prod_2$.

The preceding example illustrates a characteristic feature of dimensional analysis. Normally the modeler studying a given physical system has an intuitive idea of the variables involved and has a working knowledge of general principles and laws (such as Newton's second law) but lacks the precise laws governing the interaction of the variables. Of course, the modeler can always experiment with each independent variable separately, holding the others constant and measuring the effect on the system. Often, however, the efficiency of the experimental work can be improved through an application of dimensional analysis. Although we did not illustrate Steps 6 and 7 of the dimensional analysis process for the preceding example, these steps will be illustrated in Section 9.3.

We now make some observations concerning the dimensional analysis process. Suppose n variables have been identified in the physical problem under investigation. When determining a complete set of dimensionless products, we form a system of three linear algebraic equations by equating the exponents for M, L, and T to zero. That is, we obtain a system of three equations in n unknowns (the exponents). If the three equations are independent, we can solve the system for three of the unknowns in terms of the remaining $n - 3$ unknowns (declared to be arbitrary). In this case, we find $n - 3$ independent dimensionless products that make up the complete set we seek. For instance, in the preceding example there are five unknowns, a, b, c, d, e, and we determined three of them (b, c, and e) in terms of the remaining $(5 - 3)$ two arbitrary ones (a and d). Thus, we obtained a complete set of two dimensionless products. When choosing the $n - 3$ dimensionless products, we must be sure that the dependent variable appears in only one of them. We can then solve Equation (9.11) guaranteed by Buckingham's theorem for the dependent variable, at least under suitable assumptions on the function f in that equation. (The full story telling when such a solution is possible is the content of an important result in advanced calculus known as the implicit function theorem.)

We acknowledge that we have been rather sketchy in our presentation for solving the system of linear algebraic equations that results in the process of determining all dimensionless products. Recall how to solve simple linear systems by the method of elimination of variables. We conclude this section with another example.

EXAMPLE 2 *Automobile Gas Mileage Revisited*

Consider again the automobile gasoline mileage problem presented in Chapter 2. One of our submodels in that problem was for the force of propulsion F_p. The variables we identified that affect the propulsion force are C_r, the amount of fuel burned per unit time, the amount K of energy contained in each gallon of gasoline, and the speed v. Let's perform a dimensional analysis. The following table relates the various variables to their dimensions:

Variable	F_p	C_r	K	v
Dimension	MLT^{-2}	L^3T^{-1}	$ML^{-1}T^{-2}$	LT^{-1}

Thus, the product

$$F_p^a C_r^b K^c v^d \tag{9.16}$$

must have the dimension

$$(MLT^{-2})^a(L^3T^{-1})^b(ML^{-1}T^{-2})^c(LT^{-1})^d$$

The requirement for a dimensionless product leads to the system

$$\left.\begin{array}{r} a \quad\quad + c \quad\quad = 0 \\ a + 3b - c + d = 0 \\ -2a - b - 2c - d = 0 \end{array}\right\} \qquad (9.17)$$

Solution of the system (9.17) gives $b = -a$, $c = -a$, and $d = a$, where a is arbitrary. Choosing $a = 1$, we obtain the dimensionless product

$$\prod_1 = F_p C_r^{-1} K^{-1} v$$

From Buckingham's theorem there is a function f with $f(\prod_1) = 0$, so $\prod_1$ equals a constant. Therefore,

$$F_p \propto \frac{C_r K}{v}$$

in agreement with the conclusion reached in Chapter 2. ■ ■ ■

9.2 PROBLEMS

1. Predict the time of revolution for two bodies of mass m_1 and m_2 in empty space revolving about each other under their mutual gravitational attraction.

2. A projectile is fired with initial velocity v at an angle θ with the horizon. Predict the range R.

3. Consider an object that is falling under the influence of gravity. Assume that air resistance is negligible. Using dimensional analysis, find the speed v of the object after it has fallen a distance s. Let $v = f(m, g, s)$, where m is the mass of the object and g is the acceleration due to gravity. Does you answer agree with your knowledge of the physical situation?

4. Using dimensional analysis, find a proportionality relationship for the centrifugal force F of a particle in terms of its mass m, velocity v, and radius r of the curvature of its path.

5. One would like to know the nature of the drag forces experienced by a sphere as it passes through a fluid. It is assumed that the sphere has a low speed. Therefore, the drag force is highly dependent on the viscosity of the fluid. The fluid density is to be neglected. Use the dimensional analysis process to develop a model for drag force F as a function of the radius r and velocity m of the sphere and the viscosity μ of the fluid.

6. The volume flow rate q for laminar flow in a pipe depends on the pipe radius r, the viscosity μ of the fluid, and the pressure drop per unit length dp/dz. Develop a model for the flow rate q as a function of r, μ, and dp/dz.

7. In fluid mechanics, the Reynolds number is a dimensionless number involving the fluid velocity v, density ρ, viscosity μ, and a characteristic length r. Use dimensional analysis to find the Reynolds number.

8. The power P delivered to a pump depends on the specific weight w of the fluid pumped, the height h to which the fluid is pumped, and the fluid flow rate q in cubic feet per second. Use dimensional analysis to determine an equation for power.

9. Find the volume flow rate dV/dt of blood flowing in an artery as a function of the pressure drop per unit length of artery, the radius r of the artery, the blood density ρ, and the blood viscosity μ.

10. The speed of sound in a gas depends on the pressure and the density. Use dimensional analysis to find the speed of sound in terms of pressure and density.

11. The lift force F on a missile depends on its length r, velocity v, diameter δ, and initial angle θ with the horizon; it also depends on the density ρ, viscosity μ, gravity g, and speed of sound s of the air. Show that

$$F = \rho v^2 r^2 h \left(\frac{\delta}{r}, \theta, \frac{\mu}{\rho v r}, \frac{s}{v}, \frac{rg}{v^2} \right)$$

12. The height h that a fluid will rise in a capillary tube decreases as the diameter D of the tube increases. Use dimensional analysis to determine how h varies with D and the specific weight w and surface tension σ of the liquid.

9.3   A Damped Pendulum

In Section 9.1 we investigated the pendulum problem under the assumptions that the hinge is frictionless, the mass is concentrated at one end of the pendulum, and the drag force is negligible. Suppose we are not satisfied with the results predicted by the constructed model. Then we can refine the model by incorporating drag forces. If F represents the total drag force, the problem now is to determine the function

$$t = f(r, m, g, \theta, F)$$

Let's consider a submodel for the drag force. As we have seen in previous examples, the modeler is usually faced with a trade-off between simplicity and accuracy. For the pendulum it might seem reasonable to expect the drag force to be proportional to some positive power of the velocity. To keep our model simple, we assume that F is proportional to either v or v^2, as depicted in Figure 9.7.

Now we can experiment to determine directly the nature of the drag force. However, we will first perform a dimensional analysis because we expect it to reduce our experimental effort. Assume F is proportional to v so that $F = kv$. For convenience we choose to work with the dimensional constant $k = F/v$, which has dimension MLT^{-2}/LT^{-1}, or simply MT^{-1}. Notice that the dimensional constant captures the assumption about the drag force.

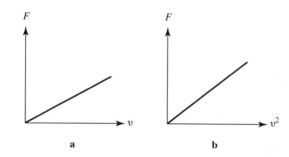

Thus, we apply dimensional analysis to the model

$$t = f(r, m, g, \theta, k)$$

An analysis of the dimensions of the variables gives

Variable	t	r	m	g	θ	k
Dimension	T	L	M	LT^{-2}	$M^0L^0T^0$	MT^{-1}

Any product of the variables must be of the form

$$t^a r^b m^c g^d \theta^e k^f \tag{9.18}$$

and hence must have dimension

$$(T)^a (L)^b (M)^c (LT^{-2})^d (M^0 L^0 T^0)^e (MT^{-1})^f$$

Therefore, a product of the form (9.18) is dimensionless if and only if

$$\left.\begin{array}{r} c \qquad\qquad + f = 0 \\ b \quad + d \qquad\quad = 0 \\ a \qquad - 2d - f = 0 \end{array}\right\} \tag{9.19}$$

The equations in the system (9.19) are independent, so we know we can solve for three of the variables in terms of the remaining $(6 - 3)$ three variables. We would like to choose the solutions in such a way that t appears in only one of the dimensionless products. Thus, we choose a, e, and f as the arbitrary variables with

$$c = -f, \; b = -d = \frac{-a}{2} + \frac{f}{2}, \; d = \frac{a}{2} - \frac{f}{2}$$

Setting $a = 1$, $e = 0$, and $f = 0$, we obtain $c = 0$, $b = -1/2$, and $d = 1/2$ with the corresponding dimensionless product $t\sqrt{g/r}$. Similarly, choosing $a = 0$, $e = 1$, and $f = 0$, we get $c = 0$, $b = 0$, and $d = 0$, corresponding to the dimensionless product θ. Finally, choosing $a = 0$, $e = 0$, and $f = 1$, we obtain $c = -1$, $b = 1/2$, and $d = -1/2$, corresponding to the dimensionless product $k\sqrt{r}/m\sqrt{g}$. Notice that t appears in only the first of these products. From Buckingham's theorem, there is a function h with

$$h\left(t\sqrt{g/r},\theta,\frac{k\sqrt{r}}{m\sqrt{g}}\right)=0$$

Assuming we can solve this last equation for $t\sqrt{g/r}$, we obtain

$$t=\sqrt{r/g}\,H\left(\theta,\frac{k\sqrt{r}}{m\sqrt{g}}\right)$$

for some function H of two arguments.

Testing the Model (Step 6)

Given $t=\sqrt{r/g}\,H(\theta,k\sqrt{r}/m\sqrt{g})$, our model predicts that $t_1/t_2=\sqrt{r_1/r_2}$ if the parameters of the function H (namely, θ and $k\sqrt{r}/m\sqrt{g}$) could be held constant. Now there is no difficulty with keeping θ and k constant. However, varying r while simultaneously keeping $k\sqrt{r}/m\sqrt{g}$ constant is more complicated. Because g is constant, we could try to vary r and m in such a manner that $\sqrt{r}/m$ remains constant. This might be done using a pendulum with a hollow mass to vary m without altering the drag characteristics. Under these conditions we would expect the plot in Figure 9.8.

■ **Figure 9.8**

A plot of t versus $\sqrt{r}$ keeping the variables k, θ, and $\sqrt{r}/m$ constant

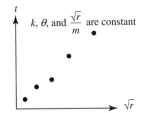

Presenting the Results (Step 7)

As was suggested in predicting the period of the undamped pendulum, we can plot $t\sqrt{g/r}=H(\theta,k\sqrt{r}/m\sqrt{g})$. However, because H is here a function of two arguments, this would yield a three-dimensional figure that is not easy to use. An alternative technique is to plot $t\sqrt{g/r}$ versus $k\sqrt{r}/m\sqrt{g}$ for various values of θ. This is illustrated in Figure 9.9. To be safe

■ **Figure 9.9**

Presenting the results

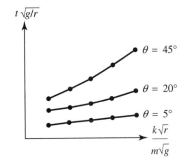

in predicting t over the range of interest for representative values of θ, it would be necessary to conduct sufficient experiments at various values of $k\sqrt{r}/m\sqrt{g}$. Note that once data are collected, various empirical models could be constructed using an appropriate interpolating scheme for each value of θ.

Choosing Among Competing Models

Because dimensional analysis involves only algebra, it is tempting to develop several models under different assumptions before proceeding with costly experimentation. In the case of the pendulum, under different assumptions, we can develop the following three models (see Problem 1 in the Section 9.3 problem set):

$$\textbf{A:} \quad t = \sqrt{r/g}\, h\,(\theta) \qquad\qquad\qquad \text{No drag forces}$$

$$\textbf{B:} \quad t = \sqrt{r/g}\, h\left(\theta,\, \frac{k\sqrt{r}}{m\sqrt{g}}\right) \qquad \text{Drag forces proportional to } v: F = kv$$

$$\textbf{C:} \quad t = \sqrt{r/g}\, h\left(\theta,\, \frac{k_1 r}{m}\right) \qquad\quad \text{Drag forces proportional to } v^2: F = k_1 v^2$$

Because all the preceding models are approximations, it is reasonable to ask which, if any, is suitable in a particular situation. We now describe the experimentation necessary to distinguish among these models, and we present some experimental results.

Model A predicts that when the angle of displacement θ is held constant, the period t is proportional to $\sqrt{r}$. Model B predicts that when θ and $\sqrt{r}/m$ are both held constant, while maintaining the same drag characteristics k, t is proportional to $\sqrt{r}$. Finally, Model C predicts that if θ, r/m, and k_1 are held constant, then t is proportional to $\sqrt{r}$.

The following discussion describes our experimental results for the pendulum.[1] Various types of balls were suspended from a string in such a manner as to minimize the friction at the hinge. The kinds of balls included tennis balls and various types and sizes of plastic balls. A hole was made in each ball to permit variations in the mass without altering appreciably the aerodynamic characteristics of the ball or the location of the center of mass. The models were then compared with one another. In the case of the tennis ball, Model A proved to be superior. The period was independent of the mass, and a plot of t versus $\sqrt{r}$ for constant θ is shown in Figure 9.10.

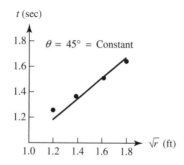

_____

[1] Data collected by Michael Jaye.

Having decided that $t = \sqrt{r/g}h(\theta)$ is the best of the models for the tennis ball, we isolated the effect of θ by holding r constant to gain insight into the nature of the function h. A plot of t versus θ for constant r is shown in Figure 9.11.

■ **Figure 9.11**

Isolating the effect of θ

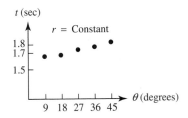

Note from Figure 9.11 that for small angles of initial displacement θ, the period is virtually independent of θ. However, the displacement effect becomes more noticeable as θ is increased. Thus, for small angles we might hypothesize that $t = c\sqrt{r/g}$ for some constant c. If one plots t versus $\sqrt{r}$ for small angles, the slope of the resulting straight line should be constant.

For larger angles, the experiment demonstrates that the effect of θ needs to be considered. In such cases, one may desire to estimate the period for various angles. For example, if $\theta = 45°$ and we know a particular value of $\sqrt{r}$, we can estimate t from Figure 9.10. Although not shown, plots for several different angles can be graphed in the same figure.

Dimensional Analysis in the Model-Building Process

Let's summarize how dimensional analysis assists in the model-building process. In the determination of a model we must first decide which factors to neglect and which to include. A dimensional analysis provides additional information on how the included factors are related. Moreover, in large problems, we often determine one or more submodels before dealing with the larger problem. For example, in the pendulum problem we had to develop a submodel for drag forces. A dimensional analysis helps us choose among the various submodels.

A dimensional analysis is also useful for obtaining an initial test of the assumptions in the model. For example, suppose we hypothesize that the dependent variable y is some function of five variables, $y = f(x_1, x_2, x_3, x_4, x_5)$. A dimensional analysis in the MLT system in general yields $\prod_1 = h(\prod_2, \prod_3)$, where each $\prod_i$ is a dimensionless product. The model predicts that $\prod_1$ will remain constant if $\prod_2$ and $\prod_3$ are held constant, even though the components of $\prod_2$ and $\prod_3$ may vary. Because there are, in general, an infinite number of ways of choosing $\prod_i$, we should choose those that can be controlled in laboratory experiments. Having determined that $\prod_1 = h(\prod_2, \prod_3)$, we can isolate the effect of $\prod_2$ by holding $\prod_3$ constant and vice versa. This can help explain the functional relationship among the variables. For instance, we say in our example that the period of the pendulum did not depend on the initial displacement for small displacements.

Perhaps the greatest contribution of dimensional analysis is that it reduces the number of experiments required to predict the behavior. If we wanted to conduct experiments to predict values of y for the assumed relationship $y = f(x_1, x_2, x_3, x_4, x_5)$ and it was decided that 5 data points would be necessary over the range of each variable, 5^5, or 3125, experiments would be necessary. Because a two-dimensional chart is required to interpolate conveniently,

y might be plotted against x_1 for five values of x_1, holding x_2, x_3, x_4, x_5 constant. Because x_2, x_3, x_4, and x_5 must vary as well, 5^4, or 625, charts would be necessary. However, after a dimensional analysis yields $\prod_1 = h(\prod_2, \prod_3)$, only 25 data points are required. Moreover, $\prod_1$ can be plotted versus $\prod_2$, for various values of $\prod_3$, on a single chart. Ultimately, the task is far easier after dimensional analysis.

Finally, dimensional analysis helps in presenting the results. It is usually best to present experimental results using those $\prod_i$ that are classical representations within the field of study. For instance, in the field of fluid mechanics there are eight factors that might be significant in a particular situation: velocity v, length r, mass density ρ, viscosity μ, acceleration of gravity g, speed of sound c, surface tension σ, and pressure p. Thus, a dimensional analysis could require as many as five independent dimensionless products. The five generally used are the Reynolds number, Froude number, Mach number, Weber number, and pressure coefficient. These numbers, which are discussed in Section 9.5, are defined as follows:

Reynolds number	$\dfrac{vr\rho}{\mu}$
Froude number	$\dfrac{v^2}{rg}$
Mach number	$\dfrac{v}{c}$
Weber number	$\dfrac{\rho v^2 r}{\sigma}$
Pressure coefficient	$\dfrac{p}{\rho v^2}$

Thus, the application of dimensional analysis becomes quite easy. Depending on which of the eight variables are considered in a particular problem, the following steps are performed.

1. Choose an appropriate subset from the preceding five dimensionless products.
2. Apply Buckingham's theorem.
3. Test the reasonableness of the choice of variables.
4. Conduct the necessary experiments and present the results in a useful format.

We illustrate an application of these steps to a fluid mechanics problem in Section 9.5.

9.3   PROBLEMS

1. For the damped pendulum,
    a. Assume that F is proportional to v^2 and use dimensional analysis to show that $t = \sqrt{r/g}h(\theta, rk_1/m)$.
    b. Assume that F is proportional to v^2 and describe an experiment to test the model $t = \sqrt{r/g}h(\theta, rk_1/m)$.

2. Under appropriate conditions, all three models for the pendulum imply that t is proportional to $\sqrt{r}$. Explain how the conditions distinguish among the three models by considering how m must vary in each case.

3. Use a model employing a differential equation to predict the period of a simple frictionless pendulum for small initial angles of displacement. (*Hint*: Let $\sin\theta = \theta$.) Under these conditions, what should be the constant of proportionality? Compare your results with those predicted by Model A in the text.[2]

9.4 Examples Illustrating Dimensional Analysis

EXAMPLE 1 *Explosion Analysis*[3]

In excavation and mining operations, it is important to be able to predict the size of a crater resulting from a given explosive such as TNT in some particular soil medium. Direct experimentation is often impossible or too costly. Thus, it is desirable to use small laboratory or field tests and then scale these up in some manner to predict the results for explosions far greater in magnitude.

We may wonder how the modeler determines which variables to include in the initial list. Experience is necessary to intelligently determine which variables can be neglected. Even with experience, however, the task is usually difficult in practice, as this example will illustrate. It also illustrates that the modeler must often change the list of variables to get usable results.

Problem Identification   *Predict the crater volume V produced by a spherical explosive located at some depth d in a particular soil medium.*

Assumptions and Model Formulation   Initially, let's assume that the craters are geometrically similar (see Chapter 2), where the crater size depends on three variables: the radius r of the crater, the density ρ of the soil, and the mass W of the explosive. These variables are composed of only two primary dimensions, length L and mass M, and a dimensional analysis results in only one dimensionless product (see Problem 1a in the Section 9.4 problem set):

$$\prod_r = r\left(\frac{\rho}{W}\right)^{1/3}$$

According to Buckingham's theorem, $\prod_r$ must equal a constant. Thus, the crater dimensions of radius or depth vary with the cube root of the mass of the explosive. Because the crater

---

[2]For students who have studied differential equations.

[3]This example is adapted with permission from R. M. Schmidt, "A Centrifuge Cratering Experiment: Development of a Gravity-Scaled Yield Parameter." In *Impact and Explosion Cratering*, edited by D. J. Roddy, R. O. Pepin, and R. B. Merrill (New York: Pergamon, 1977), pp. 1261–1276.

volume is proportional to r^3, it follows that the volume of the crater is proportional to the mass of the explosive for constant soil density. Symbolically, we have

$$V \propto \frac{W}{\rho} \tag{9.20}$$

Experiments have shown that the proportionality (9.20) is satisfactory for small explosions (less than 300 lb of TNT) at zero depth in soils, such as moist alluvium, that have good cohesion. For larger explosions, however, the rule proves unsatisfactory. Other experiments suggest that gravity plays a key role in the explosion process, and because we want to consider extraterrestial craters as well, we need to incorporate gravity as a variable.

If gravity is taken into account, then we assume crater size to be dependent on four variables: crater radius r, density of the soil ρ, gravity g, and charge energy E. Here, the charge energy is the mass W of the explosive times its specific energy. Applying a dimensional analysis to these four variables again leads to a single dimensionless product (see Problem 1b in the 9.4 problem set):

$$\prod_{rg} = r \left(\frac{\rho g}{E} \right)^{1/4}$$

Thus, $\prod_{rg}$ equals a constant and the linear crater dimensions (radius or depth of the crater) vary with the one-fourth root of the energy (or mass) of the explosive for a constant soil density. This leads to the following proportionality known as the quarter-root scaling and is a special case of *gravity scaling*:

$$V \propto \left(\frac{E}{\rho g} \right)^{3/4} \tag{9.21}$$

Experimental evidence indicates that gravity scaling holds for large explosions (more than 100 tons of TNT) where the stresses in the cratering process are much larger than the material strengths of the soil. The proportionality (9.21) predicts that crater volume decreases with increased gravity. The effect of gravity on crater formation is relevant in the study of extraterrestial craters. Gravitational effects can be tested experimentally using a centrifuge to increase gravitational accelerations.[4]

A question of interest to explosion analysts is whether the material properties of the soil do become less important with increased charge size and increased gravity. Let's consider the case in which the soil medium is characterized only by its density ρ. Thus, the crater volume V depends on the explosive, soil density ρ, gravity g, and the depth of burial d of the charge. In addition, the explicit role of material strength or cohesion has been tested and the strength–gravity transition is shown to be a function of charge size and soil strength.

We now describe our explosive in more detail than in previous models. To characterize an explosive, three independent variables are needed: size, energy field, and explosive density δ. The size can be given as charge mass W, as charge energy E, or as the radius α of the spherical explosive. The energy yield can be given as a measure of the specific

---

[4]See the papers by R. M. Schmidt (1977, 1980) and by Schmidt and Holsapple (1980), cited in Further Reading, which discuss the effects when a centrifuge is used to perform explosive cratering tests under the influence of gravitational acceleration up to 480 G, where 1 G is the terrestrial gravity field strength of 981 cm/sec^2.

energy Q_e or the energy density per unit volume Q_V. The following equations relate the variables:

$$W = \frac{E}{Q_e}$$

$$Q_V = \delta Q_e$$

$$\alpha^3 = \left(\frac{3}{4\pi}\right)\left(\frac{W}{\delta}\right)$$

One choice of these variables leads to the model formulation

$$V = f(W, Q_e, \delta, \rho, g, d)$$

Because there are seven variables under consideration and the MLT system is being used, a dimensional analysis generally will result in four $(7-3)$ dimensionless products. The dimensions of the variables are shown in the following table.

Variable	V	W	Q_e	δ	ρ	g	d
Dimension	L^3	M	L^2T^{-2}	ML^{-3}	ML^{-3}	LT^{-2}	L

Any product of the variables must be of the form

$$V^a W^b Q_e^c \delta^e \rho^f g^k d^m \qquad (9.22)$$

and hence have dimensions

$$(L^3)^a (M^b)(L^2T^{-2})^c (ML^{-3})^{e+f} (LT^{-2})^k (L)^m$$

Therefore, a product of the form (9.22) is dimensionless if and only if the exponents satisfy the following homogeneous system of equations:

$$
\begin{array}{lrcl}
M: & b & + e + f & = 0 \\
L: & 3a & + 2c - 3e - 3f + k + m & = 0 \\
T: & & -2c \qquad\qquad -2k & = 0
\end{array}
$$

Solution to this system produces

$$b = \frac{k-m}{3} - a, \quad c = -k, \quad e = a - f + \frac{k-m}{3}$$

where a, f, k, and m are arbitrary. By setting one of these arbitrary exponents equal to 1 and the other three equal to 0, in succession, we obtain the following set of dimensionless products:

$$\frac{V\delta}{W}, \quad \left(\frac{g}{Q_e}\right)\left(\frac{W}{\delta}\right)^{1/3}, \quad d\left(\frac{\delta}{W}\right)^{1/3}, \quad \frac{\rho}{\delta}$$

(Convince yourself that these are dimensionless.) Because the dimensions of ρ and δ are equal, we can rewrite these dimensionless products as follows:

$$\Pi_1 = \frac{V\rho}{W}$$

$$\Pi_2 = \left(\frac{g}{Q_e}\right)\left(\frac{W}{\delta}\right)^{1/3}$$

$$\Pi_3 = d\left(\frac{\rho}{W}\right)^{1/3}$$

$$\Pi_4 = \frac{\rho}{\delta}$$

so Π_1 is consistent with the dimensionless product implied by Equation (9.20). Then, applying Buckingham's theorem, we obtain the model

$$h\left(\Pi_1, \Pi_2, \Pi_3, \Pi_4\right) = 0 \tag{9.23}$$

or

$$V = \frac{W}{\rho} H\left(\frac{gW^{1/3}}{Q_e\delta^{1/3}}, \frac{d\delta^{1/3}}{W^{1/3}}, \frac{\rho}{\delta}\right)$$

Presenting the Results For oil-base clay, the value of ρ is approximately 1.53 g/cm^2; for wet sand, 1.65; and for desert alluvium, 1.60. For TNT, δ has the value 2.23 g/cm^3. Thus, $0.69 < \Pi_4 < 0.74$, so for simplicity we can assume that for these soils and TNT, Π_4 is constant. Then, Equation (9.23) becomes

$$h\left(\Pi_1, \Pi_2, \Pi_3\right) = 0 \tag{9.24}$$

■ **Figure 9.12**

A plot of the surface $h(\Pi_1, \Pi_2, \Pi_3) = 0$, showing the crater volume parameter Π_1 as a function of gravity-scaled yield Π_2 and depth of burial parameter Π_3 (reprinted by permission of R. M. Schmidt)

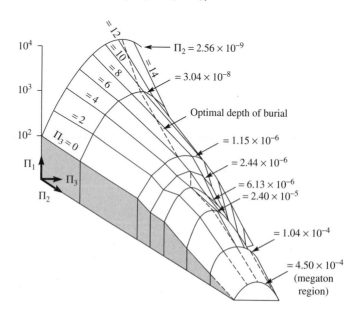

■ Figure 9.13

Data values for a cross section of the surface depicted in Figure 9.12 (data reprinted by permission from R. M. Schmidt)

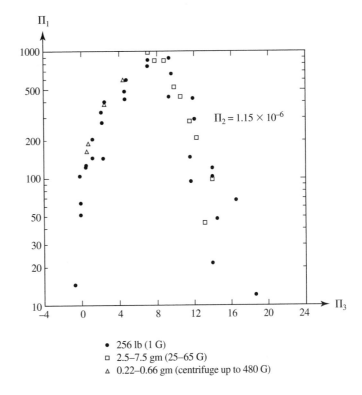

- 256 lb (1 G)
- □ 2.5–7.5 gm (25–65 G)
- ▵ 0.22–0.66 gm (centrifuge up to 480 G)

R. M. Schmidt gathered experimental data to plot the surface described by Equation (9.24). A plot of the surface is depicted in Figure 9.12, showing the crater and volume parameter Π_1 as a function of the scaled energy charge Π_2 and the depth of the burial parameter Π_3. Cross-sectional data for the surface parallel to the Π_1 Π_3 plane when $\Pi_2 = 1.15 \times 10^{-6}$ are depicted in Figure 9.13.

Experiments have shown that the physical effect of increasing gravity is to reduce crater volume for a given charge yield. This result suggests that increased gravity can be compensated for by increasing the size of the charge to maintain the same cratering efficiency. Note also that Figures 9.12 and 9.13 can be used for prediction once an empirical interpolating model is constructed from the data. Holsapple and Schmidt (1982) extend these methods to impact cratering, and Housen, Schmidt, and Holsapple (1983) extend them to crater ejecta scaling.   ■ ■ ■

EXAMPLE 2   *How Long Should You Roast a Turkey?*

One general rule for roasting a turkey is the following: Set the oven to 400 °F and allow 20 min per pound for cooking. How good is this rule?

Assumptions Let t denote the cooking time for the turkey. Now, on what variables does t depend? Certainly the size of the turkey is a factor that must be considered. Let's assume that the turkeys are geometrically similar and use l to denote some characteristic dimension of the uncooked meat; specifically, we assume that l represents the length of the turkey. Another factor is the difference between the temperature of the raw meat and the oven ΔT_m. (We know from experience that it takes longer to cook a bird that is nearly frozen than it does to cook one that is initially at room temperature.) Because the turkey will have to reach a certain interior temperature before it is considered fully cooked, the difference ΔT_c between the temperature of the cooked meat and the oven is a variable determining the cooking time. Finally, we know that different foods require different cooking times independent of size; it takes only 10 min or so to bake a pan of cookies, whereas a roast beef or turkey requires several hours. A measure of the factor representing the differences between foods is the *coefficient of heat conduction* for a particular uncooked food. Let k denote the coefficient of heat conduction for a turkey. Thus, we have the following model formulation for the cooking time:

$$t = f(\Delta T_m, \Delta T_c, k, l)$$

Dimensional Analysis Consider the dimensions of the independent variables. The temperature variables ΔT_m and ΔT_c measure the energy per volume and therefore have the dimension ML^2T^{-2}/L^3, or simply $ML^{-1}T^{-2}$. Now, what about the heat conduction variable k? **Thermal conductivity** k is defined as the amount of energy crossing one unit cross-sectional area per second divided by the gradient perpendicular to the area. That is,

$$k = \frac{\text{energy}/(\text{area} \times \text{time})}{\text{temperature}/\text{length}}$$

Accordingly, the dimension of k is $(ML^2T^{-2})(L^{-2}T^{-1})/(ML^{-1}T^{-2})(L^{-1})$, or simply L^2T^{-1}. Our analysis gives the following table:

Variable	ΔT_m	ΔT_c	k	l	t
Dimension	$ML^{-1}T^{-2}$	$ML^{-1}T^{-2}$	L^2T^{-1}	L	T

Any product of the variables must be of the form

$$\Delta T_m^a \Delta T_c^b k^c l^d t^e \tag{9.25}$$

and hence have dimension

$$(ML^{-1}T^{-2})^a (ML^{-1}T^{-2})^b (L^2T^{-1})^c (L)^d (T)^e$$

Therefore, a product of the form (9.25) is dimensionless if and only if the exponents satisfy

$$
\begin{aligned}
M: \quad & a + b && = 0 \\
L: \quad & -a - b + 2c + d && = 0 \\
T: \quad & -2a - 2b - c + e && = 0
\end{aligned}
$$

Solution of this system of equations gives

$$a = -b, \quad c = e, \quad d = -2e$$

where b and e are arbitrary constants. If we set $b = 1, e = 0$, we obtain $a = -1, c = 0$, and $d = 0$; likewise, $b = 0, e = 1$ produces $a = 0, c = 1$, and $d = -2$. These independent solutions yield the complete set of dimensionless products:

$$\prod_1 = \Delta T_m^{-1} \Delta T_c \quad \text{and} \quad \prod_2 = k l^{-2} t$$

From Buckingham's theorem, we obtain

$$h \left(\prod_1, \prod_2 \right) = 0$$

or

$$t = \left(\frac{l^2}{k} \right) H \left(\frac{\Delta T_c}{\Delta T_m} \right) \tag{9.26}$$

The rule stated in our opening remarks gives the roasting time for the turkey in terms of its weight w. Let's assume the turkeys are geometrically similar, or $V \propto l^3$. If we assume the turkey is of constant density (which is not quite correct because the bones and flesh differ in density), then, because weight is density times volume and volume is proportional to l^3, we get $w \propto l^3$. Moreover, if we set the oven to a constant baking temperature and specify that the turkey must initially be near room temperature ($65 \,^\circ$F), then $\Delta T_c / \Delta T_m$ is a dimensionless constant. Combining these results with Equation (9.26), we get the proportionality

$$t \propto w^{2/3} \tag{9.27}$$

because k is constant for turkeys. Thus, the required cooking time is proportional to weight raised to the two-thirds power. Therefore, if t_1 hours are required to cook a turkey weighing w_1 pounds and t_2 is the time for a weight of w_2 pounds,

$$\frac{t_1}{t_2} = \left(\frac{w_1}{w_2} \right)^{2/3}$$

it follows that a doubling of the weight of a turkey increases the cooking time by the factor $2^{2/3} \approx 1.59$.

How does our result (9.27) compare to the rule stated previously? Assume that ΔT_m, ΔT_c, and k are independent of the length or weight of the turkey, and consider cooking a 23-lb turkey versus an 8-lb bird. According to our rule, the ratio of cooking times is given by

$$\frac{t_1}{t_2} = \left(\frac{20 \cdot 23}{20 \cdot 8} \right) = 2.875$$

On the other hand, from dimensionless analysis and Equation (9.27),

$$\frac{t_1}{t_2} = \left(\frac{23}{8} \right)^{2/3} \approx 2.02$$

Thus, the rule predicts it will take nearly three times as long to cook a 23-lb bird as it will to cook an 8-lb turkey. Dimensional analysis predicts it will take only twice as long. Which rule is correct? Why have so many cooks overcooked a turkey?

Testing the Results   Suppose that turkeys of various sizes are cooked in an oven pre-heated to 325 °F. The initial temperature of the turkeys is 65 °F. All the turkeys are removed from the oven when their internal temperature, measured by a meat thermometer, reaches 195 °F. The (hypothetical) cooking times for the various turkeys are recorded as shown in the following table.

w (lb)	5	10	15	20
t (hr)	2	3.4	4.5	5.4

A plot of t versus $w^{2/3}$ is shown in Figure 9.14. Because the graph approximates a straight line through the origin, we conclude that $t \propto w^{2/3}$, as predicted by our model.

■ ■ ■

■ **Figure 9.14**

Plot of cooking times versus weight to the two-thirds power reveals the predicted proportionality.

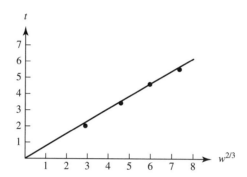

PROBLEMS

1. a. Use dimensional analysis to establish the cube-root law

$$r\left(\frac{\rho}{W}\right)^{1/3} = \text{constant}$$

for scaling of explosions, where r is the radius or depth of the crater, ρ is the density of the soil medium, and W the mass of the explosive.

b. Use dimensional analysis to establish the one-fourth-root law

$$r\left(\frac{\rho g}{E}\right)^{1/4} = \text{constant}$$

for scaling of explosions, where r is the radius or depth of the crater, ρ is the density of the soil medium, g is gravity, and E is the charge energy of the explosive.

2. a. Show that the products $\prod_1, \prod_2, \prod_3, \prod_4$ for the refined explosion model presented in the text are dimensionless products.

b. Assume ρ is essentially constant for the soil being used and restrict the explosive to a specific type, say TNT. Under these conditions, ρ/δ is essentially constant, yielding $\prod_1 = f(\prod_2, \prod_3)$. You have collected the following data with $\prod_2 = 1.5 \times 10^{-6}$.

Π_3	0	2	4	6	8	10	12	14
Π_1	15	150	425	750	825	425	250	90

i. Construct a scatterplot of Π_1 versus Π_3. Does a trend exist?

ii. How accurate do you think the data are? Find an empirical model that captures the *trend* of the data with accuracy commensurate with your appraisal of the accuracy of the data.

iii. Use your empirical model to predict the volume of a crater using TNT in desert alluvium with (CGS system) $W = 1500$ g, $\rho = 1.53$ g/cm³, and $\Pi_3 = 12.5$.

3. Consider a zero-depth burst, spherical explosive in a soil medium. Assume the value of the crater volume V depends on the explosive size, energy yield, and explosive energy, as well as on the strength Y of the soil (considered a resistance to pressure with dimensions $ML^{-1}T^{-2}$), soil density ρ, and gravity g. In this problem, assume

$$V = f(W, Q_e, \delta, Y, \rho, g)$$

and use dimensional analysis to produce the following *mass set* of dimensionless products.

$$\Pi_1 = \frac{V\rho}{W} \qquad \Pi_2 = \left(\frac{g}{Q_e}\right)\left(\frac{W}{\delta}\right)^{1/3}$$

$$\Pi_3 = \frac{Y}{\delta Q_e} \qquad \Pi_4 = \frac{\rho}{\delta}$$

4. For the explosion process and material characteristics discussed in Problem 3, consider

$$V = f(E, Q_v, \delta, Y, \rho, g)$$

and use dimensional analysis to produce the following *energy set* of dimensionless products.

$$\overline{\Pi}_1 = \frac{VQ_v}{E} \qquad \overline{\Pi}_2 = \frac{\rho g E^{1/3}}{Q_v^{4/3}}$$

$$\overline{\Pi}_3 = \frac{Y}{Q_v} \qquad \overline{\Pi}_4 = \frac{\rho}{\delta}$$

5. Repeat Problem 4 for

$$V = f(E, Q_e, \delta, Y, \rho, g)$$

and use dimensional analysis to produce the following *gravity set* of dimensionless products.

$$\overline{\overline{\Pi}}_1 = V\left(\frac{\rho g}{E}\right)^{3/4} \qquad \overline{\overline{\Pi}}_2 = \left(\frac{1}{Q_e}\right)\left(\frac{g^3 E}{\delta}\right)^{1/4}$$

$$\overline{\overline{\Pi}}_3 = \frac{Y}{\delta Q_e} \qquad \overline{\overline{\Pi}}_4 = \frac{\rho}{\delta}$$

6. An experiment consists of dropping spheres into a tank of heavy oil and measuring the times of descent. It is desired that a relationship for the time of descent be determined and verified by experimentation. Assume the time of descent is a function of mass m, gravity g, radius r, viscosity μ, and distance traveled d. Neglect fluid density. That is,

$$t = f(m, g, r, \mu, d)$$

   a. Use dimensional analysis to find a relationship for the time of descent.

   b. How will the spheres be chosen to verify that the time of descent relationship is independent of fluid density? Assuming you have verified the assumptions on fluid density, describe how you would determine the nature of your function experimentally.

   c. Using differential equations techniques, find the velocity of the sphere as a function of time, radius, mass, viscosity, gravity, and fluid density. Using this result and that found in part (a), predict under what conditions fluid density may be neglected. (*Hints:* Use the results of Problem 5 in Section 9.2 as a submodel for drag force. Consider the buoyant force.)[5]

7. A windmill is rotated by air flow to produce power to pump water. It is desired to find the power output P of the windmill. Assume that P is a function of the density of the air ρ, viscosity of the air μ, diameter of the windmill d, wind speed v, and the rotational speed of the windmill ω (measured in radians per second). Thus,

$$P = f(\rho, \mu, d, v, \omega)$$

   a. Using dimensional analysis, find a relationship for P. Be sure to check your products to make sure that they are dimensionless.

   b. Do your results make common sense? Explain.

   c. Discuss how you would design an experiment to determine the nature of your function.

8. For a sphere traveling through a liquid, assume that the drag force F_D is a function of the fluid density ρ, fluid viscosity μ, radius of the sphere r, and speed of the sphere v. Use dimensional analysis to find a relationship for the drag force

$$F_D = f(\rho, \mu, r, v)$$

   Make sure you provide some justification that the given independent variables influence the drag force.

9.4 PROJECT

1. Complete the requirements for the module, "Listening to the Earth: Controlled Source Seismology," by Richard G. Montgomery, UMAP 292-293. This module develops the elementary theory of wave reflection and refraction and applies it to a model of the earth's subsurface. The model shows how information on layer depth and sound velocity may

---

[5]For students who have studied differential equations.

be obtained to provide data on width, density, and composition of the subsurface. This module is a good introduction to controlled seismic methods and requires no previous knowledge of either physics or geology.

9.4 Further Reading

Holsapple, K. A., & R. M. Schmidt. "A Material-Strength Model for Apparent Crater Volume." *Proc. Lunar Planet Sci. Conf.* 10 (1979): 2757–2777.

Holsapple, K. A., & R. M. Schmidt. "On Scaling of Crater Dimensions–2: Impact Process." *J. Geophys. Res.* 87 (1982): 1849–1870.

Housen, K. R., K. A. Holsapple, & R. M. Schmidt." Crater Ejecta Scaling Laws 1: Fundamental Forms Based on Dimensional Analysis." *J. Geophys. Res.* 88 (1983): 2485–2499.

Schmidt, R. M. "A Centrifuge Cratering Experiment: Development of a Gravity-Scaled Yield Parameter." In *Impact and Explosion Cratering*, edited by D. J. Roddy et al., pp. 1261–1278. New York: Pergamon, 1977.

Schmidt, R. M. "Meteor Crater: Energy of Formation—Implications of Centrifuge Scaling." *Proc. Lunar Planet Sci. Conf.* 11 (1980): 2099–2128.

Schmidt, R. M., & K. A. Holsapple. "Theory and Experiments on Centrifuge Cratering." *J. Geophys. Res.* 85 (1980): 235–252.

9.5 Similitude

Suppose we are interested in the effects of wave action on a large ship at sea, heat loss of a submarine and the drag force it experiences in its underwater environment, or the wind effects on an aircraft wing. Quite often, because it is physically impossible to duplicate the actual phenomenon in the laboratory, we study a scaled-down model in a simulated environment to predict accurately the performance of the physical system. The actual physical system for which the predictions are to be made is called the **prototype**. How do we scale experiments in the laboratory to ensure that the effects observed for the model will be the same effects experienced by the prototype?

Although extreme care must be exercised in using simulations, the dimensional products resulting from dimensional analysis of the problem can provide insight into how the scaling for a model should be done. The idea comes from Buckingham's theorem. If the physical system can be described by a dimensionally homogeneous equation in the variables, then it can be put into the form

$$f\left(\prod_1, \prod_2, \ldots, \prod_n\right) = 0$$

for a complete set of dimensionless products. Assume that the independent variable of the problem appears only in the product $\prod_n$ and that

$$\prod_n = H\left(\prod_1, \prod_2, \ldots, \prod_{n-1}\right)$$

For the solution to the model and the prototype to be the same, it is sufficient that the value of all independent dimensionless products $\prod_1, \prod_2, \ldots, \prod_{n-1}$ be the same for the model and the prototype.

For example, suppose the Reynolds number $vr\rho/\mu$ appears as one of the dimensionless products in a fluid mechanics problem, where v represents fluid velocity, r a characteristic dimension (such as the diameter of a sphere or the length of a ship), ρ the fluid density, and μ the fluid viscosity. These values refer to the prototype. Next, let v_m, r_m, ρ_m, and μ_m denote the corresponding values for the scaled-down model. For the effects on the model and the prototype to be the same, we want the two Reynolds numbers to agree so that

$$\frac{v_m r_m \rho_m}{\mu_m} = \frac{vr\rho}{\mu}$$

The last equation is referred to as a *design condition* to be satisfied by the model. If the length of the prototype is too large for the laboratory experiments so that we have to scale down the length of the mode, say $r_m = r/10$, then the same Reynolds number for the model and the prototype can be achieved by using the same fluid ($\rho_m = \rho$ and $\mu_m = \mu$) and varying the velocity, $v_m = 10v$. If it is impractical to scale the velocity by the factor of 10, we can instead scale it by a lesser amount $0 < k < 10$ and use a different fluid so that the equation

$$\frac{k\rho_m}{10\mu_m} = \frac{\rho}{\mu}$$

is satisfied. We do need to be careful in generalizing the results from the scaled-down model to the prototype. Certain factors (such as surface tension) that may be negligible for the prototype may become significant for the model. Such factors would have to be taken into account before making any predictions for the prototype.

EXAMPLE 1   *Drag Force on a Submarine*

We are interested in the drag forces experienced by a submarine to be used for deep-sea oceanographic explorations. We assume that the variables affecting the drag D are fluid velocity v, characteristic dimension r (here, the length of the submarine), fluid density ρ, the fluid viscosity μ, and the velocity of sound in the fluid c. We wish to predict the drag force by studying a model of the prototype. How will we scale the experiments for the model?

A major stumbling block in our problem is in describing shape factors related to the physical object being modeled—in this case, the submarine. Let's consider submarines that are ellipsoidal in shape. In two dimensions, if a is the length of the major axis and b is the length of the minor axis of an ellipse, we can define $r_1 = a/b$ and assign a characteristic dimension such as r, the length of the submarine. In three dimensions, define also $r_2 = a/b'$, where a is the original major axis and b' is the second minor axis. Then r, r_1, and r_2 describe the shape of the submarine. In a more irregularly shaped object, additional shape factors would be required. The basic idea is that the object can be described using a characteristic dimension and an appropriate collection of shape factors. In the case of our three-dimensional ellipsoidal submarine, the shape factors r_1 and r_2 are needed. These shape factors are dimensionless constants.

Returning to our list of six fluid mechanics variables D, v, r, ρ, μ, and c, notice that we are neglecting surface tension (because it is small) and that gravity is not being considered. Thus, it is expected that a dimensionless analysis will produce three $(6-3)$ independent dimensionless products. We can choose the following three products for convenience:

$$\text{Reynolds number} \qquad R = \frac{vr\rho}{\mu}$$

$$\text{Mach number} \qquad M = \frac{v}{c}$$

$$\text{Pressure coefficient} \qquad P = \frac{p}{\rho v^2}$$

The added shape factors are dimensionless so that Buckingham's theorem gives the equation

$$h(P, M, R, r_1, r_2) = 0$$

Assuming that we can solve for P yields

$$P = H(M, R, r_1, r_2)$$

Substituting $P = p/\rho v^2$ and solving for p gives

$$p = \rho v^2 H(R, M, r_1, r_2)$$

Remembering that the total drag force is the pressure (force per unit area) times the area (which is proportional to r^2 for geometrically similar objects) and gives the proportionality $D \propto pr^2$, or

$$D = kp v^2 r^2 H(R, M, r_1, r_2) \tag{9.28}$$

Now a similar equation must hold to give the proportionality for the model

$$D_m = kp_m v_m^2 r_m^2 H(R_m, M_m, r_{1m}, r_{2m}) \tag{9.29}$$

Because the prototype and model equations refer to the same physical system, both equations are identical in form. Therefore, the design conditions for the model require that

$$\text{Condition (a)} \qquad R_m = R$$
$$\text{Condition (b)} \qquad M_m = M$$
$$\text{Condition (c)} \qquad r_{1m} = r_1$$
$$\text{Condition (d)} \qquad r_{2m} = r_2$$

Note that if conditions (a)–(d) are satisfied, then Equations (9.28) and (9.29) give

$$\frac{D_m}{D} = \frac{\rho_m v_m^2 r_m^2}{\rho v^2 r^2} \tag{9.30}$$

Thus, D can be computed once D_m is measured. Note that the design conditions (c) and (d) imply geometric similarity between the model and the prototype submarine

$$\frac{a_m}{b_m} = \frac{a}{b} \quad \text{and} \quad \frac{a_m}{b'_m} = \frac{a}{b'}$$

If the velocities are small compared to the speed of sound in a fluid, then v/c can be considered constant in accordance with condition (b). If the same fluid is used for both the model and prototype, then condition (a) is satisfied if

$$v_m r_m = vr$$

or

$$\frac{v_n}{v} = \frac{r}{r_m}$$

which states that the velocity of the model must increase inversely as the scaling factor r_m/r. Under these conditions, Equation (9.30) yields

$$\frac{D_m}{D} = \frac{\rho_m v_m^2 r_m^2}{\rho v^2 r^2} = 1$$

If increasing the velocity of the scaled model proves unsatisfactory in the laboratory, then a different fluid may be considered for the scaled model ($\rho_m \neq \rho$ and $\mu_m \neq \mu$). If the ratio v/c is small enough to neglect, then both v_m and r_m can be varied to ensure that

$$\frac{v_m r_m \rho_m}{\mu_m} = \frac{vr\rho}{\mu}$$

in accordance with condition (a). Having chosen values that satisfy design condition (a), and knowing the drag on the scaled model, we can use Equation (9.30) to compute the drag on the prototype. Consider the additional difficulties if the velocities are sufficiently great that we must satisfy condition (b) as well.

A few comments are in order. One distinction between the Reynolds number and the other four numbers in fluid mechanics is that the Reynolds number contains the viscosity of the fluid. Dimensionally, the Reynolds number is proportional to the ratio of the inertia forces of an element of fluid to the viscous force acting on the fluid. In certain problems the numerical value of the Reynolds number may be significant. For example, the flow of a fluid in a pipe is virtually always parallel to the edges of the pipe (giving *laminar flow*) if the Reynolds number is less than 2000. Reynolds numbers in excess of 3000 almost always indicate turbulent flow. Normally, there is a critical Reynolds number between 2000 and 3000 at which the flow becomes turbulent.

The design condition (a) mentioned earlier requires the Reynolds number of the model and the prototype to be the same. This requirement precludes the possibility of laminar flow in the prototype being represented by turbulent flow in the model, and vice versa. The equality of the Reynolds number for a model and prototype is important in all problems in which viscosity plays a significant role.

The Mach number is the ratio of fluid velocity to the speed of sound in the fluid. It is generally important for problems involving objects moving with high speed in fluids, such as projectiles, high-speed aircraft, rockets, and submarines. Physically, if the Mach number is the same in model and prototype, the effect of the compressibility force in the fluid relative to the inertia force will be the same for model and prototype. This is the situation that is required by condition (b) in our example on the submarine. ■ ■ ■

9.5 PROBLEMS

1. A model of an airplane wing is tested in a wind tunnel. The model wing has an 18-in. chord, and the prototype has a 4-ft chord moving at 250 mph. Assuming the air in the wind tunnel is at atmospheric pressure, at what velocity should wind tunnel tests be conducted so that the Reynolds number of the model is the same as that of the prototype?

2. Two smooth balls of equal weight but different diameters are dropped from an airplane. The ratio of their diameters is 5. Neglecting compressibility (assume constant Mach number), what is the ratio of the terminal velocities of the balls? Are the flows completely similar?

3. Consider predicting the pressure drop Δp between two points along a smooth horizontal pipe under the condition of steady laminar flow. Assume

$$\Delta p = f(s, d, \rho, \mu, v)$$

where s is the control distance between two points in the pipe, d is the diameter of the pipe, ρ is the fluid density, μ is the fluid viscosity, and v is the velocity of the fluid.

a. Determine the design conditions for a scaled model of the prototype.

b. Must the model be geometrically similar to the prototype?

c. May the same fluid be used for model and prototype?

d. Show that if the same fluid is used for both model and prototype, then the equation is

$$\Delta p = \frac{\Delta p_m}{n^2}$$

   where $n = d/d_m$.

4. It is desired to study the velocity v of a fluid flowing in a smooth open channel. Assume that

$$v = f(r, \rho, \mu, \sigma, g)$$

where r is the characteristic length of the channel cross-sectional area divided by the wetted perimeter, ρ is the fluid density, μ is the fluid viscosity, σ is the surface tension, and g is the acceleration of gravity.

a. Describe the appropriate pair of shape factors r_1 and r_2.

b. Show that

$$\frac{v^2}{gr} = H\left(\frac{\rho v r}{\mu}, \frac{\rho v^2 r}{\sigma}, r_1, r_2\right)$$

Discuss the design conditions required of the model.

    c. Will it be practical to use the same fluid in the model and the prototype?

    d. Suppose the surface tension σ is ignored and the design conditions are satisfied. If $r_m = r/n$, what is the equation for the velocity of the prototype? When is the equation compatible with the design conditions?

    e. What is the equation for the velocity v if gravity is ignored? What if viscosity is ignored? What fluid would you use if you were to ignore viscosity?

9.5 Further Reading

Massey, Bernard S. *Units, Dimensional Analysis and Physical Similarity.* London: Van Nostrand Reinhold, 1971.

10 Graphs of Functions as Models

10.1 An Arms Race

You might ask, Why study the arms race? One reason is that almost all modern wars are preceded by unstable arms races. Strong evidence suggests that an unstable arms race between great powers, characterized by a sharp acceleration in military capability, is an early warning indicator of war. In a 1979 article, Michael Wallace of the University of British Columbia studied 99 international disputes during the 1816–1965 period.[1] He found that disputes preceded by an unstable arms race escalated to war 23 out of 28 times, whereas disputes *not* preceded by an arms race resulted in war only 3 out of 71 times. Wallace calculated an arms race index for the two nations involved in each dispute that correctly predicted war or no war in 91 out of the 99 cases studied. His findings do not mean that an arms race between the powers necessarily results in war or that there is a causal link between arms races and conflict escalation. They do establish, however, that rapid competitive military growth is strongly associated with the propensity to war. Thus, by studying the arms race, we have the potential for predicting war. If we can predict war, then there is hope that we can learn to avoid it.

There is another reason for studying the arms race. If the arms race can be approximated by a mathematical model, then it can be understood more concretely. You will see that the answers to such questions as, Will civil defense dampen the arms race? and Will the introduction of mobile missile launching pads help to reduce the arms requirements? are not simply matters of political opinion. There is an objective reality to the arms race that the mathematical model intends to capture.

The former Soviet Union and the United States were engaged in a nuclear arms race during the Cold War. At that time political and military strategists asked how the United States should react to changes in numbers and sophistication of the Soviet nuclear arsenal. To answer the difficult question, How many weapons are enough?, several factors had to be considered, including American objectives, Soviet objectives, and weapon technology. A former chairman of the Joint Chiefs of Staff, General Maxwell D. Taylor, suggested the following nuclear deterrence objectives for the American strategic forces:

> The strategic forces, having the single capability of inflicting massive destruction, should have the single task of deterring the Soviet Union from resorting to any form of strategic warfare. To

---

[1] Michael Wallace, "Arms Races and Escalation: Some New Evidence." In *Explaining War*, edited by J. David Singer, pp. 240–252. Beverly Hills, CA: Sage, 1979.

maximize their deterrent effectiveness they must be able to survive a massive first strike and still be able to destroy sufficient enemy targets to eliminate the Soviet Union as a viable government, society, and economy, responsive to the national leaders who determine peace or war.[2]

Note especially that Taylor's deterrence strategy assumes the worst possible case: the Soviet's launching a preemptive first attack to destroy America's nuclear force.

How many weapons would be necessary to accomplish the objectives General Taylor suggests? After describing an appropriate system of Soviet targets (generally population and industrial centers), he states the following:

> The number of weapons we shall need will be those required to destroy the specific targets within this system of which few will be hardened silos calling for the accuracy and short flight time of ICBMs. As a safety factor, we should add extra weapons to compensate for losses that may be suffered in a first strike and for uncertainties in weapon performance. The total weapons requirement should be substantially less than the numbers available to us in our present arsenal.[3]

Thus, a minimum number of missiles would be required to destroy specific enemy targets (generally population and industrial centers) chosen to inflict unacceptable damage on the enemy. Additional missiles would be required to compensate for losses incurred in the Soviet's presumed first strike. Implicitly, the number of such additional missiles depends on the size and effectiveness of the Soviet missile forces. Taylor concluded that meeting these objectives would allow for a reduction in America's nuclear arsenal.

In response to a question on expenditures for national defense, Admiral Hyman G. Rickover testified before a congressional committee as follows:

> For example, take the number of nuclear submarines; I'll hit right close to home. I see no reason why we have to have just as many as the Russians do. At a certain point you get where it's sufficient. What's the difference whether we have 100 nuclear submarines or 200? I don't see what difference it makes. You can sink everything on the ocean several times over with the number we have and so can they. That's the point I'm making.[4]

Again, Admiral Rickover concluded that a reduction in arms would be possible.

On July 14, 2001, in Genoa, Italy, President George W. Bush and Russian President Vladimir Putin agreed to seek cuts in their nuclear arsenals. Putin said he would accept a U.S. antimissile shield if it were linked to deep cuts in offensive nuclear weapons. He suggested that both nations could reduce their nuclear arsenals to approximately 1500 strategic weapons. (At that time the United States had approximately 7000 strategic weapons, and Russia had approximately 6500.)

We are going to develop a graphical model of the nuclear arms race based on the preceding remarks. The model will help answer the question, How many weapons are enough? Although the model applies to any kind of arms race, for purposes of discussion and illustration we focus on nuclear weapons delivered by long-range intercontinental ballistic missiles (ICBMs).

---

[2]M. D. Taylor, "How to Avoid a New Arms Race." *The Monterey Peninsula Herald*, January 24, 1982, p. 3c.

[3]Ibid.

[4]Hyman C. Rickover, testimony before Joint Economic Committee. *The New York Review*, March 18, 1982, p. 13.

Developing the Graphical Model

Suppose that two countries, Country X and Country Y, are engaged in a nuclear arms race and that *each* country adopts the following strategies.

Friendly Strategy: To survive a massive first strike and inflict unacceptable damage on the enemy.

Enemy Strategy: To conduct a massive first strike to destroy the friendly missile force.

That is, *each* country follows the friendly strategy when determining its own missile force and presumes the enemy strategy for the opposing country. Note especially that the friendly strategy implies targeting population and industrial centers, whereas the enemy strategy implies targeting missile sites. This was the policy of **nuclear deterrence** advocated during the Cold War.

Now let's define the following variables:

$$x = \text{the number of missiles possessed by Country } X$$
$$y = \text{the number of missiles possessed by Country } Y$$

Next, let $y = f(x)$ denote the function representing the minimum number of missiles required by Country Y to accomplish its strategies when Country X has x missiles. Similarly, let $x = g(y)$ represent the minimum number of missiles required by Country X to accomplish its objectives. When Country Y determines the required size of its missile force, it assumes that it has the friendly strategy and that Country X is following the enemy strategy. On the other hand, Country X has the friendly strategy when determining the size of its missile force and presumes Country Y is following the enemy strategy.

We begin by investigating the nature of the curve $y = f(x)$. Because a certain number of missiles y_0 are required by Country Y to destroy the selected population and industrial centers of Country X, y_0 is the intercept when $x = 0$. That is, Country Y considers that it needs y_0 missiles even if Country X has none (basically, a psychological defense in the sense that Y fears attack or invasion by X). As Country X increases its missile force, Country Y must then add additional missiles because it assumes Country X is following the enemy strategy and targeting its missile force. Let's assume that the weapons technology is such that Country X can destroy no more than one of Country Y's missiles with each missile fired. Then the number of additional missiles Country Y needs for each missile added by Country X depends on the effectiveness of Country X's missiles. Convince yourself that the curve $y = f(x)$ must lie between the limiting lines shown in Figure 10.1. Line A, having slope 0, represents a state of absolute invulnerability of Country Y's missiles to any attack. At the other extreme, line B, having slope 1, indicates that Country Y must add one new missile for each missile added by Country X.

To determine more precisely the shape of the graph of $y = f(x)$, we will analyze what happens for various cases relating the relative sizes of the two missile forces. To determine the cases, we subdivide the region between lines A and B into smaller subregions defined by the lines $x = y, x = 2y, x = 3y$, and so forth, as shown in Figure 10.2. We then approximate $y = f(x)$ in each of these subregions. Remember that when Country Y determines the number of missiles it needs to deter Country X for the graph of $y = f(x)$, Country Y is presumed to follow the friendly strategy, whereas Country X follows the enemy strategy.

■ **Figure 10.1**

Bounding the function
$y = f(x)$

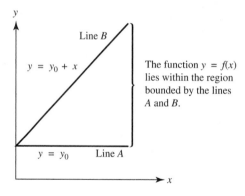

The function $y = f(x)$
lies within the region
bounded by the lines
A and B.

■ **Figure 10.2**

The subregions between
Lines A and B

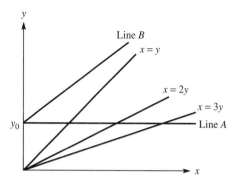

Case 1: $x < y$ If Country X attacks in this situation, it fires all of its x missiles at the same number of Country Y's missiles. Because the number $(y - x)$ of Country Y's missiles could not be attacked, at least that many would survive. Of the number x of Country Y's missiles that were fired on, a percentage s would survive, where $0 < s < 1$. Thus, the total number of missiles surviving the attack is $y - x + sx$. Now Country Y must have y_0 missiles survive to inflict unacceptable damage on Country X. Hence,

$$y_0 = y - x + sx \quad \text{for} \quad 0 < s < 1$$

or, solving for y,

$$y = y_0 + (1 - s)x \tag{10.1}$$

Equation (10.1) gives the minimum number of missiles Country Y must have to be confident that y_0 missiles will survive an attack by Country X.

Case 2: $y = x$ In firing all of its missiles, Country X fires exactly one of its missiles at each of Country Y's missiles. Assuming the percentage s will survive the attack, the number $sx = sy$ survive, in which case Country Y needs

$$y = \frac{y_0}{s} \tag{10.2}$$

missiles to inflict unacceptable damage on Country X.

Case 3: $y < x < 2y$ Country X, in firing all its missiles, targets each of Country Y's missiles once and a portion of them twice, as illustrated in Figure 10.3.

■ Figure 10.3

An example of $y < x < 2y$

Country X:

Country Y:

Convince yourself that $x - y$ of Country Y's missiles would be targeted twice and $y - (x - y) = 2y - x$ would be targeted once. Of those targeted once, a percentage $s(2y - x)$ will survive as before. Of those targeted twice, the percentage $s(x - y)$ will survive the first round. Of those that survive the first round, the percentage $s[s(x - y)] = s^2(x - y)$ will survive the second round. Hence, Country Y must have

$$y_0 = s^2(x - y) + s(2y - x)$$

missiles survive, or, solving for y,

$$y = \frac{y_0 + x(s - s^2)}{2s - s^2} \tag{10.3}$$

is the minimum number of missiles required by Country Y.

Case 4: $x = 2y$ Country X will fire exactly two missiles at each of Country Y's missiles. If we reason as in Case 2, the number $s^2 y$ survive, so

$$y = \frac{y_0}{s^2} \tag{10.4}$$

is the minimum number of missiles required by Country Y.

Now let's combine all of the preceding scenarios into a single graph. For convenience, we are going to assume that the discrete situation just discussed giving the minimum number of missiles can be represented by a continuous model (giving rise to fractions of missiles). First, observe that Equations (10.1) and (10.3) both represent straight-line segments: the first segment for $x < y$ and the second segment for $y < x < 2y$. In Case 1, when $x < y$, we obtained the equation

$$y_0 = y - x + sx$$

As x approaches y, this last equation becomes (in the limit) $y_0 = sy$. In Case 3, when $y < x < 2y$, we obtained the equation

$$y_0 = s^2(x - y) + s(2y - x) \tag{10.5}$$

Again, as x approaches y, the equation becomes $y_0 = sy$. Thus, the two line segments meet at $x = y$ with the common value $y = y_0/s$. Finally, as x approaches $2y$, Equation (10.5) becomes $y_0 = s^2 y$.

These observations mean that the two line segments defined by Equations (10.1) and (10.3) form a continuous curve meeting the lines $y = x$ and $2y = x$. Moreover, the slope $\frac{1-s}{2-s}$ for the line segment represented by Equation (10.3) is less than the slope $1 - s$ of the line segment represented by Equation (10.1) because $2 - s > 1$. Thus, the curve is *piecewise linear with decreasing slopes*. The graphical model is depicted in Figure 10.4. Note the graph lies within the cone-shaped region between lines A and B as discussed previously.

■ Figure 10.4

A graphical model relating the number of missiles for Country Y to the number of missiles for Country X when $0 \leq x \leq 2y$

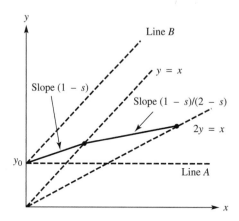

We could continue to analyze additional cases, such as what happens when $2y < x < 3y$. Because we are interested only in qualitative information, however, let's see if we can determine the general shape of the curves more simply.

To simplify the analysis, let's replace our piecewise linear approximation by a single continuous smooth curve (one without corners) that passes through each of the points $(0, y_0), (x, x), (x, \frac{x}{2}), \ldots$ shown in Figure 10.4. (These are the points where the piecewise linear approximation crosses the y axis and the lines $y = x$ and $2y = x$). We want a curve given by a single equation rather than one represented by a different equation in each subregion. Generalizing from our analysis in Cases 2 and 4, one such curve is given by the following model:

$$y = \frac{y_0}{s^{x/y}} \quad \text{for} \quad 0 < s < 1 \tag{10.6}$$

An inspection of Equation (10.6) reveals that for every ratio x/y we can find y. Thus, the curve $y = f(x)$ crosses each line $x = y, x = 2y, \ldots, x = ny$, as illustrated in Figure 10.5, at the same points as did our piecewise linear approximation.

■ Figure 10.5

The curve $y = f(x)$ must cross every line $x = ny$.

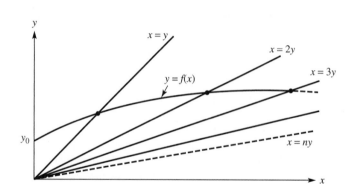

The situation for Country X is entirely symmetrical. (In determining its curve, Country X is assumed to have the friendly strategy to deter Country Y, and Country Y is presumed to have the enemy strategy.) Its minimum number of missiles is represented by a continuous curve $x = g(y)$ that crosses every line $y = x$, $y = 2x$, ..., $y = nx$. Thus, the two curves must intersect.

The preceding discussion leads us to consider two idealized continuously differentiable curves such as those drawn in Figure 10.6. Because the curve $y = f(x)$ represents the minimum number of missiles required by Country Y, the region above the curve represents missile levels satisfactory to Country Y. Likewise, the region to the right of the curve $x = g(y)$ represents missile levels satisfactory to Country X. Thus, the darkest region in Figure 10.6 represents missile levels satisfactory to both countries.

■ **Figure 10.6**

Regions of satisfaction to Country X and Country Y

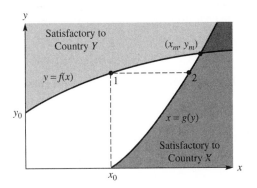

The intersection point of the curves $y = f(x)$ and $x = g(y)$ represents the minimum level at which both sides are satisfied. To see that this is so, assume Country Y has y_0 missiles and observes that Country X has x_0 missiles. To meet its objectives, Country Y will have to add sufficient missiles to reach point 1 in Figure 10.6. In turn, Country X will have to add sufficient missiles to reach point 2. This process will continue until both sides are satisfied simultaneously. Notice that any point in the darkest region will suffice to satisfy both countries, and there are many points in the darkest region that are quite likely to occur. The intersection point (x_m, y_m) in Figure 10.6 represents the minimum force levels required of both countries to meet their objectives.

Uniqueness of the Intersection Point

We would like to know if the intersection point is unique. Note from Equation (10.6) that as the ratio x/y increases, y must increase; likewise, $x = g(y)$ increases. Because both curves are increasing, it is tempting to conclude that the intersection point is unique. Consider Figure 10.7, however. In the figure both curves are steadily increasing: The curve $y = f(x)$ crosses every line $x = ny$, and $x = g(y)$ crosses every line $y = nx$. However, the curves have multiple intersection points. How can we ensure a unique intersection point? Notice that the slope of the curve $y = f(x)$ in Figure 10.7 is steadily decreasing until the point $x = x_1$, when it begins to increase. Thus, the first derivative changes from a decreasing to an increasing function at $x = x_1$. That is, the tangent line changes from continuously turning in a clockwise direction to turning in a counterclockwise direction as x advances. In other words, the second derivative changes sign. If we can show such a sign change is

■ Figure 10.7

Steadily increasing curves
with multiple intersection
points

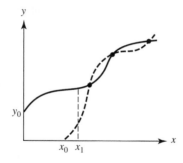

impossible, then we can conclude that the intersection point is unique. In fact, we will show
that the second derivative of $y = f(x)$ is always negative.

Taking the logarithm of Equation (10.6) yields

$$\ln y = \ln y_0 - \frac{x}{y} \ln s$$

Multiplying both sides of the equation by y and simplifying give

$$y \ln y - y \ln y_0 = -x \ln s$$

Differentiating implicitly with respect to x and simplifying yield

$$y'(1 + \ln y - \ln y_0) = -\ln s$$

or

$$y' = \frac{-\ln s}{1 + \ln y - \ln y_0}$$

Differentiating this last equation for the second derivative gives

$$y'' = \frac{-(-\ln s)\frac{1}{y} y'}{(1 + \ln y - \ln y_0)^2}$$

Next, we determine the sign of y'. Rewrite y' as

$$y' = \frac{\ln s}{-1 + \ln \frac{y_0}{y}}$$

Because $0 < s < 1$, $\ln s$ is negative.

Now, for the cases we are considering, $y > y_0$, which implies that $\ln(y_0/y) < 0$. Thus,
$y' > 0$ everywhere, in which case $y'' < 0$ everywhere. Therefore, we can conclude that
a unique intersection point does in fact exist. The model has the general shape shown in
Figure 10.8.

Graphical Behavior of $y = f(x)$

The ways in which the graph of $y = y_0/s^{x/y}$ behaves depends on three factors: the constant
y_0, which is the minimum number of missiles required by Country Y after a preemptive first
strike; the survivability percentage s, which is determined by the technology and weapon
effectiveness of Country X's missiles as well as by how securely Country Y's missiles

■ Figure 10.8

A graphical model of the nuclear arms race

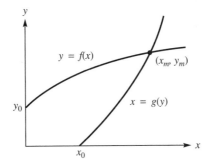

are protected; and the exchange ratio $e = x/y$. If y_0 increases, then the curve $y = f(x)$ shifts upward and also has a larger slope at each point than before (see Problem 5). If the survivability factor s increases, the curve rotates downward toward the horizontal line A: $y = y_0$ and has a smaller slope at each point than before.

If the exchange ratio e increases, then Country X can target Country Y's missiles more than once, requiring Country Y to need more of them. This results in an increase in the slope and upward rotation of the curve toward the line B: $y = y_0 + x$.

If you have access to a computer graphing package, you can see these effects by plotting the graph $y = y_0/s^{x/y}$ for various values of the three factors.

Although the curve $x = g(y)$ for Country X displays similar behavior, we note that its constant x_0, survivability, and exchange ratio factors are generally different from the values for Country Y. Let's consider several situations in which we use these ideas and the graphical model to analyze the effects on the intersection point for different political and military strategies likely to be entertained by the two countries.

When analyzing the graphical effects of a particular strategy, we first consider how the strategy changes each of the factors y_0, s, and e for Country Y. Does the factor increase, decrease, or remain unchanged? Here we assume that Country Y has the friendly strategy, whereas Country X is following the enemy strategy. How each factor changes determines how the curve $y = f(x)$ changes its position.

Next we consider how the strategy changes each of the factors x_0, s, and e for Country X. (Again, we stress that these factors are generally not the same as, and are independent of, their corresponding counterparts for Country Y.) Again, we ask if each factor increases, decreases, or remains unchanged due to the particular strategy. In answering this question, we assume that Country X has the friendly strategy, whereas Country Y follows the enemy strategy. How the factors change determines how the curve $x = g(y)$ changes its position.

The combined changes of the two curves move the original intersection point (x_m, y_m) to a new position (x'_m, y'_m), where the shifted curves (resulting from our analysis of their factors) intersect.

Model Interpretation

EXAMPLE 1 *Civil Defense*

Suppose Country X decides to double its annual budget for civil defense. Presumably, Country Y will need more missiles to inflict an unacceptable level of damage on Country X's population centers. Thus, y_0 increases. Because the effectiveness of Country X's weapons has not changed, nor has Country Y done anything to improve protection of its missiles in their silos, no change occurs in the survivability of Country Y's missiles. Also, no change occurs in the exchange ratio for Country Y: One X missile can still destroy at most one Y missile. The net effect is that the curve $y = f(x)$ shifts vertically upward with increasing slope at every x.

For Country X, the factor x_0 does not change because increased protection of its population and industrial centers does not affect the minimum number of missiles it will need to retaliate against a preemptive first strike on the part of Country Y. Also, the survivability factor of Country X's missiles does not change because civil defense does not improve the protection of its missiles typically located in silos in remote geographic regions, nor is it the case that the effectiveness of Country Y's missiles has changed. Finally, one Y missile can still destroy at most one X missile, so the exchange ratio for Country X is unchanged. The net effect is that the curve $x = g(y)$ does not change position at all.

The overall effect of Country X increasing its civil defense budget is shown in Figure 10.9. The dashed curve is the new position of the function $y = f(x)$ resulting from the civil defense of Country X. The point (x'_m, y'_m) is the new intersection point. Note that although the course of action seemed fairly passive, the effect is to increase the minimum number of missiles required by both sides because $x'_m > x_m$ and $y'_m > y_m$. ■ ■ ■

■ Figure 10.9

Country X increases its civil defense posture

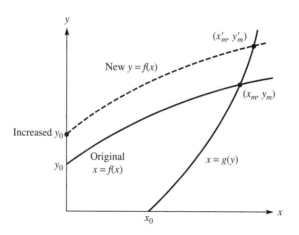

EXAMPLE 2 *Mobile Launching Pads*

For this scenario, assume that Country X puts its missiles on mobile launching pads, which can be relocated during times of international crisis. The factor y_0 does not change because Country Y must still target the same number of population and industrial centers in retaliation for a first strike by Country X. Moreover, the fact that Country X's missiles are launched from mobile pads does not alter the effectiveness of the missiles or improve the protection of Country Y's silos. So the survivability of Country Y's missiles is unchanged. No change in the exchange ratio for Country Y occurs because one X missile, though launched from a mobile launching pad, can still destroy at most one Y missile in a first strike. Thus, the curve $y = f(x)$ does not change.

Regarding the factors for Country X, there is no change in x_0 because Country X still requires the same number of missiles to inflict unacceptable damage on Country Y's population and industrial centers. Country X's missiles are less vulnerable than before because Country Y would not know their exact locations in executing a first strike. Thus, the survivability of Country X's missiles is increased. Finally, the placement of a missile on a mobile pad does not alter the exchange ratio for Country X: One Y missile can still hit and destroy at most one X missile. The net effect of these changes is to flatten the curve $x = g(y)$ toward the y axis, as shown by the new dashed curve in Figure 10.10.

■ **Figure 10.10**

Country X uses mobile launching pads.

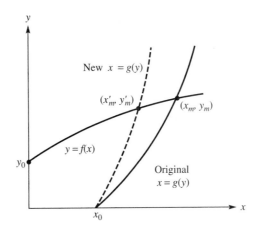

The overall effect of Country X placing its missiles on mobile launching pads is a movement of the intersection point so that $x'_m < x_m$ and $y'_m < y_m$, as depicted in Figure 10.10. Thus, the minimum number of missiles needed by *both* countries to deter the other side from engaging in hostile attack is reduced. ■ ■ ■

EXAMPLE 3 *Multiple Warheads*

Suppose now that Country X and Country Y both employ multiple warheads that can be targeted independently (MIRVs). In our development of the model we assumed each missile was armed with only one warhead, so counting missiles and counting warheads would be the same. With MIRVs this correspondence is no longer true. Let's continue to count the numbers of missiles (not warheads) required by each country. Assume that each missile is armed with 16 smaller missiles, each possessing its own warhead. Because it still takes the same number of warheads to destroy the opponent's population and industrial centers, it is reasonable to expect the number of larger missiles, x_0 and y_0, to be reduced by the factor 16.

Let's consider the survivability factor s for Country Y. If we assume a warhead released independently from an in-flight missile from Country X is just as effective in its destructive power as before, then there is no change in the survival possibility of the targeted missile in Country Y. That is, the factor s is unchanged.

However, when one missile from Country X is headed for Country Y in a preemptive first strike, it carries 16 warheads, each of which can independently target a *missile* in Country Y. Thus, 1 X missile can destroy up to 16 Y missiles, and the exchange ratio factor for Country Y increases significantly. This means the curve $y = f(x)$ must rise more sharply than before to compensate for the increased destruction if it is to meet its friendly strategy objectives.

Because both countries have MIRVed, the same argument reveals that the survivability factor for Country X remains unchanged but its exchange ratio factor is increased, causing a rise in the steepness of the curve $x = g(y)$ (away from the y axis).

■ **Figure 10.11**

Both countries use multiple independently targeted warheads on each missile.

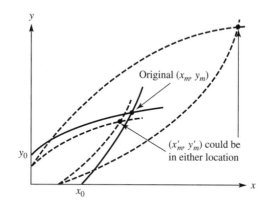

The new curves are represented in Figure 10.11. Because the reduction in values of the intercepts x_0 and y_0, and the changes in the slopes of the curves, give different effects on the new location of the intersection point, it is difficult to determine from a graphical analysis whether the minimum number of missiles actually increases or decreases. This analysis demonstrates a limitation of graphical models. To determine the location of the equilibrium point (x'_m, y'_m) would require a more detailed analysis and more exact information concerning the weapon effectiveness and technological capabilities of both countries (along with other factors, such as military intelligence). ■ ■ ■

EXAMPLE 4 *MIRVs Revisited: Counting Warheads*

Although we were unable to predict the effect of multiple warheads on the minimum number of *missiles* required by each side in Example 3, we can analyze the total number of warheads in this example's strategy. Let x and y now represent the *number of warheads* possessed by Country X and Country Y, respectively. The number of warheads needed by each country to inflict unacceptable damage on the opponent remain at the levels x_0 and y_0, as before. Also, because each warhead is located on a missile, its chance for survivability is the same as that of the missile, so the survivability factors are unchanged for each country.

Let's examine what happens to the exchange ratio factor for Country Y. A single warhead released from an incoming missile from Country X now has the capability of destroying 16 of Country Y's *warheads* instead of just 1, because they are all clustered on a single missile targeted by the incoming *warhead*. This increase in the exchange ratio for Country Y causes a sharp rise in the steepness of its curve $y = f(x)$. The same argument applies to the exchange ratio factor for Country X, so the curve $x = g(y)$ also increases in steepness. The new curves are displayed in Figure 10.12. Note that both countries require

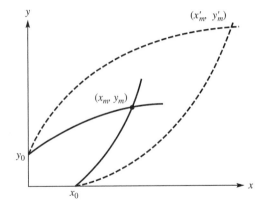

■ Figure 10.12

Multiple warheads on each missile increase the total number of warheads required by each side.

more warheads if multiple warheads are introduced on each missile because $x'_m > x_m$ and $y'_m > y_m$ in Figure 10.12.   ■ ■ ■

10.1 PROBLEMS

1. Analyze the effect on the arms race of each of the following strategies:

   a. Country X increases the accuracy of its missiles by using a better guidance system.

   b. Country X increases the payload (destructive power) of its missiles without sacrificing accuracy.

   c. Country X is able to retarget its missiles in flight so that it can aim for missiles that previous warheads have failed to destroy.

   d. Country Y employs sea-launched ballistic missiles.

   e. Country Y adds long-range intercontinental bombers to its arsenal.

   f. Country X develops sophisticated jamming devices that dramatically increase the probability of neutralizing the guidance systems of Country Y's missiles.

2. Discuss the appropriateness of the assumptions used in developing the nuclear arms race model. What is the effect on the number of missiles if each country believes the other country is also following the friendly strategy? Is disarmament possible?

3. Develop a graphical model based on the assumption that each side is following the enemy strategy. That is, each side desires a first-strike capability for destroying the missile force of the opposing side. What is the effect on the arms race if Country X now introduces antiballistic missiles?

4. Discuss how you might go about validating the nuclear arms race model. What data would you collect? Is it possible to obtain the data?

5. Use the polar coordinate substitution $x = r\cos\theta$ and $y = r\sin\theta$ in Equation (10.6) to show that a doubling of y_0 causes a doubling of r for every fixed θ. Show that if y_0 increases, then $y = f(x)$ shifts upward with an increasing slope.

10.2   Modeling an Arms Race in Stages

Let's again assume that two countries, Country X and Country Y, are engaged in an arms race. Each country follows a *deterrent strategy* that requires them to have a given number of weapons to deter the enemy (inflict unacceptable damage) even if the enemy has no weapons. Under this strategy, as the enemy adds weapons, the friendly force increases its arms inventory by some percentage of the number of attacking weapons that depends on how effective the friendly force perceives the enemy's weapons to be.

Suppose Country Y believes it needs 120 weapons to deter the enemy. Furthermore, for every 2 weapons possessed by Country X, Country Y believes it needs to add 1 additional weapon (to ensure 120 weapons remain after a strike by Country X). Thus, the number of weapons needed by Country Y (y weapons) as a function of the number of weapons it believes Country X has (x weapons) is

$$y = 120 + \frac{1}{2}x$$

Now suppose Country X is following a similar strategy, believing it needs 60 weapons even if Country Y has no weapons. Furthermore, for every 3 weapons that it believes Country Y possesses, X believes that it must add one weapon. Thus, the number x of weapons needed by Country X as a function of the number y of weapons it believes Country Y has is

$$x = 60 + \frac{1}{3}y$$

How does the arms race proceed?

A Graphical Solution

Suppose that **initially** (stage $n = 0$) Countries Y and X do not think the other side has arms. Then (stage $n = 1$) they build 120 weapons and 60 weapons, respectively. Now assume each has perfect intelligence; that is, each knows the other has built weapons. In the next stage (stage $n = 2$), Country Y increases its inventory to 150 weapons:

$$y = 120 + \frac{1}{2}(60) = 150 \text{ weapons}$$

Similarly, Country X notes that Y had 120 weapons during the previous stage and increases its inventory to 100 weapons:

$$x = 60 + \frac{1}{3}(120) = 100 \text{ weapons}$$

The arms race would proceed *dynamically*—that is, in successive stages. At each stage a country adjusts its inventory based on the strength of the enemy during the previous stage. In stage $n = 3$, Country Y realizes that Country X now has 100 weapons and reacts by increasing its inventory to $y = 120 + \frac{1}{2}(100) = 170$. Similarly, Country X increases its inventory to $x = 60 + \frac{1}{3}(150) = 110$. If we let n represent the stage of the arms race, convince yourself that the following table represents the growth of the arms race under the assumptions we have made:

Stage n	0	1	2	3	4	5
Country Y	0	120	150	170	175	178
Country X	0	60	100	110	117	118

Note that the growth in the arms race appears to be diminishing. The number of weapons needed by Country Y appears to be approaching approximately 180 weapons, whereas X appears to be approaching approximately 120 weapons (Figures 10.13 and 10.14). Does

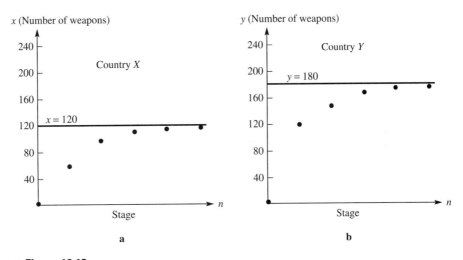

Figure 10.13

Dynamics of an arms race

Figure 10.14

Arms race curves for each country

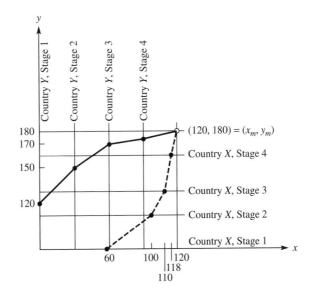

this model actually predict that an equilibrium value will be reached as suggested in the model developed earlier in this chapter? Is the equilibrium position *stable* in the sense that small changes in the number of weapons *initially* possessed by either side have little change on the final outcome? Is the outcome sensitive to changes in the coefficients of the model? Next, we build a dynamical system to answer these questions.

Numerical Solution of an Arms Race as a Dynamical System Model

Using the notation introduced in Section 10.1,

Let $n =$ stage (years, decades, fiscal periods, etc.)

$x_n =$ number of weapons possessed by X in stage n

$y_n =$ number of weapons possessed by Y in stage n

Then our assumptions imply that at stage $n + 1$,

$$\left.\begin{array}{l} y_{n+1} = 120 + \dfrac{1}{2}x_n \\[2mm] x_{n+1} = 60 + \dfrac{1}{3}y_n \end{array}\right\} \tag{10.7}$$

with

$$x_0 = 0$$
$$y_0 = 0$$

Recall that the values x_0 and y_0 are called **initial values**. Along with the coefficients $\frac{1}{2}$ and $\frac{1}{3}$, they are **parameters** we ultimately would like to vary to determine the sensitivity of the predictions. In Table 10.1, we display a numerical solution for the model and initial conditions in Equations (10.7).

What happens if both countries start off with more than the minimum number of missiles? For example, what if Country X starts with 100 and Country Y with 200? Our model becomes

$$\left.\begin{array}{l} y_{n+1} = 120 + \dfrac{1}{2}x_n \\[2mm] x_{n+1} = 60 + \dfrac{1}{3}y_n \end{array}\right\} \tag{10.8}$$

with

$$x_0 = 100$$
$$y_0 = 200$$

Is the equilibrium reached, or is there uncontrolled growth? In the problems that follow this section, we ask you to explore the long-term behavior predicted by Equations (10.8) in terms of the stability of the arms race for different initial values and other parameters (the survival coefficients of Countries X and Y).

Table 10.1 A numerical solution to Equations (10.7)

n	x_n	y_n
0	0	0
1	60	120
2	100	150
3	110	170
4	116.6667	175
5	118.3333	178.3333
6	119.4444	179.1667
7	119.7222	179.7222
8	119.9074	179.8611
9	119.9537	179.9537
10	119.9846	179.9769
11	119.9923	179.9923
12	119.9974	179.9961
13	119.9987	179.9987
14	119.9996	179.9994
15	119.9998	179.9998
16	119.9999	179.9999
17	120	180

10.2 PROBLEMS

1. Build a numerical solution to Equations (10.8).

   a. Graph your results.

   b. Is an equilibrium value reached?

   c. Try other starting values. Do you think the equilibrium value is stable?

   d. Explore other values for the survival coefficients of Countries X and Y. Describe your results.

2. Recall from Section 10.1 that an equilibrium value for the arms race requires that $x_{n+1} = x_n$ and $y_{n+1} = y_n$ simultaneously. Is there an equilibrium value for Equations (10.7)? If so, find it.

10.2 PROJECTS

For Projects 1–4, complete the requirements in the referenced UMAP module (see enclosed CD), and prepare a short summary for classroom discussion.

1. "The Distribution of Resources," by Harry M. Schey, UMAP 60–62 (one module). The author investigates a graphical model that can be used to measure the distribution of resources. The module provides an excellent review of the geometric interpretation of the derivative as applied to the economics of the distribution of a resource. Numerical calculation of the derivative and definite integral is also discussed.

2. "Nuclear Deterrence," by Harvey A. Smith, UMAP 327. The author analyzes the stability of the arms race, assuming objectives similar to those suggested by General Taylor. The module develops analytic models using probabilistic arguments. An understanding of elementary probability is required.

3. "The Geometry of the Arms Race," by Steven J. Brams, Morton D. Davis, and Philip D. Straffin, Jr., UMAP 311. This module analyzes the possibilities of both parties disarming by introducing elementary game theory. Interesting conclusions are based on Country X's ability to detect Country Y's intentions, and vice versa.

4. "The Richardson Arms Race Model," by Dina A. Zinnes, John V. Gillespie, and G. S. Tahim, UMAP 308. A model is constructed on the basis of the classical assumptions of Lewis Fry Richardson. Difference equations are introduced.

10.2   Further Reading

Saaty, Thomas L. *Mathematical Models of Arms Control and Disarmament*. New York: Wiley, 1968.
Schrodt, Philip A. "Predicting Wars with the Richardson Arms-Race Model." BYTE 7, no. 7 (July 1982): 108–134.
Wallace, Michael D. "Arms Races and Escalation; Some New Evidence." *Explaining War.* Edited by J. David Singe. Beverly Hills, CA: Sage, 1979; 240–252.

10.3   Managing Nonrenewable Resources: The Energy Crisis

During the past century, the United States has shifted into nearly complete dependence on nonrenewable energy sources. Petroleum and natural gas now constitute about three-fourths of the nation's fuel, and nearly half of our crude oil comes from foreign sources. The rise of the Organization of Petroleum-Exporting Countries cartel (OPEC) has caused some analysts to fear for supply security, especially during periods of political unrest when we are threatened by constraints on the supply of foreign oil, such as oil embargoes. Thus, there are significant attempts to conserve energy so as to reduce our long-term oil consumption. There is also interest in more drastic short-term reductions to survive a crisis situation.

Various solutions have been proposed to address these long- and short-term needs. One solution is gas rationing. Another is to place a surcharge tax on each gallon of gasoline sold at the local pump. Basically, the idea behind this solution is that gasoline companies will pass the tax on to the consumer by increasing the price per gallon by the amount of the tax. Accordingly, it is supposed the consumer will reduce consumption because of the higher price. Let's study this proposal by constructing a graphical model and qualitatively addressing the following questions:

1. What is the effect of the surcharge tax on short- and long-term consumer demand?

2. Who actually pays the tax—the consumer or the oil companies?

3. Does the tax contribute to inflation?

Stated more succinctly, the problem is to determine the effect of a surcharge tax on the market price of, and consumer demand for, gasoline. In the following analysis we are concerned with gaining a qualitative understanding of the principal factors involved with the problem. A graphical analysis is appropriate to gain this understanding, especially because precise data would be difficult to obtain. We begin by graphically analyzing some pertinent general economic principles. In the ensuing sections we interpret the conclusions of the graphical model as they apply to the oil situation.

Constructing a Graphical Model

Suppose a firm within a large competitive industry produces a single product. A question facing the firm is how many units to produce to maximize profits. Assume that the industry in question is so large that any particular firm's production has no appreciable effect on the market price. Hence, the firm may assume the price of the product is constant and need consider only the difference between the price and the firm's costs in producing the product. Individual firms encounter *fixed costs*, which are independent of the amount produced over a wide range of production levels. These costs include rent and utilities, equipment capitalization costs, and management costs. The *variable costs* depend on the quantity produced. Variable costs include the cost of raw materials, taxes, and labor. When the fixed costs are divided by the quantity produced, the share apportioned to each unit is obtained. This per-unit share is relatively high when production levels are low. However, as production levels increase, not only does the per-unit share of the fixed costs diminish but also economies of scale (such as buying raw materials in large quantities at reduced rates) often reduce some of the variable cost rates. Eventually, production levels are reached that strain the capabilities of the firm. At this point the firm is faced with hiring additional employees, paying overtime, or capitalizing additional machinery or similar costs. Because the per-unit costs tend to be relatively high when production levels are either very low or very high, one intuitively expects the existence of a production level q^* that yields a maximum profit over the range of production levels being considered. This idea is illustrated in Figure 10.15. Next, consider the characteristics of q^* mathematically.

■ **Figure 10.15**

Profit is maximized at q^*.

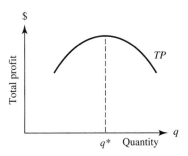

At a given level of production q, total profit $TP(q)$ is the difference between total revenue $TR(q)$ and total cost $TC(q)$. That is,

$$TP(q) = TR(q) - TC(q)$$

A necessary condition for a relative maximum to exist is that the derivative of TP with respect to q must be 0:

$$TP' = TR' - TC' = 0$$

or, at the level q^* of maximum profit,

$$TR'(q^*) = TC'(q^*) \tag{10.9}$$

Thus, at q^* it is necessary that the slope of the total revenue curve equal the slope of the total cost curve. This condition is depicted in Figure 10.16.

■ **Figure 10.16**

At q^* the slopes of the total revenue and total cost curves are equal.

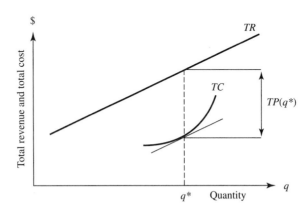

Let's interpret economically the meaning of the derivatives TR' and TC'. From the definition of the derivative,

$$TR'(q) \approx \frac{TR(q + \Delta q) - TR(q)}{\Delta q}$$

for Δq small. Thus, if $\Delta q = 1$, you can see that $TR'(q)$ approximates $TR(q + 1) - TR(q)$, which is the revenue generated by the next unit sold, or the **marginal revenue** (MR) of the $q + 1$st unit. Because total revenue is the price per unit times the number of units, it follows that the marginal revenue of the $q + 1$st unit is the price of that unit less the revenue lost on previous units resulting from price reductions (see Problem 4). Similarly, $TC'(q)$ represents the **marginal cost** (MC) of the $q + 1$st unit; that is, the extra cost in changing output to include one additional unit. If Equation (10.9) is interpreted in these new terms, a necessary condition for maximum profit to occur at q^* is that marginal revenue equal marginal cost:

$$MR(q^*) = MC(q^*) \tag{10.10}$$

For the critical point defined by Equations (10.8) to be a relative maximum, it is sufficient that the second derivative TP'' be negative. Because $TP' = MR - MC$, we have

$$TP''(q^*) = MR'(q^*) - MC'(q^*) < 0$$

or

$$MR'(q^*) < MC'(q^*) \tag{10.11}$$

This means that at the level q^* of maximum profit the slope of the marginal revenue curve is less than the slope of the marginal cost curve. The results (10.10) and (10.11) together imply that the marginal revenue and marginal cost curves intersect at q^*, with the marginal cost curve rising more rapidly. These results are illustrated in Figure 10.17.

■ **Figure 10.17**

At q^*, $MR = MC$ and $MR' < MC$

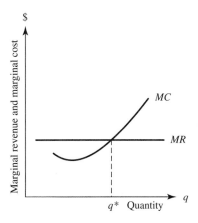

Interpreting the Graphical Model

Now let's interpret the graphical model represented by Figure 10.17. The MR curve represents the revenue generated by the next unit sold. The curve is drawn horizontally because in a large competitive industry, the amount one particular firm produces seldom influences the market price so there is no loss in revenue on previous units resulting from price reduction. Thus, the MR curve represents the (constant) price of the product. Given a market price determined by the entire industry and aggregate consumer demand, a firm attempting to maximize profits will continue to produce units until the cost of the next unit produced exceeds its market price. Verify that this situation is suggested by the graphical model in Figure 10.17.

10.3 PROBLEMS

1. Justify mathematically, and interpret economically, the graphical model for the theory of the firm given in Figure 10.18. What are the major assumptions on which the model is based?

2. Show that for total profit to reach a relative minimum, MR = MC and MC' < MR'.

3. Suppose the large competitive industry is the oil industry, and the firm within that industry is a gasoline station. How well does the model depicted in Figure 10.17 reflect the reality of that situation? How would you adjust the graphical model to make improvements?

4. Verify the result that the marginal revenue of the $q + 1$st unit equals the price of that unit minus the loss in revenue on previous units resulting from price reduction.

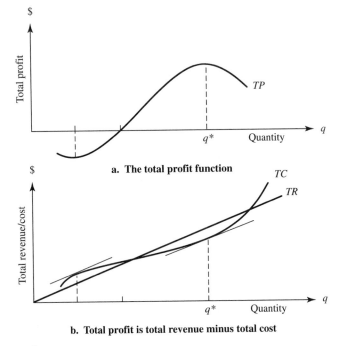

a. **The total profit function**

b. **Total profit is total revenue minus total cost**

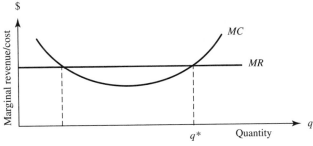

c. **Marginal revenue and marginal cost curves**

10.4   Effects of Taxation on the Energy Crisis

Let's suppose a firm is currently maximizing its profits; that is, given a market price MR, it is producing q^* units as suggested by Figure 10.17. Assume further that a tax is added to each unit sold. Because the firm must pay the government the amount of the tax for each unit sold, the marginal cost to the firm of each unit increases by the amount of the tax. Geometrically, that means the marginal cost curve shifts upward by the amount of the tax. Assume for the moment that the entire industry is able to increase the market price by simply adding on the amount of the tax to the price of each unit. Under this condition the MR curve also shifts upward by the amount of the tax. This situation is depicted in Figure 10.19. Note from the figure that the optimal production quantity is still q^*. Hence the model predicts no change in production as a result of the tax. Rather, the firm will produce the same amount

■ Figure 10.19

Both the marginal revenue
and marginal cost curves
shift upward by the amount
of the tax, leaving the
optimal production at the
same level q^*.

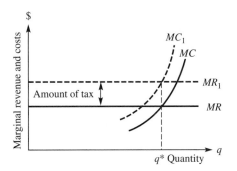

but charge a higher price, thereby contributing to inflation. Note too that is it the consumer
who pays the full amount of the tax in the form of a price increase.

The shortcoming with the model in Figure 10.19 is that it does not reveal whether the
entire industry can in fact continue to sell the same quantity at the higher price. To find
out, we need to construct a model for the industry. Thus, for each firm in the industry,
consider the intersection of the firm's various marginal revenue curves with each MC curve.
(Remember that each horizontal MR curve corresponds to a price of a unit.) This situation
is depicted for one firm in Figure 10.20a.

For each price, calculate the total that all firms in the industry would optimally produce.
This summing procedure yields a curve for the entire industry. Because this curve represents
the amount the industry would supply at various price levels, it is called a **supply curve**, an
example of which is depicted in Figure 10.20b. Qualitatively, as the market price increases,
the industry is willing to produce greater quantities.

■ Figure 10.20

The industry's supply curve
is obtained by summing
together the amounts the
firms would produce at
each price level.

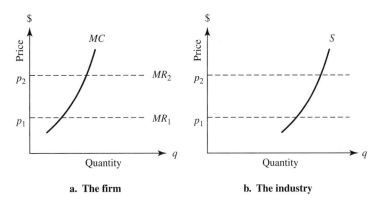

a. The firm   **b. The industry**

Next, consider aggregate consumer demand for the product at various market price
levels. From a consumer's point of view, the quantity demanded is a function of the market
price. However, it is traditional to plot market price as a function of quantity (Figure 10.21a).
Conceptually, for each price level, individual consumer demands could be summed as in the
procedure for obtaining the industry's supply curve. This summation is depicted graphically
in Figure 10.21. Qualitatively, as the price increases, we expect the aggregate demand for
the product to decrease as consumers begin to use less or substitute cheaper alternative
products (Figure 10.21b).

Finally, consider the industry's supply and demand curves together. Suppose the two
curves intersect at a unique point (q^*, p^*) as depicted in Figure 10.22. If the industry supplies

■ **Figure 10.21**

The industry's demand curve *D* represents the aggregate demand for the product at various price levels and is obtained by summing individual consumer demands at those levels. Notice that we plot price versus quantity for demand curves rather than vice versa.

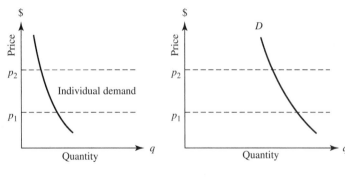

a. A consumer   **b. All consumers**

■ **Figure 10.22**

The intersection of the supply and demand curves gives a market price and a market quantity that satisfy both consumers and suppliers alike.

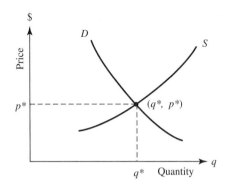

q^* and charges p^* (supply curve), then consumers are willing to buy the amount q^* at the price p^* (demand curve). Thus, there is equilibrium in the sense that no excess supply exists at that price and both consumers and suppliers are satisfied.

Obviously, an industry does not know the precise demand curve for its product. Therefore, it is important to determine what occurs if the industry supplies an amount other than q^*. For example, suppose the industry supplies an amount q_1 greater than q^* (Figure 10.23a). Then consumers are willing to buy an amount as large as q_1 only if the price is as low as p_1, forcing a reduction in price. However, if the market price drops to p_1, the industry is willing to supply only q_2 units and will cut production back to that level. Then at q_2 the unsatisfied consumers would drive the price up to p_2. (Convince yourself that the process converges to (q^*, p^*) in Figure 10.23a, where the supply curve is steeper than the demand curve.) In that situation, market forces actually drive supply and demand to the equilibrium point. On the other hand, consider Figure 10.23b, in which the supply curve is more horizontal than the demand curve. In this case, the equilibrium point (q^*, p^*) will not be achieved by the iterative process just described. Instead, there is likely to be wild fluctuation in the amount supplied and the market price as the industry and consumers search for the equilibrium point. (Convince yourself from Figure 10.23b that the equilibrium point is difficult to achieve when the supply curve is not as steep as the demand curve.)

The demand curve is steep at q^* when consumers cannot switch in the short run to an alternative product after the price p^* increases. Water and electricity are examples of such products essential to today's consumers. The supply curve is steep at q^* when industry cannot supply more of the product unless it incurs significant additional cost, causing a sharp rise in p^*. This situation occurs when the industry is operating at full supply capacity

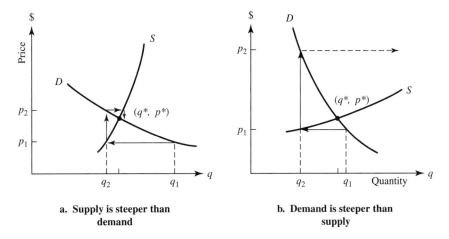

**a. Supply is steeper than
demand**

**b. Demand is steeper than
supply**

■ **Figure 10.23**

The ease with which the equilibrium point of supply and demand is achieved depends on the relative slopes of the supply and demand curves.

and demand for the product increases sharply. At this point, the industry must increase its costs (e.g., from capitalizing additional machinery or hiring additional employees), and these costs are passed on to the consumers in the form of price increases. The power crisis in California during 2000 and 2001 is such an example.

Now consider the effect of a tax on the supply and demand curves. Suppose that a particular industry is in an equilibrium market position (q^*, p^*) when a tax is added to each unit sold. Because each firm has to pay the tax to the government, each marginal cost curve shifts upward by the amount of the tax (Figure 10.19). These individual shifts cause the aggregate supply curve for the industry to shift upward by the amount of the tax as well. This phenomenon is depicted in Figure 10.24. If there is no reason for a shift in the demand curve, the intersection of the demand curve with the new supply curve shifts upward toward the left to a new equilibrium point (q_1, p_1), indicating an increase in the equilibrium market price with a corresponding decrease in market quantity. Furthermore,

■ **Figure 10.24**

A tax added to each item sold causes a decrease in the quantity produced and an increase in the price.

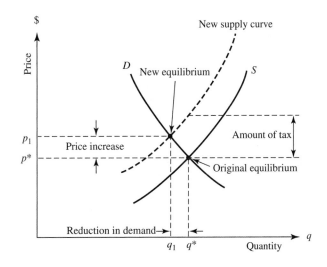

notice from Figure 10.24 that the increase in price, $p_1 - p^*$, is less than the tax. Thus, the model predicts that the consumer and industry share the tax. Study Figure 10.24 carefully and convince yourself that the proportion of the tax that the consumer pays and the relative reduction in the quantity supplied at equilibrium depend on the slopes of the supply and demand curves at the time the tax is imposed.

10.4 PROBLEMS

1. Show that when the demand curve is very steep, a tax added to each item sold will fall primarily on consumers. Now show that when the demand curve is more nearly horizontal, the tax is paid mostly by the industry. What if the supply curve is very steep? What if the supply curve is nearly horizontal?

2. Consider the oil industry. Discuss the conditions for which the demand curve will be steep near the equilibrium. What are the situations for which the demand curve will be more horizontal (or flat)?

3. Criticize the following quotation:

> The effect of a tax on a commodity might seem at first sight to be an advance in price to the consumer. But an advance in price will diminish the demand, and a reduced demand will send the price down again. It is not certain, therefore, after all, that the tax will really raise the price.[5]

4. Suppose the government pays producers a subsidy for each unit produced instead of levying a tax. Discuss the effect on the equilibrium point of the supply and demand curves. What happens to the new price and the new quantity? Discuss how the proportion of the benefits to the consumer and to the industry depends on the slopes of the supply and demand curves at the time the subsidy is given (see Problem 1).

10.5 A Gasoline Shortage and Taxation

Now let's consider the energy crisis. Suppose a shortage of oil imports exists at a time when the vast majority of the population depends on the automobile to get to work and that no alternative mass transportation system is immediately available. Assume also that in the short term, most people cannot switch to more fuel-efficient cars because they are not readily available or easily affordable. These assumptions suggest qualitatively a demand curve that is steep over a wide range of values because to get to work, consumers will suffer a high increase in price before significantly cutting back on demand. Of course, eventually price levels are reached at which it no longer pays for consumers to go to work. The demand curve is portrayed in Figure 10.25. Note that as q increases, consumers enter regions where the use of additional gasoline is for leisure. In such flat regions the consumer is most sensitive to price changes.

[5]H. D. Henderson, *Supply and Demand*, p. 22. Chicago: University of Chicago Press, 1958.

■ Figure 10.25

In the short term, the demand curve for gasoline is likely to be quite steep.

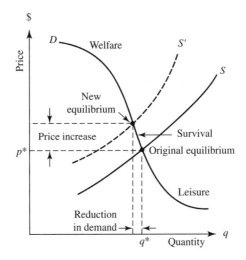

Next, consider the oil industry's supply curve. If the shortage of foreign oil catches the industry by surprise, most likely it will seek to find and develop alternative sources of oil. Possibly, the industry will be forced to turn to more expensive sources to provide the same quantities as before the shortage. Furthermore, in the short term, the oil industry will be more sensitive to price because it will be very difficult for the industry to provide immediate increases in supply. These arguments suggest an upward shift of the supply curve, which may become more vertical as well. Study Figure 10.25 and convince yourself that if the demand curve is steep, a significant price increase may result, but that it will take an appreciable shift in the supply curve to reduce demand significantly. Decide whether the new equilibrium point can easily be attained in Figure 10.25.

Now consider the supply–demand curves depicted in Figure 10.26. Suppose the government is dissatisfied with the reduction in demand for oil resulting from the shift in the supply curve, so it imposes a tax on each gallon of gasoline to reduce demand further. As discussed in the previous section, the tax causes the supply curve to shift upward (Figure 10.26). If the consumers are less sensitive to price than the industry, the new equilibrium point will be difficult to achieve. Furthermore, the new equilibrium point suggests that consumers will pay most of the tax in the form of a price increase. In summary, the graphical model

■ Figure 10.26

If the demand curve is relatively steep, the consumer pays the lion's share of tax, and reductions in demand are modest.

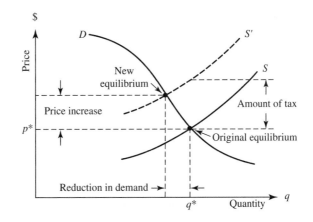

suggests that large fluctuations in price are probable, only modest reductions in demand are achievable, and the consumer bears the large portion of the tax burden.

What about the long-term effects of the foreign oil shortage? After the crisis, the oil industry will again have the foreign oil as well as the new sources that were developed during the crisis. These two sources cause the supply curve to shift downward and perhaps become more nearly horizontal than during the foreign oil shortage. Meanwhile, a change in consumer demand has occurred as well. Carpools have been formed, mass transportation systems are in place, and a larger proportion of the people have switched to fuel-efficient cars. These changes in supply and demand effectively transform the x axis for the consumers. That is, for the same amount of gasoline, the consumer is operating closer to the leisure range, where the demand curve is flat (Figure 10.27). The effect of these shifts in the supply and demand curves promises lower prices, but it is difficult to determine whether a significant reduction in demand will occur from the qualitative model depicted in Figure 10.27 (see Problem 1).

■ **Figure 10.27**

After the crisis both the demand and supply curves shift.

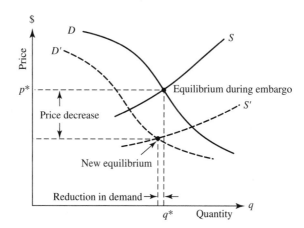

Finally, suppose the government is still dissatisfied with the level of demand and imposes a tax to reduce demand further. The supply curve shifts upward by the amount of the tax as before (Figure 10.28). Notice from Figure 10.28 that because the demand curve is

■ **Figure 10.28**

If the demand curve is relatively flat, the industry pays the larger portion of the tax, and the reduction in demand is more significant.

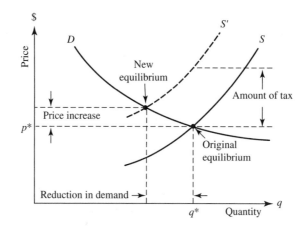

more horizontal than the supply curve at the original equilibrium, the increase in price due to the shift in the supply curve caused by the tax is small compared to the amount of the tax. In essence, the oil industry suffers the burden of the tax. Moreover, notice that the reduction in demand is more significant with this flatter demand curve. Finally, the new equilibrium position is more easily obtained.

10.5 PROBLEMS

1. Consider the graphical model in Figure 10.27. Argue that if the demand curve fails to shift significantly to the left, an *increase* in the equilibrium quantity could occur after the crisis.

2. Consider the situation in which demand is a fixed curve but there is an increase in supply, so the supply curve shifts downward. Discuss how the slope of the demand curve affects the change in price and the change in quantity: How does the price change, and when does it change the most? When does it change the least? Answer similar questions for the quantity.

3. Criticize the graphical model of the oil industry. Name some major factors that have been neglected. Which of the underlying assumptions are not satisfied by the crisis situation? Did the graphical model help you identify some of the key factors and their interactions? How could you adjust the model?

10.5 PROJECTS

For Projects 1–4, complete the requirements in the referenced UMAP module (see enclosed CD) and prepare a short summary for classroom discussion.

1. "Differentiation, Curve Sketching, and Cost Functions," by Christopher H. Nevison, UMAP 376. In this module costs and revenue for a firm are discussed using elementary calculus. The author discusses several of the economic ideas presented in this chapter.

2. "Price Discrimination and Consumer Surplus: An Application of Calculus to Economics," by Christopher H. Nevison, UMAP 294. The topics in this module are analyzed in a competitive market, and two-tier price discrimination is also discussed. The module examines several of the economic ideas presented in this chapter.

3. "Economic Equilibrium: Simple Linear Models," by Philip M. Tuchinsky, UMAP 208. In this module linear supply and demand functions are constructed, and the equilibrium market position is analyzed for an industry producing one product. The result is then extended to n products. The author concludes by briefly considering nonlinear and discontinuous functions.

4. "I Will If You Will ... A Critical Mass Model," by Jo Anne S. Growney, UMAP 539. A graphical model is presented to treat the problems of individual behavior in a group when the individual makes a choice dependent on his or her perception of the behavior of fellow group members. The model can provide insight into paradoxical situations

in which members of a group prefer one type of behavior but actually engage in the opposite behavior (such as not cheating versus cheating in a class).

10.5 Further Reading

Asimakopulos, A. *An Introduction to Economic Theory: Microeconomics*. New York: Oxford University Press, 1978.

Cohen, Kalman J., & Richard M. Cyert. *Theory of the Firm*. Englewood Cliffs, NJ: Prentice-Hall, 1975.

Thompson, Arthur A., Jr. *Economics of the Firm: Theory and Practice*. Englewood Cliffs, NJ: Prentice-Hall, 1973.

Modeling with a Differential Equation

Introduction

Quite often we have information relating a rate of change of a dependent variable with respect to one or more independent variables and are interested in discovering the function relating the variables. For example, if P represents the number of people in a large population at some time t, then it is reasonable to assume that the rate of change of the population with respect to time depends on the current size of P as well as other factors that are discussed in Section 11.1. For ecological, economical, and other important reasons, it is desirable to determine a relationship between P and t to make predictions about P. If the present population size is denoted by $P(t)$ and the population size at time $t + \Delta t$ is $P(t + \Delta t)$, then the change in population ΔP during that time period Δt is given by

$$\Delta P = P(t + \Delta t) - P(t) \tag{11.1}$$

The factors affecting the population growth are developed in detail in Section 11.1. For now, let's assume a simple proportionality: $\Delta P \propto P$. For example, if immigration, emigration, age, and gender are all neglected, we can assume that during a unit time period, a certain percentage of the population reproduces while a certain percentage dies. Suppose the constant of proportionality k is expressed as a percentage per unit time. Then our proportionality assumption gives

$$\Delta P = P(t + \Delta t) - P(t) = kP\Delta t \tag{11.2}$$

Equation (11.2) is a **difference equation** in which we are treating a discrete set of time periods rather than allowing t to vary *continuously* over some interval. In this situation the discrete set of times may give the population in future years at those distinct times (perhaps after the spring spawn in a fish population). Referring to Figure 11.1, observe that the horizontal distance between the points $(t_0, P(t_0))$ and $(t_0 + \Delta t, P(t_0 + \Delta t))$ is Δt, which may represent the time between spawning periods in a fish population growth problem or the length of a fiscal period in a budget growth problem. The time t_0 refers to a particular time. The vertical distance, ΔP in this case, represents the change in the dependent variable.

Assume that t does vary continuously so that we can take advantage of the calculus. Division of Equation (11.2) by Δt gives

$$\frac{\Delta P}{\Delta t} = \frac{P(t + \Delta t) - P(t)}{\Delta t} = kP \tag{11.3}$$

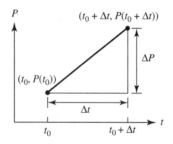

We can interpret $\Delta P/\Delta t$ physically as the *average rate of change* in P during the time period Δt. For example, $\Delta P/\Delta t$ may represent the average daily growth of the budget. In other scenarios, however, it may have no physical interpretation: If fish spawn only in the spring, it is somewhat meaningless to talk about the average daily growth in the fish population. Again in Figure 11.1, $\Delta P/\Delta t$ can be interpreted geometrically as the slope of the line segment connecting the points $(t_0, P(t_0))$ and $(t_0 + \Delta t, P(t_0 + \Delta t))$. Next, allow Δt to approach zero. The definition of the derivative gives the *differential equation*

$$\lim_{\Delta t \to 0} \frac{\Delta P}{\Delta t} = \frac{dP}{dt} = kP$$

where dP/dt represents the *instantaneous rate of change*. In many situations, the instantaneous rate of change has an identifiable physical interpretation, such as in the flow of heat from a space capsule after entering the ocean or the reading of a car speedometer as the car accelerates. However, in the case of a fish population that has a discrete spawning period or the budget process that has a discrete fiscal period, the instantaneous change may be somewhat meaningless. These latter scenarios are more appropriately modeled using difference equations, but it is occasionally advantageous to approximate a difference equation with a differential equation.

The derivative is used in two distinct roles:

1. To represent the instantaneous rate of change in *continuous* problems.

2. To approximate an average rate of change in *discrete* problems.

The advantage of approximating an average rate of change by a derivative is that the calculus often helps in uncovering a functional relationship between the variables under investigation. For instance, the solution to the Model (11.3) is $P = P_0 e^{kt}$, where P_0 is the population at time $t = 0$. However, many differential equations cannot be solved so easily using analytic techniques. In such cases the solutions are approximated using discrete methods. An introduction to numerical techniques is presented in Section 11.5. In cases in which the solution being approximated is to a differential equation that is an approximation to a difference equation, the modeler should consider using a discrete method with the finite difference equation directly (see Chapter 1).

The interpretation of the derivative as an instantaneous rate of change is useful in many modeling applications. The geometric interpretation of the derivative as the slope of the line tangent to the curve is useful for constructing numerical solutions. Let's briefly review these important concepts from the calculus.

The Derivative as a Rate of Change

The origins of the derivative lie in humankind's curiosity about motion and our need to develop a deeper understanding of motion. The search for the laws governing planetary motion, the study of the pendulum and its application to clock building, and the laws governing the flight of a cannonball were the kinds of problems stimulating the minds of mathematicians and scientists in the sixteenth and seventeenth centuries. Such problems motivated the development of the calculus.

To remind ourselves of one interpretation of the derivative, consider a particle whose distance s from a fixed position depends on time t. Let the graph in Figure 11.2 represent the distance s as a function of time t, and let (t_1, s_1) and (t_2, s_2) denote two points on the graph.

■ **Figure 11.2**

Graph of distance s as a function of time t

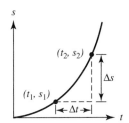

Define $\Delta t = t_2 - t_1$ and $\Delta s = s_2 - s_1$, and form the ratio $\Delta s / \Delta t$. Note that this ratio represents a rate: an increment of distance traveled Δs over some increment of time Δt. That is, the ratio $\Delta s / \Delta t$ represents the average velocity during the time period in question. Now remember how the derivative ds/dt evaluated at $t = t_1$ is defined:

$$\left. \frac{ds}{dt} \right|_{t=t_1} = \lim_{\Delta t \to 0} \frac{\Delta s}{\Delta t} \tag{11.4}$$

Physically, what occurs as $\Delta t \to 0$? Using the interpretation of average velocity, we can see that at each state of using a smaller Δt, we are computing the average velocity over smaller and smaller intervals with left endpoint at t_1 until, in the limit, we have the instantaneous velocity at $t = t_1$. If we think of the motion of a moving vehicle, this instantaneous velocity would correspond to the exact reading of its (perfect) speedometer at the instant t_1.

More generally, if $y = f(x)$ is a differentiable function, then the derivative dy/dx at any given point can be interpreted as the *instantaneous rate of change* of y with respect to x at that point. Interpreting the derivative as an instantaneous rate of change is useful in many modeling applications.

The Derivative as the Slope of the Tangent Line

Let's consider another interpretation of the derivative. As scholars sought knowledge about the laws of planetary motion, their chief need was to observe and measure the heavenly bodies. However, the construction of lenses for use in telescopes was a difficult task. To grind a lens to the correct curvature to achieve the desired light refraction requires knowing the tangent to the curve describing the lens surface.

■ **Figure 11.3**

The slope of each secant line approximates the slope of the tangent line to the curve at the point A.

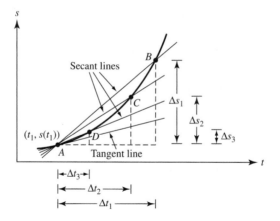

Let's examine the geometrical implications of the limit in Equation (11.4). We consider $s(t)$ simply as a curve. Let's examine a set of secant lines each emanating from the point $A = (t_1, s(t_1))$ on the curve. To each secant there corresponds a pair of increments $(\Delta t_i, \Delta s_i)$ as shown in Figure 11.3. The lines AB, AC, and AD are secant lines. As $\Delta t \to 0$, these secant lines approach the line tangent to the curve at the point A. Because the slope of each secant is $\Delta s / \Delta t$, we may interpret the derivative as *the slope of the line tangent to the curve $s(t)$ at the point A.* The interpretation of the derivative evaluated at a point as the slope of the line tangent to the curve at that point is useful in constructing numerical approximations to solutions of differential equations. Numerical approximations are discussed in Section 11.5.

11.1 Population Growth

Interest in how populations tend to grow was stimulated in the late eighteenth century when Thomas Malthus (1766–1834) published *An Essay on the Principle of Population as It Affects the Future Improvement of Society.* In his book, Malthus put forth an exponential growth model for human population and concluded that eventually the population would exceed the capacity to grow an adequate food supply. Although the assumptions of the Malthusian model leave out factors important to population growth (so the model has proven to be inaccurate for technologically developed countries), it is instructive to examine this model as a basis for later refinement.

Problem Identification   Suppose we know the population at some given time, for example, P_0 at time $t = t_0$, and we are interested in predicting the population P at some future time $t = t_1$. In other words, we want to find a population function $P(t)$ for $t_0 \leq t \leq t_1$ satisfying $P(t_0) = P_0$.

Assumptions   Consider some factors that pertain to population growth. Two obvious ones are the *birthrate* and the *death rate.* The birthrate and death rate are determined by different factors. The birthrate is influenced by infant mortality rate, attitudes toward and availability of contraceptives, attitudes toward abortion, health care during pregnancy, and so forth.

The death rate is affected by sanitation and public health, wars, pollution, medicines, diet, psychological stress and anxiety, and so forth. Other factors that influence population growth in a given region are immigration and emigration, living space restrictions, availability of food and water, and epidemics. For our model, let's neglect all these latter factors. (If we are dissatisfied with our results, we can include these factors later in a more refined model, possibly in a simulation model.) Now we'll consider only the birthrate and death rate. Because knowledge and technology have helped humankind diminish the death rate below the birthrate, human populations have tended to grow.

Let's begin by assuming that during a small unit time period, a percentage b (given as a decimal equivalent) of the population is newly born. Similarly, a percentage c of the population dies. In other words, the new population $P(t + \Delta t)$ is the old population $P(t)$ plus the number of births minus the number of deaths during the time period Δt. Symbolically,

$$P(t + \Delta t) = P(t) + bP(t)\Delta t - cP(t)\Delta t$$

or

$$\frac{\Delta P}{\Delta t} = bP - cP = kP$$

From our assumptions the average rate of change of the population over an interval of time is proportional to the size of the population. Using the instantaneous rate of change to approximate the average rate of change, we have the following differential equation model:

$$\frac{dP}{dt} = kP, \qquad P(t_0) = P_0, \qquad t_0 \le t \le t_1 \tag{11.5}$$

where (for growth) k is a positive constant.

Solving the Model We can separate the variables and rewrite Equation (11.5) by moving all terms involving P and dP to one side of the equation and all terms in t and dt to the other. This gives

$$\frac{dP}{P} = k\,dt$$

Integration of both sides of this last equation yields

$$\ln P = kt + C \tag{11.6}$$

for some constant C. Applying the condition $P(t_0) = P_0$ to Equation (11.6) to find C results in

$$\ln P_0 = kt_0 + C$$

or

$$C = \ln P_0 - kt_0$$

Then, substitution for C into Equation (11.6) gives

$$\ln P = kt + \ln P_0 - kt_0$$

or, simplifying algebraically,

$$\ln \frac{P}{P_0} = k(t - t_0)$$

Finally, by exponentiating both sides of the preceding equation and multiplying the result by P_0, we obtain the solution

$$P(t) = P_0 e^{k(t-t_0)} \tag{11.7}$$

Equation (11.7), known as the **Malthusian model of population growth**, predicts that population grows exponentially with time.

Verifying the Model   Because $\ln(P/P_0) = k(t - t_0)$, our model predicts that if we plot $\ln P/P_0$ versus $t - t_0$, a straight line passing through the origin with slope k should result. However, if we plot the population data for the United States for several years, the model does not fit very well, especially in the later years. In fact, the 1990 census for the population of the United States was 248,710,000, and in 1970 it was 203,211,926. Substituting these values into Equation (11.7) and dividing the first result by the second gives

$$\frac{248,710,000}{203,211,926} = e^{k(1990-1970)}$$

Thus,

$$k = \left(\frac{1}{20} \right) \ln \frac{248,710,000}{203,211,926} \approx 0.01$$

That is, during the 20-year period from 1970 to 1990, population in the United States was increasing at the average rate of 1.0% per year. We can use this information together with Equation (11.7) to predict the population for 2000. In this case, $t_0 = 1990$, $P_0 = 248,710,000$, and $k = 0.01$ yields

$$P(2000) = 248,710,000 e^{0.01(2000-1990)} = 303,775,080$$

The 2000 census for the population of the United States was 281,400,000 (rounded to the nearest hundred thousand). Thus our prediction is off the mark by approximately 8%. We can probably live with that magnitude of error, but let's look into the distant future. Our model predicts that the population of the United States will be 55,209 billion in the year 2300, a population that far exceeds current estimates of the maximum sustainable population of the entire planet! We are forced to conclude that our model is unreasonable over the long term.

Some populations do grow exponentially, provided that the population is not too large. In most populations, however, individual members eventually compete with one another for food, living space, and other natural resources. Let's refine our Malthusian model of population growth to reflect this competition.

Refining the Model to Reflect Limited Growth   Let's consider that the proportionality factor k, measuring the rate of population growth in Equation (11.5), is now no longer constant but a function of the population. As the population increases and gets closer to the maximum population M, the rate k decreases. One simple submodel for k is the linear one

$$k = r(M - P), \qquad r > 0$$

where r is a constant. Substitution into Equation (11.5) leads to

$$\frac{dP}{dt} = r(M - P)P \tag{11.8}$$

or

$$\frac{dP}{P(M - P)} = r \, dt \tag{11.9}$$

Again we assume the *initial condition* $P(t_0) = P_0$. (Model (11.8) was first introduced by the Dutch mathematical biologist Pierre-Francois Verhulst, 1804–1849, and is referred to as **logistic growth**.) It follows from elementary algebra that

$$\frac{1}{P(M - P)} = \frac{1}{M}\left(\frac{1}{P} + \frac{1}{M - P}\right)$$

Thus, Equation (11.9) can be rewritten as

$$\frac{dP}{P} + \frac{dP}{M - P} = rM \, dt$$

which integrates to

$$\ln P - \ln | M - P | = rMt + C \tag{11.10}$$

for some arbitrary constant C. Using the initial condition, we evaluate C in the case $P < M$:

$$C = \ln \frac{P_0}{M - P_0} - rMt_0$$

Substituting into Equation (11.10) and simplifying give

$$\ln \frac{P}{M - P} - \ln \frac{P_0}{M - P_0} = rM(t - t_0)$$

or

$$\ln \frac{P(M - P_0)}{P_0(M - P)} = rM(t - t_0)$$

Exponentiating both sides of this equation gives

$$\frac{P(M - P_0)}{P_0(M - P)} = e^{rM(t-t_0)}$$

or

$$P_0(M - P)e^{rM(t-t_0)} = P(M - P_0)$$

Then,

$$P_0 M e^{rM(t-t_0)} = P(M - P_0) + P_0 P e^{rM(t-t_0)}$$

so solving for the population P gives

$$P(t) = \frac{P_0 M e^{rM(t-t_0)}}{M - P_0 + P_0 e^{rM(t-t_0)}}$$

To estimate P as $t \to \infty$, we rewrite this last equation as

$$P(t) = \frac{M P_0}{\left[P_0 + (M - P_0)e^{-rM(t-t_0)} \right]} \tag{11.11}$$

Notice from Equation (11.11) that $P(t)$ approaches M as t tends to infinity. Moreover, from Equation (11.8) we calculate the second derivative

$$P'' = rMP' - 2rPP' = rP'(M - 2P)$$

so that $P'' = 0$ when $P = M/2$. This means that when the population P reaches half the limiting population M, the growth dP/dt is most rapid and then starts to diminish toward zero. One advantage of recognizing that the maximum rate of growth occurs at $P = M/2$ is that the information can be used to estimate M. In a situation in which the modeler is satisfied that the growth involved is essentially logistic, if the point of maximum rate of growth has been reached, then $M/2$ can be estimated. The graph of the limited growth Equation (11.11) is depicted in Figure 11.4 for the case $P < M$ (see Problem 2 for the case $P > M$). Such a curve is called a **logistic curve**.

Figure 11.4

Graph of the limited growth model

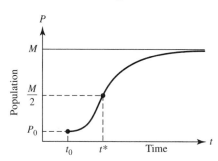

Verifying the Limited Growth Model   Let's test our model (11.11) against some real-world data. Equation (11.10) suggests a straight-line relationship of $\ln[P/(M-P)]$ versus t. Let's test this model using the data given in Table 11.1 for the growth of yeast in a culture. To plot $\ln[P/(M-P)]$ versus t, we need an estimate for the limiting population M. From the data in Table 11.1 we see that the population never exceeds 661.8. We estimate $M \approx 665$ and plot $\ln[P/(665-P)]$ versus t. The graph is shown in Figure 11.5 and does approximate a straight line. Thus, we accept the assumptions of logistic growth for bacteria. Now Equation (11.10) gives

$$\ln \frac{P}{M - P} = rMt + C$$

and from the graph in Figure 11.5 we can estimate the slope $rM \approx 0.55$ so that $r \approx 0.0008271$ from our estimate for $M \approx 665$.

Table 11.1 Growth of yeast in a culture

Time (hr)	Observed yeast biomass	Biomass calculated from logistic equation (11.13)	Percent error
0	9.6	8.9	−7.3
1	18.3	15.3	−16.4
2	29.0	26.0	−10.3
3	47.2	43.8	−7.2
4	71.1	72.5	2.0
5	119.1	116.3	−2.4
6	174.6	178.7	2.3
7	257.3	258.7	0.5
8	350.7	348.9	−0.5
9	441.0	436.7	−1.0
10	513.3	510.9	−4.7
11	559.7	566.4	1.2
12	594.8	604.3	1.6
13	629.4	628.6	−0.1
14	640.8	643.5	0.4
15	651.1	652.4	0.2
16	655.9	657.7	0.3
17	659.6	660.8	0.2
18	661.8	662.5	0.1

Data from R. Pearl, "The Growth of Population," *Quart. Rev. Biol.* 2 (1927): 532–548.

■ **Figure 11.5**

Plot of $\ln[P/(665 - P)]$ versus t for the data in Table 11.1

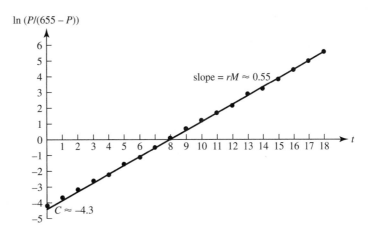

It is often convenient to express the logistic equation (11.11) in another form. To this end, let t^* denote the time when the population P reaches half the limiting value; that is, $P(t^*) = M/2$. It follows from Equation (11.11) that

$$t^* = t_0 - \frac{1}{rM} \ln \frac{P_0}{M - P_0}$$

(see Problem 1a at the end of this section). Solving this last equation for t_0, substituting the result into Equation (11.11), and simplifying algebraically give

$$P(t) = \frac{M}{1 + e^{-rM(t-t^*)}} \tag{11.12}$$

(see Problem 1b at the end of this section).

We can estimate t^* for the yeast culture data presented in Table 11.1 using Equation (11.10) and our graph in Figure 11.5:

$$t^* = -\frac{C}{rM} \approx \frac{4.3}{0.55} \approx 7.82$$

This calculation gives the logistic equation

$$P(t) = \frac{665}{1 + 73.8e^{-0.55t}} \tag{11.13}$$

by substituting $M = 665$, $r = 0.0008271$, and $t^* = 7.82$ in Equation (11.12).

The logistic model is known to agree quite well for populations of organisms that have very simple life histories—for instance, yeast growing in a culture where space is limited. Table 11.1 shows the calculations for the logistic equation (11.13), and we can see from the calculated error that there is very good agreement with the original data. A plot of the curve is shown in Figure 11.6.

■ **Figure 11.6**

Logistic curve showing the growth of yeast in a culture based on the data from Table 11.1 and Model (11.13); the small dots indicate the observed values.

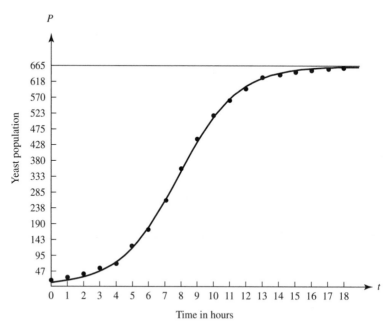

Next let's consider some data for human populations. A logistic equation for the population growth of the United States was formulated by Pearl and Reed in 1920. One form of their logistic curve is given by

$$P(t) = \frac{197{,}273{,}522}{1 + e^{-0.03134(t-1914.32)}} \tag{11.14}$$

where $M = 197,273,522$, $r = 1.5887 \times 10^{-10}$, and $t^* = 1914.32$ were determined using the census figures for the years 1790, 1850, and 1910 (we ask you to estimate M, r, and t^* in Problem 4 at the end of this section).

Table 11.2 compares the values predicted in 1920 by the logistic equation (11.14) with the observed values of the population of the United States. The predicted values agree quite well with the observations up to the year 1950, but the predicted values are much too small for the years 1970, 1980, 1990, and 2000. This should not be too surprising because our model fails to take into account such factors as immigration into the United States, wars, and advances in medical technology. In populations of higher plants and animals, which have complicated life histories and long periods of individual development, there are likely to be numerous responses that greatly modify the population growth.

Table 11.2 Population of the United States from 1790 to 2000, with predictions from Equation (11.14)

Year	Observed population	Predicted population	Percent error
1790	3,929,000	3,929,000	0.0
1800	5,308,000	5,336,000	0.5
1810	7,240,000	7,227,000	−0.2
1820	9,638,000	9,756,000	1.2
1830	12,866,000	13,108,000	1.9
1840	17,069,000	17,505,000	2.6
1850	23,192,000	23,191,000	−0.0
1860	31,443,000	30,410,000	−3.3
1870	38,558,000	39,370,000	2.1
1880	50,156,000	50,175,000	0.0
1890	62,948,000	62,767,000	−0.3
1900	75,995,000	76,867,000	1.1
1910	91,972,000	91,970,000	−0.0
1920	105,711,000	107,393,000	1.6
1930	122,755,000	122,396,000	−0.3
1940	131,669,000	136,317,000	3.5
1950	150,697,000	148,677,000	−1.3
1960	179,323,000	159,230,000	−11.2
1970	203,212,000	167,943,000	−17.4
1980	226,505,000	174,941,000	−22.8
1990	248,710,000	180,440,000	−27.5
2000	281,416,000	184,677,000	−34.4

11.1 PROBLEMS

1. a. Show that the population P in the logistic equation reaches half the maximum population M at time t^* given by

$$t^* = t_0 - (1/rM) \ln[P_0/(M - P_0)]$$

   b. Derive the form given by Equation (11.12) for population growth according to the logistic law.

   c. Derive the equation $\ln[P/(M - P)] = rMt - rMt^*$ from Equation (11.12).

2. Consider the solution of Equation (11.8). Evaluate the constant C in Equation (11.10) in the case that $P > M$ for all t. Sketch the solutions in this case. Also sketch a solution curve for the case that $M/2 < P < M$.

3. The following data were obtained for the growth of a sheep population introduced into a new environment on the island of Tasmania (adapted from J. Davidson, "On the Growth of the Sheep Population in Tasmania," *Trans. R. Soc. S. Australia* 62(1938): 342–346).

t (year)	1814	1824	1834	1844	1854	1864
$P(t)$	125	275	830	1200	1750	1650

   a. Make an estimate of M by graphing $P(t)$.

   b. Plot $\ln[P/(M - P)]$ against t. If a logistic curve seems reasonable, estimate rM and t^*.

4. Using the data for the U.S. population in Table 11.2, estimate M, r, and t^* using the same technique as in the text. Assume you are making the prediction in 1951 using a previous census. Use the data from 1960 to 2000 to check your model.

5. The modern philosopher Jean-Jacques Rousseau formulated a simple model of population growth for eighteenth-century England based on the following assumptions:

   The birthrate in London is less than that in rural England.

   The death rate in London is greater than that in rural England.

   As England industrializes, more and more people migrate from the countryside to London.

Rousseau then reasoned that because London's birthrate was lower and its death rate higher and rural people tend to migrate there, the population of England would eventually decline to zero. Criticize Rousseau's conclusion.

6. Consider the spreading of a highly communicable disease on an isolated island with population size N. A portion of the population travels abroad and returns to the island infected with the disease. You would like to predict the number of people X who will have been infected by some time t. Consider the following model, where $k > 0$ is constant:

$$\frac{dX}{dt} = kX(N - X)$$

   a. List two major assumptions implicit in the preceding model. How reasonable are your assumptions?

   b. Graph dX/dt versus X.

   c. Graph X versus t if the initial number of infections is $X_1 < N/2$. Graph X versus t if the initial number of infections is $X_2 > N/2$.

   d. Solve the model given earlier for X as a function of t.

e. From part (d), find the limit of X as t approaches infinity.

f. Consider an island with a population of 5000. At various times during the epidemic the number of people infected was recorded as follows:

t (days)	2	6	10
X (people infected)	1887	4087	4853
$\ln(X/(N - X))$	−.5	1.5	3.5

Do the collected data support the given model?

g. Use the results in part (f) to estimate the constants in the model, and predict the number of people who will be infected by $t = 12$ days.

7. Assume we are considering the survival of whales and that if the number of whales falls below a minimum survival level m, the species will become extinct. Assume also that the population is limited by the carrying capacity M of the environment. That is, if the whale population is above M, then it will experience a decline because the environment cannot sustain that high a population level.

a. Discuss the following model for the whale population

$$\frac{dP}{dt} = k(M - P)(P - m)$$

where $P(t)$ denotes the whale population at time t and k is a positive constant.

b. Graph dP/dt versus P and P versus t. Consider the cases in which the initial population $P(0) = P_0$ satisfies $P_0 < m$, $m < P_0 < M$, and $M < P_0$.

c. Solve the model in part (a), assuming that $m < P < M$ for all time. Show that the limit of P as t approaches infinity is M.

d. Discuss how you would test the model in part (a). How would you determine M and m?

e. Assuming that the model reasonably estimates the whale population, what implications are suggested for fishing? What controls would you suggest?

8. Sociologists recognize a phenomenon called *social diffusion*, which is the spreading of a piece of information, a technological innovation, or a cultural fad among a population. The members of the population can be divided into two classes: those who have the information and those who do not. In a fixed population whose size is known, it is reasonable to assume that the rate of diffusion is proportional to the number who have the information times the number yet to receive it. If X denotes the number of individuals who have the information in a population of N people, then a mathematical model for social diffusion is given by $dX/dt = kX(N - X)$, where t represents time and k is a positive constant.

a. Solve the model and show that it leads to a logistic curve.

b. At what time is the information spreading fastest?

c. How many people will eventually receive the information?

11.1   PROJECTS (See enclosed CD for UMAP modules.) ——

1. Complete the requirements of the UMAP module "The Cobb–Douglas Production Function," by Robert Geitz, UMAP 509. A mathematical model relating the output of an economic system to labor and capital is constructed from the assumptions that (a) marginal productivity of labor is proportional to the amount of production per unit of labor, (b) marginal productivity of capital is proportional to the amount of production per unit of capital, and (c) if either labor or capital tends to zero, then so does production.

2. Complete the UMAP module "The Diffusion of Innovation in Family Planning," by Kathryn N. Harmon, UMAP 303. This module gives an interesting application of finite difference equations to study the process through which public policies are diffused to understand how national governments might adopt family planning policies.

3. Complete the UMAP module "Difference Equations with Applications," by Donald R. Sherbert, UMAP 322. This module presents a good introduction to solving first- and second-order linear difference equations, including the method of undetermined coefficients for nonhomogeneous equations. Applications to problems in population and economic modeling are included.

11.1   Further Reading

Frauenthal, James C. *Introduction to Population Modeling.* Lexington, MA: COMAP, 1979.

Hutchinson, G. Evelyn. *An Introduction to Population Ecology.* New Haven, CT: Yale University Press, 1978.

Levins, R. "The Strategy of Model Building in Population Biology." *American Scientist* 54 (1966): 421–431.

Pearl, R., & L. J. Reed. "On the Rate of Growth of the Population of the United States since 1790." *Proceedings of the National Academy of Science* 6 (1920): 275–288.

11.2   Prescribing Drug Dosage[1]

The problem of how much of a drug dosage to prescribe and how often the dosage should be administered is an important one in pharmacology. For most drugs there is a concentration below which the drug is ineffective and a concentration above which the drug is dangerous.

Problem Identification   *How can the doses and the time between doses be adjusted to maintain a safe but effective concentration of the drug in the blood?*

The concentration in the blood resulting from a single dose of a drug normally decreases with time as the drug is eliminated from the body (Figure 11.7). We are interested in what happens to the concentration of the drug in the blood as doses are given at regular intervals.

---

[1]This section is adapted from UMAP Unit 72, based on the work of Brindell Horelick and Sinan Koont. The adaptation is presented with the permission of COMAP, 57 Bedford St., Lexington, MA 02173.

■ **Figure 11.7**

The concentration of a drug in the bloodstream decreases with time.

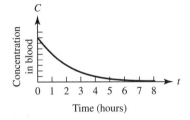

If H denotes the highest safe level of the drug and L its lowest effective level, it would be desirable to prescribe a dosage C_0 with time T between doses so that the concentration of the drug in the bloodstream remains between L and H over each dose period.

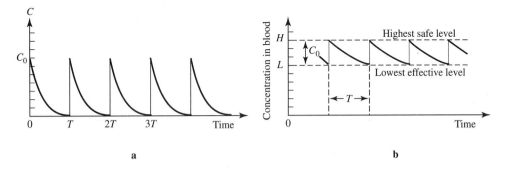

a

b

■ **Figure 11.8**

Residual buildup depends on the time interval between administration of drug doses.

Let's consider several ways in which the drugs might be administered. In Figure 11.8a the time between doses is such that effectively there is no buildup of the drug in the system. In other words, the residual concentration from previous doses is approximately zero. On the other hand, in Figure 11.8b the interval between doses relative to the amount administered and the decay rate of the concentration is such that a residual concentration exists at each time the drug is taken (after the first dose). Furthermore, as depicted in the graph, this residual level seems to be approaching a limit. We will be concerned with determining whether this is indeed the case and, if so, what that limit must be. Our ultimate goal in prescribing drugs is to determine dose *amounts* and *intervals* between doses such that the lowest effective level L is reached quickly and thereafter the concentration is maintained between the lowest effective level L and the highest safe level H, as depicted in Figure 11.9. We begin by determining the limiting residual level, which depends on our assumptions for the rate of assimilation of the drug in the bloodstream and the rate of decay after assimilation.

■ **Figure 11.9**

Safe but effective levels of the drug in the blood; C_0 is the change in concentration produced by one dose, and T is the time interval between doses.

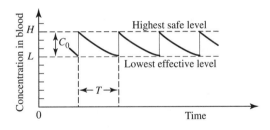

Assumptions   To solve the problem we have identified, let's consider the factors that determine the concentration $C(t)$ of the drug in the bloodstream at any time t. We begin with

$$C(t) = f(\text{decay rate, assimilation rate, dosage amount, dosage interval,} \dots)$$

and various other factors, including body weight and blood volume. To simplify our assumptions, let's assume that body weight and blood volume are constants (e.g., an average over some specific age group), and that concentration level is the critical factor in determining the effect of a drug. Next we determine submodels for decay rate and assimilation rate.

Submodel for Decay Rate   Consider the elimination of the drug from the bloodstream. This is probably a discrete phenomenon, but let's approximate it by a continuous function. Clinical experiments have revealed that the decrease in the concentration of a drug in the bloodstream will be proportional to the concentration. Mathematically, this assumption means that if we assume the concentration of the drug in the blood at time t is a differentiable function $C(t)$, then

$$C'(t) = -kC(t) \tag{11.15}$$

In this formula k is a positive constant called the **elimination constant** of the drug. Note that $C'(t)$ is negative, as it should be if it is to describe a decreasing concentration. Usually the quantities in Equation (11.15) are measured as follows: the time t is given in hours, $C(t)$ is milligrams per milliliter of blood (mg/ml), $C'(t)$ is mg ml^{-1}hr^{-1}, and k is hr^{-1}.

Assume that the concentrations H and L can be determined experimentally for a given population, such as an age group. (We will say more about this assumption in the ensuing discussion.) Then set the drug concentration for a single dose at the level

$$C_0 = H - L \tag{11.16}$$

If we assume that C_0 is the concentration at $t = 0$, then we have the model

$$\frac{dC}{dt} = -kC, \qquad C(0) = C_0 \tag{11.17}$$

The variables can be separated in Equation (11.17) and the model solved in the same way as the Malthusian model of population growth presented in the preceding section. Solution of the model gives

$$C(t) = C_0 e^{-kt} \tag{11.18}$$

To obtain the concentration at time $t > 0$, multiply the *initial concentration* C_0 by e^{-kt}. The graph of $C(t)$ looks like the one in Figure 11.10.

■ **Figure 11.10**

Exponential model for
decay of drug concentration
with time

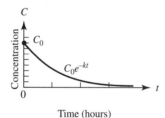

Submodel for Assimilation Rate   Having made an assumption about how drug concentrations decrease with time, let's consider how they increase again when drugs are administered. Our initial assumption is that when a drug is taken, it is diffused so rapidly throughout the blood that the graph of the concentration for the absorption period is, for all practical purposes, vertical. That is, we assume an instantaneous rise in concentration whenever a drug is administered. This assumption may not be as reasonable for a drug taken by mouth as it is for a drug that is injected directly into the bloodstream. Now let's see how the drug accumulates in the bloodstream with repeated doses.

Drug Accumulation with Repeated Doses   Consider what happens to the concentration $C(t)$ when a dose that is capable of raising the concentration by C_0 mg/ml each time it is given is administered regularly at fixed time intervals of length T.

Suppose the first dose is administered at time $t = 0$. According to Model (11.18), after T hours have elapsed, the residual $R_1 = C_0 e^{-kT}$ remains in the blood, and then the second dose is administered. Because of our assumption concerning the increase in drug concentration as previously discussed, the level of concentration instantaneously jumps to $C_1 = C_0 + C_0 e^{-kT}$. Then after T hours elapse again, the residual $R_2 = C_1 e^{-kT} = C_0 e^{-kT} + C_0 e^{-2kT}$ remains in the blood. This possibility of accumulation of the drug in the blood is depicted in Figure 11.11.

■ **Figure 11.11**

One possible effect of repeating equal doses

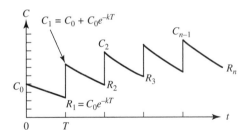

Next, we determine a formula for the nth residual R_n. If we let C_{i-1} be the concentration at the beginning of the ith interval and R_i the *residual concentration* at the end of it, we can easily obtain Table 11.3.

Table 11.3   Calculation of residual concentration of drug

i	C_{i-1}	R_i
1	C_0 —— multiply by $e^{-kT} \longrightarrow$	$C_0 e^{-kT}$
	—— add C_0 ——	
2	$C_0 + C_0 e^{-kT}$	$C_0 e^{-kT} + C_0 e^{-2kT}$
3	$C_0 + C_0 e^{-kT} + C_0 e^{-2kT}$	$C_0 e^{-kT} + C_0 e^{-2kT} + C_0 e^{-3kT}$
$\vdots$	$\vdots$	$\vdots$
n		$C_0 e^{-kT} + \cdots + C_0 e^{-nkT}$

From the table,

$$R_n = C_0 e^{-kT} + C_0 e^{-2kT} + \cdots + C_0 e^{-nkT} \tag{11.19}$$
$$= C_0 e^{-kT} (1 + r + r^2 + \cdots + r^{n-1})$$

where $r = e^{-kT}$. Algebraically, it is easy to verify that

$$1 + r + r^2 + \cdots + r^{n-1} = \frac{1 - r^n}{1 - r}$$

so substitution for r in Equation (11.19) gives the result

$$R_n = \frac{C_0 e^{-kT} (1 - e^{-nkT})}{1 - e^{-kT}} \tag{11.20}$$

Notice that the number e^{-nkT} is close to 0 when n is large. In fact, the larger n becomes, the closer e^{-nkT} gets to 0. As a result, the sequence of R_n's has a limiting value, which we call R:

$$R = \lim_{n \to \infty} R_n = \frac{C_0 e^{-kT}}{1 - e^{-kT}}$$

or

$$R = \frac{C_0}{e^{kT} - 1} \tag{11.21}$$

In summary, if a dose that is capable of raising the concentration by C_0 mg/ml is repeated at intervals of T hours, then the limiting value R of the residual concentrations is given by Equation (11.21). The number k in the formula is the elimination constant of the drug.

Determining the Dose Schedule From Table 11.3 the concentration C_{n-1} at the beginning of the nth interval is given by

$$C_{n-1} = C_0 + R_{n-1} \tag{11.22}$$

If the desired dosage level is required to approach the highest safe level H as depicted in Figure 11.9, then we want C_{n-1} to approach H as n becomes large. That is,

$$H = \lim_{n \to \infty} C_{n-1} = \lim_{n \to \infty} (C_0 + R_{n-1}) = C_0 + R$$

Combining this last result with $C_0 = H - L$ yields

$$R = L \tag{11.23}$$

A meaningful way to examine what happens to the residual concentration R for different intervals T between doses is to examine R in comparison with C_0, the change in concentration due to each dose. To make this comparison, we form the dimensionless ratio

$$\frac{R}{C_0} = \frac{1}{e^{kT} - 1} \tag{11.24}$$

Equation (11.24) states that R/C_0 will be close to 0 whenever the time T between doses is long enough to make $e^{kT} - 1$ sufficiently large. As for the intermediate values of

R_n, we can see from Table 11.3 that each R_n is obtained from the previous R_{n-1} by adding a positive quantity $C_0 e^{-nkT}$. This means that all the R_n's are positive because R_1 is positive. It also means that R is larger than each of the R_n's. In symbols,

$$0 < R_n < R, \quad \text{for all } n$$

The implication of this for drug dosage is that whenever R is small, the R_n's are even smaller. In particular, whenever T is long enough to make $e^{kT} - 1$ significantly large, the residual concentration from each dose is almost nil. The various administrations of the drug are then essentially independent, and the graph of $C(t)$ looks like the one depicted in Figure 11.12.

■ **Figure 11.12**

Drug concentration for long intervals between doses

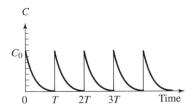

On the other hand, suppose the length of time T between doses is so short that e^{kT} is not much larger than 1 so that R/C_0 is significantly greater than 1. As R_n becomes larger, the concentration C_n after each dose becomes larger. The loss during the time period after each dose increases with larger C_n from Equation (11.17). Finally, the drop in concentration after each dose becomes imperceptibly close to the rise in concentration C_0 resulting from each dose. When this condition prevails (the loss in concentration equaling the gain), the concentration will oscillate between R at the end of each period and $R + C_0$ at the start of each period. This situation is depicted in Figure 11.13.

■ **Figure 11.13**

Buildup of drug concentration when the interval between doses is short

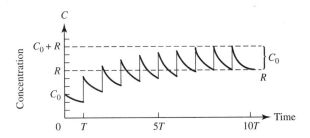

Suppose a drug is ineffective below the concentration L and harmful above some higher concentration H, as discussed previously. Assume that L and H are safe guidelines so that a person will not suffer a severe overdose if the drug concentration rises somewhat above H, and that it is not necessary to begin the buildup process all over again if the concentration falls slightly below L. Then, for patient convenience we might opt for the strategy of maximizing the time between drug doses by setting $R = L$ and $C_0 = H - L$, as we have indicated previously. Then substitution of $R = L$ and $C_0 = H - L$ in Equation (11.21) yields

$$L = \frac{H - L}{e^{kT} - 1}$$

We then solve the preceding equation for e^{kT} to obtain

$$e^{kT} = H/L$$

Taking the logarithm of both sides of this last equation and dividing the result by k give the desired dose schedule:

$$T = \frac{1}{k} \ln \frac{H}{L} \qquad (11.25)$$

To reach an effective level rapidly, administer a dose, often called a *loading dose*, that will immediately produce a blood concentration of H mg/ml. (For example, this loading dose might equal $2C_0$.) This medication can be followed every $T = (1/k) \ln(H/L)$ hours by a dose that raises the concentration by $C_0 = H - L$ mg/ml.

Verifying the Model   Our model for prescribing a safe and effective dosage of drug concentration appears to be a good one. It is in accord with the common medical practice of prescribing an initial dose several times larger than the succeeding periodic doses. Also, the model is based on the assumption that the decrease in the concentration of the drug in the bloodstream is proportional to the concentration, which has been verified clinically. Moreover, the elimination constant k, which is the positive constant of proportionality in that relationship, is an easily measured parameter (see Problem 1 in the 11.2 problem set). Equation (11.21) permits the prediction of concentration levels under varying conditions for dose rates. Thus, the drug may be tested to determine experimentally the lowest effective level L and the highest safe level H with appropriate safety factors to allow for inaccuracies in the modeling process. Then, Equations (11.16) and (11.25) can be used to prescribe a safe and effective dosage of the drug (assuming the loading dose is several times larger than C_0). Thus, our model is useful.

One deficiency in the model is the assumption of an instantaneous rise in concentration whenever a drug is administered. A drug, such as aspirin, taken orally requires a finite time to diffuse into the bloodstream; therefore, the assumption is not realistic for such a drug. For such cases, the graph of concentration versus time for a single dose might resemble Figure 11.14.

■ **Figure 11.14**

The concentration of a drug in the bloodstream for a single dose taken orally

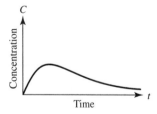

11.2   PROBLEMS

1. Discuss how the elimination constant k in Equation (11.15) could be obtained experimentally for a given drug.

2. a. If $k = 0.05$ hr^{-1} and the highest safe concentration is e times the lowest effective concentration, find the length of time between repeated doses that will ensure safe but effective concentrations.

   b. Does part (a) give enough information to determine the size of each dose?

3. Suppose $k = 0.01$ hr^{-1} and $T = 10$ hr. Find the smallest n such that $R_n > 0.5R$.

4. Given $H = 2$ mg/ml, $L = 0.5$ mg/ml, and $k = 0.02$ hr^{-1}, suppose concentrations below L are not only ineffective but also harmful. Determine a scheme for administering this drug (in terms of concentration and times of dosage).

5. Suppose that $k = 0.2$ hr^{-1} and that the smallest effective concentration is 0.03 mg/ml. A single dose that produces a concentration of 0.1 mg/ml is administered. Approximately how many hours will the drug remain effective?

6. Suggest other phenomena for which the model described in the text might be used.

7. Sketch how a series of doses might accumulate based on the concentration curve given in Figure 11.14.

8. A patient is given a dosage Q of a drug at regular intervals of time T. The concentration of the drug in the blood has been shown experimentally to obey the law

$$\frac{dC}{dt} = -ke^C$$

   a. If the first dose is administered at $t = 0$ hr, show that after T hr have elapsed, the residual

$$R_1 = -\ln (kT + e^{-Q})$$

   remains in the blood.

   b. Assuming an instantaneous rise in concentration whenever the drug is administered, show that after the second dose and T hr have elapsed again, the residual

$$R_2 = -\ln \left[kT(1 + e^{-Q}) + e^{-2Q}\right]$$

   remains in the blood.

   c. Show that the limiting value R of the residual concentrations for doses of Q mg/ml repeated at intervals of T hr is given by the formula

$$R = -\ln \frac{kT}{1 - e^{-Q}}$$

   d. Assuming the drug is ineffective below a concentration L and harmful above some higher concentration H, show that the dose schedule T for a safe and effective concentration of the drug in the blood satisfies the formula

$$T = \frac{1}{k}(e^{-L} - e^{-H})$$

   where k is a positive constant.

11.2   PROJECTS

1. Write a summary report on the article "Case Studies in Cancer and Its Treatment by Radiotherapy," by J. R. Usher and D. A. Abercrombie, *International Journal of Mathematics Education in Science and Technology* 12, no. 6 (1981), pp. 661–682. Present your report to the class.

In Projects 2–5, complete the requirements of the designated UMAP module (see enclosed CD for UMAP modules).

2. "Selection in Genetics," by Brindell Horelick and Sinan Koont, UMAP 70. This module introduces genetic terminology and basic results about genotype distribution in successive generations. A recurrence relationship is obtained from which the nth-generation frequency of a recessive gene can be determined. Calculus is used to derive a technique for approximating the number of generations required for this frequency to fall below any given positive value.

3. "Epidemics," by Brindell Horelick and Sinan Koont, UMAP 73. This unit poses two problems: (1) At what rate must infected persons be removed from a population to keep an epidemic under control? (2) What portion of a community will become infected during an epidemic? Threshold removal rate is discussed, and the extent of an epidemic is discussed when the removal rate is slightly below threshold.

4. "Tracer Methods in Permeability," by Brindell Horelick and Sinan Koont, UMAP 74. This module describes a technique for measuring the permeability of red corpuscle surfaces to K^{42} ions, using radioactive tracers. Students learn how radioactive tracers can be used to monitor substances in the body and learn some of the limitations and strengths of the model described in this unit.

5. "Modeling the Nervous System: Reaction Time and the Central Nervous System," by Brindell Horelick and Sinan Koont, UMAP 67. The module models the process by which the central nervous system reacts to a stimulus, and it compares the predictions of the model with experimental data. Students learn what conclusions can be drawn from the model about reaction time and are given an opportunity to discuss the merits of various assumptions about the relation between intensity of excitation and stimulus intensity.

11.3   Braking Distance Revisited

In our model for vehicular total stopping distance (see Section 2.2), one of the submodels is braking distance:

$$\text{braking distance} = h(\text{weight, speed})$$

Using an argument based on the result that the work done by the braking system must equal the change in kinetic energy, we found that the braking distance d_b is proportional to the square of the velocity. We now use an argument based on the derivative to establish that same result.

Let's assume that the braking system is designed in such a way that the maximum braking force increases in proportion to the mass of the car. Basically, this means that if the force per unit area applied by the braking hydraulic system remains constant, the surface area in contact with the brakes would have to increase in proportion to the mass of the car. From an engineering standpoint, this assumption seems reasonable.

The implication of the assumption is that the deceleration felt by the passengers is constant, which is probably a plausible design criterion. If it is further assumed that under a panic stop the maximum braking force F is applied continuously, then we obtain

$$F = -km$$

for some positive proportionality constant k. Because F is the only force acting on the car under our assumptions, this gives

$$ma = m \frac{dv}{dt} = -km$$

(the negative sign signals deceleration). Thus,

$$\frac{dv}{dt} = -k$$

which integrates to

$$v = -kt + C_1$$

If v_0 denotes the velocity at $t = 0$ when the brakes are initially applied, substitution gives $C_1 = v_0$ so that

$$v = -kt + v_0 \tag{11.26}$$

If t_s denotes the time it takes for the car to stop after the brakes have been applied, then $v = 0$ when $t = t_s$, substituted into Equation (11.26) gives

$$t_s = \frac{v_0}{k} \tag{11.27}$$

If x represents the distance traveled by the car after the brakes are applied, then x is the integral of $v = dx/dt$. Thus, from Equation (11.26),

$$x = -0.5kt^2 + v_0 t + C_2$$

When $t = 0$, $x = 0$, which implies $C_2 = 0$, so

$$x = -0.5kt^2 + v_0 t \tag{11.28}$$

Next, let d_b denote the braking distance; that is, $x = d_b$ when $t = t_s$. Substitution of these results into Equation (11.28) yields

$$d_b = -0.5kt_s^2 + v_0 t_s$$

Using Equation (11.27) in this last equation, we have

$$d_b = \frac{-v_0^2}{2k} + \frac{v_0^2}{k} = \frac{v_0^2}{2k} \tag{11.29}$$

Therefore, d_b is proportional to the square of the velocity, in accordance with the submodel obtained in Section 2.2.

In Chapter 2 we tested the submodel $d_b \propto v^2$ against some data and found reasonable agreement. The constant of proportionality was estimated to be 0.054 ft·hr²/mi², which corresponds to a value of k in Equation (11.29) of approximately 19.9 ft/sec² (see Problem 1 in the problem set). If we interpret k as the deceleration felt by a passenger in the vehicle (because $F = -km$ by assumption), we will find it useful to interpret this constant as $0.6g$ (where g is the acceleration of gravity).

11.3 PROBLEMS

1. **a.** Using the estimate that $d_b = 0.054v^2$, where 0.054 has dimension ft·hr²/mi², show that the constant k in Equation (11.29) has the value 19.9 ft/sec².

    b. Using the data in Table 4.4, plot d_b in ft versus $v^2/2$ in ft²/sec² to estimate $1/k$ directly.

2. Consider launching a satellite into orbit using a single-stage rocket. The rocket is continuously losing mass, which is being propelled away from it at significant speeds. We are interested in predicting the maximum speed the rocket can attain.[2]

    a. Assume the rocket of mass m is moving with speed v. In a small increment of time Δt it loses a small mass Δm_p, which leaves the rocket with speed u in a direction opposite to v. Here, Δm_p is the small propellant mass. The resulting speed of the rocket is $v + \Delta v$. Neglect all external forces (gravity, atmospheric drag, etc.) and assume Newton's second law of motion:

    $$\text{force} = \frac{d}{dt}(\text{momentum of system})$$

    where momentum is mass times velocity. Derive the model

    $$\frac{dv}{dt} = \left(\frac{-c}{m}\right)\frac{dm}{dt}$$

    where $c = u + v$ is the relative exhaust speed (the speed of the burnt gases relative to the rocket).

    b. Assume that initially, at time $t = 0$, the velocity $v = 0$ and the mass of the rocket is $m = M + P$, where P is the mass of the payload satellite and $M = \epsilon M + (1 - \epsilon)M$ $(0 < \epsilon < 1)$ is the initial fuel mass ϵM plus the mass $(1 - \epsilon)M$ of the rocket casings and instruments. Solve the model in part (a) to obtain the speed

    $$v = -c \ln \frac{m}{M + P}$$

---

[2]This problem was suggested by D. N. Burghes and M. S. Borrie, *Modelling with Differential Equations*. West Sussex, UK: Horwood, 1981.

c. Show that when all the fuel is burned, the speed of the rocket is given by

$$v_f = -c \ln \left[1 - \frac{\epsilon}{1+\beta} \right]$$

where $\beta = P/M$ is the ratio of the payload mass to the rocket mass.

d. Find v_f if $c = 3$ km/sec, $\epsilon = 0.8$, and $\beta = 1/100$. (These are typical values in satellite launchings.)

e. Suppose scientists plan to launch a satellite in circular orbit h km above the earth's surface. Assume that the gravitational pull toward the center of the earth is given by Newton's inverse square law of attraction:

$$\frac{\gamma m M_e}{(h + R_e)^2}$$

where γ is the universal gravitational constant, m is the mass of the satellite, M_e is the earth's mass, and R_e is the radius of the earth. Assume that this force must be balanced by the centrifugal force $mv^2/(h + R_e)$, where v is the speed of the satellite. What speed must be attained by a rocket to launch a satellite into an orbit 100 km above the earth's surface? From your computation in part (d), can a single-stage rocket launch a satellite into an orbit of that height?

3. The gross national product (GNP) represents the sum of consumption purchases of goods and services, government purchases of goods and services, and gross private investment (which is the increase in inventories plus buildings constructed and equipment acquired). Assume that the GNP is increasing at the rate of 3% per year and that the national debt is increasing at a rate proportional to the GNP.

a. Construct a system for two ordinary differential equations modeling the GNP and national debt.

b. Solve the system in part (a), assuming the GNP is M_0 and the national debt is N_0 at year 0.

c. Does the national debt eventually outstrip the GNP? Consider the ratio of the national debt to the GNP.

11.3 PROJECTS

Complete the requirements of the indicated UMAP module. (See enclosed CD for UMAP modules.)

1. "Kinetics of Single Reactant Reactions," by Brindell Horelick and Sinan Koont, UMAP 232. The unit discusses reaction orders of irreversible single reactant reactions. The equation $a'(t) = -k(a(t))^n$ is solved for selected values of n; reaction orders of various reactions are found from experimental data, and the notion of half-life is discussed. Some background knowledge of chemistry is required.

2. "Radioactive Chains: Parents and Daughters," by Brindell Horelick and Sinan Koont, UMAP 234. When a radioactive substance A decays into a substance B, A and B are called parent and daughter. It may happen that B is radioactive and is the parent of a new daughter C, and so on. There are three radioactive chains that together account for all naturally occurring radioactive substances beyond thallium on the periodic table. This unit develops models for calculating the amounts of the substances in radioactive chains and discusses transient and secular states of equilibrium between parent and daughter.

3. "The Relationship between Directional Heading of an Automobile and Steering Wheel Deflection," by John E. Prussing, UMAP 506. This unit develops a model relating the compass heading and the steering wheel deflection using basic geometric and kinematic principles.

Graphical Solutions of Autonomous Differential Equations

The models developed in this chapter are first-order differential equations of the form

$$\frac{dy}{dx} = g(x, y)$$

relating the derivative dy/dx to some function g of the independent and dependent variables. In some cases, either of the variables x or y may not appear explicitly.

Slope Fields: Viewing Solution Curves

Each time we specify an initial condition $y(x_0) = y_0$ for the solution of a differential equation $y' = g(x, y)$, the **solution curve** (graph of the solution) is required to pass through the point (x_0, y_0) and to have slope $g(x_0, y_0)$ there. We can picture these slopes graphically by drawing short line segments of slope $g(x, y)$ at selected points (x, y) in the region of the xy-plane that constitutes the domain of g. Each segment has the same slope as the solution curve through (x, y) and thus is tangent to the curve there. We see how the curves behave by following these tangents (Figure 11.15).

Constructing a slope field with pencil and paper can be quite tedious. All our examples were generated by a computer. Let's build on our knowledge from calculus of how derivatives determine the shape of a graph to solve differential equations graphically.

Equilibrium Values and Phase Lines

We saw in Chapter 10 the important role that the critical points play in determining how a function behaves and in finding its extreme points. Let's investigate what happens when the derivative of a function is zero from a slightly different point of view. In this case, the derivative dy/dx will be a function of y only (the dependent variable). For example, differentiating the equation

$$y^2 = x + 1$$

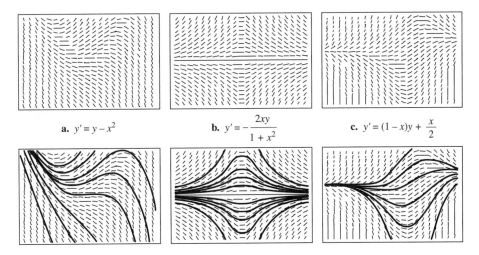

a. $y' = y - x^2$ **b.** $y' = -\dfrac{2xy}{1 + x^2}$ **c.** $y' = (1 - x)y + \dfrac{x}{2}$

■ **Figure 11.15**

Slope fields (top row) and selected solution curves (bottom row). In computer renditions, slope segments are sometimes portrayed with arrows, as they are here. This is not to be taken as an indication that slopes have directions, however, for they do not.

implicitly gives

$$2y\,\frac{dy}{dx} = 1 \quad \text{or} \quad \frac{dy}{dx} = \frac{1}{2y}$$

A differential equation for which dy/dx is a function of y only is called an **autonomous** differential equation.

Definition

> If $dy/dx = g(y)$ is an autonomous differential equation, then the values of y for which $dy/dx = 0$ are called **equilibrium values** or **rest points**.

Thus, equilibrium values are those at which no change occurs in the dependent variable, so y is at *rest*. The emphasis is on the value of y where $dy/dx = 0$, not on the value of x, as in the preceding section. For example, the equilibrium values for $dy/dx = (y + 1)(y - 2)$ are $y = -1$ and $y = 2$.

To construct a graphical solution to an autonomous differential equation, we first make a **phase line** for the equation, which is a plot on the y axis that shows the equation's equilibrium values along with the intervals where dy/dx and d^2y/dx^2 are positive and negative. Then we know where the solutions are increasing and decreasing and the concavity of the solution curves. These are the essential features needed to determine the shapes of the solution curves without having to find formulas for them.

EXAMPLE 1   *Drawing a Phase Line and Sketching Solution Curves*

Draw a phase line for the equation

$$\frac{dy}{dx} = (y+1)(y-2)$$

and use it to sketch solutions to the equation.

Solution

STEP 1   DRAW A NUMBER LINE FOR *y* AND MARK THE EQUILIBRIUM VALUES *y* = −1 AND *y* = 2, WHERE *dy/dx* = 0.

STEP 2   IDENTIFY AND LABEL THE INTERVALS WHERE *y′* > 0 AND *y′* < 0. This step resembles what we do in calculus, only now we are marking the *y* axis instead of the *x* axis.

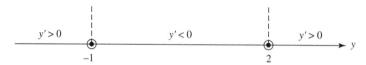

We can encapsulate the information about the sign of *y′* on the phase line. Because *y′* > 0 on the interval to the left of *y* = −1, a solution of the differential equation with a *y* value less than −1 will increase from there toward *y* = −1. We display this information by drawing an arrow on the interval pointing to −1.

Similarly, *y′* < 0 between *y* = −1 and *y* = 2, so any solution with a value in this interval will decrease toward *y* = −1.

For *y* > 2, we have *y′* > 0, so a solution with a *y* value greater than 2 will increase from there without bound.

In short, solution curves below the horizontal line *y* = −1 in the *xy*-plane rise toward *y* = −1. Solution curves between the lines *y* = −1 and *y* = 2 fall away from *y* = 2 toward *y* = −1. Solution curves above *y* = 2 rise away from *y* = 2 and keep rising.

STEP 3   CALCULATE *y″* AND MARK THE INTERVALS WHERE *y″* > 0 AND *y″* < 0. To find *y″*, we differentiate *y′* *with respect to x* using implicit differentiation.

$$y' = (y+1)(y-2) = y^2 - y - 2$$
$$y'' = \frac{d}{dx}(y') = \frac{d}{dx}(y^2 - y - 2)$$
$$= 2yy' - y'$$
$$= (2y-1)y'$$
$$= (2y-1)(y+1)(y-2)$$

From this formula, we see that y'' changes sign at $y = -1$, $y = 1/2$, and $y = 2$. We add the sign information to the phase line.

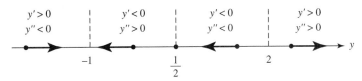

STEP 4   SKETCH AN ASSORTMENT OF SOLUTION CURVES IN THE *xy*-PLANE. The horizontal lines $y = -1$, $y = 1/2$, and $y = 2$ partition the plane into horizontal bands in which we know the signs of y' and y''. In each band, this information tells us whether the solution curves rise or fall and how they bend as x increases (Figure 11.16).

■ Figure 11.16

Solution curves for the autonomous differential equation

$dy/dx = (y+1)(y-2)$

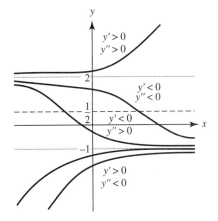

The equilibrium lines $y = -1$ and $y = 2$ are also solution curves. (The constant functions $y = -1$ and $y = 2$ satisfy the differential equation.) Solution curves that cross the line $y = 1/2$ have an inflection point there. The concavity changes form concave down (above the line) to concave up (below the line).

As predicted in Step 2, solutions in the middle and lower bands approach the equilibrium value $y = -1$ as x increases. Solutions in the upper band rise steadily away from the value $y = 2$.   ■ ▨ ■

Stable and Unstable Equilibria

Look at Figure 11.16 once more, particularly at the behavior of the solution curves near the equilibrium values. Once a solution curve has a value near $y = -1$, it tends steadily toward that value; $y = -1$ is a **stable equilibrium**. The behavior near $y = 2$ is just the opposite: All solutions except the equilibrium solution $y = 2$ move away from it as x increases. We call $y = 2$ an **unstable equilibrium**. If the solution is *at* that value, it stays, but if it is off by any amount, no matter how small, it moves away. (Sometimes an equilibrium value is unstable because a solution moves away from it only on one side of the point.)

Now that we know what to look for, we can already see this behavior on the initial phase line. The arrows lead away from $y = 2$ and, once to the left of $y = 2$, toward $y = -1$.

Isaac Newton postulated that the rate of change in the temperature of a cooled or heated object is proportional to the difference in temperature between the object and its surrounding medium. We can use this idea to describe how the object's temperature will change over time.

EXAMPLE 2 *Cooling Soup*

What happens to the temperature of a cup of hot soup when it is placed on a table in a room? We know the soup cools down, but what does a typical temperature curve look like as a function of time?

Solution We assume that the soup's Celsius temperature H is a differentiable function of time t. Choose a suitable unit for t (e.g., minutes) and start measuring time at $t = 0$. We also assume that the volume of the surrounding medium is large enough so that the heat of the soup has a negligible effect on its surrounding temperature.

Suppose that the surrounding medium has a constant temperature of $15\,°C$. We can then express the difference in temperature as $H(t) - 15$. According to Newton's law of cooling, there is a constant of proportionality $k > 0$ such that

$$\frac{dH}{dt} = -k(H - 15) \tag{11.30}$$

(*minus k to give a negative derivative when $H > 15$*).

Because $dH/dt = 0$ at $H = 15$, the temperature $15\,°C$ is an equilibrium value. If $H > 15$, Equation (11.30) tells us that $(H - 15) > 0$ and $dh/dt < 0$. If the object is hotter than the room, it will get cooler. Similarly, if $H < 15$, then $(H - 15) < 0$ and $dH/dt > 0$. An object cooler than the room will warm up. Thus, the behavior described by Equation (11.30) agrees with our intuition of how temperature should behave. These observations are captured in the initial phase line diagram in Figure 11.17.

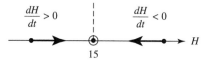

■ Figure 11.17

First step in constructing phase line for Newton's Law of Cooling. The temperature tends toward the equilibrium (surrounding-medium) value in the long run.

We determine the concavity of the solution curves by differentiating both sides of Equation (11.30) with respect to t:

$$\frac{d}{dt}\left(\frac{dH}{dt}\right) = \frac{d}{dt}(-k(H - 15))$$

$$\frac{d^2 H}{dt^2} = -k\frac{dH}{dt}$$

Because $-k$ is negative, we see that d^2H/dt^2 is positive when $dH/dt < 0$ and negative when $dH/dt > 0$. Figure 11.18 adds this information to the phase line.

■ **Figure 11.18**

The complete phase line for Newton's Law of Cooling model

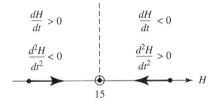

The completed phase line shows that if the temperature of the object is above the equilibrium value of $15\,°C$, the graph of $H(t)$ will be decreasing and concave upward. If the temperature is below $15\,°C$ (the temperature of the surrounding medium), the graph of $H(t)$ will be increasing and concave downward. We use this information to sketch typical solution curves (Figure 11.19).

■ **Figure 11.19**

Temperature versus time. Regardless of initial temperature, the object's temperature $H(t)$ tends toward $15\,°C$, the temperature of the surrounding medium.

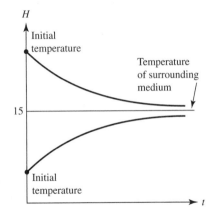

From the upper solution curve in Figure 11.19, we see that as the object cools down, the rate at which it cools slows down because dH/dt approaches zero. This observation is implicit in Newton's law of cooling and contained in the differential equation, but the flattening of the graph as time advances gives an immediate visual representation of the phenomenon. The ability to discern physical behavior from graphs is a powerful tool in understanding real-world systems.   ■ ■ ■

EXAMPLE 3 *Logistic Growth Revisited*

Let's apply our phase line techniques to obtain solution curves for the logistic growth equation

$$\frac{dP}{dt} = r(M - P)P \tag{11.31}$$

which we studied in Section 11.1.

The equilibrium values for autonomous Equation (11.31) are $P = M$ and $P = 0$, and we can see that $dP/dt > 0$ if $0 < P < M$ and if $dP/dt < 0$ if $P > M$. These observations are recorded on the phase line in Figure 11.20.

■ Figure 11.20

The initial phase line for logistic growth

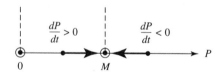

We determine the concavity features of the population curves by differentiating the equation

$$\frac{dP}{dt} = r(M - P)P = rMP - rP^2$$

Thus,

$$\frac{d^2P}{dt^2} = \frac{d}{dt}(rMP - rP^2)$$

$$= rM\frac{dP}{dt} - 2rP\frac{dP}{dt}$$

$$= r(M - 2P)\frac{dP}{dt}$$

If $P = M/2$, then $d^2P/dt^2 = 0$. If $P < M/2$, then $(M - 2P)$ and dP/dt are positive and $d^2P/dt^2 > 0$. If $M/2 < P < M$, then $(M - 2P) < 0$, $dP/dt > 0$, and $d^2P/dt^2 < 0$. If $P > M$, then $(M - 2P)$ and dP/dt are both negative and $d^2P/dt^2 > 0$. We add this information to the phase line (Figure 11.21).

■ Figure 11.21

The completed phase line for logistic growth

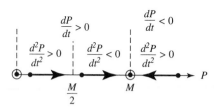

The lines $P = M/2$ and $P = M$ divide the first quadrant of the tP-plane into horizontal bands in which we know the signs of both dP/dt and d^2P/dt^2. In each band, we know how the solution curves rise and fall and how they bend as time passes. The equilibrium lines $P = 0$ and $P = M$ are both population curves. Population curves crossing the line $P = M/2$ have an inflection point there, giving them a **sigmoid** shape (curved in two directions like a letter S). Figure 11.22 displays typical population curves. They closely resemble the growth of yeast in a culture displayed in Figure 11.6 in Section 11.1. ■ ■ ■

■ **Figure 11.22**

Population curves
representing logistic growth

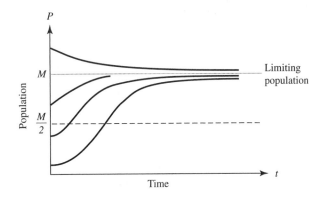

11.4   PROBLEMS

1. Construct a direction field and sketch a solution curve for the following differential equations:

   a. $dy/dx = y$

   b. $dy/dx = x$

   c. $dy/dx = x + y$

   d. $dy/dx = x - y$

   e. $dy/dx = xy$

   f. $dy/dx = 1/y$

For Problems 2–5,

(a) Identify the equilibrium values. Which are stable and which are unstable?

(b) Construct a phase line. Identify the signs of y' and y''.

(c) Sketch several solution curves.

2. $dy/dx = (y + 2)(y - 3)$

3. $dy/dx = y^2 - 2y$

4. $dy/dx = (y - 1)(y - 2)(y - 3)$

5. $dy/dx = y - \sqrt{y}, y > 0$

The autonomous differential equations in Problems 6–9 represent models for population growth. For each problem, use a phase line analysis to sketch solution curves for $P(t)$, selecting different starting values $P(0)$ (as in Example 3). Which equilibria are stable, and which are unstable?

6. $dP/dt = 1 - 2P$

7. $dP/dt = P(1 - 2P)$

8. $dP/dt = 2P(P - 3)$

9. $dP/dt = 3P(1 - P)(P - \frac{1}{2})$

10. *Catastrophic continuation of Example 3*—Suppose that a healthy population of some species is growing in a limited environment and that the current population P_0 is fairly close to the carrying capacity M_0. You might imagine a population of fish living in a freshwater lake in a wilderness area. Suddenly, a catastrophe such as the Mount St. Helens volcanic eruption contaminates the lake and destroys a significant part of the food and oxygen on which the fish depend. The result is a new environment with a carrying capacity M_1 considerably less than M_0 and, in fact, less than the current population P_0. Starting at some time before the catastrophe, sketch a "before-and-after" curve that shows how the fish population responds to the change in the environment.

11. *Controlling a population*—The fish and game department in a certain state is planning to issue hunting permits to control the deer population (one deer per permit). It is known that if the deer population falls below a certain level m, the deer will become extinct. It is also known that if the deer population rises above the carrying capacity M, the population will decrease back to M through disease and malnutrition.

   a. Discuss the reasonableness of the following model for the growth rate of the deer population as a function of time:

$$\frac{dP}{dt} = rP(M - P)(P - m)$$

   where P is the population of the deer and r is a positive constant of proportionality. Include a phase line.

   b. Explain how this model differs from the logistic model $dP/dt = rP(M - P)$. Is it better or worse than the logistic model?

   c. Show that if $P > M$ for all t, then $\lim_{t \to \infty} P(t) = M$.

   d. What happens if $P < M$ for all t?

   e. Discuss the solutions to the differential equation. What are the equilibrium points of the model? Explain the dependence of the steady-state value of P on the initial values of P. About how many permits should be issued?

11.5   Numerical Approximation Methods

In the models developed in earlier sections of this chapter, we found an equation relating a derivative to some function of the independent and dependent variables; that is,

$$\frac{dy}{dx} = g(x, y)$$

where g is some function in which either x or y may not appear explicitly. Moreover, we were given some starting value; that is, $y(x_0) = y_0$. Finally, we were interested in the values of y for a specific set of x values; that is, $x_0 \leq x \leq b$. In summary, we determined models of the general form

$$\frac{dy}{dx} = g(x, y), \quad y(x_0) = y_0, \quad x_0 \leq x \leq b$$

We call first-order ordinary differential equations with the preceding conditions **first-order initial value problems**. As seen from our previous models, they constitute an important class of problems. We now discuss the three parts of the model.

First-Order Initial Value Problems

The Differential Equation $dy/dx = g(x, y)$ As discussed in our models, we are interested in finding a function $y = f(x)$ whose derivative satisfies an equation $dy/dx = g(x, y)$. Although we do not know f, we can compute its derivative given particular values of x and y. As a result, we can find the slope of the tangent line to the solution curve $y = f(x)$ at specified points (x, y).

The Initial Value $y(x_0) = y_0$ The initial value equation states that at the initial point x_0, we know the y value is $f(x_0) = y_0$. Geometrically, this means that the point (x_0, y_0) lies on the solution curve (Figure 11.23). Thus, we know where our solution curve begins. Moreover, from the differential equation $dy/dx = g(x, y)$, we know that the slope of the solution curve at (x_0, y_0) is the number $g(x_0, y_0)$. This is also depicted in Figure 11.23.

■ **Figure 11.23**

The solution curve passes through the point (x_0, y_0) and has slope $g(x_0, y_0)$.

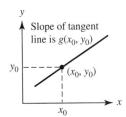

The Interval $x_0 \leq x \leq b$ The condition $x_0 \leq x \leq b$ gives the particular interval of the x axis with which we are concerned. Thus, we would like to relate y with x over the interval $x_0 \leq x \leq b$ by finding the solution function $y = f(x)$ passing through the point (x_0, y_0) with slope $g(x_0, y_0)$ (Figure 11.24). Note that the function $y = f(x)$ is continuous over $x_0 \leq x \leq b$ because its derivative exists there.

■ **Figure 11.24**

The solution $y = f(x)$ to the initial value problem is a continuous function over the interval from x_0 to b.

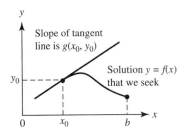

Approximating Solutions to Initial Value Problems

If we do not require an exact solution function to an initial value problem $dy/dx = g(x, y)$, $y(x_0) = y_0$, we can probably use a computer to generate a table of approximate numerical values of y for values of x in an appropriate interval $x_0 \leq x \leq b$. Such a table is called a **numerical solution** of the problem, and the method by which we generate the table is a **numerical method**. One such method, called Euler's method, is described as follows.

Given a differential equation $dy/dx = g(x, y)$ and an initial condition $y(x_0) = y_0$, we can approximate the solution by its tangent line

$$T(x) = y_0 + g(x_0, y_0)(x - x_0)$$

The value $g(x_0, y_0)$ is the slope of the solution curve and its tangent line at point (x_0, y_0). The function $T(x)$ gives a good approximation to the solution $y(x)$ in a short interval about x_0 (Figure 11.25). The basis of Euler's method is to patch together a string of tangent line approximations to the curve over a longer stretch. Here's how the method works.

■ **Figure 11.25**

The tangent line $T(x)$ is a good approximation to the solution curve $y(x)$ in a short interval about x_0.

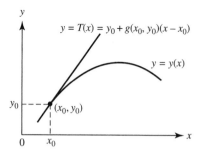

We know the point (x_0, y_0) lies on the solution curve. Suppose that we specify a new value for the independent variable to be $x_1 = x_0 + \Delta x$. If the increment Δx is small, then

$$y_1 = T(x_1) = y_0 + g(x_0, y_0)\Delta x$$

is a good approximation to the exact solution value $y = y(x_1)$. Thus, from the point (x_0, y_0), which lies *exactly* on the solution curve, we have obtained the point (x_1, y_1), which lies very close to the point $(x_1, y(x_1))$ on the solution curve (Figure 11.26).

■ **Figure 11.26**

The first Euler step approximates $y(x_1)$ with the value $T(x_1)$ on the tangent line.

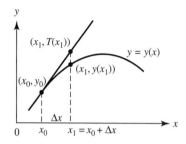

Using the point (x_1, y_1) and the slope $g(x_1, y_1)$ of the solution curve through (x_1, y_1), we take a second step. Setting $x_2 = x_1 + \Delta x$, we use the tangent linearization of the solution curve through (x_1, y_1) to calculate

$$y_2 = y_1 + g(x_1, y_1)\Delta x$$

This gives the next approximation (x_2, y_2) to values along the solution curve $y = y(x)$ (Figure 11.27). Continuing in this fashion, we take a third step from the point (x_2, y_2) with slope $g(x_2, y_2)$ to obtain the third approximation

$$y_3 = y_2 + g(x_2, y_2)\Delta x$$

and so on. We are literally building an approximation to one of the solutions by following the direction of the slope field of the differential equation.

■ **Figure 11.27**

Three steps in the Euler approximation to the solution of the initial value problem $y' = g(x, y)$, $y(x_0) = y_0$. As we take more steps, the errors involved usually accumulate, but not in the exaggerated way shown here.

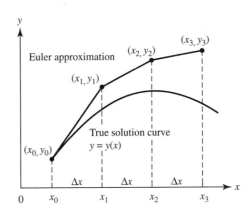

EXAMPLE 1 *Using Euler's Method*

Find the first three approximations y_1, y_2, y_3 using Euler's method for the initial value problem

$$y' = 1 + y, \quad y(0) = 1$$

starting at $x_0 = 0$, with $\Delta x = 0.1$.

Solution   We have $x_0 = 0$, $y_0 = 1$, $x_1 = x_0 + \Delta x = 0.1$, $x_2 = x_0 + 2\Delta x = 0.2$, and $x_3 = x_0 + 3\Delta x = 0.3$.

First: $y_1 = y_0 + g(x_0, y_0)\Delta x$
            $= y_0 + (1 + y_0)\Delta x$
            $= 1 + (1 + 1)(0.1) = 1.2$

Second: $y_2 = y_1 + g(x_1, y_1)\Delta x$
            $= y_1 + (1 + y_1)\Delta x$
            $= 1.2 + (1 + 1.2)(0.1) = 1.42$

Third: $y_3 = y_2 + g(x_2, y_2)\Delta x$
            $= y_2 + (1 + y_2)\Delta x$
            $= 1.42 + (1 + 1.42)(0.1) = 1.662$

This step-by-step process can be easily continued. Using equally spaced values for the independent variable in the table and generating n of them, set

$$x_1 = x_0 + \Delta x$$
$$x_2 = x_1 + \Delta x$$
$$\vdots$$
$$x_n = x_{n-1} + \Delta x$$

Then calculate the approximations to the solution

$$y_1 = y_0 + g(x_0, y_0)\Delta x$$
$$y_2 = y_1 + g(x_1, y_1)\Delta x$$
$$\vdots$$
$$y_n = y_{n-1} + g(x_{n-1}, y_{n-1})\Delta x$$

The number of steps n can be as large as we like, but errors can accumulate if n is too large.

Euler's method is easy to program on a computer or programmable calculator. A program generates a table of numerical solutions to an initial value problem, allowing us to input x_0 and y_0, the number of steps n, and the step size Δx. It then calculates the approximate solution values $y_1, y_2, \ldots, y_n$ in iterative fashion, as just described.

EXAMPLE 2 A Savings Certificate Revisited

Let's discuss again the savings certificate example investigated in Chapter 1 as a discrete dynamical system. Here, we consider the value of a certificate initially worth $1000 that accumulates annual interest at 12% **compounded continuously** (rather than 1% each month as in Example 1 of Section 1.1). We would like to know the value of the certificate in 10 years. If $Q(t)$ represents the value of the certificate at any time t, we have the model

$$\frac{dQ}{dt} = 0.12Q, \quad Q(0) = 1000$$

Use Euler's method to approximate the value in 10 years if

(a) $\Delta t = 1$ year   (b) $\Delta t = 1$ month   (c) $\Delta t = 1$ week

Compare your results with the analytic solution when $t = 10$ years.

Solution   First, let's find the analytic solution. Then we will use Euler's method for each of the three cases and compare the results at each year $t = 1, 2, 3, \ldots, 10$. Just as we did for the population problem in Section 11.1, we can separate the variables and integrate the initial value problem

$$\frac{dQ}{dt} = 0.12Q, \quad Q(0) = 1000, \quad 0 \le t \le 10$$

to obtain

$$Q = C_1 e^{0.12t}$$

We apply the initial condition $Q(0) = 1000$ to evaluate $C_1 = 1000$. The exact solution is

$$Q(t) = 1000e^{0.12t}$$

Table 11.4 Euler solutions of $dQ/dt = 0.12Q, Q(0) = 1000$

Year	(a) $\Delta t = 1$	(b) $\Delta t = 1/12$	(c) $\Delta t = 1/52$	Exact solution
0	1000.00	1000.00	1000.00	1000.00
1	1120.00	1126.83	1127.34	1127.50
2	1254.40	1269.73	1270.90	1271.25
3	1404.93	1430.77	1432.74	1433.33
4	1573.52	1612.23	1615.18	1616.07
5	1762.34	1816.70	1820.86	1822.12
6	1973.82	2047.10	2052.73	2054.43
7	2210.68	2306.72	2314.13	2316.37
8	2475.96	2599.27	2608.81	2611.70
9	2773.08	2928.93	2941.02	2944.68
10	3105.85	3300.39	3315.53	3320.12

Using a computer, we generated the Euler approximation values in Table 11.4 for the three cases (a) $\Delta t = 1$, (b) $\Delta t = 1/12$, and (c) $\Delta t = 1/52$. All entries are rounded to the nearest cent. Note that the Euler approximations become more accurate as Δt decreases. The Euler approximation with $\Delta t = 1/52$ (representing weeks) gives $Q(10) = \$3315.53$ for the value of the savings certificate after 10 years, an error of $\$4.59$ from the analytical solution value. Figure 11.28 displays the Euler approximations plotted with the exact solution curve.

■ **Figure 11.28**

A plot of the exact solution and Euler approximations in Table 11.4 for the savings certificate in Example 2

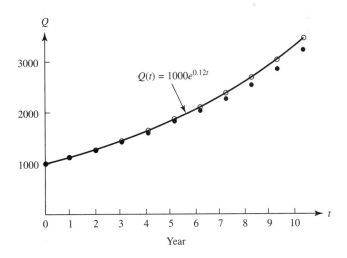

It might be tempting to reduce the step size even further to obtain greater accuracy. However, each additional calculation not only requires additional computer time but also, more important, introduces round-off error. Because these errors accumulate, an ideal method would improve the accuracy of the approximations but minimize the number of calculations. For this reason, Euler's method may prove unsatisfactory. More refined numerical methods for solving the initial value problem are investigated in courses in the numerical methods of differential equations. We will not pursue those methods here.   ▨ ▦ ▧

Separation of Variables

The method of direct integration employed in this chapter works only in those cases in which the dependent and independent variables can be algebraically separated. Any such differential equation can be written in the form

$$p(y)\,dy = q(x)\,dx$$

For example, given the differential equation

$$u(x)v(y)\,dx + q(x)p(y)\,dy = 0$$

we can arrange the equation in the following form:

$$\frac{p(y)}{v(y)}\,dy = \frac{-u(x)}{q(x)}\,dx$$

Because the right-hand side is a function of x and the left-hand side is a function of y only, the solution is obtained by simply integrating both sides directly. Remember from calculus that even if we are successful in separating the variables, it may not be possible to find the integrals in closed form. The method just described is called *separation of variables*. We will study the method in the next section.

11.5   PROBLEMS

In Problems 1–4, use Euler's method to calculate the first three approximations to the given initial value problem for the specified increment size. Round your results to four decimal places.

1. $y' = x(1 - y)$, $y(1) = 0$, $\Delta x = 0.2$

2. $y' = 1 - \dfrac{y}{x}$, $y(2) = -1$, $\Delta x = 0.5$

3. $y' = 2xy + 2y$, $y(0) = 3$, $\Delta x = 0.2$

4. $y' = y^2(1 + 2x)$, $y(-1) = 1$, $\Delta x = 0.5$

5. When interest is compounded, the interest earned is added to the principal amount so that it may also earn interest. For a 1-year period, the principal amount Q is given by

$$Q = \left(1 + \frac{i}{n}\right)^n Q(0)$$

where i is the annual interest rate (given as a decimal) and n is the number of times during the year that the interest is compounded.

To lure depositors, banks offer to compound interest at different intervals: semiannually, quarterly, or daily. A certain bank advertises that it compounds interest continuously. If $100 is deposited initially, formulate a mathematical model describing the growth of the initial deposit during the first year. Assume an annual interest rate of 10%.

6. Use the differential equation model formulated in the preceding problem to answer the following:

   a. From the derivative evaluated at $t = 0$, determine an equation of the tangent line T passing through the point $(0, 100)$.

   b. Estimate $Q(1)$ by finding $T(1)$, where $Q(t)$ denotes the amount of money in the bank at time t (assuming no withdrawals).

   c. Estimate $Q(1)$ using a step size of $\Delta t = 0.5$

   d. Estimate $Q(1)$ using a step size of $\Delta t = 0.25$.

   e. Plot the estimates you obtained for $\Delta t = 1.0, 0.5$, and 0.25 to approximate the graph of $Q(t)$.

7. For the differential equation model obtained in Problem 5, find $Q(t)$ by separating the variables and integrating.

   a. Evaluate $Q(1)$.

   b. Compare your previous estimates of $Q(1)$ with its actual value.

   c. Find the effective annual interest rate when an annual rate of 10% is compounded continuously.

   d. Compare the effective annual interest rate computed in part (c) with interest compounded

      i. Semiannually: $(1 + 0.1/2)^2$

      ii. Quarterly: $(1 + 0.1/4)^4$

      iii. Daily: $(1 + 0.1/365)^{365}$

      iv. Estimate the limit of $(1 + 0.1/n)^n$ as $n \to \infty$ by evaluating the expression for $n = 1000;\ 10,000;\ 100,000$.

      v. What is $\lim_{n \to \infty} (1 + 0.1/n)^n$?

11.5 PROJECTS

Complete the requirements of the indicated UMAP modules. (See enclosed CD for UMAP modules.)

1. "Feldman's Model," by Brindell Horelick and Sinan Koont, UMAP 75. This unit develops a version of G. A. Feldman's model of growth in a planned economy in which all the means of production are owned by the state. Originally, the model was developed by Feldman in connection with planning the economy of the former Soviet Union. Students compute numerical values for rates of output, national income, their rates of change, and the propensity to save, and they discuss the effects of changes in the parameters of the model and in the units of measurement.

2. "The Digestive Process of Sheep," by Brindell Horelick and Sinan Koont, UMAP 69. This unit introduces a differential equation model for the digestive processes of sheep. The model is tested and fit using collected data and the least-squares criterion.

11.5 Further Reading

Fox, William P., & George Schnibben. "Using Euler's Method in Autonomous Ordinary Differential Equations: The Importance of Step Size." *COED Journal* (January–March 2000): 44–50.

11.6 Separation of Variables

Suppose an object is dropped from a tall building. Initially the object has zero velocity ($v(0) = 0$), but we wish to know the object's velocity at any time $t > 0$. If we neglect all resistance, Newton's second law leads to

$$m \frac{dv}{dt} = mg$$

where m is the mass of the object and g is the acceleration due to gravity. In this formulation, positive displacement is being measured downward from the top of the building. The differential equation simply integrates directly to yield

$$v = gt + C$$

Knowing $v(0) = 0$ enables us to evaluate C:

$$v(0) = 0 = g \cdot 0 + C$$

or

$$C = 0$$

Thus the solution function $v = gt$ predicts the velocity of an object dropped with zero initial velocity if gravity is the only force considered.

This very simple falling-body example illustrates two features typical of first-order differential equations. First, an integration process is required to obtain y from its derivative y'. Second, the integration process introduces a single arbitrary constant of integration, which can then be evaluated if an initial condition is known.

We can always integrate (at least theoretically) a first-order differential equation of the form

$$y' = f(x)$$

whenever f is a continuous function. However, consider the differential equation

$$\frac{dy}{dx} = f(x, y) \tag{11.32}$$

where the derivative is a function of both the variables x and y. We may be able to factor $f(x, y)$ into factors containing only x or y terms, but not both:

$$f(x, y) = p(x)q(y)$$

We allow for the possibility that either p or q may be a constant function. When the variables are separable in this way, differential Equation (11.32) becomes

$$\frac{dy}{dx} = p(x)q(y)$$

which can be rewritten as

$$\frac{dy}{q(y)} = p(x)\,dx \tag{11.33}$$

To solve separable Equation (11.33), simply integrate both sides (with respect to the same variable x).

Integration of both sides is permitted because we are assuming y is a function of x. Thus the left side of Equation (11.33) is

$$\frac{dy}{q(y)} = \frac{y'(x)}{q(y(x))}\,dx$$

Substitution of this expression into Equation (11.33) gives

$$\frac{y'(x)}{q(y(x))}\,dx = p(x)\,dx$$

If we set $u = y(x)$ and $du = y'(x)\,dx$, then integration of both sides leads to the solution

$$\int \frac{du}{q(u)} = \int p(x)\,dx + C \tag{11.34}$$

Of course, we have to be concerned about any values of x where $q(y(x))$ is zero. Let us consider several examples.

EXAMPLE 1

Solve $y' = 3x^2 e^{-y}$.

Solution   We separate the variables and write

$$e^y dy = 3x^2 dx$$

Integration of each side yields

$$e^y = x^3 + C$$

Applying the natural logarithm to each side results in

$$y = \ln(x^3 + C) \tag{11.35}$$

Let's verify that y does solve the given differential equation. Differentiating Equation (11.35), we find

$$y' = \frac{3x^2}{x^3 + C}$$

Substitution of y and y' into the original equation then gives

$$\frac{3x^2}{x^3 + C} = 3x^2 e^{-\ln(x^3 + C)} \tag{11.36}$$

Because

$$e^{-\ln(x^3 + C)} = e^{\ln(x^3 + C)^{-1}}$$
$$= \frac{1}{x^3 + C}$$

Equation (11.36) is valid for all values of x satisfying $x^3 + C > 0$, and the differential equation is satisfied. ■ ■ ■

EXAMPLE 2

Solve $y' = 2(x + y^2 x)$.

Solution   The differential equation can be written as

$$\frac{dy}{dx} = 2x(1 + y^2)$$

and separating the variables gives

$$\frac{dy}{1 + y^2} = 2x\, dx$$

Integration of both sides leads to

$$\tan^{-1} y = x^2 + C$$

or

$$y = \tan(x^2 + C). \tag{11.37}$$

To verify that Equation (11.37) is a solution to the given equation, we differentiate y:

$$y' = 2x \sec^2(x^2 + C)$$
$$= 2x[1 + \tan^2(x^2 + C)]$$
$$= 2x(1 + y^2)$$

Substitution into the original differential equation gives the identity

$$2x(1 + y^2) = 2(x + y^2 x) \qquad\qquad ■ ■ ■$$

From now on, to save space, we will not always verify that the function we find by a solution method is in fact a solution to the differential equation (as we have done in the examples so far). However, it is good practice to do so, especially if the solution method is fairly involved.

The Differential Form

In many cases the first-order differential equation appears in **differential form** as

$$M(x, y)\, dx + N(x, y)\, dy = 0 \tag{11.38}$$

For example, the equations

$$ye^{-x}\, dy + x\, dx = 0 \tag{11.39}$$

$$\sec x\, dy - x \cot y\, dx = 0 \qquad \text{and} \tag{11.40}$$

$$(xe^{y} - e^{2y})\, dy + (e^{y} + x)\, dx = 0 \tag{11.41}$$

all have differential form. If separation of variables is to apply to Equation (11.38), first write the equation in the form

$$\frac{dy}{dx} = -\frac{M(x, y)}{N(x, y)}$$

Next look for cancellation of common terms in the numerator and denominator, and then separate the variables, if possible. Finally, integrate each side as before. Observe that Equations (11.39) and (11.40) are indeed separable, whereas Equation (11.41) is not.

EXAMPLE 3

Solve $\sec x\, dy - x \cot y\, dx = 0$.

Solution   After division by $\sec x \cot y$, the equation becomes

$$\tan y\, dy - x \cos x\, dx = 0$$

Integration then gives

$$\ln |\cos y| + x \sin x + \cos x = C \qquad \blacksquare \blacksquare \blacksquare$$

Using separation of variables involves no new ideas, only the ability to recognize factors and integrate. If the correct factoring can be found, the solution technique is simply to integrate. You studied a variety of integration techniques in integral calculus and may have forgotten some of the more important ones. Appendix D provides a brief review of the main integration techniques needed for this text. These techniques are simple u-substitution, integration by parts, and integration of rational functions requiring partial fraction decomposition. We call your attention especially to the convenient *tableau method* for integration by parts and to the *Heaviside method*, which is convenient for certain partial fraction decompositions. The following examples illustrate these integration techniques in the context of solving separable differential equations.

EXAMPLE 4

Solve $e^{-x}y' = x$.

Solution Separating the variables yields (see Appendix D for a review of integration techniques)

$$dy = xe^x \, dx.$$

Integrating the right side by the tableau method, we have

Sign	Derivatives	Integrals
+ →	x	e^x
− →	1	e^x
+ →	0	e^x

Thus, interpreting the tableau, we get

$$\int xe^x \, dx = +xe^x - 1 \cdot e^x + \int 0 \cdot e^x dx + C$$
$$= (x - 1)e^x + C$$

Therefore,

$$y = (x - 1)e^x + C$$

EXAMPLE 5

Solve $e^{x+y}y' = x$.

Solution The equation can be written as

$$e^x e^y \, dy = x \, dx$$

Separating the variables leads to

$$e^y \, dy = xe^{-x} dx$$

Integrating the right side by parts with e^x replaced by e^{-x} in the tableau in Example 4 gives us

$$\int xe^{-x} \, dx = -xe^{-x} - e^{-x} + \int 0 \cdot e^{-x} \, dx$$

Thus the separable equation integrates to

$$e^y = -xe^{-x} - e^{-x} + C$$
$$= -(x + 1)e^{-x} + C$$

We now solve for y by taking the logarithm of each side to obtain

$$y = \ln[C - (x + 1)e^{-x}] \tag{11.42}$$

Let us verify that y does solve the original differential equation. Differentiating Equation (11.42) yields

$$\begin{aligned} y' &= [C - (x + 1)e^{-x}]^{-1} \cdot [-e^{-x} + (x + 1)e^{-x}] \\ &= e^{-y} \cdot xe^{-x} \\ &= xe^{-(x+y)} \end{aligned}$$

Substituting this result into the differential equation yields the identity

$$e^{(x+y)}xe^{-(x+y)} = x$$

EXAMPLE 6

Solve the differential equation $dy/dx = \ln x$, where $x > 0$.

Solution   Integration of the right side gives

Sign	Derivatives	Integrals
$+$	$\ln x$	1
$-$	$1/x$	x

Thus,

$$\begin{aligned} y &= \int \ln x \, dx \\ &= x \ln x - \int \left(\frac{1}{x}\right) x \, dx + C \\ &= x \ln x - x + C \end{aligned}$$

EXAMPLE 7

Solve the initial value problem $x^2 y y' = e^y$, where $y(2) = 0$.

Solution   Separating the variables and integrating each side leads to

$$ye^{-y} \, dy = x^{-2} \, dx$$

$$-(y + 1)e^{-y} = -\frac{1}{x} + C$$

To evaluate the arbitrary constant, substitute $y = 0$ and $x = 2$, to obtain

$$-(0 + 1)1 = -\frac{1}{2} + C$$

or

$$C = -\frac{1}{2}$$

Thus

$$-(y+1)e^{-y} = -\frac{1}{x} - \frac{1}{2}$$

or

$$2x(y+1) = (2+x)e^y \qquad \blacksquare \blacksquare \blacksquare$$

EXAMPLE 8

Solve $y' = x(1 - y^2)$, where $-1 < y < 1$.

Solution   Separating the variables, we have

$$\frac{dy}{1-y^2} = x\,dx.$$

Partial fraction decomposition yields

$$\frac{1}{1-y^2} = \frac{1}{(1+y)(1-y)}$$

$$= \frac{1/2}{1+y} + \frac{1/2}{1-y}$$

Integration then results in

$$\frac{1}{2}\ln|1+y| - \frac{1}{2}\ln|1-y| = \frac{x^2}{2} + C$$

or

$$\ln\left(\frac{1+y}{1-y}\right) = x^2 + C_1$$

where $C_1 = 2C$. Exponentiating each side leads to

$$\left(\frac{1+y}{1-y}\right) = e^{x^2}e^{C_1}$$

Setting $e^{C_1} = C_2$ and solving algebraically for y, we have

$$1 + y = C_2 e^{x^2} - y C_2 e^{x^2}$$

or

$$y = \frac{C_2 e^{x^2} - 1}{C_2 e^{x^2} + 1}$$

where $C_2 = e^{C_1}$ is an arbitrary constant. $\blacksquare \blacksquare \blacksquare$

EXAMPLE 9 *Newton's Law of Cooling Revisited*

Consider the following model for the cooling of a hot cup of soup:

$$\frac{dT_m}{dt} = -k(T_m - \beta), \quad k > 0$$

where $T_m(0) = \alpha$. Here T_m is the temperature of the soup at any time $t > 0$, β is the constant temperature of the surrounding medium, α is the initial temperature of the soup, and k is a constant of proportionality depending on the thermal properties of the soup. We will solve the above differential equation for T_m.

Solution   After separating the variables, we obtain

$$\frac{dT_m}{T_m - \beta} = -k \, dt$$

Integration yields

$$\ln |T_m - \beta| = -kt + C$$

Exponentiating both sides, we get

$$|T_m - \beta| = e^{-kt+C} = e^{-kt} e^C$$

Since e^C is a constant, we substitute $C_1 = e^C$ into the above equation:

$$|T_m - \beta| = C_1 e^{-kt}$$

From the initial condition $T_m(0) = \alpha$, we evaluate C_1:

$$|\alpha - \beta| = C_1$$

Substitution of this result into the solution produces

$$|T_m - \beta| = |\alpha - \beta| e^{-kt}$$

Assuming the object is initially warmer than the surrounding medium, we have

$$T_m = \beta + (\alpha - \beta)e^{-kt}$$

The graph of $T_m(t)$ is shown in Figure 11.29. As $t \to \infty$, you can see that $T_m \to \beta$, in agreement with the graphical analysis presented in Section 11.4. ■ ■ ■

EXAMPLE 10 *Population Growth with Limited Resources Revisited*

In Section 11.1 we developed the following model for population growth in a limited environment:

$$\frac{dP}{dt} = r(M - P)P = rMP - rP^2$$

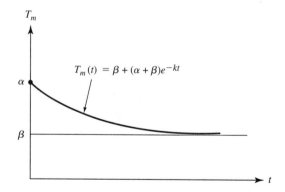

Figure 11.29

This graph of $T_m(t)$ assumes that the initial temperature $T_m(0) = \alpha$ is greater than the temperature β of the surrounding medium.

where $P(t_0) = P_0$. Here P denotes the population at any time $t > 0$, M is the carrying capacity of the environment, and r is a proportionality constant. Let us solve this model.

Solution   Separating the variables, we get

$$\frac{dP}{P(M - P)} = r\,dt$$

Using partial fraction decomposition on the left side gives us

$$\frac{1}{M}\left(\frac{dP}{P} + \frac{dP}{M - P}\right) = r\,dt$$

Multiplying by M and integrating then yield

$$\ln P - \ln|M - P| = rMt + C$$

for some arbitrary constant C. Note that since $P > 0$, the absolute value symbol in the expression $\ln|P|$ is not necessary. Using the initial condition, we evaluate C in the case where $0 < P_0 < M$:

$$C = \ln\left(\frac{P_0}{M - P_0}\right) - rMt_0$$

Substitution for C into the solution and algebraic simplification give us

$$\ln\left[\frac{P(M - P_0)}{P_0(M - P)}\right] = rM(t - t_0)$$

Exponentiating both sides of this equation, we obtain

$$\frac{P(M - P_0)}{P_0(M - P)} = e^{rM(t - t_0)}$$

or

$$P_0(M - P)e^{rM(t - t_0)} = P(M - P_0)$$

Finally, solving this equation algebraically for population P yields the **logistic curve**

$$P(t) = \frac{P_0 M e^{rM(t-t_0)}}{M - P_0 + P_0 e^{rM(t-t_0)}}$$

If you divide the numerator and denominator of this last expression by $e^{rM(t-t_0)}$ and then take the limit as $t \to \infty$, you will find that $P(t) \to M$. That is, the population tends toward the maximum sustainable population. Our analytic result agrees with our graphical analysis for the case $0 < P_0 < M$, which was depicted in Figure 11.4. ▪ ▪ ▪

Uniqueness of Solutions

Often the solutions to a first-order differential equation can be expressed in the explicit form $y = f(x)$, where each solution is distinguished by a different value of the arbitrary constant of integration. For instance, if we separate the variables in the equation

$$\frac{dy}{dx} = \frac{2y}{x}, \quad x \neq 0, \tag{11.43}$$

we obtain

$$\frac{dy}{y} = 2 \frac{dx}{x}.$$

Notice, however, that the algebra is not valid when $y = 0$. Next, integration of both sides of this last result gives us

$$\ln |y| = 2 \ln |x| + C_1 \tag{11.44}$$

Exponentiating both sides of Equation (11.44) yields

$$|y| = C_2 x^2$$

where $C_2 = e^{C_1} > 0$ is a constant. Finally, by applying the definition of absolute value, we have

$$y = Cx^2 \tag{11.45}$$

where $C = \pm C_2$ is positive or negative according to whether y is positive or negative. However, there are still other solutions not given by Equation (11.45). Equation (11.45) represents a *family of parabolas,* each parabola distinguished by a different value of the constant C. If $C > 0$, the curves $y = Cx^2$ open upward, and if $C < 0$, they open downward (see Figure 11.30). Exactly one of these parabolas passes through each point in the plane excluding the origin. Thus, by specifying an initial condition $y(x_0) = y_0$, a unique solution curve from the family denoted by Equation (11.45) is selected that passes through the point (x_0, y_0).

The question of whether more than one solution curve can pass through a specific point (x_0, y_0) in the plane is an important one. For uniqueness to occur, certain conditions must be met by the function $f(x, y)$ defining the differential equation $y' = f(x, y)$. Geometrically, the condition of uniqueness means that two solution curves cannot cross at the point in

■ **Figure 11.30**

The family of parabolas
$y = Cx^2$

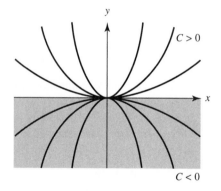

question. In Figure 11.30 there is a unique solution parabola passing through each point in the plane with the exception of the origin. Through the origin pass infinitely many solution parabolas, along with other solution curves as well. For instance, notice that the constant function $y \equiv 0$ is a solution to the differential equation. This fact is readily seen when Equation (11.43) is written in the form

$$xy' = 2y, \tag{11.46}$$

which is clearly satisfied when $y \equiv 0$. However, the solution $y \equiv 0$ is *not* a member of the family of solutions represented by Equation (11.45), since C_2 is always positive. The difficulty lies with the lack of continuity of the function $f(x, y) = 2y/x$ at the origin. Nevertheless, if we let $C = 0$ in Equation (11.45), we do pick up the solution $y \equiv 0$. Even so, there are still other solutions not given by Equation (11.45) for any value of C.

The issue of uniqueness of solutions is rather involved for nonlinear first-order equations. However, in the very important case of the linear first-order equation the matter is easily settled. This points up one of the significant differences between linear and nonlinear differential equations. In the next section we will investigate the nature of the linear first-order equation.

11.6 PROBLEMS

In Problems 1–8, solve the separable differential equation using u-substitution.

1. $\dfrac{dy}{dx} = y^2 - 2y + 1$

2. $\dfrac{dy}{dx} = \sqrt{y} \cos^2 \sqrt{y}$

3. $y' = \dfrac{3y(x+1)^2}{y-1}$

4. $yy' = \sec y^2 \sec^2 x$

5. $y \cos^2 x \, dy + \sin x \, dx = 0$

6. $y' = \left(\dfrac{y}{x}\right)^2$

7. $y' = xe^y \sqrt{x-2}$

8. $y' = xye^{x^2}$

In Problems 9–16, solve the separable differential equation using integration by parts.

9. $\sec x\, dy + x \cos^2 y\, dx = 0$

10. $2x^2\, dx - 3\sqrt{y} \csc x\, dy = 0$

11. $y' = \dfrac{e^y}{xy}$

12. $y' = xe^{x-y}\csc y$

13. $y' = e^{-y} \ln\left(\dfrac{1}{x}\right)$

14. $y' = y^2 \tan^{-1} x$

15. $y' = y \sin^{-1} x$

16. $\sec(2x+1)\, dy + 2xy^{-1}\, dx = 0$

In Problems 17–24, solve the separable differential equation using partial fractions.

17. $(x^2 + x - 2)\, dy + 3y\, dx = 0$

18. $y' = (y^2 - 1)x^{-1}$

19. $x(x-1)\, dy - y\, dx = 0$

20. $y' = \dfrac{(y+1)^2}{x^2 + x - 2}$

21. $9y\, dx - (x-1)^2(x+2)\, dy = 0$

22. $e^x\, dy + (y^3 - y^2)\, dx = 0$

23. $\sqrt{1 - y^2}\, dx + (x^2 - 2x + 2)\, dy = 0$

24. $(2x - x^2)\, dy + e^{-y}\, dx = 0$

In Problems 25–32, solve the separable differential equation.

25. $\sqrt{2xy}\, \dfrac{dy}{dx} = 1$

26. $(\ln x)\dfrac{dx}{dy} = xy$

27. $x^2\, dy + y(x-1)\, dx = 0$

28. $ye^x\, dy - (e^{-y} + e^{2x-y})\, dx = 0$

29. $(x \ln y)y' = \left(\dfrac{x+1}{y}\right)^2$

30. $y' = \dfrac{\sin^{-1} x}{2y \ln y}$

31. $y' = e^{-y} - xe^{-y} \cos x^2$

32. $(1 + x + xy^2 + y^2)\, dy = (1 - x)^{-1}\, dx$

In Problems 33–39, solve the initial value problem.

33. $y^{-2}\dfrac{dx}{dy} = \dfrac{e^x}{e^{2x} + 1}$, $y(0) = 1$

34. $\dfrac{dy}{dx} + xy = x$, $y(1) = 2$

35. $y' - 2y = 1$, $y(2) = 0$

36. $2(y^2 - 1)\, dx + \sec x \csc x\, dy = 0$, $y\left(\dfrac{\pi}{4}\right) = 0$

37. $\dfrac{dP}{dt} + P = Pte^t, \quad P(0) = 1$

38. $\dfrac{dP}{dt} = (P^2 - P)t^{-1}, \quad P(1) = 2$

39. $x\,dy - (y + \sqrt{y})\,dx = 0, \quad y(1) = 1$

11.7   Linear Equations

Serious problems of water pollution face the world's industrialized nations. If the polluted body of water is a river, it can clean itself fairly rapidly once the pollution is stopped, provided excessive damage has not already occurred. But the problem of pollution in a lake or reservoir is not so easily overcome. A polluted lake, such as one of the Great Lakes, contains a large amount of water that must somehow be cleaned. Presently government and industry still rely on natural processes for this cleanup. In this section we shall model how long a given lake might take to return to an acceptable level of pollution through natural processes alone; modeling this problem will enable us to examine some characteristics of linear equations.

Let's make some simplifying assumptions to model the situation. Imagine the lake to be a large container or tank holding a volume $V(t)$ of water at any time t. Assume that when water enters the lake, perfect mixing occurs, so that the pollutants are uniformly distributed throughout the lake at any time. Assume also that the pollutants are not removed from the lake by sedimentation, decay, or any other natural mechanism except the outflow of water from the lake. Moreover, the pollutants flow freely from the lake (unlike DDT, which tends to concentrate in the fatty tissues of animals and thus be retained in biological systems). Let $p(t)$ denote the amount of the pollutant in the lake at time t. Then the **concentration** of the pollutant is the ratio $c(t) = p(t)/V(t)$.

Over the time interval $[t, t + \Delta t]$, the change in the amount of pollutant Δp is the amount of pollutant that enters the lake minus the amount that leaves:

$$\Delta p = \text{amount input} - \text{amount output}.$$

If water enters the lake with a constant concentration of c_{in} in grams per liter at a rate of r_{in} in liters per second, then

$$\text{amount input} \approx r_{in} c_{in}\, \Delta t = \alpha\, \Delta t$$

where $\alpha = r_{in} c_{in}$ is constant. For example, if polluted water with a concentration of 9 g/L of pollutant enters the lake at 10 L/sec, then $\alpha = 90$ g of pollutant enters the lake each second.

If water leaves the lake at a constant rate of r_{out} L/sec, then, since the concentration of pollutant in the lake is given by p/V, we have

$$\text{amount output} \approx r_{out} \frac{p}{V}\, \Delta t$$

Thus

$$\Delta p \approx \left(\alpha - \frac{p r_{out}}{V}\right) \Delta t$$

Dividing by Δt and passing to the limit as $\Delta t \to 0$ result in

$$\frac{dp}{dt} = \alpha - \frac{r_{\text{out}}}{V} p. \tag{11.47}$$

Suppose $V(0) = V_0$ is the volume of the lake initially. Then $V(t) = V_0 + (r_{\text{in}} - r_{\text{out}})t$ represents the volume at any time t. Substituting for V in Equation (11.47) and rearranging terms give us

$$\frac{dp}{dt} + \frac{p r_{\text{out}}}{V_0 + (r_{\text{in}} - r_{\text{out}})t} = \alpha \tag{11.48}$$

where α, r_{out}, V_0, and $(r_{\text{in}} - r_{\text{out}})$ are all constants. Equation (11.48) is an example of a *linear first-order differential equation*. Note that if $r_{\text{in}} - r_{\text{out}} = 0$, Equation (11.48) is a separable equation. However, if $r_{\text{in}} - r_{\text{out}} \neq 0$, Equation (11.48) represents a new type of first-order equation.

At the end of this section we will return to the pollution problem to find the amount $p(t)$ of pollutant in the lake at any time t. In order to solve this model, we now take up the general question of solving linear first-order equations.

First-Order Linear Equations

The **first-order linear equation** is an equation of the form

$$a_1(x)y' + a_0(x)y = b(x) \tag{11.49}$$

where $a_1(x)$, $a_0(x)$, and $b(x)$ depend only on the independent variable x, not on y. For example,

$$2xy' - y = xe^{-x} \tag{11.50}$$
$$(x^2 + 1)y' + xy = x \quad \text{and} \tag{11.51}$$
$$y' + (\tan x)y = \cos^2 x \tag{11.52}$$

are all first-order linear equations. The equation

$$y' - x^2 e^{-2y} = 0 \tag{11.53}$$

is not linear, although it is separable. The equation

$$(2x - y^2)y' + 2y = x \tag{11.54}$$

is neither linear nor separable.

We assume in Equation (11.49) that the functions $a_1(x)$, $a_0(x)$, and $b(x)$ are continuous on an interval and that $a_1(x) \neq 0$ on that interval. Division of both sides of (11.49) by $a_1(x)$ gives the **standard form** of the linear equation,

$$y' + P(x)y = Q(x) \tag{11.55}$$

where $P(x)$ and $Q(x)$ are continuous on the interval. The solution method we present for the linear equation proceeds from the standard form.

To provide some insight into the solution form, we are going to solve Equation (11.55) in three stages. The first stage is the case when $P(x) \equiv$ constant and $Q(x) \equiv 0$. (The symbol $\equiv$ means "is identically equal to." Thus $P(x) \equiv$ constant says that $P(x)$ is constant for all values of x.) Then the case is considered for $P(x) \equiv$ constant and $Q(x) \neq 0$. Finally, we consider the general case given by Equation (11.55).

Case 1: $y' + ky = 0$, $k =$ Constant   The differential equation is separable with

$$\frac{dy}{y} = -k\, dx$$

Then

$$y = Ce^{-kx}$$

is a solution for any constant C. If we write the last equation as

$$e^{kx}y = C$$

and differentiate implicitly, we obtain

$$e^{kx}y' + ke^{kx}y = 0$$

or

$$e^{kx}(y' + ky) = 0$$

That is, multiplication of each side of the equation $y' + ky = 0$ by the exponential function e^{kx} results in

$$\frac{d}{dx}(e^{kx}y) = \frac{d}{dx}(C)$$

The solution is now readily obtained by integrating each side. Armed with this insight, we consider the second case.

Case 2: $y' + ky = Q(x)$, $k =$ Constant   From our observation in Case 1, if we multiply both sides of Equation (11.55) by e^{kx}, we get

$$e^{kx}(y' + ky) = e^{kx}Q(x)$$

or

$$\frac{d}{dx}(e^{kx}y) = e^{kx}Q(x) \tag{11.56}$$

Integration then gives us

$$e^{kx}y = \int e^{kx}Q(x)\, dx + C \tag{11.57}$$

where C is an arbitrary constant.

Let us pause to reflect on this procedure. We multiplied the equation $y' + ky = Q(x)$ by a function e^{kx} of the independent variable. The left side was then transformed into the derivative of a product:

$$e^{kx} y' + ke^{kx} y = \frac{d}{dx} (e^{kx} y)$$

Then all that was needed to arrive at a solution was to integrate resulting Equation (11.56), obtaining Equation (11.57).

Case 3: The General Linear Equation $y' + P(x)y = Q(x)$ Let us try the same idea as in Case 2: Multiply both sides of the equation by some function $\mu(x)$ so that the left side is the derivative of the product μy. That is,

$$\mu(x)[y' + P(x)y] = \frac{d}{dx} [\mu(x)y]$$
$$= \mu(x)y' + \mu'(x)y$$

The function $\mu(x)$ is not yet known, but from the last equation, it must satisfy

$$\mu(x)P(x)y = \mu'(x)y$$

or

$$\frac{\mu'(x)}{\mu(x)} = P(x) \tag{11.58}$$

We seek only *one* function $\mu(x)$ in our procedure, so assume that $\mu(x)$ is positive over the interval. Then integrating Equation (11.58) gives us

$$\ln \mu(x) = \int P(x)\,dx$$

or, exponentiating both sides,

$$\mu(x) = e^{\int P(x)\,dx} \tag{11.59}$$

That is, Equation (11.59) determines precisely a function $\mu(x)$ that will work for our procedure. Note that $\mu(x)$ defined by Equation (11.59) is indeed positive. The function $\mu(x)$ is called an **integrating factor** for linear first-order Equation (11.55).

Now, multiplying Equation (11.55) through by the integrating factor (11.59) results in

$$\mu(x)[y' + P(x)y] = \mu(x)Q(x)$$

or

$$\frac{d}{dx} [\mu(x)y] = \mu(x)Q(x) \tag{11.60}$$

In order to solve Equation (11.60), simply integrate both sides:

$$\mu(x)y = \int \mu(x)Q(x)\,dx + C \tag{11.61}$$

where $\mu(x)$ is given by Equation (11.59). We can then solve explicitly for the solution y by dividing each side of Equation (11.61) by the integrating factor $\mu(x)$.

The method of solution presented for linear first-order equations requires two integrations. The first integration in Equation (11.59) produces the integrating factor $\mu(x)$; the second leads to the general solution y from Equation (11.61). Both integrations are possible because $P(x)$ and $Q(x)$ are assumed to be continuous over the interval. Using the Fundamental Theorem of Calculus, we could easily verify that the function y defined by Equation (11.61) does satisfy the original linear Equation (11.55) (see Problem 23 at the end of this section). We will now summarize the solution method.

Solving a Linear First-Order Equation

STEP 1 Write the linear first-order equation in standard form:

$$y' + P(x)y = Q(x) \tag{11.62}$$

STEP 2 Calculate the integrating factor:

$$\mu(x) = e^{\int P(x)\,dx} \tag{11.63}$$

STEP 3 Multiply the right-hand side of Equation (11.62) by μ and integrate:

$$\int \mu(x)Q(x)\,dx + C \tag{11.64}$$

STEP 4 Write the general solution:

$$\mu(x)y = \int \mu(x)Q(x)\,dx + C \tag{11.65}$$

$$\underset{\substack{\text{integrating factor} \\ \text{from Step 2}}}{\uparrow} \qquad \underbrace{}_{\text{result of Step 3}}$$

Observe that in Step 2, we introduce no arbitrary constant of integration when determining the integrating factor μ. The reason for this is that only a *single* function is sought as an integrating factor, not an entire family of functions. Now let us apply the method to several examples.

EXAMPLE 1

Find the general solution of

$$xy' + y = e^x, \quad x > 0$$

Solution

STEP 1 We write the linear equation in standard form:

$$y' + \left(\frac{1}{x}\right)y = \left(\frac{1}{x}\right)e^x$$

Thus $P(x) = 1/x$ and $Q(x) = e^x/x$.

STEP 2 The integrating factor is

$$\mu(x) = e^{\int P(x)\,dx} = e^{\int dx/x}$$
$$= e^{\ln x} = x$$

STEP 3 We multiply the right-hand side of the equation in Step 1 by $\mu = x$ and integrate the results to get

$$\int \mu(x)Q(x)\,dx = \int x \cdot \left(\frac{1}{x}\right) e^x\,dx$$
$$= \int e^x\,dx$$
$$= e^x + C$$

STEP 4 The general solution is given by Equation (11.65):

$$xy = e^x + C$$

or

$$y = \frac{e^x + C}{x}, \quad x > 0$$

Let us verify that y does indeed solve the original equation. Differentiation of y gives us

$$y' = \frac{-1}{x^2}(e^x + C) + \frac{1}{x}e^x$$

Then

$$xy' + y = \left[\frac{-1}{x}(e^x + C) + e^x\right] + \left(\frac{1}{x}\right)(e^x + C)$$
$$= e^x$$

so the differential equation is satisfied. ■ ■ ■

EXAMPLE 2

Find the general solution of

$$y' + (\tan x)y = \cos^2 x$$

over the interval $-\pi/2 < x < \pi/2$.

Solution

STEP 1 The equation is in standard form with $P(x) = \tan x$ and $Q(x) = \cos^2 x$.

STEP 2 The integrating factor is

$$\mu(x) = e^{\int P(x)\,dx} = e^{\int \tan x\,dx} = e^{-\ln|\cos x|} = \sec x$$

since $\cos x > 0$ over the interval $-\pi/2 < x < \pi/2$.

STEP 3 Next we integrate the product $\mu(x)Q(x)$:

$$\int \sec x \cos^2 x\,dx = \int \cos x\,dx = \sin x + C$$

STEP 4 The general solution is given by

$$(\sec x)y = \sin x + C$$

or

$$y = \sin x \cos x + C \cos x$$

EXAMPLE 3

Find the solution of

$$3xy' - y = \ln x + 1, \quad x > 0$$

satisfying $y(1) = -2$.

Solution   In this example we shall omit the designation of the steps. With $x > 0$, we rewrite the equation in the standard form as

$$y' - \frac{1}{3x}y = \frac{\ln x + 1}{3x}$$

Then the integrating factor is given by

$$\mu = e^{\int -dx/3x} = e^{(-1/3)\ln x} = x^{-1/3}$$

Thus

$$x^{-1/3}y = \frac{1}{3}\int (\ln x + 1)x^{-4/3}\,dx$$

Integration by parts results in the following (with the details left to you to figure out):

$$x^{-1/3}y = -x^{-1/3}(\ln x + 1) + \int x^{-4/3}\,dx + C$$

Therefore,

$$x^{-1/3}y = -x^{-1/3}(\ln x + 1) - 3x^{-1/3} + C$$

or

$$y = -(\ln x + 4) + Cx^{1/3}$$

When $x = 1$ and $y = -2$ are substituted into the general solution, the arbitrary constant C is evaluated:

$$-2 = -(0 + 4) + C$$

or

$$C = 2$$

Thus

$$y = 2x^{1/3} - \ln x - 4$$

is the particular solution we seek.

EXAMPLE 4 *Water Pollution*

We now return to the problem of water pollution of a large lake introduced at the beginning of this section. Suppose a large lake that was formed by damming a river holds initially 100 million gallons of water. Because a nearby agricultural field was sprayed with a pesticide, the water has become contaminated. The concentration of the pesticide has been measured and is equal to 35 ppm (parts per million), or 35×10^{-6}. The river continues to flow into the lake at a rate of 300 gal/min. The river is only slightly contaminated with pesticide and has a concentration of 5 ppm. The flow of water over the dam can be controlled and is set at 400 gal/min. Assume that no additional spraying causes the lake to become even more contaminated. How long will it be before the water reaches an acceptable level of concentration equal to 15 ppm?

Solution From the opening discussion in this section, recall that

$$V(t) = V_0 + (r_{in} - r_{out})t$$

For the particular lake at hand, we are given that $V_0 = 100 \times 10^6$ and $r_{in} - r_{out} = 300 - 400 = -100$ gal/min. Thus,

$$V(t) = 100 \times 10^6 - 100t$$

represents the volume of the lake at time t. Since $r_{in} - r_{out} = -100$, note that the lake will be empty when $V(t) = 0$, or $t = 10^6$ min ≈ 1.9 yr. It is hoped that the contamination in the lake can be reduced to the acceptable level of $15/10^6$ before the lake is empty.

Using the notation introduced in the opening discussion, we have $\alpha = r_{in}c_{in} = 300(5/10^6)$. Thus, from Equation (11.48), the differential equation governing the change in pollution is given by

$$\frac{dp}{dt} + \frac{400p}{100 \times 10^6 - 100t} = 15 \times 10^{-4} \tag{11.66}$$

The integrating factor for Equation (11.66) is

$$\mu = e^{\int 4dt/(10^6 - t)} = e^{-4\ln(10^6 - t)} = (10^6 - t)^{-4}$$

assuming $t < 10^6$. Thus the solution satisfies

$$(10^6 - t)^{-4}p(t) = \int 15 \times 10^{-4}(10^6 - t)^{-4} \, dt$$
$$= 5 \times 10^{-4}(10^6 - t)^{-3} + C$$

Therefore,

$$p(t) = 5 \times 10^{-4}(10^6 - t) + C(10^6 - t)^4 \tag{11.67}$$

From the initial condition, when $t = 0$ the concentration is $c_0 = p(0)/V_0 = 35 \times 10^{-6}$. Hence

$$p(0) = (35 \times 10^{-6}) \times 100 \times 10^6 = 3500$$

By substituting this result into Equation (11.67), we evaluate the constant of integration C:

$$3500 = 5 \times 10^{-4} \times 10^6 + C \times 10^{24}$$

or $C = 3 \times 10^{-21}$. The particular solution for the level of pollution at any time $t < 10^6$ is therefore

$$p(t) = 5 \times 10^{-4}(10^6 - t) + 3 \times 10^{-21}(10^6 - t)^4 \tag{11.68}$$

The problem asks for the time t when the concentration level $c(t) = p(t)/V(t) = 15 \times 10^{-6}$. Here t is measured in minutes. Division of Equation (11.68) by $V(t)$ and application of this condition yield

$$15 \times 10^{-6} = \frac{5 \times 10^{-4}(10^6 - t) + 3 \times 10^{-21}(10^6 - t)^4}{100(10^6 - t)}$$

Simplifying algebraically, we get

$$3 \times 10^{-18}(10^6 - t)^3 - 1 = 0$$

Using a calculator or computer, we find that solving this last equation for t gives

$$t \approx 306{,}650 \text{ min} \approx 7 \text{ months}$$

We end this section with a brief discussion of the initial value problem for the linear first-order equation.

Uniqueness of Solutions

Unlike the difficulties we can encounter when solving (nonlinear) separable equations, the linear first-order equation always has one and only one solution satisfying a specified initial condition. This result is stated precisely in the following theorem.

Theorem 1 _____

Suppose that $P(x)$ and $Q(x)$ are continuous functions over the interval $\alpha < x < \beta$. Then there is one and only one function $y = y(x)$ satisfying the first-order linear equation

$$y' + P(x)y = Q(x)$$

on the interval and the initial condition

$$y(x_0) = y_0$$

at the specified point x_0 in the interval.

Theorem 1 is known as the **existence and uniqueness theorem** for the linear first-order equation. Any real value whatsoever may be assigned to y_0 and the theorem will be satisfied. Thus the particular solution found in Example 3 is the only function satisfying the differential equation and the initial condition specified there. Problems 23 and 24 outline a proof of the existence and uniqueness theorem based on the Fundamental Theorem of Calculus.

11.7 PROBLEMS _____

In Problems 1–15, find the general solution of the given first-order linear differential equation. State an interval over which the general solution is valid.

1. $y' + 2xy = x$

2. $y' - 3y = e^x$

3. $2y' - y = xe^{x/2}$

4. $\dfrac{y'}{2} + y = e^{-x} \sin x$

5. $xy' + 2y = 1 - x^{-1}$

6. $xy' - y = 2x \ln x$

7. $y' = y - e^{2x}$

8. $y' = \dfrac{2y}{x} + x^3 e^x - 1$

9. $x^2 \dfrac{dy}{dx} + xy = 2$

10. $(1 + x)\dfrac{dy}{dx} + y = \sqrt{x}$

11. $x^2\, dy + xy\, dx = (x - 1)^2\, dx$

12. $(1 + e^x)\, dy + (ye^x + e^{-x})\, dx = 0$

13. $e^{-y} \, dx + (e^{-y} x - 4y) \, dy = 0$

14. $(x + 3y^2) \, dy + y \, dx = 0$

15. $y \, dx + (3x - y^{-2} \cos y) \, dy = 0, \quad y > 0$

In Problems 16–20, solve the initial value problem.

16. $y' + 4y = 1, \quad y(0) = 1$

17. $\dfrac{dy}{dx} + 3x^2 y = x^2, \quad y(0) = -1$

18. $x \, dy + (y - \cos x) \, dx = 0, \quad y\left(\dfrac{\pi}{2}\right) = 0$

19. $xy' + (x - 2)y = 3x^3 e^{-x}, \quad y(1) = 0$

20. $y \, dx + (3x - xy + 2) \, dy = 0, \quad y(2) = -1, \, y < 0$

21. Oxygen flows through one tube into a liter flask filled with air, and the mixture of oxygen and air (considered well stirred) escapes through another tube. Assuming that air contains 21% oxygen, what percentage of oxygen will the flask contain after 5 L have passed through the intake tube?

22. If the average person breathes 20 times per minute, exhaling each time 100 in.^{3} of air containing 4% carbon dioxide. Find the percentage of carbon dioxide in the air of a 10,000-ft^3 closed room 1 hr after a class of 30 students enters. Assume that the air is fresh at the start, that the ventilators admit 1000 ft^3 of fresh air per minute, and that the fresh air contains 0.04% carbon dioxide.

23. *Existence*—Assume the hypothesis of Theorem 1.

   a. From the Fundamental Theorem of Calculus, we have

   $$\frac{d}{dx}\left[\int \mu(x)Q(x)\,dx\right] = \mu(x)Q(x)$$

   Use this fact to show that any function y given by Equation (11.61) solves linear first-order Equation (11.55). *Hint:* Differentiate both sides of Equation (11.61).

   b. If the constant C is given by

   $$C = y_0 \mu(x_0) - \int_{x_0}^{x} \mu(t)Q(t)\,dt$$

   in Equation (11.61), show that the resulting function y defined by Equation (11.61) satisfies the initial condition $y(x_0) = y_0$.

24. *Uniqueness*—Assume the hypothesis of Theorem 1, and assume that $y_1(x)$ and $y_2(x)$ are both solutions to the linear first-order equation satisfying the initial condition $y(x_0) = y_0$.

a. Verify that $y(x) = y_1(x) - y_2(x)$ satisfies the initial value problem

$$y' + P(x)y = 0, \quad y(x_0) = 0$$

b. For the integrating factor $\mu(x)$ defined by Equation (11.63), show that

$$\frac{d}{dx}(\mu(x)[y_1(x) - y_2(x)]) = 0$$

Conclude that $\mu(x)[y_1(x) - y_2(x)] \equiv$ constant.

c. From part (a), we have $y_1(x_0) - y_2(x_0) = 0$. Since $\mu(x) > 0$ for $\alpha < x < \beta$, use part (b) to establish that $y_1(x) - y_2(x) \equiv 0$ on the interval (α, β). Thus $y_1(x) = y_2(x)$ for all $\alpha < x < \beta$.

12 Modeling with Systems of Differential Equations

Introduction

Interactive situations occur in the study of economics, ecology, electrical circuits, mechanical systems, celestial mechanics, control systems, and so forth. For example, the study of the dynamics of population growth of various plants and animals is an important ecological application of mathematics. Different species interact in a variety of ways. One animal may serve as the primary food source for another, commonly referred to as a predator–prey relationship. Two species may depend on one another for mutual support, such as a bee's using a plant's nectar as food while simultaneously pollinating that plant; such a relationship is referred to as mutualism. Another possibility occurs when two or more species compete against one another for a common food source or even compete for survival. In this chapter, we develop some elementary models to explain these interactive situations, and we analyze the models using graphical techniques.

In modeling interactive situations involving the dynamics of population growth, we are interested in the answers to certain questions concerning the species under investigation. For instance, will one species eventually dominate the other and drive it to extinction? Can the species coexist? If so, will their populations reach equilibrium levels, or will they vary in some predictable fashion? Moreover, how sensitive are the answers to the preceding questions relative to the initial population levels or to external perturbations (such as natural disasters, development of chemical or biological agents used to control the populations, and the like)?

Because we are modeling the rates of change with respect to time, the models invariably involve differential equations (or, in a discrete analysis, difference equations). Even with very simple assumptions, these equations are often nonlinear and generally cannot be solved analytically, although numerical techniques exist. Nevertheless, qualitative information about the behavior of the variables can often be obtained by simple graphical analysis. We will demonstrate how graphical analysis can be used to answer questions such as those posed in the preceding paragraph. We will also point out limitations for such an analysis and conditions requiring a more sophisticated mathematical analysis. Our graphical analysis for systems of differential equations extends the graphical procedures we applied in Section 11.4 to two dimensions.

12.1 Graphical Solutions of Autonomous Systems of First-Order Differential Equations

In Chapter 11, we solved differential equations by separating the variables and integrating. Usually, however, it is not so easy to solve a system of differential equations; in fact, it is rare that we can find an analytic solution when the equations are nonlinear, although numerical solution methods exist. Therefore, it is worthwhile to consider a qualitative graphical analysis for solutions to a system of differential equations analogous to our development for single equations. We restrict our discussion to special systems involving only two first-order differential equations.

The system

$$\frac{dx}{dt} = f(x, y)$$
$$\frac{dy}{dt} = g(x, y)$$

(12.1)

is called an **autonomous** system of differential equations. In such a system, the independent variable t is absent (i.e., t does not appear explicitly on the right side of Equation (12.1)). To emphasize the physical significance of autonomous systems, think of the independent variable t as denoting time and the dependent variables as giving position (x, y) in the Cartesian plane. Thus, autonomous systems are not time dependent. In order that the system be suitably well behaved, we assume throughout our discussion that the functions f and g, together with their first partial derivatives $\partial f/\partial x, \partial f/\partial y, \partial g/\partial x$, and $\partial g/\partial y$, are all continuous over a suitable region of the xy-plane.

It is useful to think of a solution to the autonomous system (12.1) as a curve in the xy-plane. That is, a **solution** to Equation (12.1) is a pair of parametric equations, $x = x(t)$ and $y = y(t)$, whose derivatives satisfy the system. The solution curve whose coordinates are $(x(t), y(t))$, as t varies over time, is called a **trajectory**, **path**, or **orbit** of the system. The xy-plane is referred to as the **phase plane**. It is convenient to think of a trajectory as the path of a moving particle and we appeal to this idea throughout this chapter. Note that as the particle moves through the phase plane with increasing t, the direction it moves from a point (x, y) depends only on the coordinates (x, y) and not on the time of arrival.

If (x, y) is a point in the phase plane for which $f(x, y) = 0$ and $g(x, y) = 0$ simultaneously, then both the derivatives dx/dt and dy/dt are zero. Hence, there is no motion in either the x or the y direction, and the particle is stationary. Such a point is called a **rest point**, or **equilibrium point**, of the system. Notice that whenever (x_0, y_0) is a rest point of the system (12.1), the equations $x = x_0$ and $y = y_0$ give a solution to the system. In fact, this constant solution is the only one passing through the point (x_0, y_0) in the phase plane. The trajectory associated with this solution is simply the rest point (x_0, y_0). A trajectory $x = x(t)$, $y = y(t)$ is said to approach the rest point (x_0, y_0) if $x(t) \to x_0$ and $y(t) \to y_0$ as $t \to \infty$. In applications it is of interest to see what happens to a trajectory when it comes near a rest point.

The idea of stability is central to any discussion of the behavior of trajectories near a rest point. Roughly, the rest point (x_0, y_0) is **stable** if any trajectory that starts close to the point stays close to it for all future time. It is **asymptotically stable** if it is stable and if any trajectory that starts close to (x_0, y_0) approaches that point as t tends to infinity. If it is not stable, the rest point is said to be **unstable**. These notions will be clarified when we examine specific modeling applications later in the chapter. Our goal here is to have a language with which to discuss qualitatively our differential equations models. It is not our intent to study the theoretical aspects of stability, which would require greater mathematical precision than we have presented here.

The following results are useful in investigating solutions to the autonomous system (12.1) We offer these results without proof:

1. There is at most one trajectory through any point in the phase plane.

2. A trajectory that starts at a point other than a rest point cannot reach a rest point in a finite amount of time.

3. No trajectory can cross itself unless it is a closed curve. If it is a closed curve, it is a periodic solution.

The implications of these three properties are that from a starting point that is not a rest point, the resulting motion

a. will move along the same trajectory regardless of the starting time;

b. cannot return to the starting point unless the motion is periodic;

c. can never cross another trajectory; and

d. can only approach (never reach) a rest point.

Therefore, the resulting motion of a particle along a trajectory behaves in one of three possible ways: (1) The particle approaches a rest point; (2) the particle moves along or approaches asymptotically a closed path; or (3) at least one of the trajectory components, $x(t)$ or $y(t)$, becomes arbitrarily large as t tends to infinity.

EXAMPLE 1   *A Linear Autonomous System*

The pair of functions

$$x = e^{-t} \sin t$$
$$y = e^{-t} \cos t$$

solve the linear autonomous system

$$\frac{dx}{dt} = -x + y$$

$$\frac{dy}{dt} = -x - y$$

They are readily verified to be solutions by differentiating x and y and showing that the differential equations in the system are satisfied:

$$\frac{dx}{dt} = \frac{d}{dt}(e^{-t} \sin t) = -e^{-t} \sin t + e^{-t} \cos t$$

$$= -x + y$$

and

$$\frac{dy}{dt} = \frac{d}{dt}(e^{-t}\cos t) = -e^{-t}\cos t - e^{-t}\sin t$$

$$= -y - x$$

If simultaneously $dx/dt = 0$ and $dy/dt = 0$, then $x = y = 0$. Thus, the origin $(0, 0)$ is the only rest point of the system. Because

$$x^2 + y^2 = e^{-2t}\sin^2 t + e^{-2t}\cos^2 t = e^{-2t}$$

each trajectory is a circular spiral of decreasing radius around and approaching the origin as t approaches plus infinity. Therefore, $(0, 0)$ is an asymptotically stable rest point. A typical trajectory to the system, starting from the initial position $x(0) = x_0$ and $y(0) = y_0$ in the phase plane, is shown in Figure 12.1. ■ ■ ■

■ **Figure 12.1**

The origin is an asymptotically stable rest point (Example 1).

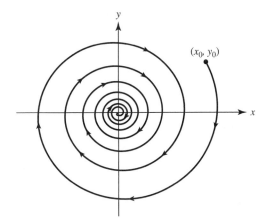

A spiral around the rest point $(0, 0)$ for the system
$dx/dt = -x + y$
$dy/dt = -x - y$

EXAMPLE 2 A Nonlinear Autonomous System

For $-\infty < t < \infty$, the two function pairs

$$x = -a \text{ csch } at \tag{12.2}$$
$$y = -a \text{ coth } at$$

and

$$x = a \text{ csch } at \tag{12.3}$$
$$y = -a \text{ coth } at$$

satisfy the nonlinear autonomous system

$$dx/dt = xy \tag{12.4}$$
$$dy/dt = x^2$$

These results are easily verified from the differentiation formulas

$$\frac{d}{dt}(a \text{ csch } at) = -a^2 \text{ csch } at \text{ coth } at$$

$$\frac{d}{dt}(-a \text{ coth } at) = a^2 \text{ csch}^2 at$$

Each of the function pairs (12.2) and (12.3) satisfy the equation

$$y^2 - x^2 = a^2$$

because of the property that

$$\coth^2 u - \text{csch}^2 u = 1$$

Thus, function pairs (12.2) and (12.3) represent upper and lower branches of the hyperbolas displayed in Figure 12.2. The straight half-lines $y = x$ and $y = -x$, $x \neq 0$, are also solutions to system (12.4).

■ **Figure 12.2**

Several trajectories for the nonlinear autonomous system $dx/dt = xy$ and $dy/dt = x^2$ (Example 2)

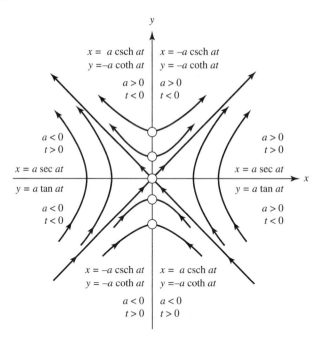

Let us find the rest points for system (12.4). Because $dx/dt = 0$ and $dy/dt = 0$ hold simultaneously whenever $x = 0$, all points along the y axis are rest points for the system. We now classify these rest points.

Consider first the origin. If we select an initial point $(a, 0)$ close to the origin such that $a > 0$, for instance, then as t increases, the branch of the hyperbola trajectory passing through that point is traversed upward and to the right, as indicated by the arrows in Figure 12.2. In fact, note that as t approaches $\pi/2a$, both x and y approach plus infinity. Therefore, the rest point $(0, 0)$ is unstable.

If $(0, a)$ is any point along the y axis for $a > 0$, the point $(0, a)$ is unstable. For instance, whenever a particle starts along a trajectory near $(0, a)$, the trajectory is traversed upward and away from the point $(0, a)$, as indicated by the trajectories shown in Figure 12.2; whether the trajectory moves to the left or to the right depends on whether the second or first quadrant is selected for the starting point.

The rest points $(0, b)$, where $b < 0$, are of an entirely different nature. If a starting point in the phase plane is near $(0, b)$, the trajectory through that point approaches some point $(0, a)$ on the negative y axis as t approaches plus infinity. For instance, a particle starting very close to and to the left of the point $(0, b)$ will move along some trajectory given by

$$y = -a \text{ csch } at$$

$$y = -a \text{ coth } at$$

where $a < 0$ and $t > 0$ (Figure 12.2). As $t \to +\infty$, we have $at \to -\infty$. Then (csch at) $\to 0^-$, so $x \to 0^-$. Also, (coth at) $\to -1$, so $y \to a$. As time advances, the moving particle gets increasingly close to but never reaches the rest point $(0, a)$. If $(0, a)$ is close to $(0, b)$, then a particle that starts near $(0, b)$ stays near it for all future time (because the particle will also be near the point $(0, a)$). Thus, the rest points along the negative y axis are stable. They are not asymptotically stable, however, because *not every* trajectory that starts near $(0, b)$ approaches $(0, b)$. Only two trajectories have that property; namely, the left and right branches of the lower hyperbola that approach that rest point, as indicated in Figure 12.2. It is important to observe that no trajectory can ever cross the y axis, because every point $(0, b)$ along the y axis is a rest point and hence a solution to the autonomous system. We apply the ideas discussed previously to the models we develop in the next several sections. ■ ■ ■

12.1 PROBLEMS

In Problems 1–4, verify that the given function pair is a solution to the first-order system.

1. $x = -e^t, \quad y = e^t$

$$\frac{dx}{dt} = -y, \quad \frac{dy}{dt} = -x$$

2. $x = -\dfrac{1}{2} + \dfrac{e^{2t}}{2}, \quad y = -\dfrac{3}{4} + \dfrac{3e^{2t}}{8} + \dfrac{3e^{-2t}}{8}$

$$\frac{dx}{dt} = 2x + 1, \quad \frac{dy}{dt} = 3x - 2y$$

3. $x = e^{2t}, \quad y = e^t$

$$\frac{dx}{dt} = 2y^2, \quad \frac{dy}{dt} = y$$

4. $x = b \tanh bt, y = b \text{ sech } bt, b = $ any real number

$$\frac{dx}{dt} = y^2, \quad \frac{dy}{dt} = xy$$

In Problems 5–8, find and classify the rest points of the given autonomous system.

5. $\dfrac{dx}{dt} = 2y, \quad \dfrac{dy}{dt} = -3x$

6. $\dfrac{dx}{dt} = -(y - 1), \quad \dfrac{dy}{dt} = x - 2$

7. $\dfrac{dx}{dt} = -y(y-1), \quad \dfrac{dy}{dt} = (x-1)(y-1)$

8. $\dfrac{dx}{dt} = \dfrac{1}{y}, \quad \dfrac{dy}{dt} = \dfrac{1}{x}$

9. Sketch a number of trajectories corresponding to the following autonomous systems, and indicate the direction of motion for increasing t. Identify and classify any rest points as being stable, asymptotically stable, or unstable.

   a. $dx/dt = x, \quad dy/dt = y$

   b. $dx/dt = -x, \quad dy/dt = 2y$

   c. $dx/dt = y, \quad dy/dt = -2x$

   d. $dx/dt = -x + 1, \quad dy/dt = -2y$

12.1 PROJECT

1. Complete the requirements of the UMAP module "Whales and Krill: A Mathematical Model," by Raymond N. Greenwell, UMAP 610. This module models a predator–prey system involving whales and krill by a system of differential equations. Although the equations are not solvable, information is extracted using dimensional analysis and the study of equilibrium points. The concept of maximum sustainable yield is introduced and used to draw conclusions about fishing strategies. You will learn to construct a differential equations model, remove dimensions from a set of equations, find equilibrium points of a system of differential equations and learn their significance, and practice manipulative skills in algebra and calculus.

12.2 A Competitive Hunter Model[1]

Up to now we have seen how single species growth can be modeled as the Malthusian model or the limited growth model. Let's turn our attention to how two different species might compete for common resources.

Problem Identification   Imagine a small pond that is mature enough to support wildlife. We desire to stock the pond with game fish, say trout and bass. Let $x(t)$ denote the population of the trout at any time t, and let $y(t)$ denote the bass population. *Is coexistence of the two species in the pond possible? If so, how sensitive is the final solution of population levels to the initial stockage levels and external perturbations?*

Assumptions   The level of the trout population $x(t)$ depends on many variables: the initial level x_0, the amount of competition for limited resources, the existence of predators, and so forth. Initially, we assume that the environment can support an unlimited number of trout

---

[1]This section is adapted from UMAP Unit 628, based on the work of Stanley C. Leja and one of the authors. The adaptation is presented with the permission of COMAP, Inc., 57 Bedford St., Lexington, MA 02420.

so that in isolation

$$\frac{dx}{dt} = ax \quad \text{for } a > 0$$

(Later we may find it desirable to refine the model and use a limited growth assumption.) Next, we modify the preceding differential equation to take into account the competition of the trout with the bass population for living space and a common food supply. The effect of the bass population is to decrease the growth rate of the trout population. This decrease is approximately proportional to the number of possible interactions between the two species, so one submodel is to assume that the decrease is proportional to the product of x and y. These considerations are modeled by the equation

$$\frac{dx}{dt} = ax - bxy = (a - by)x \tag{12.5}$$

The **intrinsic growth rate** $k = a - by$ decreases as the level of the bass population increases. The constants a and b indicate the degrees of self-regulation of the trout population and its competition with the bass population, respectively. These coefficients must be determined experimentally or by analyzing historical data.

The situation for the bass population is analyzed in the same manner. Thus, we obtain the following autonomous system of two first-order differential equations for our model:

$$\frac{dx}{dt} = (a - by)x \qquad \frac{dy}{dt} = (m - nx)y \tag{12.6}$$

where $x(0) = x_0$, $y(0) = y_0$, and a, b, m, and n are all positive constants. This model is useful in studying the growth patterns of species exhibiting competitive behavior such as the trout and bass.

Graphical Analysis of the Model One of our concerns is whether the trout and bass populations reach equilibrium levels. If so, then we will know whether coexistence of the two species in the pond is possible. The only way such a state can be achieved is for both populations to stop growing; that is, $dx/dt = 0$ and $dy/dt = 0$. Thus, we seek the rest points or equilibrium points of the system (12.6).

Setting the right sides of Equations (12.6) equal to zero and solving for x and y simultaneously, we find the rest points $(x, y) = (0, 0)$ and $(x, y) = (m/n, a/b)$ in the phase plane. Along the vertical line $x = m/n$ and the x axis in the phase plane, the growth dy/dt in the bass population is zero; along the horizontal line $y = a/b$ and the y axis, the growth dx/dt in the trout population is zero. If the initial stockage were at these rest point levels, there would be no growth in either population. These features are depicted in Figure 12.3.

Considering the approximations necessary in any model, it is inconceivable that we would estimate precisely the values for the constants a, b, m, and n in our system (12.6). Therefore, the pertinent behavior we need to investigate is what happens to the solution trajectories in the vicinity of the rest points $(0, 0)$ and $(m/n, a/b)$. Specifically, are these points stable or unstable?

To investigate this question graphically, let's analyze the signs of dx/dt and dy/dt in the phase plane. (Although $x(t)$ and $y(t)$ represent the trout and bass populations, respectively,

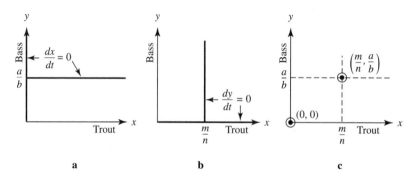

■ Figure 12.3

Rest points of the competitive hunter model given by the System (12.6)

it is helpful to think of the trajectories as paths of a moving particle, in accord with our discussion in the preceding section.) Whenever dx/dt is positive, the horizontal component $x(t)$ of the trajectory is increasing and the particle is moving toward the right; whenever dx/dt is negative, the particle is moving to the left. Likewise, if dy/dt is positive, the component $y(t)$ is increasing and the particle is moving upward; if dy/dt is negative, the particle is moving downward. In our system (12.6), the vertical line $x = m/n$ divides the phase plane into two half-planes. In the left half-plane dy/dt is positive, and in the right half-plane it is negative. The directions of the associated trajectories are indicated in Figure 12.4. Likewise, the horizontal line $y = a/b$ determines the half-planes where dx/dt is positive or negative. The directions of the associated trajectories are indicated in Figure 12.5. Along the line $y = a/b$, $dx/dt = 0$. Therefore, any trajectory crossing this line will do so vertically. Similarly, along the line $x = m/n$, $dy/dt = 0$, so the line will be crossed horizontally. Finally, along the y axis, motion must be vertical, and along the x axis, motion must be horizontal. Combining all this information into a single graph gives the four distinct regions A–D with their respective trajectory directions as depicted in Figure 12.6.

■ Figure 12.4

To the left of $x = m/n$ the trajectories move upward; to the right they move downward.

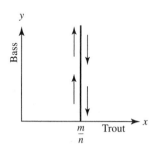

■ Figure 12.5

Above the line $y = a/b$ the trajectories move to the left; below the line they move to the right.

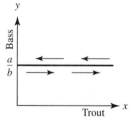

■ Figure 12.6

Composite graphical
analysis of the trajectory
directions in the four
regions determined by
$x = m/n$ and $y = a/b$

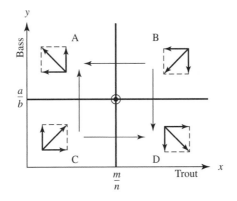

Now let us analyze the motion in the vicinity of the rest points. For $(0, 0)$, we can see that all motion is away from it—upward and toward the right. In the vicinity of the rest point $(m/n, a/b)$, the behavior depends on the region in which the trajectory begins. If the trajectory starts in region B, for instance, then it will move downward and leftward toward the rest point. However, as it gets nearer to the rest point, the derivatives dx/dt and dy/dt approach zero. Depending on where the trajectory begins and on the relative sizes of the constants a, b, m, and n, either the trajectory will continue moving downward and into region D as it swings past the rest point or it will move leftward into region A. Once it enters either one of these latter two regions, it will move away from the rest point. Thus, both rest points are unstable. These features are suggested in Figure 12.7.

■ Figure 12.7

Motion along the
trajectories near the rest
points (0,0) and (m/n, a/b)

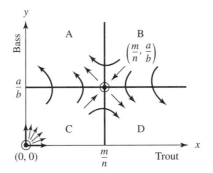

Model Interpretation   Now let us consider the half-planes $y < a/b$ and $y > a/b$. In each half-plane there is exactly one trajectory approaching the rest point $(m/n, a/b)$. A proof of this fact is outlined in Problem 7. Above these two trajectories the bass population increases, and below them the bass population decreases. The trajectory for $y < a/b$ is shown as a line joining $(0, 0)$ to $(m/n, a/b)$ in Figure 12.8 for simplicity, but it is not likely to be a line.

The graphical analysis conducted so far leads us to the preliminary conclusion that under the assumptions of our model, it is highly unlikely for both species to reach equilibrium levels. Furthermore, the initial stockage levels turn out to be important in determining which of the two species might survive. Perturbations of the system may also affect the outcome of the competition. Thus, mutual coexistence of the species is highly improbable. This phenomenon is known as the **principle of competitive exclusion**, or Gause's

■ **Figure 12.8**

Qualitative results of analyzing the competitive hunter model. There are exactly two trajectories approaching the point $(m/n, a/b)$.

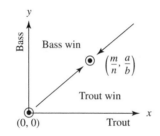

principle.[2] Moreover, the initial conditions completely determine the outcome, as depicted in Figure 12.8. We can see from that graph that any perturbation causing a switch from one region (e.g., below the two trajectories approaching the rest point $(m/n, a/b)$) to the other region (above the two trajectories) would change the outcome. One of the limitations of our graphical analysis is that we have not determined those separating trajectories precisely. If we are satisfied with our model, we may very well want to determine that separating boundary for the two regions.

■ **Figure 12.9**

Trajectory direction near a rest point

Limitations of a Graphical Analysis   It is not always possible to determine the nature of the motion near a rest point using only graphical analysis. To understand this limitation, consider the rest point and the direction of motion of the trajectories shown in Figure 12.9. The information given in the figure is insufficient to distinguish between the three possible motions shown in Figure 12.10. Moreover, even if we have determined by some other means that Figure 12.10c correctly portrays the motion near the rest point, we might be tempted to deduce that the motion will grow without bound in both the x and y directions. However, consider the system given by

$$\frac{dx}{dt} = y + x - x(x^2 + y^2)$$
$$\frac{dy}{dt} = -x + y - y(x^2 + y^2)$$

(12.7)

■ **Figure 12.10**

Three possible trajectory motions: (a) periodic motion, (b) motion toward an asymptotically stable rest point, and (c) motion near an unstable rest point

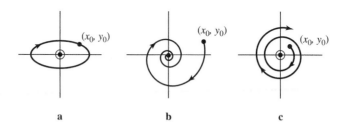

It can be shown that $(0, 0)$ is the only rest point for Equations (12.7). However, any trajectory starting on the unit circle $x^2 + y^2 = 1$ will traverse the unit circle in a periodic solution

---

[2]Named after G. F. Gause, who furthered the work of Joseph Grinnel, Alfred Lotka, and Vita Volterra in population ecology with his book *The Struggle for Existence* (Baltimore: Williams & Wilkins, 1934). Actually, it was Grinnel who first expressed the exclusion principle in 1904. For an interesting historical account, see the article by G. Hardin, "The Competitive Exclusion Principle," *Science* 131(1960):1291–1297.

because in that case $dy/dx = -x/y$ (see Problem 2 in the problem set). Moreover, if a trajectory starts inside the circle (provided it does not start at the origin), it will spiral outward asymptotically, getting increasingly closer to the circular path as t tends to infinity. Likewise, if the trajectory starts outside the circular region, it will spiral inward and again approach the circular path asymptotically. The solution $x^2 + y^2 = 1$ is called a **limit cycle**. The trajectory behavior is sketched in Figure 12.11. Thus, if the system (12.7) models population behavior for two competing species, we would have to conclude that the population levels will eventually be periodic. This example illustrates that the results of a graphical analysis are useful for determining the motion in the immediate vicinity of an equilibrium point only. (Here we have assumed that negative values for x and y have a physical meaning in Figure 12.11, or that the point $(0, 0)$ represents the translation of a rest point from the first quadrant to the origin.)

■ **Figure 12.11**

The solution $x^2 + y^2 = 1$ to (12.7) is a limit cycle.

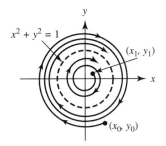

12.2 PROBLEMS

1. List three important considerations that are ignored in the development of the competitive hunter model presented in this section.

2. For the system (12.7), show that any trajectory starting on the unit circle $x^2 + y^2 = 1$ will traverse the unit circle in a periodic solution. First introduce polar coordinates and rewrite the system as $dr/dt = r(1 - r^2)$ and $d\theta/dt = 1$.

3. Develop a model for the growth of trout and bass, assuming that in isolation trout demonstrate exponential decay (so that $a < 0$ in Equation (12.6)) and that the bass population actually grows logistically with a population limit M. Analyze graphically the motion in the vicinity of the rest points in your model. Is coexistence possible?

4. How might the competitive hunter model (Equations (12.6)) be validated? Include a discussion of how the various constants a, b, m, and n might be estimated. How could state conservation authorities use the model to ensure the survival of both species?

5. Consider the competitive hunter model defined by

$$\frac{dx}{dt} = a(1 - x/k_1)x - bxy$$

$$\frac{dy}{dt} = m(1 - y/k_2)y - nxy$$

where x represents the trout population and y the bass population.

a. What assumptions are implicitly being made about the growth of trout and bass in the absence of competition?

b. Interpret the constants a, b, m, n, k_1, and k_2 in terms of the physical problem.

c. Perform a graphical analysis and answer the following questions:

    i. What are the possible equilibrium levels?

    ii. Is coexistence possible?

    iii. Pick several typical starting points and sketch typical trajectories in the phase plane.

    iv. Interpret the outcomes predicted by your graphical analysis in terms of the constants a, b, m, n, k_1, and k_2.

Note: When you get to step (i), you should realize that at least five cases exist. You will need to analyze all five cases. One case is when the lines are coincident.

6. Consider the following economic model: Let P be the price of a single item on the market. Let Q be the quantity of the item available on the market. Both P and Q are functions of time. If we consider price and quantity as two interacting species, the following model might be proposed:

$$\frac{dP}{dt} = aP(b/Q - P)$$

$$\frac{dQ}{dt} = cQ(fP - Q)$$

where a, b, c, and f are positive constants. Justify and discuss the adequacy of the model.

a. If $a = 1$, $b = 20{,}000$, $c = 1$, and $f = 30$, find the equilibrium points of this system. Classify each equilibrium point with respect to its stability, if possible. If a point cannot be readily classified, explain why.

b. Perform a graphical stability analysis to determine what will happen to the levels of P and Q as time increases.

c. Give an economic interpretation of the curves that determine the equilibrium points.

7. Show that the two trajectories leading to $(m/n, a/b)$ shown in Figure 12.8 are unique.

a. From system (12.6) derive the following equation:

$$\frac{dy}{dx} = \frac{(m - nx)y}{(a - by)x}$$

b. Separate variables, integrate, and exponentiate to obtain

$$y^a e^{-by} = K x^m e^{-nx}$$

where K is a constant of integration.

c. Let $f(y) = y^a/e^{by}$ and $g(x) = x^m/e^{nx}$. Show that $f(y)$ has a unique maximum of $M_y = (a/eb)^a$ when $y = a/b$ as shown in Figure 12.12. Similarly, show that $g(x)$ has a unique maximum $M_x = (x/en)^m$ when $x = m/n$, also shown in Figure 12.12.

Figure 12.12

Graphs of the functions
$f(y) = y^a/e^{by}$ and
$g(x) = x^m/e^{nx}$

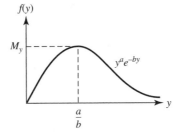

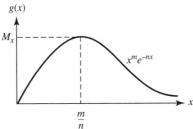

d. Consider what happens as (x, y) approaches $(m/n, a/b)$. Take limits in part (b) as $x \to m/n$ and $y \to a/b$ to show that

$$\lim_{\substack{y \to a/b \\ x \to m/n}} \left[\left(\frac{y^a}{e^{by}} \right) \left(\frac{e^{nx}}{x^m} \right) \right] = K$$

or $M_y/M_x = K$. Thus, any solution trajectory that approaches $(m/n, a/b)$ must satisfy

$$\frac{y^a}{e^{by}} = \left(\frac{M_y}{M_x} \right) \left(\frac{x^m}{e^{nx}} \right)$$

e. Show that only one trajectory can approach $(m/n, a/b)$ from below the line $y = a/b$. Pick $y_0 < a/b$. From Figure 12.12 you can see that $f(y_0) < M_y$, which implies that

$$\frac{M_y}{M_x} \left(\frac{x^m}{e^{nx}} \right) = y_0^a/e^{by_0} < M_y$$

This in turn implies that

$$\frac{x^m}{e^{nx}} < M_x$$

Figure 12.12 tells you that for $g(x)$ there is a unique value $x_0 < m/n$ satisfying this last inequality. That is, for each $y < a/b$ there is a unique value of x satisfying the equation in part (d). Thus, there can exist only one trajectory solution approaching $(m/n, a/b)$ from below, as shown in Figure 12.13.

■ **Figure 12.13**

For any $y < a/b$, only one solution trajectory leads to the rest point $(m/n, a/b)$.

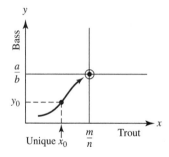

f. Use a similar argument to show that the solution trajectory leading to $(m/n, a/b)$ is unique if $y_0 > a/b$.

12.2 PROJECTS

Complete the requirements of the referenced UMAP modules (see enclosed CD).

1. "The Budgetary Process: Incrementalism," UMAP 332: "The Budgetary Process: Competition," UMAP 333, by Thomas W. Likens. The politics of budgeting revolve around the allocation of limited resources to agencies and groups competing for them. UMAP 332 develops a model to explain how levels of appropriation change from one period to the next if Congress and federal agencies determine new budgets by making marginal adjustments in the status quo. The model assumes that the share received by one agency will not affect or depend on the share received by another agency. In UMAP 333 the model is refined to address the conflictive nature of politics and the necessary interdependence of budgetary decisions.

2. "The Growth of Partisan Support I: Model and Estimation," UMAP 304; "The Growth of Partisan Support II: Model Analytics," UMAP 305, by Carol Weitzel Kohfeld. UMAP 304 presents a simple model of political mobilization, refined to include the interaction between supporters of a particular party and recruitable nonsupporters. UMAP 305 investigates the mathematical properties of the first-order quadratic difference equation model. The model is tested using data from three U.S. counties. An understanding of linear first-order difference equations with constant coefficients is required.

3. "Random Walks: An Introduction to Stochastic Processes," by Ron Barnes, UMAP 520. This module introduces random walks by an example of a gambling game. It develops and solves the associated finite difference equation, while introducing the concept of expected gain. Generalizations to Markov chains and continuous processes are discussed. Applications in the life sciences and genetics are noted.

12.2 Further Reading

Tuchinsky, Philip M. *Man in Competition with the Spruce Budworm*, UMAP Expository Monograph. The population of tiny caterpillars periodically explodes in the evergreen forests of eastern Canada and Maine. They devour the trees' needles and cause great damage to forests that are central to the economy of the region. The province of New Brunswick is using mathematical models of the budworm–forest interaction in an effort to plan for and control the damage. The monograph surveys the ecological situation and examines the computer simulation and differential equation models that are currently in use.

12.3 A Predator–Prey Model

In this section we study a model of population growth of two species in which one species is the primary food source for the other. One example of such a situation occurs in the Southern Ocean, where the baleen whales eat the fish known as the Antarctic krill, *Euphausia superboa*, as their principal food source. Another example is wolves and rabbits in a closed forest: The wolves eat the rabbits for their principal food source, and the rabbits eat vegetation in the forest. Still other examples include sea otters as predators and abalone as prey and the ladybird beetle, *Novius cardinalis*, as predator and the cottony cushion insect, *Icerya purchasi*, as prey.

Problem Identification   Let's take a closer look at the situation of the baleen whales and Antarctic krill. The whales eat the krill, and the krill live on the plankton in the sea. If the whales eat so many krill that the krill cease to be abundant, the food supply of the whales is greatly reduced. Then the whales will starve or leave the area in search of a new supply of krill. As the population of baleen whales dwindles, the krill population makes a comeback because not so many of them are being eaten. As the krill population increases, the food supply for the whales grows and, consequently, so does the baleen whale population. Also, more baleen whales are eating increasingly more krill again. *In the pristine environment, does this cycle continue indefinitely or does one of the species eventually die out*? The baleen whales in the Southern Ocean have been overexploited to the extent that their current population is approximately one-sixth its estimated pristine level. Thus, there appears to be a surplus of Antarctic krill. (Already, approximately 100,000 tons of krill are being harvested annually.) What effect does exploitation of the whales have on the balance between the whale and krill populations? What are the implications that a krill fishery may hold for the depleted stocks of baleen whales and for other species, such as seabirds, penguins, and fish, that depend on krill for their main source of food? The ability to answer such questions is important to management of multispecies fisheries. Let's see what answers can be obtained from a graphical modeling approach.

Assumptions   Let $x(t)$ denote the Antarctic krill population at any time t, and let $y(t)$ denote the population of baleen whales in the Southern Ocean. The level of the krill population depends on a number of factors, including the ability of the ocean to support them, the existence of competitors for the plankton they ingest, and the presence and levels of predators. As a rough first model, let's start by assuming that the ocean can support an unlimited number of krill so that

$$\frac{dx}{dt} = ax \quad \text{for } a > 0$$

(Later we may want to refine the model with a limited growth assumption. This refinement is presented in UMAP 610 described in the 12.1 Projects section.) Second, assume the krill are eaten primarily by the baleen whales (so neglect any other predators). Then the growth rate of the krill is diminished in a way that is proportional to the number of interactions between them and the baleen whales. One interaction assumption leads to the differential equation

$$\frac{dx}{dt} = ax - bxy = (a - by)x \tag{12.8}$$

Notice that the **intrinsic growth rate** $k = a - by$ decreases as the level of the baleen whale population increases. The constants a and b indicate the degrees of self-regulation of the krill population and the predatoriness of the baleen whales, respectively. These coefficients must be determined experimentally or from historical data. So far, Equation (12.8) governing the growth of the krill population looks just like either of the equations in the competitive hunter model presented in the preceding section.

Next, consider the baleen whale population $y(t)$. In the absence of krill the whales have no food, so we will assume that their population declines at a rate proportional to their numbers. This assumption produces the exponential decay equation

$$\frac{dy}{dt} = -my \quad \text{for } m > 0$$

However, in the presence of krill the baleen whale population increases at a rate proportional to the interactions between the whales and their krill food supply. Thus, the preceding equation is modified to give

$$\frac{dy}{dt} = -my + nxy = (-m + nx)y \tag{12.9}$$

Notice from Equation (12.9) that the **intrinsic growth rate** $r = -m + nx$ of the whales increases as the level of the krill population increases. The positive coefficients m and n would be determined experimentally or from historical data. Putting results (12.8) and (12.9) together gives the following autonomous system of differential equations for our predator–prey model:

$$\frac{dx}{dt} = (a - by)x \tag{12.10}$$

$$\frac{dy}{dt} = (-m + nx)y$$

where $x(0) = x_0$, $y(0) = y_0$, and a, b, m, and n are all positive constants. The system (12.10) governs the interaction of the baleen whales and Antarctic krill populations under our unlimited growth assumptions and in the absence of other competitors and predators.

Graphical Analysis of the Model   Let's determine whether the krill and whale populations reach equilibrium levels. The rest points or equilibrium levels occur when $dx/dt = dy/dt = 0$. Setting the right side of Equations (12.10) equal to zero and solving for x and y simultaneously give the rest points $(x, y) = (0, 0)$ and $(x, y) = (m/n, a/b)$. Along the vertical line $x = m/n$ and the x axis in the phase plane, the growth dy/dt in the baleen whale population is zero; along the horizontal line $y = a/b$ and the y axis, the growth dx/dt in the krill population is zero. These features are depicted in Figure 12.14.

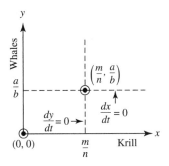

■ **Figure 12.14**

Rest points of the predator–prey model given by the system (12.10)

Because the values for the constants a, b, m, and n in system (12.10) will be only estimates, we need to investigate the behavior of the solution trajectories near the two rest points $(0, 0)$ and $(m/n, a/b)$. Thus, we analyze the signs of dx/dt and dy/dt in the phase plane. In system (12.10), the vertical line $x = m/n$ divides the phase plane into two half-planes. In the left half-plane dy/dt is negative, and in the right half-plane it is positive. In a similar way, the horizontal line $y = a/b$ determines two half-planes. In the upper half-plane dx/dt is negative, and in the lower half-plane it is positive.

The directions of the associated trajectories are indicated in Figure 12.15. Along the y axis, motion must be vertical and toward the rest point $(0, 0)$, and along the x axis, motion must be horizontal and away from the rest point $(0, 0)$.

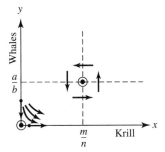

■ **Figure 12.15**

Trajectory directions in the predator–prey model

From Figure 12.15 you can see that the rest point $(0, 0)$ is unstable. Entrance to the rest point is along the line $x = 0$, where there is no krill population. Thus, the whale population declines to zero in the absence of its primary food supply. All other trajectories recede from the rest point. The rest point $(m/n, a/b)$ is more complicated to analyze. The information the figure gives is insufficient to distinguish between the three possible motions shown in Figure 12.10 of the preceding section. We cannot tell whether the motion is periodic, asymptotically stable, or unstable. Thus, we must perform a further analysis.

An Analytic Solution of the Model   Because the number of baleen whales depends on the number of Antarctic krill available for food, we assume that y is a function of x. Then from the chain rule for derivatives, we have

$$\frac{dy}{dx} = \frac{dy/dt}{dx/dt}$$

or

$$\frac{dy}{dx} = \frac{(-m + nx)y}{(a - by)x} \tag{12.11}$$

Equation (12.11) is a separable first-order differential equation and may be rewritten as

$$\left(\frac{a}{y} - b\right) dy = \left(n - \frac{m}{x}\right) dx \tag{12.12}$$

Integration of each side of Equation (12.12) yields

$$a \ln y - by = nx - m \ln x + k_1$$

or

$$a \ln y + m \ln x - by - nx = k_1$$

where k_1 is a constant.

Using properties of the natural logarithm and exponential functions, this last equation can be rewritten as

$$\frac{y^a x^m}{e^{by+nx}} = K \tag{12.13}$$

where K is a constant. Equation (12.13) defines the solution trajectories in the phase plane. We now show that these trajectories are closed and represent periodic motion.

Periodic Predator–Prey Trajectories   Equation (12.13) can be rewritten as

$$\left(\frac{y^a}{e^{by}}\right) = K\left(\frac{e^{nx}}{x^m}\right) \tag{12.14}$$

Let's determine the behavior of the function $f(y) = y^a/e^{by}$. Using the first derivative test (see Problem 1 in the problem set), we can easily show that $f(y)$ has a relative maximum at $y = a/b$ and no other critical points. For simplicity of notation, call this maximum value M_y. Moreover, $f(0) = 0$, and from l'Hôpital's rule, $f(y)$ approaches 0 as y tends to infinity. Similar arguments apply to the function $g(x) = x^m/e^{nx}$, which achieves its maximum value M_x at $x = m/n$. The graphs of the functions f and g are depicted in Figure 12.16.

■ **Figure 12.16**

Graphs of the functions
$f(y) = y^a/e^{by}$ and
$g(x) = x^m/e^{nx}$

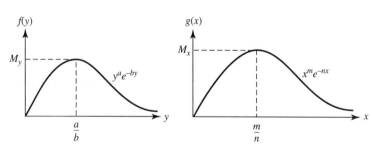

From Figure 12.16, the largest value for $y^a e^{-by} x^m e^{-nx}$ is $M_y M_x$. That is, Equation (12.14) has no solutions if $K > M_y M_x$ and exactly one solution, $x = m/n$ and $y = a/b$, when $K = M_y M_x$. Let's consider what happens when $K < M_y M_x$.

Suppose $K = s M_y,$ where $s < M_x$ is a positive constant. Then the equation

$$x^m e^{-nx} = s$$

Figure 12.17

The equation $x^m e^{-nx} = s$ has exactly two solutions for $s < M_x$.

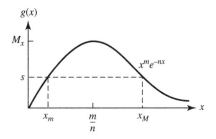

has exactly two solutions: $x_m < m/n$ and $x_M > m/n$ (Figure 12.17). Now if $x < x_m$, then $x^m e^{-nx} < s$ so that $s e^{nx} x^{-m} > 1$ and

$$f(y) = y^a e^{-by} = K e^{nx} x^{-m} = s M_y e^{nx} x^{-m} > M_y$$

Therefore, there is no solution for y in Equation (12.14) when $x < x_m$. Likewise, there is no solution when $x > x_M$. If $x = x_m$ or $x = x_M$, Equation (12.14) has exactly the one solution $y = a/b$.

Finally, if x lies between x_m and x_M, Equation (12.14) has exactly two solutions. The smaller solution $y_1(x)$ is less than a/b, and the larger solution $y_2(x)$ is greater than a/b. This situation is depicted in Figure 12.18. Moreover, as x approaches either x_m or x_M, $f(y)$ approaches M_y so that both $y_1(x)$ and $y_2(x)$ approach a/b. It follows that the trajectories defined by Equation (12.14) are periodic and have the form depicted in Figure 12.19.

Figure 12.18

When $x_m < x < x_M$, there are exactly two solutions for y in Equation (12.14).

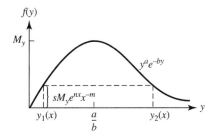

Figure 12.19

Trajectories in the vicinity of the rest point $(m/n, a/b)$ are periodic.

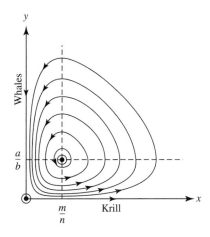

Model Interpretation   What conclusions can be drawn from the trajectories in Figure 12.19? First, because the trajectories are closed curves, they predict that under the assumptions of our Model (12.10), neither the baleen whales nor the Antarctic krill will become extinct. (Remember, the model is based on the pristine situation.) The second observation is that along a single trajectory the two populations fluctuate between their maximum and minimum values. That is, starting with populations in the region where $x > m/n$ and $y > a/b$, the krill population will decline and the whale population will increase until the krill population reaches the level $x = m/n$, at which point the whale population also begins to decline. Both populations continue to decline until the whale population reaches the level $y = a/b$ and the krill population begins to increase, and so on, counterclockwise around the trajectory. Recall from our discussion in Section 12.1 that the trajectories never cross. A sketch of the two population curves is shown in Figure 12.20. In the figure, we can see that the krill population fluctuates between its maximum and minimum values over one complete cycle. Notice that when the krill are plentiful, the whale population has its maximum rate of increase but that the whale population reaches its maximum value after the krill population is on the decline. The predator lags behind the prey in a cyclic fashion.

■ Figure 12.20

The whale population lags behind the krill population as both populations fluctuate cyclically between their maximum and minimum values.

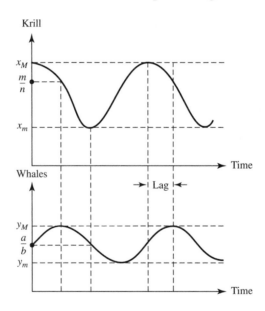

Effects of Harvesting   For given initial population levels $x(0) = x_0$ and $y(0) = y_0$, the whale and krill populations will fluctuate with time around one of the closed trajectories depicted in Figure 12.19. Let T denote the time it takes to complete one full cycle and return to the starting point. The average levels of the krill and baleen whale populations over the time cycle are defined, respectively, by the integrals

$$\bar{x} = \frac{1}{T} \int_0^T x(t)\, dt \quad \text{and} \quad \bar{y} = \frac{1}{T} \int_0^T y(t)\, dt$$

Now, from Equation (12.8)

$$\left(\frac{1}{x}\right)\left(\frac{dx}{dt}\right) = a - by$$

so that integration of both sides from $t = 0$ to $t = T$ leads to

$$\int_0^T \left(\frac{1}{x}\right) \left(\frac{dx}{dt}\right) dt = \int_0^T (a - by) \, dt$$

or

$$\ln x(T) - \ln x(0) = aT - b \int_0^T y(t) \, dt$$

Because of the periodicity of the trajectory, $x(T) = x(0)$, this last equation gives the average value

$$\bar{y} = \frac{a}{b}$$

In an analogous manner, it can be shown that

$$\bar{x} = \frac{m}{n}$$

(see Problem 2 in the problem set). Therefore, the *average* levels of the predator and prey populations are in fact their *equilibrium* levels. Let's see what this means in terms of harvesting krill.

Let's assume that the effect of fishing for krill is to decrease its population level at a rate $rx(t)$. The constant r indicates the intensity of fishing and includes such factors as the number of fishing vessels at sea and the number of crew members on those vessels casting nets for krill. Because less food is now available for the baleen whales, assume the whale population also decreases at a rate $ry(t)$. Incorporating these fishing assumptions into our model, we obtain the refined model

$$\frac{dx}{dt} = (a - by)x - rx = [(a - r) - by]x$$
$$\frac{dy}{dt} = (-m + nx)y - ry = [-(m + r) + nx]y$$

(12.15)

The autonomous system (12.15) is of the same form as Equations (12.10) (provided that $a - r > 0$) with a replaced by $a - r$ and m replaced by $m + r$. Thus, the new average population levels will be

$$\bar{x} = \frac{m + r}{n} \quad \text{and} \quad \bar{y} = \frac{a - r}{b}$$

Consequently, a moderate amount of harvesting krill (so that $r < a$) actually *increases* the average level of krill and *decreases* the average baleen whale population (under our assumptions for the model). The increase in krill population is beneficial to other species in the Southern Ocean (seals, seabirds, penguins, and fish) that depend on the krill as their main food source. The fact that some fishing increases the number of krill is known as **Volterra's principle**. The autonomous system (12.10) was first proposed by Lotka (1925) and Volterra (1931) as a simple model of predator–prey interaction.

The Lotka–Volterra model can be modified to reflect the situation in which both the predator and the prey are diminished by some kind of depleting force, such as the application of insecticide treatments that destroy both the insect predator and its insect prey. An example is given in Problem 3 in the problem set.

Some biologists and ecologists argue that the Lotka–Volterra Model (12.10) is unrealistic because the system is not asymptotically stable, whereas most observable natural
predator–prey systems tend to equilibrium levels over time. Nevertheless, regular population
cycles, as suggested by the trajectories in Figure 12.19, do occur in nature. Some scientists
have proposed models other than the Lotka–Volterra model that do exhibit oscillations that
are asymptotically stable (so that the trajectories approach equilibrium solutions). One such
model is given by

$$\frac{dx}{dt} = ax + bxy - rx^2$$

$$\frac{dy}{dt} = -my + nxy - sy^2$$

In this last autonomous system, the term rx^2 indicates the degree of internal competition
of the prey for their limited resource (such as food and space), and the term sy^2 indicates
the degree of competition among the predators for the finite amount of available prey. An
analysis of this model is more difficult than that presented for the Lotka–Volterra model, but
it can be shown that the trajectories of the model are not periodic and tend to equilibrium
levels. The constants r and s are positive and would be determined by experimentation or
historical data.

12.3 PROBLEMS

1. Apply the first and second derivative tests to the function $f(y) = y^a / e^{by}$ to show that
   $y = a/b$ is a unique critical point that yields the relative maximum $f(a/b)$. Show also
   that $f(y)$ approaches zero as y tends to infinity.

2. Derive the result that the average value $\bar{x}$ of the prey population modeled by the Lotka–
   Volterra system (12.10) is given by the population level m/n.

3. In 1868 the accidental introduction into the United States of the cottony cushion insect
   (*Icerya purchasi*) from Australia threatened to destroy the American citrus industry.
   To counteract this situation, a natural Australian predator, a ladybird beetle (*Novius
   cardinalis*) was imported. The beetles kept the scale insects down to a relatively low
   level. When DDT was discovered to kill scale insects, farmers applied it in the hope of
   reducing even further the scale insect population. However, DDT turned out to be fatal
   to the beetle as well, and the overall effect of using the insecticide was to increase the
   numbers of the scale insect.

   Modify the Lotka–Volterra model to reflect a predator–prey system of two insect
   species where farmers apply (on a continuing basis) an insecticide that destroys both
   the insect predator and the insect prey at a common rate proportional to the numbers
   present. What conclusions do you reach concerning the effects of the application of the
   insecticide? Use graphical analysis to determine the effect of using the insecticide once
   on an irregular basis.

4. In a 1969 study, E. R. Leigh concluded that the fluctuations in the numbers of Canadian
   lynx and its primary food source, the hare, trapped by the Hudson's Bay Company

between 1847 and 1903 were periodic. The actual population levels of both species differed greatly from the predicted population levels obtained from the Lotka–Volterra predator–prey model.

Use the entire model-building process to modify the Lotka–Volterra model to arrive at a more realistic model for the growth rates of both species. Answer the following questions at the appropriate times in the model-building process:

a. How have you modified the basic assumptions of the predator–prey model?

b. Why are your modifications an improvement to the basic model?

c. What are the equilibrium points for your model?

d. Is it possible to classify each equilibrium point as either stable or unstable? If so, classify, them.

e. Based on your equilibrium analysis, what values will the population levels of lynx and hare approach as t tends to infinity?

f. How would you use your revised model to suggest hunting policies for Canadian lynx and hare? *Hint*: You are introducing a second predator—the human—into the system.

5. Consider two species whose survival depends on their mutual cooperation. Let's take as an example a species of bee that feeds primarily on the nectar of one plant species and simultaneously pollinates that plant. One simple model of this **mutualism** is given by the autonomous system

$$\frac{dx}{dt} = -ax + bxy$$

$$\frac{dy}{dt} = -my + nxy$$

a. What assumptions are implicitly being made about the growth of each species in the absence of cooperation?

b. Interpret the constants a, b, m, and n in terms of the physical problem.

c. What are the equilibrium levels?

d. Perform a graphical analysis and indicate the trajectory directions in the phase plane.

e. Find an analytic solution and sketch typical trajectories in the phase plane.

f. Interpret the outcomes predicted by your graphical analysis. Do you believe the model is realistic? Why?

12.3 PROJECTS

1. Complete the requirements of the UMAP module (see enclosed CD), "Graphical Analysis of Some Difference Equations in Biology," by Martin Eisen, UMAP 553. Difference equations model the growth of many biological populations. This module predicts the behavior of the solutions to certain equations by graphical techniques.

2. Prepare a summary of one of the papers by May (or May et al.) listed in Further Reading for this section.

12.3   Further Reading

Clark, Colin W. *Mathematical Bioeconomics: The Optimal Management of Renewable Resources.* New York: Wiley, 1976.

May, R. M. *Stability and Complexity in Model Ecosystems.* Monographs in Population Biology VI. Princeton, NJ: Princeton University Press, 1973.

May, R. M., ed. *Theoretical Ecology: Principles and Applications.* Philadelphia: Saunders, 1976.

May, R. M. *Stability and Complexity in Model Ecosystems.* Princeton, NJ: Princeton University Press, 2001.

May, R. M., J. R. Beddington, C. W. Clark, S. J. Holt, & R. M. Lewis. "Management of Multispecies Fisheries." *Science* 205 (July 1979): 267–277.

12.4   Two Military Examples

In Section 1.4, we investigated the Battle of Trafalgar in 1805 as a system of difference equations. Here, we study two military forces in combat as a system of differential equations. Our graphical analysis then reveals the condition for which one of the forces may eventually win out over the other.

EXAMPLE 1   *Lanchester Combat Models*

Consider the situation of combat between two homogeneous forces: a homogeneous X force (e.g., tanks) opposed by another homogeneous Y force (e.g., antitank weapons). We want to know if one force will eventually win out over the other, or will the combat end in a draw? Other questions of interest include the following: How do the force levels decrease over time in battle? How many survivors will the winner have? How long will the battle last? How do changes in the initial force levels and weapon-system parameters affect the battle's outcome? In this example, we consider one basic combat model and several of its refinements.

Assumptions   Let $x(t)$ and $y(t)$ denote the strengths of the forces X and Y at time t, respectively. Usually t is measured in hours or days from the beginning of the combat. Let's examine what is meant by the *strengths* of the two forces X and Y. The strength $x(t)$, for example, includes a number of factors. If X is a homogeneous tank force, its strength depends on the number of tanks in operation, the level of technology used in the tank design, the quality of workmanship in the manufacturing process, the level of training and the skills of the individuals operating the tanks, and so forth. For our purposes, let's assume that the strength $x(t)$ is simply the number of tanks in operation at time t. Likewise, the strength $y(t)$ is the number of antitank weapons operational at time t.

In the actual state of affairs, the numbers $x(t)$ and $y(t)$ are nonnegative integers. However, it is convenient to idealize the situation and assume that $x(t)$ and $y(t)$ are continuous functions of time. For instance, if there are 500 tanks at 1400 hours (2 o'clock in the afternoon in military time) and 487 tanks at 1500 hours, then it is reasonable to assume that there are 497.4 tanks at 1412 hours (if we perform a linear interpolation between the data points). That is, 2.6 of the 13 tanks lost in the 1 hr of combat were lost in the first 12 min.

We also assume that $x(t)$ and $y(t)$ are differentiable functions of t so that they are smooth functions without any corners or cusps on their graphs. These idealizations enable us to model the strength functions by differential equations.

What can we assume about how the force levels change as a result of combat? For the basic combat model, let's assume that the combat *casualty rate* for the X force is proportional to the strength of the Y force. Other factors affect the change in the X force over time, such as reinforcements of the force by bringing in additional tanks (or troops if it is a troop force), tank losses due to mechanical or electronic failures, or losses due to operator errors or desertions. (Can you name some additional factors?) We will ignore all these other factors in our initial model so that the rate of change in $x(t)$ is

$$\frac{dx}{dt} = -ay, \qquad a > 0 \tag{12.16}$$

The positive constant a in Equation (12.16) is called the *antitank weapon kill rate or attrition-rate coefficient* and reflects the degree to which a single antitank weapon can destroy tanks. Thus, we begin our analysis with the simplest assumption: that the loss rate is proportional to the number of firers. Later we assume an interaction is necessary between firer and target (the firer must locate the target before firing), and we refine our model accordingly. In the refined model the attrition-rate coefficient is proportional to the number of targets.

Under similar assumptions, the rate of change in the Y force is given by

$$\frac{dy}{dt} = -bx, \qquad b > 0 \tag{12.17}$$

Here, the constant b indicates the degree to which a single tank can destroy antitank weapons. The autonomous system given by Equations (12.16) and (12.17), together with the initial strength levels $x(0) = x_0$ and $y(0) = y_0$, is called a **Lanchester-type combat model** after F. W. Lanchester, who investigated air combat situations during World War I. Equations (12.16) and (12.17) constitute our basic model subject to the assumptions we have made. We assume throughout that $x \geq 0$ and $y \geq 0$ because negative force levels have no physical meaning.

Analysis of the Model   Setting the right sides of Equations (12.16) and (12.17) equal to zero, we see that $(0, 0)$ is a rest point for the basic combat model. The trajectory directions in the phase plane are determined from the observations that $dx/dt < 0$ and $dy/dt < 0$, when $x > 0$ and $y > 0$. Moreover, if $x = 0$, then $dy/dt = 0$ (and we assume also that $dx/dt = 0$ because $x < 0$ has no physical meaning). These considerations lead to the trajectory directions depicted in Figure 12.21. Note that our assumptions imply that a

■ **Figure 12.21**

The rest point $(0, 0)$ for the basic Lanchester combat model

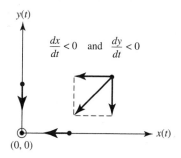

$y(t)$

$$\frac{dx}{dt} < 0 \quad \text{and} \quad \frac{dy}{dt} < 0$$

$x(t)$

$(0, 0)$

trajectory terminates when it reaches either coordinate axis. (Otherwise, the rest point $(0, 0)$ would be unstable.)

It is easy to find an analytic solution to the basic model. From the chain rule

$$\frac{dy}{dx} = \frac{dy/dt}{dx/dt}$$

and substitution from Equations (12.16) and (12.17), we get

$$\frac{dy}{dx} = \frac{-bx}{-ay}$$

Separating the variables in this last equation yields

$$-ay\,dy = -bx\,dx \tag{12.18}$$

Integration of each side of Equation (12.18) and employment of the initial force levels $x(0) = x_0$ and $y(0) = y_0$ produces the **Lanchester square law model**

$$a(y^2 - y_0^2) = b(x^2 - x_0^2) \tag{12.19}$$

Setting $C = ay_0^2 - bx_0^2$, we obtain the equation

$$ay^2 - bx^2 = C \tag{12.20}$$

Typical trajectories in the phase plane represented by Equation (12.20) are depicted in Figure 12.22. The trajectories for $C \neq 0$ are hyperbolas, and when $C = 0$, the trajectory is the straight line $y = \sqrt{b/a}\,x$. When $C < 0$, the trajectory intersects the x axis at $x = \sqrt{-C/b}$; then the X (tank) force wins because the Y force has been totally eliminated. On the other hand, if $C > 0$, the Y force wins with a final strength level of $y = \sqrt{C/a}$. These considerations are shown in Figure 12.22.

■ **Figure 12.22**

Trajectories of the basic Lanchester combat model: the trajectories are hyperbolas satisfying the Lanchester square law (12.19).

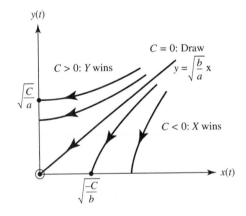

Let's investigate the situation in which the Y (antitank) force wins. Then the constant C must be positive, so

$$\left(\frac{y_0}{x_0}\right)^2 > \frac{b}{a} \tag{12.21}$$

The inequality (12.21) gives a necessary and sufficient condition that the Y force wins under the assumptions of our model (e.g., so that reinforcements are not permitted). From the inequality we can see that a doubling of the initial Y force level results in a fourfold advantage for that force, assuming the X force remains at the same initial level x_0 for constants a and b. This means Country X must increase b (its technology) by a factor of 4 to keep pace with the increase in the size of Country Y's force if the strength x_0 is kept at the same level. Figure 12.23 depicts a typical graph showing the force level curves $x(t)$ and $y(t)$ when the inequality (12.21) is satisfied. Observe from the figure that it is not necessary for the initial Y force level y_0 to exceed the level x_0 of the X force to ensure victory for Y. The crucial relationship is given by the inequality (12.21).

■ **Figure 12.23**

Force level curves $x(t)$ and $y(t)$ for the basic Lanchester combat model when $C > 0$ and the Y force wins

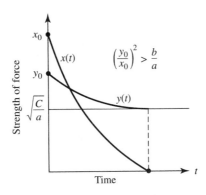

The model given by Equations (12.16) and (12.17) can be converted to a single second-order differential equation in the dependent variable y as follows: Differentiate Equation (12.17) to obtain

$$\frac{d^2y}{dt^2} = -b\frac{dx}{dt}$$

Then substitute Equation (12.16) into this last equation to get

$$\frac{d^2y}{dt^2} = aby$$

or

$$\frac{d^2y}{dt^2} - aby = 0 \tag{12.22}$$

In the problem set you are asked to verify that the function

$$y(t) = y_0 \cosh \sqrt{ab}\, t - x_0 \sqrt{a/b} \sinh \sqrt{ab}\, t \tag{12.23}$$

satisfies the differential Equation (12.22) subject to the condition that $y(0) = y_0$. Similarly, the solution for the X force level is

$$x(t) = x_0 \cosh \sqrt{ab}\, t - y_0 \sqrt{a/b} \sinh \sqrt{ab}\, t \tag{12.24}$$

subject to the initial level $x(0) = x_0$.

Equation (12.23) can be written in the more revealing form

$$\frac{y(t)}{y_0} = \cosh \sqrt{ab}\, t - \left(\frac{x_0}{y_0}\right) \sqrt{\frac{a}{b}} \sinh \sqrt{ab}\, t \qquad (12.25)$$

Expression (12.25) states that Y's current force level divided by the initial one—that is, the *normalized force level*—depends on just two parameters: a dimensionless engagement parameter $E = (x_0/y_0)\sqrt{a/b}$ and a time parameter $T = \sqrt{ab}\, t$. The constant $\sqrt{ab}$ represents the intensity of battle and controls how quickly the battle is driven to its conclusion. The ratio a/b represents the relative effectiveness of individual combatants on the two opposing sides.[3]

Refinements of the Basic Lanchester Combat Model   In the basic Lanchester combat model

$$\frac{dx}{dt} = -ay$$
$$\frac{dy}{dt} = -bx \qquad (12.26)$$

it has been assumed that the single-weapon attrition rates a and b are constant over time. In many circumstances, however, the X force and the Y force are changing positions, and the weapons' effectiveness depends on the distance between firer and target. Thus, $a = a(t)$ and $b = b(t)$ are time dependent. In that case, the Model (12.26) is no longer an autonomous system, and it is much more difficult to extract information from the model analytically.

In some situations, the single-weapon attrition rate a depends not only on time but also on the number of targets x. This situation occurs, for example, when target detection depends on the number of targets. In this case, $a = a(t, x)$ is a function of time as well as of the number of targets. The model now becomes analytically intractable, but numerical methods can be used to generate force-level results.

We can further enrich the basic Model (12.26). For example, if $a = a(t, x/y)$, then the single-weapon attrition rate depends on time and on the force ratio x/y. In still another operational circumstance, it is possible that $a = a(t, x, y)$ so that the attrition rate coefficient depends on time, the number of targets, and the number of firers.

When a weapon system uses area fire and enemy targets defend a constant area, the corresponding Lanchester attrition-rate coefficients depend on the number of targets. Then the basic model becomes

$$\frac{dx}{dt} = -gxy$$
$$\frac{dy}{dt} = -hyx \qquad (12.27)$$

where g and h are positive constants, and $x(0) = x_0$ and $y(0) = y_0$ are the initial force levels. Assuming there are no operational losses and no reinforcements on either side, Model (12.27) reflects combat between two guerilla forces.

---

[3]For further study, see the excellent paper "An Introduction to Lanchester-Type Models of Warfare," by James G. Taylor, Naval Postgraduate School, Monterey, California (1993).

System (12.27) is easily solvable. From the chain rule we obtain

$$\frac{dy}{dx} = \frac{h}{g}$$

and separation of variables leads to the equation

$$g\,dy = h\,dx$$

Integration then yields the **linear combat law**

$$g(y - y_0) = h(x - x_0) \tag{12.28}$$

Setting $K = gy_0 - hx_0$, we obtain the equation

$$gy - hx = K \tag{12.29}$$

If $K > 0$, the Y force wins; if $K < 0$, the X force wins. The trajectories for Model (12.27) are depicted in Figure 12.24. Notice from Equation (12.29) that the Y force wins provided

$$\frac{y_0}{x_0} > \frac{h}{g} \tag{12.30}$$

In this case, a doubling of the initial Y force simply doubles the advantage of that force, assuming that the X force retains its same initial level x_0.

■ **Figure 12.24**

Trajectories when the attrition-rate coefficients are directly proportional to the number of targets

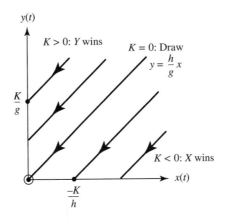

It can be shown (see Problem 3 in the problem set) that the X force level satisfying Model (12.27) is given by

$$x(t) = \begin{cases} x_0 \left[\dfrac{hx_0 - gy_0}{hx_0 - gy_0 e^{-(hx_0 - gy_0)t}} \right] & \text{for } hx_0 \neq gy_0 \\[2em] \dfrac{x_0}{1 + hx_0 t} & \text{for } hx_0 = gy_0 \end{cases} \tag{12.31}$$

A similar result holds for the Y force level.

Let's consider a few more simple, but natural, enrichments to the basic homogeneous-force combat Model (12.26). For example, if the X force were to continuously commit more combatants to the battle at the rate $q(t) \geq 0$, then the rate of change of the X force level would become

$$\frac{dx}{dt} = -ay + q(t) \tag{12.32}$$

Here, $q(t) < 0$ would mean the continuous withdrawal of forces. Now what might the function $q(t)$ look like? First, we would observe that replacements are drawn from a limited pool of combatants and weapon stocks. Then we could assume that these resources are committed to the battle at a constant rate m for as long as they last. If R denotes the total number of reserves that can commit to battle, these considerations translate into the submodel

$$q(t) = \begin{cases} m & \text{if } 0 \leq t \leq \dfrac{R}{m} \\ 0 & \text{if } t > \dfrac{R}{m} \end{cases} \tag{12.33}$$

Another consideration is that operational losses may occur. By operational losses, we mean those due to noncombat mishaps, such as disease, desertions, and breakdown of machinery. Many of the factors involved in the operational loss rate, such as psychological factors inherent in desertions, would be difficult to identify precisely. However, we might assume that the operational loss rate is proportional to the strength of the force. In that case, also assuming there are reinforcements, the rate of change of the X force level would be

$$\frac{dx}{dt} = -cx - ay + q(t) \tag{12.34}$$

where a and c are positive constants and $q(t) \geq 0$. Similar considerations apply to the Y force. ▪ ▪ ▪

EXAMPLE 2   *Economic Aspects of an Arms Race*

In this example we investigate the impact of a two-country arms race on their defense spending.

Problem Identification   Consider two countries engaged in an arms race. Let's attempt to assess qualitatively the effect of an arms race on the level of defense spending. *Specifically, we are interested in knowing whether the arms race will lead to uncontrolled spending and eventually be dominated by the country with the greatest economic assets. Or will an equilibrium level of spending eventually be reached in which each country spends a steady-state amount on defense?*

Assumptions   Define the variable x as the annual defense expenditure for Country 1 and the variable y as the annual defense expenditure for Country 2. We assume that each nation is ready to defend itself and considers a defense budget to be necessary. Let's examine the situation from the point of view of Country 1. Its rate of spending depends on several factors. In the absence of any spending on the part of Country 2 or any grievance with that country, it is reasonable to assume that the defense spending would decrease at a rate proportional to the amount being spent. The proportionality constant is indicative of the requirement to maintain the current arsenal (a percentage of the current spending) and acts as an economic restraint on defense spending (in the sense that government monies need to be spent in other areas, such as health and education). Thus,

$$\frac{dx}{dt} = -ax \quad \text{for } a > 0$$

Qualitatively, this equation states there will be no growth in defense spending. Now what happens when Country 1 perceives Country 2 engaging in defense spending? Country 1 will feel compelled to increase its defense budget to offset the defense buildup of its adversary and shore up its own security. Let's assume that the rate of increase for Country 1 is proportional to the amount Country 2 spends, where the proportionality coefficient is a measure of the perceived effectiveness of Country 2's weapons. This assumption seems reasonable, at least up to a point. As Country 2 adds weapons to its arsenal, Country 1 will perceive a need to add weapons to its own arsenal, where the numbers added are based on the assessment of the effectiveness of Country 2's weapons. Thus, the previous equation is modified to become

$$\frac{dx}{dt} = -ax + by$$

We are assuming b is a constant, although this assumption is somewhat unrealistic. We would expect some kind of diminishing return of perceived effectiveness as Country 2 continues to add weapons. That is, if Country 2 has 100 weapons, Country 1 might perceive the need to add 40 weapons. However, if Country 2 adds 200 weapons to its arsenal, Country 1 might perceive the need to add only 75 weapons. Thus, realistically, the constant b is more likely to be a decreasing function of y (and, in some instances, might be an increasing function). Later, we may wish to refine our model to account for this probable diminishing return.

Finally, let's add a constant term to reflect any underlying grievance felt by Country 1 toward Country 2. That is, even if the level of spending by both countries were zero, Country 1 would still feel compelled to be armed against Country 2, perhaps because of a fear of future aggressive action on the part of its adversary. If $y = 0$, then $c - ax$ represents a growth in defense spending for Country 1, and until $c = ax$, growth will continue to achieve deterrence. These assumptions lead to the differential equation

$$\frac{dx}{dt} = -ax + by + c$$

where a, b, and c are nonnegative constants. The constant a indicates an economic restraint on defense spending, b indicates the intensity of rivalry with Country 2, and c indicates the deterrent or grievance factor. Although we are assuming c to be constant, it is more likely to be a function of both the variables x and y.

An entirely similar argument for Country 2 yields the differential equation

$$\frac{dy}{dt} = mx - ny + p$$

where m, n, and p are nonnegative constants and are interpreted the same way as b, a, and c, respectively. The preceding equations constitute our model for arms expenditures.

Graphical Analysis of the Model   We are interested in whether the defense expenditures reach equilibrium levels. If so, we will know the arms race will not lead to uncontrolled spending. This situation means that both defense budgets must stop growing, so $dx/dt = 0$ and $dy/dt = 0$. Thus, we seek the rest points, or equilibrium points, of our model.

First, consider the case in which neither country has a grievance against the other or perceives any need for deterrence. Then $c = 0$ and $p = 0$, so our model becomes

$$\frac{dx}{dt} = -ax + by$$

$$\frac{dy}{dt} = mx - ny \tag{12.35}$$

For the autonomous system (12.35), $(x, y) = (0, 0)$ is a rest point. In this state there are no defense expenditures on either side and the two countries live in permanent peace with all conflicts resolved through nonmilitary means. (Such a peaceful state has existed between the United States and Canada since 1817.) However, if grievances that are not resolved to the mutual satisfaction of both sides do arise, the two countries will feel compelled to arm, leading to the equations

$$\frac{dx}{dt} = c \quad \text{and} \quad \frac{dy}{dt} = p$$

Thus, (x, y) will not remain at the rest point $(0, 0)$ if c and p are positive, so the rest point is unstable.

Now consider our general model

$$\frac{dx}{dt} = -ax + by + c$$

$$\frac{dy}{dt} = mx - ny + p \tag{12.36}$$

Setting the right sides equal to zero yields the linear system

$$ax - by = c$$
$$mx - ny = -p \tag{12.37}$$

Each of the equations in (12.37) represents a straight line in the phase plane. If the determinant of the coefficients, $bm - an$, is not equal to zero, then these two straight lines intersect at a unique rest point, denoted by (X, Y). It is easy to solve Equations (12.37) to obtain this rest point:

$$X = \frac{bp + cn}{an - bm}$$

$$Y = \frac{ap + cm}{an - bm}$$

Assume that $an - bm > 0$, so the rest point (X, Y) lies in the first quadrant of the phase plane. The situation is depicted in Figure 12.25. Shown in the figure are four regions labeled A–D determined by the two intersecting lines. Let's examine the trajectory directions in each of these regions.

Any point (x, y) in region A lies above both of the lines represented by Equations (12.37), so $ny - mx - p > 0$ and $by - ax + c > 0$. It follows from Equations (12.36) that for

Figure 12.25

If $an - bm > 0$, the Model (12.36) has a unique rest point (X, Y) in the first quadrant. Along the line $by = ax - c$, $dx/dt = 0$; along the line $ny = mx + p$, $dy/dt = 0$.

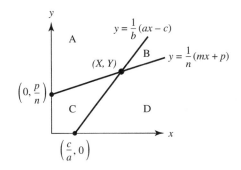

region A, $dy/dt < 0$ and $dx/dt > 0$. For a point (x, y) in region B, $ny - mx - p > 0$ and $by - ax + c < 0$ so that $dy/dt < 0$ and $dx/dt < 0$. Similarly, in region C, $dy/dt > 0$ and $dx/dt > 0$; and in region D, $dy/dt > 0$ and $dx/dt < 0$. These features are suggested in Figure 12.26.

Figure 12.26

Composite graphical analysis of the trajectory directions in the four regions determined by the intersecting lines (Equations 12.37)

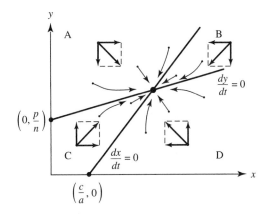

An analysis of the trajectory motion in the vicinity of the rest point (X, Y) in Figure 12.26 reveals that every trajectory in the phase plane approaches the rest point. Thus, under our assumption that $a/b > m/n$, so the intersection point (X, Y) lies in the first quadrant, the rest point (X, Y) is stable. Therefore, we conclude that defense spending for both countries will approach equilibrium (steady-state) levels of $x = X$ and $y = Y$. (In the problems for this section you are asked to investigate the case in which $a/b < m/n$; you will see that uncontrolled spending occurs.)

Let's examine the meaning of the inequality $a/b > m/n$ (which is equivalent to $an - bm > 0$). From the point of view of Country 1, the economic restraint a compared to the perceived intensity of rivalry b must be high against some specific ratio m/n for Country 2. The constant b has to do with perception and is, in part, a psychological factor. If b is lowered, the chances for the rest point to lie in the first quadrant are increased, with the beneficial result of steady-state levels of defense spending eventually likely for both countries. The value of our model is not in its capacity to make predictions but in its clarification of what can happen under different conditions regulating the parameters a, b, c, m, n, and p in the model. Certainly, we can see that it is important for nations to reduce tension levels and perceived threats through mutual cooperation, respect, and disarmament policies if a runaway arms race is to be avoided. Moreover, a permanent peace can be achieved only if grievances are resolved to the mutual satisfaction of each country. ■ ■ ■

12.4 PROBLEMS

1. Verify that the Function (12.23) satisfies the differential Equation (12.22).

   a. Solve Equation (12.29) for y and substitute the results into the differential equation $dx/dt = -gxy$ from Model (12.27).

   b. Separate the variables in the differential equation resulting from part (a) and integrate using partial fractions to obtain the force level given by Equation (12.31).

2. In the basic Lanchester model (12.26), assume the two forces are of equal effectiveness, so $a = b$. The Y force initially has 50,000 soldiers. There are two geographically separate units of 40,000 and 30,000 soldiers that make up the X force. Use the basic model and result (12.20) to show that if the commander of the Y force fights each of the X units separately, he can bring about a draw.

3. Let X denote a guerilla force and Y denote a conventional force. The autonomous system

$$\frac{dx}{dt} = -gxy$$

$$\frac{dy}{dt} = -bx$$

   is a Lanchestrian model for conventional–guerilla combat in which there are no operational loss rates and no reinforcements.

   a. Discuss the assumptions and relationships necessary to justify the model. Does the model seem reasonable?

   b. Solve the system and obtain the *parabolic law*

$$gy^2 = 2bx + M$$

   where $M = gy_0^2 - 2bx_0$.

   c. What condition must be satisfied by the initial force levels x_0 and y_0 for the conventional Y force to win? If the Y force does win, how many survivors will there be?

4. a. Assuming that the single-weapon attrition rates a and b in Equations (12.26) are constant over time, discuss the submodels

$$a = r_y p_y \quad \text{and} \quad b = r_x p_x$$

   where r_y and r_x are the respective *firing rates* (shots/combatant/day) of the Y and the X forces, and p_y and p_x are the respective probabilities that a single shot kills an opponent.

   b. How would you model the attrition-rate coefficients g and h in Model (12.27)? It may be helpful to think of Equations (12.27) as modeling guerilla–guerilla combat.

5. In our Model (12.36) for the arms race, assume that $an - bm < 0$, so the rest point lies in a quadrant other than the first one in the phase plane. Sketch the lines $dx/dt = 0$

and $dy/dt = 0$ in the phase plane, and label them and their intercepts on the coordinate axes. Perform a graphical stability analysis to respond to the following:

a. Do any potential equilibrium levels for defense spending exist? List any such points and classify them as stable or unstable.

b. Pick at least four starting points in the first quadrant, and sketch their trajectories in the phase plane.

c. What outcome for defense spending is predicted by your graphical analysis?

d. From the point of view of Country 1, interpret qualitatively the outcome predicted by your graphical analysis in terms of the relative values of the parameters in Model (12.36).

 PROJECT

1. Complete the requirements of "The Richardson Arms Race Model," by Dina A. Zinnes, John V. Gillespie, and G. S. Tahim, UMAP 308 (see enclosed CD). This unit constructs a model based on the classical assumptions of Lewis Fry Richardson and introduces difference equations. Students gain experience in analyzing the equilibrium, stability, and sensitivity properties of an interactive model.

12.4 Further Reading

Callahan, L. G. "Do We Need a Science of War?" *Armed Forces Journal* 106.36 (1969): 16–20.

Engel, J. H. "A Verification of Lanchester's Law." *Operations Research* 2 (1954): 163–171.

Lanchester, F. W. "Mathematics in Warfare." *The World of Mathematics.* Edited by J. R. Newman. Vol. 4. New York: Simon & Schuster, 1956. 2138–2157.

McQuie, R. "Military History and Mathematical Analysis." *Military Review* 50.5 (1970): 8–17.

Richardson, L. F. "Mathematics of War and Foreign Politics." *The World of Mathematics.* Edited by J. R. Newman. Vol. 4. New York: Simon & Schuster, 1956. 1240–1253.

U.S. General Accounting Office (GAO). "Models, Data, and War: A Critique of the Foundation Defense Analysis." PAD-80–21. Washington, DC; GAO, March 1980.

12.5 Euler's Method for Systems of Differential Equations

Throughout this chapter we have investigated autonomous systems of first-order differential equations. A more general form of a system of two ordinary first-order differential equations in the dependent variables x and y with independent variable t is given by

$$\frac{dx}{dt} = f(t, x, y)$$
$$\frac{dy}{dt} = g(t, x, y)$$

$$(12.38)$$

If the variable t does appear explicitly in one of the functions f or g, the system is a **nonautonomous system**; otherwise, it is autonomous. In this section we present Euler's numerical method for approximating the solution functions $x(t)$ and $y(t)$ to system (12.38) subject to the initial conditions $x(t_0) = x_0$ and $y(t_0) = y_0$. As was the case for a single differential equation in Section 11.5, Euler's method for systems patches together a string of tangent line approximations to each curve $x(t)$ and $y(t)$ over an interval $I: t_0 \leq t \leq b$.

To execute Euler's method for systems, we first subdivide the interval I for the independent variable t into n equally spaced points:

$$t_1 = t_0 + \Delta t$$
$$t_2 = t_1 + \Delta t$$
$$\vdots$$
$$t_n = t_{n-1} + \Delta t = b$$

We then calculate successive approximations to the solution functions

$$x_1 = x_0 + f(t_0, x_0, y_0)\Delta t$$
$$y_1 = y_0 + g(t_0, x_0, y_0)\Delta t$$
$$x_2 = x_1 + f(t_1, x_1, y_1)\Delta t$$
$$y_2 = y_1 + g(t_1, x_1, y_1)\Delta t$$
$$\vdots$$
$$x_n = x_{n-1} + f(t_{n-1}, x_{n-1}, y_{n-1})\Delta t$$
$$y_n = y_{n-1} + g(t_{n-1}, x_{n-1}, y_{n-1})\Delta t$$

Therefore, the method for systems is just like Euler's method studied in Section 11.5 except that we iterate *two* equations corresponding to the system (12.38) rather than one equation. As before, the number of steps n can be as large as we like, but errors do accumulate if n is too large. Here's an example to illustrate how the method works.

EXAMPLE 1   *Using Euler's Method for Systems*

Find the first three approximations (x_1, y_1), (x_2, y_2), (x_3, y_3) using Euler's method for the predator–prey system

$$\frac{dx}{dt} = 3x - xy$$

$$\frac{dy}{dt} = xy - 2y$$

subject to the initial conditions $x_0 = 1$ and $y_0 = 2$ starting at $t_0 = 0$, with $\Delta t = 0.1$.

Solution   We have $t_0 = 0$, $t_1 = t_0 + \Delta t = 0.1$, $t_2 = t_1 + \Delta t = 0.2$, $t_3 = t_2 + \Delta t = 0.3$, and $(x_0, y_0) = (1, 2)$.

First:
$$x_1 = x_0 + f(t_0, x_0, y_0)\Delta t$$
$$= x_0 + (3x_0 - x_0 y_0)\Delta t$$
$$= 1 + (3 - 2)(0.1) = 1.1$$
$$y_1 = y_0 + g(t_0, x_0, y_0)\Delta t$$
$$= y_0 + (x_0 y_0 - 2y_0)\Delta t$$
$$= 2 + (2 - 4)(0.1) = 1.8$$

Second:
$$x_2 = x_1 + f(t_1, x_1, y_1)\Delta t$$
$$= x_1 + (3x_1 - x_1 y_1)\Delta t$$
$$= 1.1 + (3.3 - (1.1)(1.8))(0.1) = 1.232$$
$$y_2 = y_1 + g(t_1, x_1, y_1)\Delta t$$
$$= y_1 + (x_1 y_1 - 2y_1)\Delta t$$
$$= 1.8 + ((1.1)(1.8) - 3.6)(0.1) = 1.638$$

Third:
$$x_3 = x_2 + f(t_2, x_2, y_2)\Delta t$$
$$= x_2 + (3x_2 - x_2 y_2)\Delta t$$
$$= 1.232 + (3.696 - (1.232)(1.638))(0.1)$$
$$= 1.3997984$$
$$y_3 = y_2 + g(t_2, x_2, y_2)\Delta t$$
$$= y_2 + (x_2 y_2 - 2y_2)\Delta t$$
$$= 1.638 + ((1.232)(1.638) - 3.276)$$
$$= 1.5122016$$

Euler's method for systems is easy to program on a computer. A program generates a table of numerical solutions (t_k, x_k, y_k) to the system (12.38) starting with the input (t_0, x_0, y_0), the number of steps n, and the step size Δt. If the system is autonomous, as in Example 1, we can then plot the points $(x_0, y_0), (x_1, y_1), (x_2, y_2), \ldots, (x_n, y_n)$ approximating the solution trajectory through the initial point (x_0, y_0) in the xy phase plane. We can also plot the points $x_0, x_1, x_2, \ldots, x_n$ approximating the solution curve $x(t)$ in the xt-plane. Likewise, we can plot $y_0, y_1, y_2, \ldots, y_n$ approximating the solution curve $y(t)$ in the yt-plane.

EXAMPLE 2 *A Trajectory and Solution Curves*

Use Euler's method to find the trajectory through the point $(1, 2)$ in the phase plane for the predator–prey model in Example 1.

Solution Through experimentation with a computer program, we found that a complete cycle of the solution trajectory through $(1, 2)$ occurs (with some overlap) if we choose $0 \le t \le 3$. Starting at $t_0 = 0$ with $\Delta t = 0.1$, we obtain the point plot of the trajectory displayed in Figure 12.27. Note that for this fairly large value of $\Delta t = 0.1$, the approximations do not cycle around counterclockwise exactly to the initial point $(1, 2)$. Thus, the approximations

■ **Figure 12.27**

A plot of the points in the phase plane approximating the solution trajectory through (1, 2) for the predator–prey model in Example 1 for $\Delta t = 0.1$, $0 \le t \le 3$

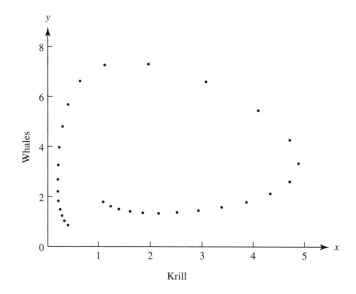

depart from the true solution trajectory, which is periodic. This failure is due to the errors inherent in the approximation process (as well as to round-off errors).

Let's plot the approximations to the trajectory using the smaller $\Delta t = 0.02$ for $0 \le t \le 3$. The trajectory is shown in Figure 12.28, and we can see that it is more nearly periodic because of the reduced error using a smaller step size.

■ **Figure 12.28**

Solution trajectory through (1, 2) for Example 2 with $\Delta t = 0.02$, $0 \le t \le 3$

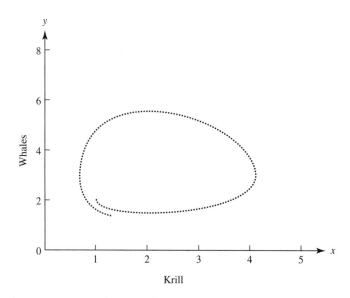

Further reducing the step size to $\Delta t = 0.003125$ (so $n = 960$ for the interval $0 \le t \le 3$), we obtain the trajectory displayed in Figure 12.29, which is very close to being truly periodic.

Finally, we plot each of the solution population curves $x(t)$ and $y(t)$ satisfying $x(0) = 1$ and $y(0) = 2$ for $0 \le t \le 9$ and $\Delta t = 0.0025$ (corresponding to $n = 3600$). The curves are

■ Figure 12.29

With $\Delta t = 0.003125$ ($n = 960$) on the interval $0 \le t \le 3$, the approximating solution trajectory through the point $(1, 2)$ is very nearly periodic, corresponding well to the actual solution.

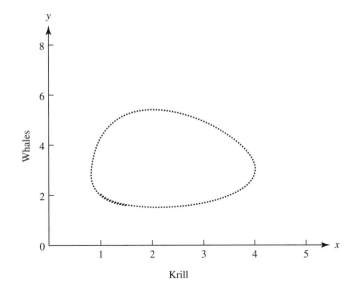

■ Figure 12.30

Periodic fluctuations of the individual population levels over time for the predator–prey model in Example 1. The maximum predator population always occurs at a later time than the occurrence of the maximum prey population (so the prey population is already in decline in its cycle when the predator reaches its zenith).

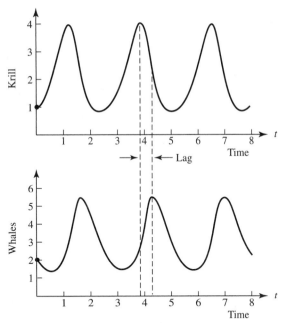

shown in Figure 12.30 and show how the predator whale population lags behind the prey krill population.

As discussed in Section 11.5 for a single first-order differential equation, more refined and accurate numerical methods for solving systems exist and are usually studied in the numerical methods portion of a differential equations course. Many of those methods are simply modifications of the basic Euler method we presented here, with fairly accurate approximations to the true solution trajectory for an appropriate step size.   ■ ■ ■

12.5 PROBLEMS

In Problems 1–4, use Euler's method to solve the first-order system subject to the specified initial condition. Use the given step size Δt and calculate the first three approximations (x_1, y_1), (x_2, y_2), and (x_3, y_3). Then repeat your calculations for $\Delta t/2$. Compare your approximations with the values of the given analytical solution.

1. $\dfrac{dx}{dt} = 2x + 3y$

$\dfrac{dy}{dt} = 3x + 2y$

$x(0) = 1, \ y(0) = 1, \ \Delta t = \dfrac{1}{4}$

$x(t) = \dfrac{1}{2}e^{-t} + \dfrac{1}{2}e^{5t}, \ y(t) = -\dfrac{1}{2}e^{-t} + \dfrac{1}{2}e^{5t}$

2. $\dfrac{dx}{dt} = x + 5y$

$\dfrac{dy}{dt} = -x - 3y$

$x(0) = 5, \ y(0) = 4, \ \Delta t = \dfrac{1}{4}$

$x(t) = 5e^{-t}(\cos t + 6 \sin t), \ y(t) = e^{-t}(4 \cos t - 13 \sin t)$

3. $\dfrac{dx}{dt} = x + 3y$

$\dfrac{dy}{dt} = x - y + 2e^t$

$x(0) = 0, \ y(0) = 2, \ \Delta t = \dfrac{1}{4}$

$x(t) = -e^{-2t} + 3e^{2t} - 2e^t, \ y(t) = e^{-2t} + e^{2t}$

4. $\dfrac{dx}{dt} = 3x + e^{2t}$

$\dfrac{dy}{dt} = -x + 3y + te^{2t}$

$x(0) = 2, \ y(0) = -1, \ \Delta t = \dfrac{1}{4}$

$x(t) = 3e^{3t} - e^{2t}, \ y(t) = e^{3t} - 3te^{2t} - 2e^{2t} - te^{2t}$

5. Assume we have a lake that is stocked with both bass and trout. Because both eat the same food sources, they are competing for survival. Let $B(t)$ and $T(t)$ denote the bass and trout populations, respectively, at time t. The rates of growth for bass and for trout are estimated by the differential equations

$$\frac{dB}{dt} = B \cdot (10 - B - T), \, B(0) = 5$$

$$\frac{dT}{dt} = T \cdot (15 - B - 3 \cdot T), \, T(0) = 2$$

Use Euler's method with step size $\Delta t = 0.1$ to estimate the solution curves from $0 \le t \le 7$ for

a. $B(t)$ versus t

b. $T(t)$ versus t

c. The solution trajectory $(B(t), T(t))$ in the phase plane

6. Repeat Problem 5 for step size $\Delta t = 1$. Discuss the differences in your plots, and explain why these differences occur.

7. The following system is a predator–prey model in which human harvesting occurs for both species. Use Euler's method with step size $\Delta t = 1$ over $0 \le t \le 4$ to numerically solve

$$\frac{dx}{dt} = x - xy - \frac{3}{4}y$$

$$\frac{dy}{dt} = xy - y - \frac{3}{4}x$$

subject to $x(0) = \frac{1}{2}$ and $y(0) = 1$.

8. Use Euler's method to solve the model in Problem 7 without harvesting. Compare and discuss the differences in your solutions to the two models.

12.5 PROJECTS

The following algorithm, known as the improved Euler's method, improves on the accuracy of the original Euler method. In the improved Euler's method we take the average of two slopes. We first estimate x_{n+1} and y_{n+1}, as in the original Euler method, but denote the estimates by x_{n+1}^* and y_{n+1}^*, respectively. We then use these points to find the average slopes. Thus, to solve the system

$$\frac{dx}{dt} = f(t, x, y) \quad \text{and} \quad \frac{dy}{dt} = g(t, x, y)$$

calculate the following updates from t_n, x_n, y_n to $t_{n+1}, x_{n+1}, y_{n+1}$:

$$x_{n+1}^* = x_n + f(t_n, x_n, y_n) \, \Delta t$$

$$y_{n+1}^* = y_n + g(t_n, x_n, y_n) \, \Delta t$$

$$t_{n+1} = t_n + \Delta t$$

$$x_{n+1} = x_n + \left[f(t_n, x_n, y_n) + f(t_{n+1}, x_{n+1}^*, y_{n+1}^*) \right] \frac{\Delta t}{2}$$

$$y_{n+1} = y_n + \left[g(t_n, x_n, y_n) + g(t_{n+1}, x_{n+1}^*, y_{n+1}^*) \right] \frac{\Delta t}{2}$$

1. Using the improved Euler's method, approximate the solution to the predator–prey problem in Example 2. Compare the new solution to that obtained by Euler's method using $\Delta t = 0.1$ over the interval $0 \le t \le 3$. Graph the solution trajectories for both solutions.

2. Using the improved Euler's method, approximate the solution to the harvesting predator–prey problem in Problem 7. Compare the new solution to the one obtained in Problem 7 using the same step size $\Delta t = 1$ over the interval $0 \le t \le 4$. Graph the solution trajectories for both solutions.

12.5 Further Reading

Burden, Richard, & Douglas Faires. *Numerical Analysis*, 6th ed. Pacific Grove, CA: Brooks/Cole, 2000.

Giordano, Frank, & Maurice Weir. *Differential Equations: A Modeling Approach*. Reading, MA: Addison-Wesley, 1991.

13 Optimization of Continuous Models

Introduction

In Chapter 7 you studied linear programming models

$$\text{Optimize } f(\mathbf{X}) \tag{13.1}$$

subject to inequality constraints

$$g_i(\mathbf{X}) \left\{ \begin{matrix} \geq \\ \leq \end{matrix} \right\} b_i \quad \text{for all } i \text{ in } I$$

For linear programming models there is only one objective function f, which is a linear function of the decision variables or components of the vector $\mathbf{X}$. The constraint functions g_i must also be linear. If the decision variables are restricted to integer values, the problem is an *integer program*.

In this chapter we consider optimization problems in which the objective function f is continuous but nonlinear. Moreover, the constraint functions g_i may be nonlinear as well, and they are *equality* constraints

$$g_i(X) = b_i \quad \text{f}$$

We restrict our attention to problems for whi   These
are the optimization models you studied in

In Section 13.1 we address a special
elementary calculus. In the illustrative
determining an optimal inventory strate
quantities and how often goods shoul
inventory. The restrictions on the va
the solutions to the assumptions is e
was constructed in Section 7.4.)
model sensitivity.

In Section 13.2 we study
using two methods: the usua'
the variables when the part
approximation technique–

In Section 13.3, we
trative example, we de

storage tank. The emphasis in Section 13.3 is on *model solution* and *model sensitivity* of this class of optimization problems. We present the method of *Lagrange multipliers* for analyzing such problems.

In Section 13.4, we present *graphical optimization*. The illustrative problem addresses the management of a fishing industry and is concerned with whether a free market can lead to a satisfactory solution for fishers, consumers, and ecologists alike or whether some type of government intervention is necessary. The graphical analysis in the example provides a qualitative analysis for the type of analytic models we developed in Chapters 11 and 12 using differential equations.

The projects in this chapter allow for a more detailed study of the optimization topics we discuss. For instance, students who wish can study Lagrange multipliers or elementary ideas in the calculus of variations using UMAP modules cited in the various project sections.

13.1 An Inventory Problem: Minimizing the Cost of Delivery and Storage

Scenario   A chain of gasoline stations has hired us as consultants to determine how often and how much gasoline should be delivered to the various stations. After some questioning, we determine that each time gasoline is delivered, stations incur a charge of d dollars, which is in addition to the cost of the gasoline and is independent of the amount delivered.

Costs are also incurred when the gasoline is stored. One such cost is capital tied up in inventory—money that is invested in the stored gasoline and thus cannot be used elsewhere. The cost is normally computed by multiplying the cost of the gasoline to the company by the current interest rate for the period the gasoline was stored. Other costs include amortization of the tanks and equipment necessary to store the gasoline, insurance, taxes, and security measures.

The gasoline stations are located near interstate highways, where demand is fairly constant throughout the week. Records indicating the gallons sold daily are available for each station.

Problem Identification   Assume that the firm wishes to maximize its profits and that demand and price are constant in the short term. Thus, because total revenue is constant, total profit can be maximized by minimizing the total costs. There are many components of total costs, such as overhead and employee costs. If these costs are affected by the amount and the timing of the deliveries, they should be considered. Let's assume the costs are not affected and focus our attention on the following problem: *Minimize the average daily delivery and storing sufficient gasoline at each station to meet consumer demand.* expect such a minimum to exist. If the delivery charge is very high and the low, we would expect large orders of gasoline delivered infrequently. On delivery charge is very low and the storage costs very high, we would line delivered frequently.

Assumptions   In the following presentation, we consider some factors important in deciding how large an inventory to maintain. Obvious factors to consider are delivery costs, storage costs, and demand rate for the product. Perishability of the stored product may also be a paramount concern. In the case of gasoline, the effect of condensation may become increasingly more appreciable as the gasoline level gets lower in the tank. Also, consider the market stability of the selling price of the product and the cost of raw materials. For example, if the market price of the product is volatile, the seller would be reluctant to store large quantities of the product. On the other hand, an expected large increase in the price of raw materials in the near future argues for large inventories. The stability of the demand for the products by the consumer is another factor. There may be seasonal fluctuations in the demand for the product, or a technological breakthrough may cause the product to become obsolete. The time horizon can also be extremely important. In the short term, contracts may be signed for warehouse space, some of which will not be needed in the long term. Another consideration is the importance to the owner of an occasional unsatisfied demand (stockouts). Some owners would opt for a more costly inventory strategy to ensure that they will never run out of stock. From this discussion we can see that the inventory decision is not an easy one, and it is not difficult to build scenarios in which any one of the preceding factors may dictate a particular strategy. We restrict our initial model here to the following variables:

$$\text{average daily cost} = f(\text{storage costs, delivery costs, demand rate})$$

The Submodels

Storage Costs   We need to consider how the storage cost per unit varies with the number of units being stored. Are we renting space and receiving a discount when storage exceeds certain levels, as suggested in Figure 13.1a? Or do we rent the cheapest storage first (adding more space as needed), as suggested by Figure 13.1b? Do we need to rent an entire warehouse or floor? If so, the per-unit price is likely to decrease as the quantity stored increases until another warehouse or floor needs to be rented, as suggested in Figure 13.1c. Does the company own its own storage facilities? If so, what alternative use can be made of them? In our model we take per-unit storage as a constant.

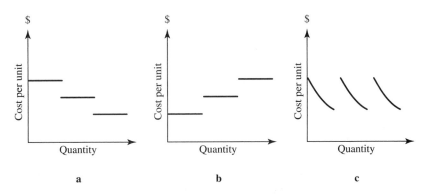

■ **Figure 13.1**
Submodels for storage costs

Delivery Costs   In many cases the delivery charge depends on the amount delivered. For example, if a larger truck or an extra flatcar is needed, there is an additional charge. In our model we consider a constant delivery charge independent of the amount delivered.

Demand   If we plot the daily demand for gasoline at a particular station, we will very likely get a graph similar to the one shown in Figure 13.2a. If we plot the frequency of each demand level over a fixed time period (e.g., 1 year), we might get a plot similar to that shown in Figure 13.2b. If demands are fairly tightly packed about the most frequently occurring demand, then we accept the daily demand as being constant. We will assume that such is the case for our model. Finally, even though we realize that the demands occur in discrete time periods, for purposes of simplification we take a continuous submodel for demand. This continuous submodel is depicted in Figure 13.2c, where the slope of the line represents the constant demand rate. Notice the importance of our assumptions in producing the linear submodel. Also, as suggested in Figure 13.2b, about half the time demand exceeds its average value. We will examine the importance of this assumption in the implementation phase when we consider the possibility of unsatisfied demands.

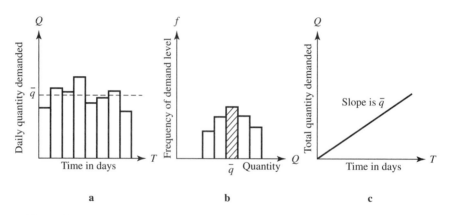

■ **Figure 13.2**

A constant demand rate

Model Formulation   We use the following notation for constructing our model:

$$s = \text{storage costs per gallon per day}$$
$$d = \text{delivery cost in dollars per delivery}$$
$$r = \text{demand rate in gallons per day}$$
$$Q = \text{quantity of gasoline in gallons}$$
$$T = \text{time in days}$$

Now suppose an amount of gasoline, say $Q = q$, is delivered at time $T = 0$, and the gasoline is used up after $T = t$ days. The same cycle is then repeated, as illustrated in Figure 13.3. The slope of each line segment in Figure 13.3 is $-r$ (the negative of the demand rate). The problem is to determine an order quantity Q^* and a time between orders T^* that minimizes the delivery and storage costs.

■ Figure 13.3

An inventory cycle consists of an order quantity *q* consumed in *t* days.

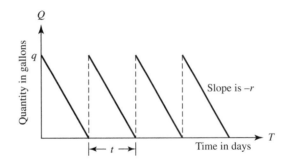

We seek an expression for the average daily cost, so consider the delivery and storage cost for a cycle of length *t* days. The delivery costs are the constant amount *d* because only one delivery is made over the single time period. To compute the storage costs, take the average daily inventory $q/2$, multiply by the number of days in storage *t*, and multiply that by the storage cost per item per day *s*. In our notation this gives

$$\text{cost per cycle} = d + s\frac{q}{2}t$$

which, upon division by *t*, yields the average daily cost:

$$c = \frac{d}{t} + \frac{sq}{2}$$

Model Solution   Apparently, the cost function to minimize has two independent variables, *q* and *t*. However, from Figure 13.3 notice that the two variables are related. For a single cyclic period, the amount delivered equals the amount demanded. This translates to $q = rt$. Substitution into the average daily cost equation yields

$$c = \frac{d}{t} + \frac{srt}{2} \tag{13.2}$$

Equation (13.2) is the sum of a hyperbola and a linear function. The situation is depicted in Figure 13.4.

■ Figure 13.4

The average daily cost *c* is the sum of a hyperbola and a linear function.

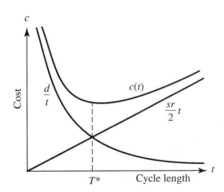

Let's find the time between orders T^* that minimizes the average daily cost. Differentiating c with respect to t and setting $c' = 0$ yield

$$c' = -\frac{d}{t^2} + \frac{sr}{2} = 0$$

This equation gives the (positive) critical point:

$$T^* = \left(\frac{2d}{sr}\right)^{1/2} \tag{13.3}$$

This critical point provides a relative minimum for the cost function because the second derivative

$$c'' = \frac{2d}{t^3}$$

is always positive for positive values of t. It is clear from Figure 13.4 that T^* gives a global minimum as well. Also, note from Equation (13.3) that $d/T^* = (sr/2)T^*$ so that T^* is the point at which the linear function and the hyperbola intersect in Figure 13.4.

Interpretation of the Model   Given a (constant) demand rate r, Equation (13.3) suggests a proportionality between the optimal period T^* and $(d/s)^{1/2}$. Intuitively, we would expect T^* to increase as the delivery cost d increases and to decrease as storage costs s increase. Thus, our model at least makes common sense. However, the nature of the relationship (13.3) is interesting.

For our submodels of demand rate, storage costs, and delivery cost, we assumed very simple relationships. To analyze models that have more complex submodels, we need to analyze what we did mathematically. Note that to determine the number of item-days of storage, we computed the area under the curve for one cycle. Thus, the storage costs for one cycle could be computed as an integral:

$$s \int_0^t (q - rx)\, dx = s\left(qt - \frac{rt^2}{2}\right) = \frac{sqt}{2}$$

The last equality in this equation follows from the substitution $r = q/t$ and agrees with our previous result for storage costs per cycle. It is important to recognize the underlying mathematical structure to facilitate generalization to other assumptions. As we will see, it is also helpful in analyzing the sensitivity of the model to changes in the assumptions.

One of our assumptions was to neglect the cost of the gasoline in the analysis. However, does the cost of gasoline actually affect the optimal order quantity and period? Because the amount purchased in each cycle is rt, if the cost per gallon is p dollars, then the constant amount $p(q/t) = pr$ would have to be added to the average daily cost. Because this amount is constant, it cannot affect T^* because the derivative of a constant is zero. Thus, we are correct in neglecting the cost of gasoline. In a more refined model the interest lost in capital invested in the inventory could be considered.

Implementation of the Model   Consider again the graph in Figure 13.3. Now the model assumes the entire inventory is used up in each cyclic period, but all demands are supposed to be satisfied immediately. Note that this assumption is based on an average daily demand of r gallons per day. This assumption means that over the long term, for approximately half of the time cycles the stations will run out of stock before the end of the period and the next delivery time, and for the other half of the time cycles the stations will still have some gasoline left in the storage tanks when the next delivery arrives. Such a situation probably won't do much for our credibility as consultants! So let's consider recommending a buffer stock to help prevent the stock-outs, as suggested in Figure 13.5. Note in Figure 13.5 that the optimal time period T^* and the optimal order quantity $Q^* = rT^*$ are indicated as labels because we know those values from our model given in Equation (13.3).

■ **Figure 13.5**

A buffer stock q_b helps prevent stock-outs.

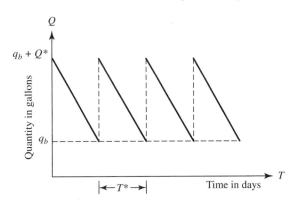

Let's examine the effect of the buffer stock on our inventory strategy. We have already observed that the storage costs for one cycle are given by the area under the curve over one period multiplied by the constant s. Now, the effect of the buffer stock is to add an additional constant area $q_b t$ to the previous area under the curve. Thus, the constant amount $q_b s$ is added to the average daily cost and consequently the value of T^* does not change from what it was without the buffer stock. Thus, the stations continue to order $Q^* = rT^*$ as before, although the maximum inventory now becomes $Q^* + q_b$. By determining the daily cost of maintaining the buffer stock, the business manager can decide how large a buffer stock to carry. What other information would be useful to the business manager in determining how large a buffer stock to carry?

Our mathematical analysis has been very straightforward and precise. However, can we obtain the necessary data to estimate r, s, and d precisely? Probably not. Moreover, how sensitive is the cost to changes in these parameters? These are issues that a consultant would want to consider. Note, too, that we would probably round T^* to an integer value. How should it be rounded? Is it better to overestimate or underestimate T^*? Of course, in a given situation, we could substitute various values of T^* for t in Equation (13.2) and see how the cost varies.

Let's use the derivative to determine the shape of the average daily cost curve more generally. We know that the average daily cost curve is minimum at $T^* = t$. The first derivative of the average daily cost represents its rate of change, or the marginal cost, and from Equation (13.2) is $-d/t^2 + sr/2$. The derivative of the marginal cost is $2d/t^3$, which for positive t is always positive and decreases as t increases. Note that the derivative of the marginal cost becomes large without bound as t approaches zero and approaches zero as t becomes large. Thus, to the left of T^* the marginal cost is negative and becomes steeper and steeper as t approaches zero. To the right of T^* the marginal cost approaches the constant value $sr/2$. Relate these results to the graph depicted in Figure 13.4 and interpret them economically.

13.1 PROBLEMS

1. Consider an industrial situation in which it is necessary to set up an assembly line. Suppose that each time the line is set up a cost c is incurred. Assume c is in addition to the cost of producing any item and is independent of the amount produced. Suggest submodels for the production rate. Now assume a constant production rate k and a constant demand rate r. What assumptions are implied by the model in Figure 13.6? Next assume a storage cost of s (in dollars per unit per day) and compute the optimal length of the production run P^* to minimize the costs. List all of your assumptions. How sensitive is the average daily cost to the optimal length of the production run?

■ **Figure 13.6**

Determine the optimal length of the production run.

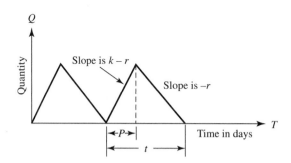

2. Consider a company that allows back ordering. That is, the company notifies customers that a temporary stock-out exists and that their order will be filled shortly. What conditions might argue for such a policy? What effect does such a policy have on storage costs? Should costs be assigned to stock-outs? Why? How would you make such an assignment? What assumptions are implied by the model in Figure 13.7? Suppose a "loss of goodwill cost" of w dollars per unit per day is assigned to each stock-out. Compute the optimal order quantity Q^* and interpret your model.

■ **Figure 13.7**

An inventory strategy that permits stock-outs

3. In the inventory model discussed in the text, we assumed a constant delivery cost that is independent of the amount delivered. Actually, in many cases, the cost varies in discrete amounts depending on the size of the truck needed, the number of platform cars required, and so forth. How would you modify the model in the text to take into account these changes? We also assumed a constant cost for raw materials. However, often bulk-order discounts are given. How would you incorporate these discount effects into the model?

4. Discuss the assumptions implicit in the two graphical models depicted in Figure 13.8. Suggest scenarios in which each model might apply. How would you determine the

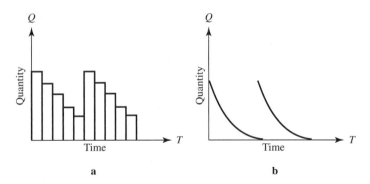

submodel for demand in Figure 13.8b? Discuss how you would compute the optimal order quantity in each case.

5. What are the optimal speed and safe following distance that allow the maximum flow rate (cars per unit time)? The solution to the problem would be useful in controlling traffic in tunnels, for roads under repair, or for other congested areas. In the following schematic, l is the length of a typical car and d is the distance between cars:

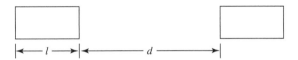

Justify that the flow rate is given by

$$f = \frac{\text{velocity}}{\text{distance}}$$

Let's assume a car length of 15 ft. In Section 2.2 the safe stopping distance

$$d = 1.1v + 0.054v^2$$

was determined, where d is measured in feet and v in miles per hour. Find the velocity in miles per hour and the corresponding following distance d that maximizes traffic flow. How sensitive is the solution to changes in v? Can you suggest a practical rule? How would you enforce it?

6. Consider an athlete competing in the shot put. What factors influence the length of his or her throw? Construct a model that predicts the distance thrown as a function of the initial velocity and angle of release. What is the optimal angle of release? If the athlete cannot maximize the initial velocity at the angle of release you propose, should he or she be more concerned with satisfying the angle of release or generating a high initial velocity? What are the trade-offs?

7. John Smith is responsible for periodically buying new trucks to replace older trucks in his company's fleet of vehicles. He is expected to determine the time a truck should be retained so as to minimize the average cost of owning the truck. Assume the purchase price of a new truck is $9000 with trade-in. Also assume the maintenance cost (in dollars)

per truck for t years can be expressed analytically by the following empirical model:

$$C(t) = 640 + 180(t + 1)t$$

where t is the time in years that the company owns the truck.

a. Determine $E(t)$, the total cost function for a single truck retained for a period of t years.

b. Determine $E_A(t)$, the *average* annual cost function for a single truck that is kept in the fleet for t years.

c. Graphically depict $E_A(t)$ as a function of t. Justify the shape of your graph.

d. Analytically determine t^*, the optimal period that a truck should be retained in the fleet. Remember that the objective is to minimize the average cost of owning a truck.

e. Suppose we have to round t^* to the nearest whole year. In general, would it be better to round up or round down? Justify your answer.

8. *Cow to Market*—A cow currently weighs 800 lb and is gaining 35 lb per week. It costs $6.50 a week to maintain the cow. The market price today is $0.95 per pound but is falling $0.01 per day. Formulate a mathematical model and find the optimal period to keep the cow until it is sold to maximize profits.

13.1 PROJECTS (See enclosed CD for UMAP modules.) ▬▬

1. "The Human Cough," by Philip M. Tuchinsky, UMAP 211. A model is developed showing how our bodies contract the windpipe during a cough to maximize the velocity of the airflow (making the cough maximally effective). Complete the module and prepare a short report for classroom discussion.

2. "An Application of Calculus in Economics: Oligopolistic Competition," by Donald R. Sherbert, UMAP 518. The author analyzes a number of mathematical models that investigate the competitive structure of a market in which a small number of firms compete. Thus, a change in price or production level by one firm will cause a reaction in the others. Complete the module and prepare a short report for classroom discussion.

3. "Five Applications of Max–Min Theory from Calculus," by Thurmon Whitley, UMAP 341. In this module several unconstrained optimization problems are solved using the calculus. Scenarios addressed include maximizing profit, minimizing cost, minimizing travel time of light as it passes through several mediums (Snell's law), minimizing the surface area of a bee's cell, and the surgeon's problem of attaching an artery in such a way as to minimize the resistance to blood flow and strain on the heart.

13.2 A Manufacturing Problem: Maximizing Profit in Producing Competing Products

In many modeling situations we are required to optimize a function of several independent variables. In this section we present a scenario with two independent variables that are unconstrained. We find the optimal solution by two methods: the usual multivariable calculus solving the equations for the variables when the partial derivatives equal zero and then by applying a gradient search algorithm.

Scenario A company manufacturing computers plans to introduce two new products. Both computers contain the same microprocessing chip, but one system is equipped with a 27-in. monitor and the other system with a 31-in. monitor. In addition to $400,000 in fixed costs, it costs the company $1950 to produce a 27-in. system and $2250 to produce a 31-in. system. The manufacturer's suggested retail price is $3390 for the 27-in. system and $3990 for the 31-in. system. In the competitive market in which the systems will be sold, the marketing staff estimates that for each additional system of a particular type sold, the selling price will fall by $0.10. Furthermore, sales of each type of system affect the sales of the other: It is estimated that the selling price for each 27-in. system is reduced by $0.03 for each 31-in. computer sold, and the selling price of each 31-in. system is reduced by $0.04 for each 27-in. computer sold. Assuming it can sell all the computers it makes, how many systems of each type should the company manufacture to maximize its profits?

Model Formulation We define the following variables for the two types of computer systems (so $i = 1$ or 2):

$$x_1 = \text{number of 27-in. systems}$$
$$x_2 = \text{number of 31-in. systems}$$
$$P_i = \text{selling price of } x_i$$
$$R = \text{revenue obtained from computer sales}$$
$$C = \text{cost to manufacture the computers}$$
$$P = \text{total profit from the sales of the computers}$$

From our previous discussion of the manufacturing and marketing situations, we obtain the following assumptions and submodels:

$$P_1 = 3390 - 0.1x_1 - 0.03x_2$$
$$P_2 = 3990 - 0.04x_1 - 0.1x_2$$
$$R = P_1 \cdot x_1 + P_2 \cdot x_2$$
$$C = 400,000 + 1950x_1 + 2250x_2$$
$$P = R - C$$

and

$$x_1, x_2 \geq 0$$

Our objective is to maximize the profit function

$$
\begin{aligned}
P(x_1, x_2) &= R - C \\
&= (3390 - 0.1x_1 - 0.03x_2)x_1 + (3990 - 0.04x_1 - 0.1x_2)x_2 \\
&\quad - (400,000 + 1950x_1 + 2250x_2) \\
&= 1440x_1 - 0.1x_1^2 + 1740x_2 - 0.1x_2^2 - 0.07x_1x_2 - 400,000
\end{aligned}
$$

The necessary conditions are

$$\frac{\partial P}{\partial x_1} = 1440 - 0.2x_1 - 0.07x_2 = 0$$

$$\frac{\partial P}{\partial x_2} = 1740 - 0.07x_1 - 0.2x_2 = 0$$

Solution of these equations gives $x_1 \approx 4736$ and $x_2 \approx 7043$ (both rounded up). That is, the company should manufacture 4736 27-in. systems and 7043 31-in. systems for a total profit of $P(4736, 7043) = \$9,136,410.25$. Figure 13.9 shows a plot of the surface represented by $P(x_1, x_2)$ verifying that the point (4736, 7043) is indeed a maximum extreme value. This fact can also be verified using the second-derivative test from multivariable calculus. At the extreme point (4736, 7043),

$$\frac{\partial^2 P}{\partial x_1^2} = -0.2 < 0$$

and

$$\frac{\partial^2 P}{\partial x_1^2} \frac{\partial^2 P}{\partial x_2^2} - \left(\frac{\partial^2 P}{\partial x_1 \partial x_2} \right)^2 = (-0.2)(-0.2) - (-0.07)^2 \approx 0.04 > 0.$$

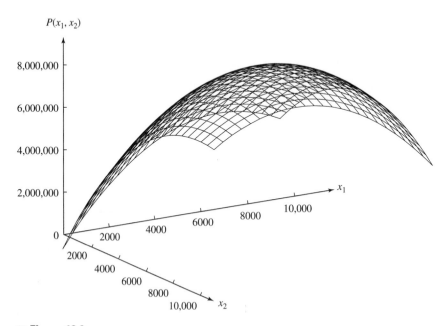

■ **Figure 13.9**

The total profit surface
$P(x_1, x_2) = 1440x_1 - 0.1x_1^2 + 1740x_2 - 0.1x_2^2 - 0.07x_1 x_2 - 400,000$

The Gradient Method of Steepest Ascent

A classical iterative method for finding an extreme point is the **gradient method of steepest ascent** (in the case of a maximum extremum) or **descent** (in the case of a minimum extremum). It is based on the fact that the gradient vector ∇f at a point in the domain of a differentiable function $f(x, y)$ always points in the direction of the maximum rate of increase of the function. Thus, starting with an initial point (x_0, y_0), we develop an iterative procedure for generating a sequence of points (x_k, y_k) obtained by moving from point to point in the direction of the gradient $\nabla f(x_k, y_k)$ such that $f(x_{k+1}, y_{k+1}) > f(x_k, y_k)$. In terms of coordinates, for some $\lambda_k > 0$,

$$
\begin{aligned}
x_{k+1} &= x_k + \lambda_k \frac{\partial f}{\partial x}(x_k, y_k) \\
y_{k+1} &= y_k + \lambda_k \frac{\partial f}{\partial y}(x_k, y_k)
\end{aligned}
\tag{13.4}
$$

Whenever f is a differentiable function throughout its domain, a theorem from advanced calculus guarantees that a $\lambda_k > 0$ does exist for which $f(x_{k+1}, y_{k+1}) > f(x_k, y_k)$. The question is, What value should be assigned to λ_k to obtain the next approximate solution (x_{k+1}, y_{k+1})?

The difficulty in implementing the gradient procedure is that as each point (x_k, y_k) gets closer to the extreme point, the length of the gradient vector becomes increasingly short (because the partial derivatives of f are zero at the extreme point). So at each stage in the process we will need a large enough value of λ_k to advance significantly toward the extreme point in the direction of the gradient. If λ_k is too small, we will never get close enough to the extreme point in a reasonably finite number of steps. On the other hand, if λ_k is too large, we overshoot the extreme point and find $f(x_{k+1}, y_{k+1}) < f(x_k, y_k)$, which is no improvement. One approach would be to determine the value of λ_k that maximizes the function value $f(x_{k+1}, y_{k+1})$ using Equations (13.4). That is, maximize the real-valued function

$$
g(\lambda) = f(x_k + \lambda \nabla f(x_k))
$$

for the optimal value $\lambda = \lambda_k$. We might try to solve this calculus problem by setting $dg/d\lambda = 0$ and solving for λ. However, in most cases we cannot easily solve the equation $dg/d\lambda = 0$; moreover, the equation may not have a unique solution.

Another approach is to apply one of the search techniques, such as the Golden Section Search Method, presented in Chapter 7, to maximize $g(\lambda)$. We will not pursue that approach here. Rather than using an elaborate search procedure to determine precisely the best $\lambda = \lambda_k$ at each stage, we will simply use a large enough λ_k to advance significantly toward the extreme point at each step. One way to accomplish this is to begin with a small $\lambda_0 > 0$ and then increase its value at each step through repeated multiplication by some fixed constant $\delta > 1$ while maintaining $f(x_{k+1}, y_{k+1}) > f(x_k, y_k)$. For a specific problem, the number $\delta > 0$ can be determined experimentally with the aid of a computer.

In the case of our scenario maximizing the profit function $P(x, y)$ for the company manufacturing two computer systems (where we renamed the variables $x_1 = x$ and $x_2 = y$ to avoid confusion with our iteration points), we selected a starting point of $(x_0, y_0) = (0, 0)$, an initial λ value equal to $1/16$, and a multiplier $\delta = 1.2$ (found experimentally). Thus,

Table 13.1   Gradient method of steepest ascent for $P(x, y) = 1440x - 0.1x^2 + 1740y - 0.1y^2 - 0.07xy - 400{,}000$ using $(x_0, y_0) = (0, 0)$, $\lambda_0 = 1/16$, and $\delta = 1.2$

k	x_k	y_k	$P(x_k, y_k)$	$\|\nabla P(x_k, y_k)\|$	λ_k
0	0.00000	0.00000	−400000.00	2258.58363	0.06250
1	90.00000	108.75000	−83852.78	2220.64513	0.07500
2	196.07906	237.14625	282264.83	2175.88751	0.09000
3	320.65562	388.24232	703216.07	2123.26645	0.10800
4	466.31434	565.35213	1183043.92	2061.65628	0.12960
5	635.72260	771.97180	1724309.91	1989.88094	0.15552
6	831.49389	1011.64446	2327249.57	1906.76591	0.18662
7	1055.98131	1287.74842	2988767.41	1811.21810	0.22395
8	1310.98315	1603.18739	3701350.51	1702.34142	0.26874
9	1597.34566	1959.96301	4452064.88	1579.59436	0.32249
10	1914.45720	2358.61835	5221908.43	1442.99306	0.38698
11	2259.64855	2797.55995	5985907.75	1293.35557	0.46438
12	2627.54965	3272.30172	6714409.45	1132.57034	0.55726
13	3009.50923	3774.73021	7375944.48	963.84679	0.66871
14	3393.25845	4292.56951	7941721.67	791.88486	0.80245
15	3763.08189	4809.31299	8391166.11	622.85723	0.96294
16	4100.81508	5304.95849	8717048.82	464.08297	1.15553
17	4387.95176	5757.86859	8928057.27	323.27676	1.38663
18	4608.92383	6147.88608	9046844.18	207.33566	1.66396
19	4755.12455	6460.37432	9103274.16	120.79222	1.99675
20	4828.50289	6690.13319	9125410.76	64.25797	2.39610
21	4842.85557	6843.43126	9132779.82	33.19815	2.87532
22	4820.97072	6936.34390	9135179.93	18.10526	3.45038
23	4787.37013	6989.00804	9136044.58	9.73259	4.14046
24	4759.61004	7018.33278	9136332.33	4.50158	4.96855
25	4743.68432	7034.03709	9136400.48	1.59395	5.96226
26	4737.00980	7040.80235	9136409.77	0.35854	7.15472
27	4735.16299	7042.58276	9136410.25	0.02579	8.58566
28	4735.04802	7042.77198	9136410.26	0.00857	10.30279

$\lambda_k = 1.2\lambda_{k-1}$ at each step in finding (x_{k+1}, y_{k+1}) according to Equations (13.4). Table 13.1 summarizes our iterative approximations to the extreme point. Figure 13.10 shows the level curves for the profit function P and the gradient vector ∇P at the 15th iteration point (3763.08, 4809.31) in the table. In this case, the actual extreme point (4735.042735, 7042.735043) was closely approximated after 28 iterations.

13.2   PROBLEMS

1. Find the local maximum value of the function

$$f(x, y) = xy - x^2 - y^2 - 2x - 2y + 4$$

2. Find the local minimum value of the function

$$f(x, y) = 3x^2 + 6xy + 7y^2 - 2x + 4y$$

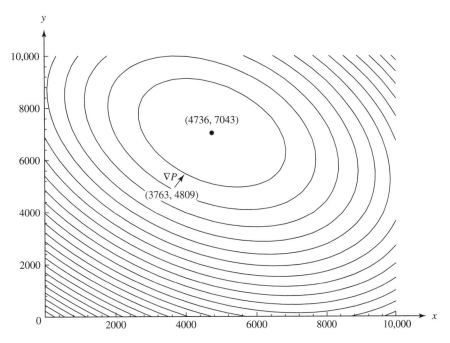

■ **Figure 13.10**

Level curves of the total profit function $P(x, y) = 1440x - 0.1x^2 + 1740y - 0.1y^2 - 0.07xy - 400,000$. The gradient vector ∇P is shown at the iteration point (3763, 4809) (rounded to integer values) and points in the direction of greatest increase for P.

3. A differentiable function $f(x, y)$ has a **saddle point** at a point (a, b) where its partial derivatives are simultaneously zero, if in every open disk centered at (a, b) there are domain points where $f(x, y) > f(a, b)$ and domain points where $f(x, y) < f(a, b)$. Find the saddle points of the following functions.

a. $f(x, y) = x^3 - y^3 - 2xy + 6$

b. $f(x, y) = 6x^2 - 2x^3 + 3y^2 + 6xy$

4. A continuous function $f(x, y)$ takes on its absolute extrema on a closed and bounded region either at an interior point or at a boundary point of the region. Find the absolute extrema of

$$f(x, y) = 48xy - 32x^3 - 24y^2$$

on the square region $0 \le x \le 1$ and $0 \le y \le 1$.

5. A company manufactures x floor lamps and y table lamps each day. The profit in dollars for the manufacture and sale of these lamps is

$$P(x, y) = 18x + 2y - 0.05x^2 - 0.03y^2 + 0.02xy - 100$$

Find the daily production level of each lamp to maximize the company's profits.

6. If x and y are the amounts of labor and capital, respectively, to produce

$$Q(x, y) = 0.54x^2 - 0.02x^3 + 1.89y^2 - 0.09y^3$$

units of output for manufacturing a product, find the values of x and y to maximize Q.

7. The total cost to manufacture one unit of product A is \$3, and for one unit of product B it is \$2. If x and y are the retail prices per unit of A and B, respectively, then marketing research has established that

$$Q_A = 2750 - 700x + 200y$$

and

$$Q_B = 2400 + 150x - 800y$$

are the quantities of each product that will be sold each day. Find a function $P(x, y)$ modeling the daily profit and the maximum daily profit.

8. An electric power-generating company charges different rates for residential and business users. (You might consider some reasons why this would be so.) The cost of producing the electricity is the same for all users and equals \$1000 in fixed costs plus an additional \$200 for each unit produced. If residential customers use x units of electricity, they pay $p = 1200 - 2x$ dollars for each unit. On the other hand, commercial customers pay $q = 1000 - y$ dollars for each of the y units of electricity they use. What price should the power company charge each type of customer to maximize profit? What is the maximum profit?

13.2 Projects

1. Write a computer code to perform the gradient method of steepest ascent algorithm using the multiplier technique $\lambda_{k+1} = \delta\lambda_k$ discussed in this section. Use your code to solve Problems 1, 5, 6, and 7 of this section.

2. Write a computer code to perform the gradient method of steepest ascent algorithm using the Golden Section Search Method (presented in Section 7.6) to maximize the function

$$g(\lambda) = f(x_k + \lambda \nabla f(x_k))$$

to obtain $\lambda = \lambda_k$ at each step determining the next point (x_{k+1}, y_{k+1}) in Equations (13.4). Use your code to solve Problems 6 and 7 of this section.

13.3 Constrained Continuous Optimization

Sometimes in modeling an optimization situation, we need to find a maximum or minimum value of a function when the independent variables are constrained to lie within some particular subset of the plane (such as a disk, along a straight line, or a closed triangular region). In this section we present two examples of such constrained problems and explore a powerful method for finding the extreme values known as the method of *Lagrange multipliers*.

EXAMPLE 1 *An Oil Transfer Company*

Consider a situation in which we are hired as consultants for a small oil transfer company. The management desires a policy of minimum cost due to restricted tank storage space.

Problem Identification *Minimize the costs associated with dispensing and holding the oil to maintain sufficient oil to satisfy demand while meeting the restricted tank storage space constraint.*

Assumptions Many factors determine the total cost of transferring oil. For our model we include the following variables: holding costs of the oil in the storage tank, withdrawal rate of the oil from the tank per unit time, the cost of the oil, and the size of the tank.

Model Formulation We define the following variables for two types of oil (so $i = 1$ or 2):

$$x_i = \text{the amount of oil type } i \text{ available}$$
$$a_i = \text{the cost of oil type } i$$
$$b_i = \text{withdrawal rate per unit time of oil type } i$$
$$h_i = \text{holding (storage) costs per unit time for oil type } i$$
$$t_i = \text{space in cubic feet to store one unit of oil type } i$$
$$T = \text{the total amount of storage space available}$$

Historical records have been studied and a formula has been derived that describes the system costs in terms of our variables. Our objective is to minimize the sum of the cost variables:

$$\left. \begin{array}{l} \text{Minimize } f(x_1, x_2) = \left(\dfrac{a_1 b_1}{x_1} + \dfrac{h_1 x_1}{2} \right) + \left(\dfrac{a_2 b_2}{x_2} + \dfrac{h_2 x_2}{2} \right) \\[2mm] \text{such that } g(x_1, x_2) = t_1 x_1 + t_2 x_2 = T \text{ (space constraint)} \end{array} \right\} \qquad (13.5)$$

We are provided the following data:

Oil type	a_i (\$)	b_i	h_i (\$)	t_i (ft^3)
1	9	3	0.50	2
2	4	5	0.20	4

We measure the storage tank and find only 24 ft^3 of available space. After substitution of the data into Equations (13.5), the formulation for our problem is

$$\text{Minimize } f(x_1, x_2) = 27/x_1 + 0.25x_1 + 20/x_2 + 0.10x_2$$

such that

$$2x_1 + 4x_2 = 24$$

Model Solution   The method commonly used for solving nonlinear optimization problems with equality constraints is known as the method of **Lagrange multipliers**. The method involves introducing a new variable λ (called a Lagrange multiplier) and setting up the function

$$L(x_1, x_2, \lambda) = f(x_1, x_2) + \lambda\,[g(x_1, x_2) - T]$$

For our problem this function is

$$L(x_1, x_2, \lambda) = 27/x_1 + 0.25x_1 + 20/x_2 + 0.10x_2 + \lambda(2x_1 + 4x_2 - 24) \qquad (13.6)$$

The solution methodology is to take the partial derivatives of Equation (13.6) with respect to the variables x_1, x_2, and λ and set them equal to zero; that is,

$$\frac{\partial L}{\partial x_1} = \frac{-27}{x_1^2} + 0.25 + 2\lambda = 0$$

$$\frac{\partial L}{\partial x_2} = \frac{-20}{x_2^2} + 0.10 + 4\lambda = 0$$

$$\frac{\partial L}{\partial \lambda} = 2x_1 + 4x_2 - 24 = 0$$

Using a computer algebra system, we find the solution to be $x_1 = 5.0968$, $x_2 = 3.4516$, $\lambda = 0.3947$, and $f(x_1, x_2) = \$12.71$. By perturbing the values of x_1 and x_2 slightly in either direction (but satisfying the constraint), we find the value of $f(x_1, x_2)$ increases. Thus, we conclude the solution is a minimum.

Model Sensitivity   The variable λ has special significance and is called a **shadow price**. The value of λ represents the amount the objective function would change for a *1-unit increase* in the constraint represented by λ. Therefore, in this problem, the value of $\lambda = 0.3947$ means that if the storage space constraint is changed from 24 to 25 ft^3, the value of the objective function would change from $\$12.71$ to approximately $12.71 + (1)(0.3947)$ or $\$13.10$. The economic interpretation is that another cubic foot of storage tank capacity increases the dispensing and holding costs by about $\$0.40$. ■ ■ ■

EXAMPLE 2 *A Space Shuttle Water Container*

Consider the space shuttle and an astronaut's water container that is stored within the shuttle's wall. The water container is formed as a sphere surmounted by a cone (like an ice cream cone), the base of which is equal to the radius of the sphere (Figure 13.11). If the radius of the sphere is restricted to exactly 6 ft and a surface area of 450 ft^2 is all that is allowed in the design, find the dimensions x_1 and x_2 such that the volume of the container is a maximum.

■ **Figure 13.11**

Space shuttle water
container

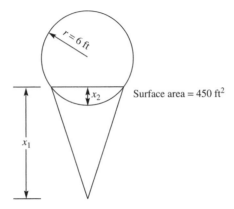

Surface area = 450 ft²

Problem Identification *Maximize the volume of the water container for the astronauts
while meeting the design restrictions.*

Assumptions There are many factors that can influence the design of a water container.
For our model we include the shape and dimensions, volume, surface area, and radius of
the sphere corresponding with Figure 13.11.

Model Formulation We define the following variables:

V_c = the volume of the conical top, which equals

$$\frac{\pi r^2}{12} x_1$$

V_s = the volume of the cut sphere, which equals

$$\frac{4}{3}\pi r^3 - \frac{1}{6}\pi x_2 \left(\frac{3r^2}{4} + x_2^2 \right)$$

$V_w = V_c + V_s$: the volume of the water container
S_c = the surface area of the cone, which equals

$$\frac{\pi r}{2} \sqrt{\frac{r^2}{4} + x_1^2}$$

S_s = the surface area of the sphere, which equals

$$4\pi r^2 - \pi \left(\frac{r^2}{4} + x_2^2 \right)$$

$S_T = S_c + S_s$: the total surface area

We want to maximize the volume V_w of the container. The total available surface area S_T
constrains the container's volume. Thus, the problem is

$$\text{Maximize } f(x_1, x_2) = \frac{\pi r^2}{12} x_1 + \frac{4}{3}\pi r^3 - \frac{1}{6}\pi x_2 \left(\frac{3r^2}{4} + x_2^2 \right) \qquad (13.7)$$

subject to

$$\frac{\pi r}{2}\sqrt{\frac{r^2}{4}+x_1^2}+4\pi r^2-\pi\left(\frac{r^2}{4}+x_2^2\right)=450$$

Model Solution   We employ the method of Lagrange multipliers to solve this equality constrained optimization problem. We set up the function

$$L(x_1, x_2, \lambda) = \frac{\pi r^2}{12}x_1 + \frac{4}{3}\pi r^3 - \frac{1}{6}\pi x_2\left(\frac{3r^2}{4}+x_2^2\right)$$

$$-\lambda\left[\frac{\pi r}{2}\sqrt{\frac{r^2}{4}+x_1^2}+4\pi r^2-\pi\left(\frac{r^2}{4}+x_2^2\right)-450\right]$$

We substitute $r = 6$ and $\pi = 3.14$ into the equation to simplify the expression:

$$L(x_1, x_2, \lambda) = 9.42x_1 - 14.13x_2 - 0.52x_2^3 + 904.32$$

$$-\lambda\left(9.42\sqrt{x_1^2+9}-3.14x_2^2-26.10\right)$$

Taking the partial derivatives of L with respect to x_1, x_2, and λ and setting them equal to zero give

$$\frac{\partial L}{\partial x_1} = 9.42 - 9.42\lambda \cdot x_1(x_1^2+9)^{-1/2} = 0$$

$$\frac{\partial L}{\partial x_2} = -14.13 - 1.57x_2^2 + 2\lambda \cdot 3.14x_2 = 0$$

$$\frac{\partial L}{\partial \lambda} = -9.42\sqrt{x_1^2+9}+3.14x_2^2+26.10 = 0$$

Using a computer algebra system, we find the solution to be

$$x_1 = 1.95 \text{ ft}, x_2 = 1.56 \text{ ft}, \lambda = 1.83, \text{ which implies } f(x_1, x_2) = 898.72 \text{ ft}^3$$

Model Sensitivity   The Lagrange multiplier value $\lambda = 1.8$ means that if the total surface area is increased by 1 unit, then the volume of the water container will increase by approximately 1.8 ft³.   ▪ ▪ ▪

13.3   PROBLEMS

1. Resolve the oil transfer problem when the storage capacity is 25 ft³. How does this result compare with our estimated value?

2. Resolve the water container problem when the surface area available is 500 ft² and the radius is 9 ft.

Use the method of Lagrange multipliers to solve Problems 3–6.

3. Find the minimum distance from the surface $x^2 + y^2 - z^2 = 1$ to the origin.

4. Find three numbers whose sum is 9 and whose sum of squares is as small as possible.

5. Find the hottest point (x, y, z) along the elliptical orbit

$$4x^2 + y^2 + 4z^2 = 16$$

where the temperature function is

$$T(x, y, z) = 8x^2 + 4yz - 16z + 600$$

6. *A Least Squares Plane*—Given the four points (x_k, y_k, z_k)

$$(0, 0, 0), (0, 1, 1), (1, 1, 1), (1, 0, -1)$$

find the values of A, B, and C to minimize the sum of squared errors

$$\sum_{k=1}^{4} (Ax_k + By_k + C - z_k)^2$$

if the points must lie in the plane

$$z = Ax + By + C$$

7. Resolve the oil transfer problem if we introduce a second storage tank that can be filled to its capacity of 30 ft^3. (*Hint:* You might reformulate with four variables x_{ij}, the amount of oil type i stored in storage tank j.)

13.3 PROJECTS (See enclosed CD for UMAP modules.)

1. "Lagrange Multipliers and the Design of Multistage Rockets," by Anthony L. Peressini, UMAP 517. The method of Lagrange multipliers is applied to compute the minimum total mass of an n-stage rocket capable of placing a given payload in an orbit at a given altitude above the earth's surface. Familiarity with elementary minimization techniques for functions of several variables, the method of Lagrange multipliers, and the concepts of linear momentum and conservation of momentum is required.

2. "Lagrange Multipliers: Applications to Economics," by Christopher H. Nevison, UMAP 270. The Lagrange multipliers method is interpreted and studied as the marginal rate of change of a utility function. Differential calculus through Lagrange multipliers is required.

3. Research the requirements for the necessary and sufficient conditions for the method of Lagrange multipliers, and prepare a 10-minute talk.

13.4   Managing Renewable Resources: The Fishing Industry

Consider the plight of the Antarctic baleen whale, which yielded a peak catch of 2.8 million tons in 1937 but only 50,000 tons in 1978. Or the case of the Peruvian anchoveta, which yielded 12.3 million tons in 1970 but only 500,000 tons just 8 years later. The anchoveta fishery was Peru's largest industry and the world's most productive fishery. The seriousness of the economic impact is realized when one considers that biologists estimate that even without fishing, it will take several years, or even decades in the case of the baleen whale, for these fisheries to reach their former levels of maximum biological productivity.[1]

Resources such as the anchoveta and baleen whale are called *renewable* (as opposed to *exhaustible*) resources. Exhaustible resources can yield only a finite total amount, whereas renewable resources can (theoretically) yield an unlimited total amount and be maintained at some positive level. The management of a renewable resource, such as a fishery, involves several critical considerations. What should the harvest rate be? How sensitive is the survival of the species to population fluctuation caused by harvesting or to natural disasters, such as a temporary alteration in the ocean currents (which contributed to the demise of the anchoveta)? The economist Adam Smith (1723–1790) proclaimed that each individual, in pursuing only his or her own selfish good, is led by an "invisible hand" to achieve the best for all. Will that invisible hand really ensure that market forces work in the best interest of humanity and the renewable resources? Or will intervention be required to improve the situation, either for humanity or for the resource? In this section we use some graphical submodels to gain a qualitative understanding of these management issues.

Scenario   Consider the harvesting of a common fish, such as haddock, in a large competitive fishing industry. *Given the population level of the fish at some time, what happens to future population levels?* Future population levels depend on, among other factors, the harvesting rate of the fish and their natural reproductive rate (births minus deaths per unit time). Let's develop submodels for harvesting and reproduction separately.

A harvesting submodel   Let's refer to the classical theory of the firm as presented in Chapter 2. The graphical models for total profit and marginal profit are reproduced in Figure 13.12. From Figure 13.12a we can see that a firm breaks even only when it can produce at least q_1 items, where total revenue equals total cost. To maximize profits the firm should continue to produce items as long as the marginal revenue exceeds the marginal cost—that is, produce up to the quantity q_2 as depicted in Figure 13.12b.

How is the theory of the firm related to the fishing industry? For a common fish such as the haddock, and a competitive fishing industry, it is probably reasonable to assume that price is a constant. If a common fish cannot be marketed at that price, consumers will simply switch to another type of fish. Likewise, in a large industry the quantity of a particular fish marketed by an individual firm should not affect the price of the fish. Hence, over a wide range of values, constant price appears to be a reasonable initial assumption.

---

[1]Data from Colin W. Clarke, "The Economics of Over-exploration." *Science* 181 (1973): 630–634.

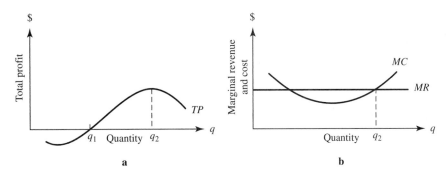

■ Figure 13.12

A graphical model for the theory of the firm

Next, let's consider the costs a fishing company encounters. These include salaries, fuel, capital tied up in equipment, processing, and refrigeration. As usual, once a time horizon is selected, these costs can be divided into two categories: fixed costs (such as capital costs), which are independent of the yield harvested, and variable costs (such as processing), which depend on the size of the yield. The basic principles underlying the theory of the firm apply reasonably well. For example, we would expect that a typical fishing company must harvest some minimum yield at a given level of effort (number of boats, person-hours of labor, and so forth) to break even.

An interesting condition exists with the fishing industry: The cost of harvesting a given number of fish depends on *the size of the fish population*. Certainly less effort is required to catch a given number of fish when they are plentiful, so at a given level of effort one would expect to catch more fish when they are plentiful. Thus, we assume the average unit harvesting cost $c(N)$ decreases as the size of the fish population N increases. This assumption is depicted in Figure 13.13, in which the harvesting cost per fish is shown as a decreasing function of the size of the fish population. It is important to note that the independent variable N is the size of the fish population, not the size of the yield. The market price p of a fish is also shown in the figure.

The submodel represented by Figure 13.13 suggests that it would be unprofitable to fish a particular species of fish unless the population level is at least N_L, where the cost of harvesting each fish equals the price paid for it by the consumer. In those instances in which it is economically feasible to harvest the species, harvesting is a force driving the population level to N_L. If the population level lies significantly above N_L, then the excess profit potential causes the fishing of the species to be intensified. On the other hand, if the

■ Figure 13.13

The (average) harvesting cost per fish decreases as the fish population level increases.

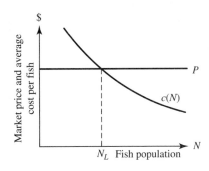

fish population falls below N_L, the fishing ceases (theoretically), causing the population level to increase.

We have reached an impasse with the harvesting submodel. Although we intended to develop the two submodels independently, we now realize that the average cost of harvesting a fish depends on the size of the fish population and hence its reproductive capacity. Even if we know the current size of the fish population, we still need to know how the magnitudes of the harvesting and reproduction rates compare to determine whether the population will increase or decrease. We turn now to the development of a simple reproduction submodel.

A Reproduction Submodel   Consider how a species of fish grows in the absence of fishing. Among the many factors affecting its net growth rate are the birthrate, death rate, and environmental conditions. Each of these factors was developed in detail in Chapter 11, in which we constructed several population models. For our purposes here, a simple qualitative graphical model will suffice. Let $N(t)$ denote the size of the fish population at any time t, and let $g(N)$ represent the rate of growth of the function $N(t)$. Assume that $g(N)$ is approximated by the continuous model $g(N) = dN/dt$. When $N = 0$, there are neither births nor deaths. Now assume there is a maximum population level N_u that can be supported by the environment. This level might be imposed by the availability of food, the effect of predators, or some similar inhibitor. At N_u we assume the births equal deaths, or $g(N_u) = 0$. Thus, under our assumptions, $g(N)$ has zeros at $N = 0$ and $N = N_u$ and positive values in between. A graphical representation of g is depicted in Figure 13.14.

■ **Figure 13.14**

A submodel for
reproduction

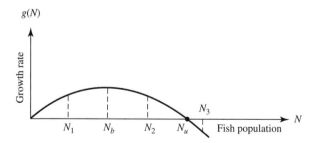

Before proceeding, we offer one analytic submodel for reproduction as an illustration. A simple quadratic function having zeros at $N = 0$ and $N = N_u$ is $g(N) = aN(N_u - N)$, where a is a positive constant. Note that the quadratic is positive for $0 < N < N_u$ and negative for $N > N_u$ as required. Because $g'(N) = a(N_u - 2N)$, $N = N_u/2$ is a critical point of g. Also, $g''(N) = -2a$ implies g is everywhere concave downward, as depicted in Figure 13.14.

Let's see what the submodel for reproduction suggests about the population level. Suppose the population level is currently at level N_1 (Figure 13.14). Because $g(N_1) > 0$ and $g(N) = dN/dt$, the function N is increasing. As N increases, $g(N) = dN/dt$ increases as well until it reaches a maximum at $N = N_b$. For $N > N_b$, the derivative remains positive but becomes smaller and approaches zero as N gets closer to N_u. Thus, the population approaches $N = N_u$ as suggested in Figure 13.15. Convince yourself that the curves suggested for starting populations of $N = N_2$ and $N = N_3$ are qualitatively correct in Figure 13.15, where $N_1 < N_b < N_2 < N_u < N_3$. Note that the ordinate in Figure 13.14

Figure 13.15

Regardless of the initial population, the population approaches $N = N_u$ without fishing.

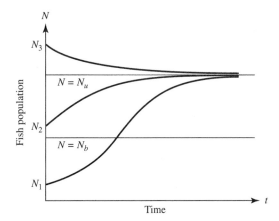

is the slope of the curves in Figure 13.15. In Chapter 11 we developed an analytic model using differential equations to represent the population growth in Figure 13.15, which is called *logistic* growth.

The Biological Optimum Population Level   We now interpret our submodel for reproduction. In the absence of fishing, the population always approaches N_u, which is called the *carrying capacity* of the environment. Moreover, a population level exists where the net biological growth is at a maximum. This population level is called *biological optimum population* and is denoted by N_b. In our particular submodel, $N_b = N_u/2$, as established earlier and depicted in Figure 13.14. (We established this result more precisely in Chapter 11.)

The Social Optimum Yield   Now that we have developed a submodel for reproduction, let's return to our harvesting submodel. Assume that the fishing industry seeks to maximize total profit, which is the product of the yield and the average profit per fish. The average profit per fish is the difference between the price p and the average cost $c(N)$ to harvest a fish. Thus, the expression for total profit TP is

$$\text{TP} = (\text{yield}) \times [p - c(N)] \tag{13.8}$$

What shall we assume about the yield? From the fishing company's point of view, the ideal situation is to have a constant yield from year to year. This situation would allow planning for efficient staffing and capitalization of the resources allocated to fishing that particular species. Otherwise, in lean years there would be resources, such as fishing vessels and staff, that would be underused. It is also reasonable to assume that consumer demand for a certain kind of fish is pretty much constant from year to year, for example, for a 3- or 4-year period. Let's see where the assumption of constant yield leads us.

First, how is constant yield obtained from year to year? One way is to harvest the difference between the births and deaths over time. Because the function $g(N)$ approximates this difference for a given population N of fish, this idea suggests that $g(N)$ equals the yield. To illustrate the idea, consider Figure 13.16. Suppose the fish population is currently at N_2. If precisely the amount of fish $g(N_2)$ are harvested, the population will remain at N_2 (zero net growth) and the fishing operation could be repeated annually indefinitely. Convince

■ **Figure 13.16**

Harvesting $g(N)$ annually
permits a sustainable yield.

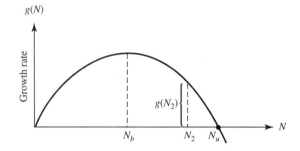

yourself that any population level N with $0 < N < N_u$ can be maintained by harvesting
the amount $g(N)$ per year.

Note that N_b, the population level for which the biological growth is a maximum, is of
particular importance under interpretation of a sustainable yield. This is true because N_b is
the population level that allows the largest sustainable yield, $g(N_b)$. Because this population
level has important social implications, we refer to N_b as the *social optimum population
level*. Note that the social and biological optimum levels are coincident in this example.

The Economic Optimum Population Level   Assuming the yield to be $g(N)$, Equation (13.8) for total profit becomes

$$TP = g(N)[p - c(N)] \tag{13.9}$$

The fishing industry wants to maximize TP. What level of fish population permits the
greatest profit? Because $g(N)$ reaches a maximum at $N = N_b$, we might be tempted to
conclude that profit is maximized there also. After all, the greatest yield occurs there.
However, the average profit continues to increase as N increases, based on our submodel
depicted in Figure 13.13. Thus, by choosing $N > N_b$, we may increase the profit while
simultaneously catching fewer fish because $g(N) < g(N_b)$. Let's see if this possibility is
indeed the case.

What does the graph of the total profit function TP look like? The factor $g(N)$ has zeros
at $N = 0$ and $N = N_u$, and the factor $[p - c(N)]$ has a zero at $N = N_L$. One representation
of a continuous function meeting these requirements is shown in Figure 13.17 and suggests
that a population level does exist at which profits can be maximized. We call that population
level the *economic optimum population level* and denote it by N_p.

■ **Figure 13.17**

A continuous total profit
function

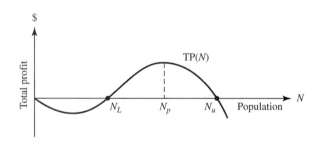

Relating All Optimal Population Levels   Up to this point several population levels of interest to us have been introduced. They are summarized as follows:

$$N_L = \text{minimum population level for economic feasibility}$$
$$N_b = \text{social and biological optimum population levels}$$
$$N_p = \text{economic optimum population level}$$
$$N_u = \text{maximum population sustainable by the environment}$$
$$\text{(the environmental carrying capacity)}$$

Next, we relate these population levels. In order that it be worthwhile to fish a species on a prolonged basis, it must be the case that $N_L < N_u$, with N_b and N_p somewhere in between. However, how are N_b and N_p related? If we believe that the fishing industry will eventually find the population level that maximizes profit, we will hope, for social and biological considerations, that $N_p = N_b$. Let's see whether Adam Smith's invisible hand is at work.

Consider Figure 13.18. For the case $N_L < N_p < N_u$, three possible locations for N_b are shown on the graph for the total profit function TP. Study the graph and convince yourself of the following situations:

1. $N_b < N_p$ if and only if TP$'(N_b) > 0$.

2. $N_b = N_p$ if and only if TP$'(N_b) = 0$.

3. $N_b > N_p$ if and only if TP$'(N_b) < 0$.

Now, taking the derivative of Equation (13.9), we have

$$\text{TP}' = g'(N)[p - c(N)] - g(N)c'(N) \tag{13.10}$$

■ **Figure 13.18**

Three possible locations for N_b are $N_b < N_p$, $N_b = N_p$, $N_b > N_p$.

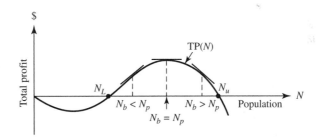

From the definition of N_b, $g'(N_b) = 0$ and $g(N_b) > 0$, and from the submodel for average cost depicted in Figure 13.13, $c'(N) < 0$ for all N. Substitution of N_b into Equation (13.10) yields TP$'(N_b) > 0$, which implies that $N_b < N_p$ from Figure 13.18. Thus, under our assumptions, by allowing the fish population to exceed N_b the fishing company can catch fewer fish but reap greater profits than is possible at N_b! Let's interpret our model to assist management in determining the alternatives for controlling the location of the various population levels N_L, N_b, and N_p.

Model Interpretation   As we saw in Chapter 11, the assumptions underlying the reproduction submodel are extremely simplistic, so we cannot expect to draw precise conclusions from the graphical models developed here. However, our purpose in constructing the model was to identify and analyze qualitatively some of the key issues in managing a renewable resource. Consider again the baleen whale and Peruvian anchoveta in terms of the models we have developed.

Several considerations must be taken into account with whales. First, the whale population displays a low natural growth rate (typically only 5–10% a year), and even a small harvest can result in a negative net growth rate when the population is low. Second, many conservationists argue that there is a minimum population N_s for the whale below which the species will not survive. If harvesting or natural disasters cause the whale population to fall below N_s, the species will be driven to extinction. These ideas are incorporated in the graphical representations depicted in Figure 13.19, showing both the reproduction and population submodels.

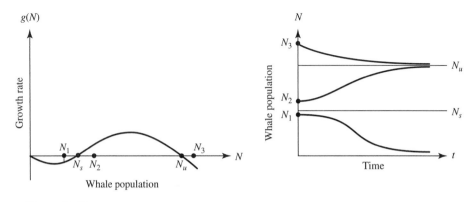

■ Figure 13.19

Reproduction and population models incorporating a minimum survival level

Note from Figure 13.19 how sensitive the future whale population is to levels in the vicinity of N_s. The relatively high market price for the whale causes the minimum population N_L for economic feasibility to be quite low (Figure 13.13) so that N_L can be quite near the level N_s. Thus, the location of N_L becomes critical. (Argue that in the case of the whale, N_L could be smaller than N_s. Management of the fishery in such a case requires separating N_s and N_L.) Ideally, one would like N_L and N_b to coincide. Taxation and the establishment of quotas are two alternatives that are discussed in the following problem set.

In the case of the Peruvian anchoveta, it is estimated that the maximum sustainable yield is 10 million tons annually. Thus, any additional catch (such as the 12.3 million tons caught in 1970) would result in population decrease, even if the population were at the biological optimum level N_b. From Figure 13.16 we suspect that as the population decreases from the biological optimum population, the sustainable yield decreases, suggesting that the harvest should be reduced. However, from a practical point of view, once the fishing fleet and staff are established to catch 12.3 million tons, there is strong economic motivation to continue harvesting at that level by fishing more intensively. This critical situation existed during the 1970s in Peru, where there were few economic alternatives, and government intervention to enforce strict regulation of the fishery was politically volatile. We can well imagine how these practical considerations become substantially more complex when more than one country is involved in managing the resource.

PROBLEMS

1. Assume that the environmental carrying capacity N_u is determined principally by the availability of food. Argue that under such an assumption, as N approaches N_u the physical condition of the average fish deteriorates as competition for the food supply becomes more severe. What does this suggest about the survival of the species when natural disasters such as storms, severe winters, and similar circumstances further restrict the food supply? Where should conservationists attempt to maintain the population level?

2. In 1981 and 1982, the deer population in the Florida Everglades was very high. Although the deer were plentiful, they were on the brink of starvation. Hunting permits were issued to thin out the herd. This action caused much furor on the part of environmentalists and conservationists. Explain the poor health of the deer and the purpose of the special hunting permits in terms of population growth and population submodels.

3. Argue that for many species, a minimum population level is required for survival. Give several examples. Call this minimum survival level N_s. Suggest a simple cubic growth submodel meeting these requirements, as depicted in Figure 13.19. Answer the questions posed in Problem 1 using your graphical submodel.

4. Suppose $N_u < N_L$. What does this inequality suggest about the economic feasibility of fishing that species? Give several examples.

Problems 5 and 6 are related to fishing regulation.

5. One of the key assumptions underlying the models developed in this section is that the harvest rate equals the growth rate for a sustainable yield. The reproduction submodels in Figures 13.16 and 13.19 suggest that if the current population levels are known, it is possible to estimate the growth rate. The implication of this knowledge is that if a quota for the season is established based on the estimated growth rate, then the fish population can be maintained, increased, or decreased as desired. This quota system might be implemented by requiring all commercial fishermen to register their catch daily and then closing the season when the quota is reached. Discuss the difficulties in determining reproduction models precise enough to be used in this manner. How would you estimate the population level? What are the disadvantages of having a quota that varies from year to year? Discuss the practical political difficulties of implementing such a procedure.

6. One of the difficulties in managing a fishery in a free-enterprise system is that excess capacity may be created through overcapitalization. This happened in 1970 when the capacity of the Peruvian anchoveta fishermen was sufficient to catch and process the maximum annual growth rate in less than 3 months. A disadvantage of restricting access to the fishery by closing the season after a quota is reached is that this excess capacity is idle during much of the season, which creates a politically and economically unsatisfying situation. An alternative is to control the capacity in some manner. Suggest several procedures for controlling the capacity that is developed. What difficulties would be involved in implementing a procedure such as restricting the number of commercial fishing permits issued?

Problems 7–9 are related to taxation.

7. Figure 13.13 suggests that market forces tend to drive the population to N_L. Use that figure to show how taxation or subsidization may be used to control the location of N_L. What forms might the taxation and subsidization take? (*Hint:* One cost to the fisherman is the various taxes he pays.) Apply your ideas to the whale fishery.

8. Taxation is appealing from a theoretical perspective because with a properly designed tax, desired goals can be achieved through normal market forces rather than by some artificial method (such as restricting the number of commercial permits). Assume a fish population is currently at N_b and you want to maintain it at that level by harvesting $g(N_b)$. How can a tax be determined for each fish caught to cause N_L, N_b, and N_p to coincide? (*Hint:* Consider Equation (13.8) and the condition required for $N_b = N_p$.)

9. A constant price has been assumed in all the models developed in this section. Suggest some fisheries for which that assumption is not realistic. How might you alter the assumption? How would you determine the appropriate tax?

13.4 Further Reading

Clark, Colin W. *Mathematical Bioeconomics: The Optimal Management of Renewable Resources.* New York: Wiley, 1976.

May, Robert M., John R. Beddington, Colin W. Clark, Sidney J. Holt, & Richard M. Laws. "Management of Multispecies Fisheries." *Science* 205 (July 1979): 267–277.

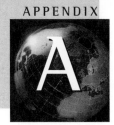

Problems from the Mathematics Contest in Modeling, 1985–2008

1985: The Animal Population Problem

Choose a fish or mammal for which appropriate data are available to model it accurately. Model the animal's natural interactions with its environment by expressing population levels of different groups in terms of the significant parameters of the environment. Then adjust the model to account for harvesting in a form consistent with the actual method by which the animal is harvested. Include any outside constraints imposed by food or space limitations that are supported by the data. Consider the value of the various quantities involved, the number harvested, and the population size to devise a numerical quantity that represents the overall value of the harvest. Find a harvesting policy in terms of population size and time that optimizes the value of the harvest over a long period of time. Check that the policy optimizes this value over a realistic range of environmental conditions.

1985: The Strategic Reserve Problem

Cobalt, which is not produced in the United States, is essential to a number of industries. (Defense accounted for 17% of the cobalt production in 1979.) Most cobalt comes from central Africa, a politically unstable region. The Strategic and Critical Materials Stockpiling Act of 1946 requires a cobalt reserve that will carry the United States through a 3-year war. The government built up a cobalt stockpile in the 1950s, sold most of it in the early 1970s, and then decided to build it up again in the late 1970s, with a stockpile goal of 85.4 million pounds. About half of this stockpile had been acquired by 1982.

Build a mathematical model for managing a stockpile of the strategic metal cobalt. You will need to consider such questions as

- How big should the stockpile be?
- At what rate should it be acquired?
- What is a reasonable price to pay for the metal?

    You will also want to consider such questions as

- At what point should the stockpile be drawn down?
- At what rate should it be drawn down?

- What is a reasonable price at which to sell the metal?
- How should sold metal be allocated?

Below we give more information on the sources, cost, demand, and recycling aspects of cobalt.

Useful Information on Cobalt

The government has projected a need of 25 million pounds of cobalt in 1985.

The United States has about 100 million pounds of proven cobalt deposits. Production becomes economically feasible when the price reaches $22/lb (as occurred in 1981). It takes 4 years to get operations rolling, and then 6 million pounds per year can be produced.

In 1980, 1.2 million pounds of cobalt were recycled, 7% of total consumption.

Please see Figures A.1–A.3, whose source is *Mineral Facts and Problems*, United States Bureau of Mines (Washington, DC: Government Printing Office, 1980).

■ Figure A.1

U.S. primary demand for cobalt, 1960–1980

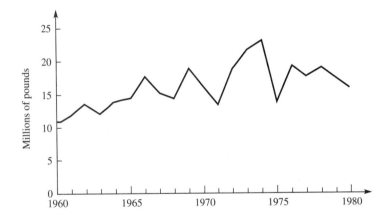

■ Figure A.2

Cobalt prices in the U.S. market, 1960–1982

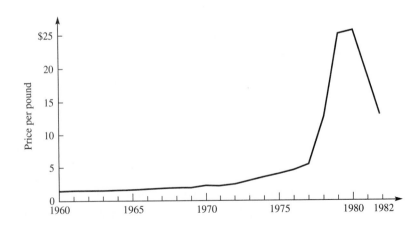

■ Figure A.3

Producers of refined metal and/or oxide 1979; an asterisk denotes a country with domestic production. Source: U.S. Bureau of Mines, *Mineral Facts and Problems* (1980)

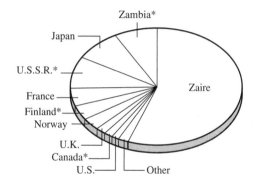

1986: The Hydrographic Data Problem

The following table gives the depth Z of water in feet for surface points with rectangular coordinates X, Y in yards. The depth measurements were taken at low tide. Your ship has a draft of 5 feet. What region should you avoid within the rectangle $(150, -50) \times (200, 75)$?

X	Y	Z
129.0	7.5	4
140.0	141.5	8
108.5	28.0	6
88.0	147.0	8
185.5	22.5	6
195.0	137.5	8
105.5	85.5	8
157.5	−6.5	9
107.5	−81.0	9
77.0	3.0	8
162.0	−66.5	9
162.0	84.0	4
117.5	−38.5	9

1986: The Emergency-Facilities Location Problem

The township of Rio Rancho has hitherto not had its own emergency facilities. It has secured funds to erect two emergency facilities in 1986, each of which will combine ambulance, fire, and police services. Figure A.4 indicates the demand, or number of emergencies per square block, for 1985. The L region in the north is an obstacle, whereas the rectangle in the south is a park with a shallow pond. It takes an emergency vehicle an average of 15 seconds

Figure A.4

A map of Rio Rancho, with number of emergencies in 1985 indicated for each block

3	1	4	2	5
3	2	3	3	2
2		3	3	2
3	0		3	1
3	4	3	3	5
2	3	4	4	0
1	2	0	1	3
0	2	0	3	2
3	0		0	4
3	1	0	4	2

N

to go one block in the N–S direction and 20 seconds in the E–W direction. Your task is to locate the two facilities so as to minimize the total response time.

- Assume that the demand is concentrated at the center of the block and that the facilities will be located on corners.

- Assume that the demand is uniformly distributed on the streets bordering each block and that the facilities may be located anywhere on the streets.

1987: The Salt Storage Problem

For approximately 15 years, a midwestern state has stored salt used on roads in the winter in circular domes. Figure A.5 shows how salt has been stored in the past. The salt is brought

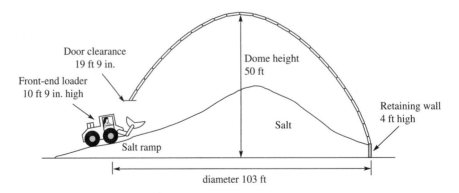

Figure A.5

Diagram of a salt storage dome

into and removed from the domes by driving front-end loaders up ramps of salt leading into the domes. The salt is piled 25–30 ft high, using the buckets on the front-end loaders.

Recently, a panel determined that this practice is unsafe. If the front-end loader got too close to the edge of the salt pile, the salt might shift, and the loader could be thrown against the retaining walls that reinforce the dome. The panel recommended that if the salt is to be piled with the use of loaders, then the piles should be restricted to a maximum height of 15 ft.

Construct a mathematical model for this situation and find a recommended maximum height for salt in the domes.

1987: The Parking Lot Problem

The owner of a paved, 100-ft-by-200-ft corner parking lot in a New England town hires you to design the layout—that is, to design how the lines are to be painted.

You realize that squeezing as many cars into the lot as possible leads to right-angle parking with the cars aligned side by side. However, inexperienced drivers have difficulty parking their cars this way, which can give rise to expensive insurance claims. To reduce the likelihood of damage to parked vehicles, the owner might then have to hire expert drivers for valet parking. On the other hand, most drivers seem to have little difficulty in parking in one attempt if there is a large enough turning radius from the access lane. Of course, the wider the access lane, the fewer cars that can be accommodated in the lot, leading to less revenue for the parking lot owner.

1988: The Railroad Flatcar Problem

Two railroad flatcars are to be loaded with seven types of packing crates. The crates have the same width and height but vary in thickness (t, in cm) and weight (w, in kg). Table A.1 gives, for each crate, the thickness, weight, and number available. Each car has 10.2 m of length available for packing the crates (like slices of toast) and can carry up to 40 metric tons. There is a special constraint on the total number of C_5, C_6, and C_7 crates because of a subsequent local trucking restriction: The total space (thickness) occupied by these crates must not exceed 302.7 cm. Load the two flatcars (Figure A.6) so as to minimize the wasted floor space.

Table A.1   The thickness, weight, and number of each kind of crate

	C_1	C_2	C_3	C_4	C_5	C_6	C_7	
t	48.7	52.0	61.3	72.0	48.7	52.0	64.0	cm
w	2000	3000	1000	500	4000	2000	1000	kg
	8	7	9	6	6	4	8	

■ Figure A.6

Diagram of loading
of a flatcar

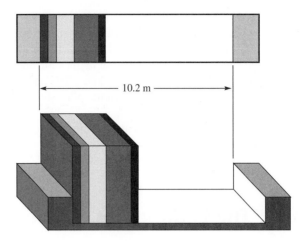

1988: The Drug Runner Problem

Two listening posts 5.43 miles apart pick up a brief radio signal. The sensing devices were
oriented at 110° and 119°, respectively, when the signal was detected (Figure A.7), and they
are accurate to within 2°. The signal came from a region of active drug exchange, and it
is inferred that there is a powerboat waiting for someone to pick up drugs. It is dusk, the
weather is calm, and there are no currents. A small helicopter leaves a pad from Post 1 and
is able to fly accurately along the 110° angle direction. The helicopter's speed is three times
the speed of the boat. The helicopter will be heard when it gets within 500 ft of the boat.
This helicopter has only one detection device, a searchlight. At 200 ft, it can just illuminate
a circular region with a radius of 25 ft.

• Describe the (smallest) region where the pilot can expect to find the waiting boat.

• Develop an optimal search method for the helicopter.

   Use a 95% confidence level in your calculations.

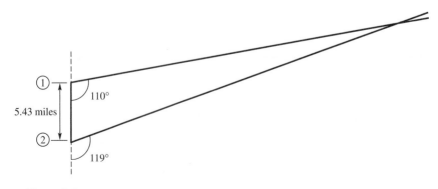

■ Figure A.7

Geometry of the problem

1989: The Aircraft Queuing Problem

A common procedure at airports is to assign aircraft (AC) to runways on a first-come-first-served basis. That is, as soon as an AC is ready to leave the gate (push back), the pilot calls ground control and is added to the queue. Suppose that a control tower has access to a fast online database with the following information for each AC:

- The time it is scheduled for pushback
- The time it actually pushes back
- The number of passengers on board
- The number of passengers who are scheduled to make a connection at the next stop, as well as the time to make that connection
- The schedule time of arrival at its next stop

Assume that there are seven types of AC with passenger capacities varying from 100 to 400 in steps of 50. Develop and analyze a mathematical model that takes into account both the travelers' and the airlines' satisfaction.

1989: The Midge Classification Problem

Two species of midges, Af and Apf, have been identified by biologists Grogan and Wirth (1981) on the basis of antenna and wing length (Figure A.8). Each of nine Af midges is denoted by □, and each of six Apf midges is denoted by ○. It is important to be able to classify a specimen as Af or Apf, given the antenna and wing length.

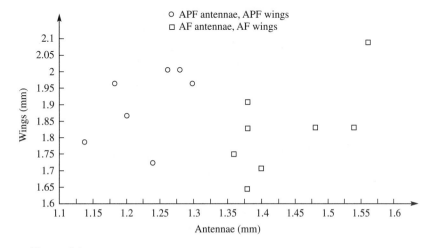

■ Figure A.8

Display of data collected by Grogan and Wirth (1981)

1. Given a midge that you know is species Af or Apf, how would you go about classifying it?

2. Apply your method to three specimens with (antenna, wing) lengths (1.24, 1.80), (1.28, 1.84), (1.40, 2.04).

3. Assume that species Af is a valuable pollinator and that species Apf is a carrier of a debilitating disease. Would you modify your classification scheme and if so, how?

1990: The Brain–Drug Problem

Researchers on brain disorders test the effects of new medical drugs (e.g., dopamine against Parkinson's disease) with intracerebral injections. To this end, they must estimate the size and the shape of the spatial distribution of the drug after the injection to estimate accurately the region of the brain that the drug has affected.

The research data consist of the measurements of the amounts of drug in each of 50 cylindrical tissue samples (Figure A.9 and Table A.2). Each cylinder has length 0.76 mm and diameter 0.66 mm. The centers of the parallel cylinders lie on a grid with mesh $1 \times 0.76 \times 1$ mm so that the cylinders touch one another on their circular bases but not along their sides, as shown in the accompanying figure. The injection was made near the center of the cylinder with the highest scintillation count. Naturally, one expects that there is drug also between the cylinders and outside the region covered by the sample.

■ Figure A.9

Orientation of the cylinders of tissue

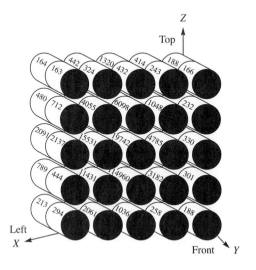

Estimate the distribution in the region affected by the drug.

One unit represents a scintillation count, or 4.753×10^{-13} mole of dopamine. For example, the table shows that the middle rear cylinder contains 28,353 units.

Table A.2   Amounts of drug in each of 50 cylindrical tissue samples

		Rear vertical section		
164	442	1,320	414	188
480	7,022	14,411	5,158	352
2,091	23,027	28,353	13,138	681
789	21,260	20,921	11,731	727
213	1,303	3,765	1,715	453
		Front vertical section		
163	324	432	243	166
712	4,055	6,098	1,048	232
2,137	15,531	19,742	4,785	330
444	11,431	14,960	3,182	301
294	2,061	1,036	258	188

1991: The Water Tank Problem

Some state water-right agencies require from communities data on the rate of water use, in gallons per hour, and the total amount of water used each day. Many communities do not have equipment to measure the flow of water into or out of the municipal tank. Instead, they can measure only the *level* of water in the tank, within 0.5% accuracy, every hour. More important, whenever the level in the tank drops below some minimum level L, a pump fills the tank up to the maximum level, H; however, there is no measurement of the pump flow either. Thus, one cannot readily relate the level in the tank to the amount of water used while the pump is working, which occurs once or twice per day, for a couple of hours each time.

Estimate the flow out of the tank $f(t)$ at all times, even when the pump is working, and estimate the total amount of water used during the day. Table A.3 gives the real data from an actual small town for one day.

The table gives the time, in seconds, since the first measurement and the level of water in the tank, in hundredths of a foot. For example, after 3316 seconds, the depth of water in

Table A.3   Water tank levels over a single day for a small town (time is in seconds and level is in 0.01 ft)

Time	Level	Time	Level	Time	Level
0	3,175	35,932	pump on	68,535	2,842
3,316	3,110	39,332	pump on	71,854	2,767
6,635	3,054	39,435	3,550	75,021	2,697
10,619	2,994	43,318	3,445	79,254	pump on
13,937	2,947	46,636	3,350	82,649	pump on
17,921	2,892	49,953	3,260	85,968	3,475
21,240	2,850	53,936	3,167	89,953	3,397
25,223	2,797	57,254	3,087	93,270	3,340
28,543	2,752	60,574	3,012		
32,284	2,697	64,554	2,927		

the tank reached 31.10 ft. The tank is a vertical circular cylinder, with a height of 40 ft and a diameter of 57 ft. Usually, the pump starts filling the tank when the level drops to about 27 ft, and the pump stops when the level rises back to about 35.50 ft.

1991: The Steiner Tree Problem

The cost for a communication line between two stations is proportional to the length of the line. The cost for conventional minimal spanning trees of a set of stations can often be cut by introducing phantom stations and then constructing a new *Steiner tree*. This device allows costs to be cut by up to 13.4% ($= 1 - \sqrt{3}/2$). Moreover, a network with n stations never requires more than $n - 2$ points to construct the cheapest Steiner tree. Two simple cases are shown in Figure A.10.

■ **Figure A.10**

Two simple cases of forming the shortest Steiner tree for a network

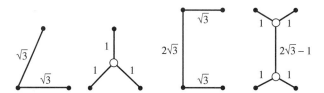

For local networks, it is often necessary to use rectilinear or checkerboard distances instead of straight Euclidean lines. Distances in this metric are computed as shown in Figure A.11.

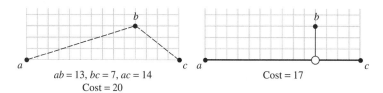

■ **Figure A.11**

Comparison of distances using straight Euclidean line distances ($ab = 13$, $bc = 7$, $ac = 14$; cost $= 20$) versus using rectilinear distances (cost $= 17$)

Suppose you wish to design a minimal-cost spanning tree for a local network with nine stations. Their rectangular coordinates are

$$a(0, 15), \quad b(5, 20), \quad c(16, 24), \quad d(20, 20), \quad e(33, 25),$$

$$f(23, 11), \quad g(35, 7), \quad h(25, 0), \quad i(10, 3)$$

You are restricted to using rectilinear lines. Moreover, all phantom stations must be located at lattice points (i.e., the coordinates must be integers). The cost for each line is its length.

1. Find a minimal-cost tree for the network.
2. Suppose each station has a cost $d^{3/2}w$, where d is the degree of the station. If $w = 1.2$, find a minimal-cost tree.
3. Try to generalize this problem.

1992: The Emergency Power-Restoration Problem

Power companies serving coastal regions must have emergency-response systems for power outages due to storms. Such systems require the input of data that allow the time and cost required for restoration to be estimated and the value of the outage judged by objective criteria. In the past, Hypothetical Electric Company (HECO) has been criticized in the media for its lack of a prioritization scheme.

You are a consultant to HECO power company. HECO possesses a computerized database with real-time access to service calls that currently require the following information:

- Time of report
- Type of requestor
- Estimated number of people affected
- Location (x, y)

Crew sites are located at coordinates $(0, 0)$ and $(40, 40)$, where x and y are in miles. The region serviced by HECO is within $-65 < x < 65$ and $-50 < y < 50$. The region is largely metropolitan with an excellent road network. Crews must return to their dispatch site only at the beginning and end of each shift. Company policy requires that no work be initiated until the storm leaves the area, unless the facility is a commuter railroad or hospital, which may be processed immediately if crews are available.

HECO has hired you to develop the objective criteria and schedule the work for the storm restoration requirements listed in Table A.5 using the work force described in Table A.4. Note that the first call was received at 4:20 A.M. and that the storm left the area at 6:00 A.M. Also note that many outages were not reported until much later in the day.

Table A.4   Crew descriptions

- Dispatch locations at $(0, 0)$ and $(40, 40)$.
- Crews consist of three trained workers.
- Crews report to the dispatch location only at the beginning and end of their shifts.
- One crew is scheduled for duty at all times on jobs assigned to each dispatch location. These crews would normally be performing routine assignments. Until the storm leaves the area, they can be dispatched for emergencies only.
- Crews work 8-hr shifts.
- There are six crew teams available at each location.
- Crews can work only one overtime shift in a work day and receive time-and-a-half for overtime.

Table A.5   Storm restoration requirements

Time (A.M.)	Location	Type	# Affected	Estimated repair time (hr for crew)
4:20	(−10, 30)	Business (cable TV)	?	6
5:30	(3, 3)	Residential	20	7
5:35	(20, 5)	Business (hospital)	240	8
5:55	(−10, 5)	Business (railroad system)	25 workers; 75,000 commuters	5
6:00	All-clear given; storm leaves area; crews can be dispatched			
6:05	(13, 30)	Residential	45	2
6:06	(5, 20)	Area	2,000	7
6:08	(60, 45)	Residential	?	9
6:09	(1, 10)	Government (city hall)	?	7
6:15	(5, 20)	Business (shopping mall)	200 workers	5
6:20	(5, −25)	Government (fire dept.)	15 workers	3
6:20	(12, 18)	Residential	350	6
6:22	(7, 10)	Area	400	12
6:25	(− 1, 19)	Industry (newspaper co.)	190	10
6:40	(−20, −19)	Industry (factory)	395	7
6:55	(− 1, 30)	Area	?	6
7:00	(−20, 30)	Government (high school)	1,200 students	3
7:00	(40, 20)	Government (elementary school)	1,700	?
7:00	(7, −20)	Business (restaurant)	25	12
7:00	(8, −23)	Government (police station & jail)	125	7
7:05	(25, 15)	Government (elementary school)	1,900	5
7:10	(−10, −10)	Residential	?	9
7:10	(− 1, 2)	Government (college)	3,000	8
7:10	(8, −25)	Industry (computer manuf.)	450 workers	5
7:10	(18, 55)	Residential	350	10
7:20	(7, 35)	Area	400	9
7:45	(20, 0)	Residential	800	5
7:50	(− 6, 30)	Business (hospital)	300	5
8:15	(0, 40)	Business (several stores)	50	6
8:20	(15, −25)	Government (traffic lights)	?	3
8:35	(−20, −35)	Business (bank)	20	5
8:50	(47, 30)	Residential	40	?
9:50	(55, 50)	Residential	?	12
10:30	(−18, −35)	Residential	10	10
10:30	(− 1, 50)	Business (civic center)	150	5
10:35	(− 7, − 8)	Business (airport)	350 workers	4
10:50	(5, −25)	Government (fire dept.)	15	5
11:30	(8, 20)	Area	300	12

HECO has asked for a technical report for its purposes and an executive summary in lay terms that can be presented to the media. Furthermore, it would like recommendations for the future. To determine your prioritized scheduling system, you will have to make additional assumptions. Detail those assumptions. In the future, you may desire additional data. If so, detail the information desired.

1992: The Air-Traffic-Control Radar Problem

You are to determine the power to be radiated by an air-traffic-control radar at a major metropolitan airport. The airport authority wants to minimize the power of the radar consistent with safety and cost.

The authority is constrained to operate with its existing antennae and receiver circuitry. The only option it is considering is upgrading the transmitter circuits to make the radar more powerful.

The question that you are to answer is what power (in watts) must be released by the radar to ensure detection of standard passenger aircraft at a distance of 100 km.

■ **Figure A.12**

Measurements for the radar system

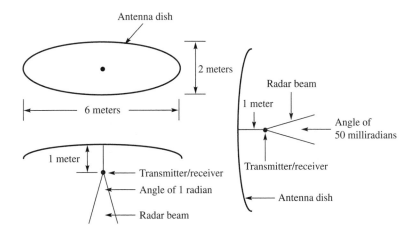

Technical specifications (see also Figure A.12):

1. The radar antenna is a section of a paraboloid of revolution with focal length of 1 meter. Its projection onto a plane tangent to its vertex is an ellipse with a major axis of 6 meters and a minor axis of 2 meters. The main lobe energy beam pattern, located at the focus, is an elliptical cone that has a major axis of 1 radian and a minor axis of 50 milliradians. The antenna and beam are sketched in the figures provided.

2. The nominal class of aircraft is one that has an effective radar reflection cross section of 75 square meters. For the purposes of this problem, this means that in your initial model, the aircraft is equivalent to a 100% reflective circular disc of 75 square meters, which is centered on the axis of the antennae and is perpendicular to it. You may want to consider alternatives or refinements to this initial model.

1993: The Coal-Tipple Operations Problem

The Aspen–Boulder Coal Company runs a loading facility consisting of a large coal tipple. When the coal trains arrive, they are loaded from the tipple. The standard coal train takes 3 hours to load, and the tipple's capacity is 1.5 standard trainloads of coal. Each day, the railroad sends three standard trains to the loading facility, and they arrive at any time between 5 A.M. and 8 P.M. local time. Each of the trains has three engines. If a train arrives and sits idle while waiting to be loaded, the railroad charges a special fee, called a *demurrage*. The fee is $5000 per engine per hour. In addition, a high-capacity train arrives once a week every Thursday between 11 A.M. and 1 P.M. This special train has five engines and holds twice as much coal as a standard train. An empty tipple can be loaded directly from the mine to its capacity in 6 hours by a single loading crew. This crew (and its associated equipment) costs $9000 per hour. A second crew can be called out to increase the loading rate by conducting an additional tipple-loading operation at the cost of $12,000 per hour. Because of safety requirements, during tipple loading no trains can be loaded. Whenever train loading is interrupted to load the tipple, demurrage charges are in effect.

The management of the coal company has asked you to determine the expected annual costs of this tipple's loading operations. Your analysis should include the following considerations:

- How often should the second crew be called out?
- What are the expected monthly demurrage costs?
- If the standard trains could be scheduled to arrive at precise times, what daily schedule would minimize loading costs?
- Would a third tipple-loading crew at $12,000 per hour reduce annual operations costs?
- Can this tipple support a fourth standard train every day?

1993: The Optimal Composting Problem

An environmentally conscious institutional cafeteria is recycling customers' uneaten food into compost by means of microorganisms. Each day, the cafeteria blends the leftover food into a slurry, mixes the slurry with crisp salad wastes from the kitchen and a small amount of shredded newspaper, and feeds the resulting mixture to a culture of fungi and soil bacteria, which digest slurry, greens, and paper into usable compost. The crisp greens provide pockets of oxygen for the fungi culture, and the paper absorbs excess humidity. At times, however, the fungi culture appears unable or unwilling to digest as much of the leftovers as customers leave; the cafeteria does not blame the chef for the fungi culture's lack of appetite. Also, the cafeteria has received offers for the purchase of large quantities of its compost. Therefore, the cafeteria is investigating ways to increase its production of compost. Because it cannot yet afford to build a new composting facility, the cafeteria seeks methods to accelerate the fungi culture's activity—for instance, by optimizing the fungi culture's environment (currently held at about 120° F and 100% humidity), by optimizing the composition of the mixture fed to the fungi culture, or both.

Determine whether any relation exists between the proportions of slurry, greens, and paper in the mixture fed to the fungi culture and the rate at which the fungi culture composts the mixture. If no relation exists, state so. Otherwise, determine what proportions would accelerate the fungi culture's activity.

In addition to the technical report following the format prescribed in the contest instructions, provide a 1-page nontechnical recommendation for implementation for the cafeteria manager.

Table A.6 shows the composition of various mixtures, in pounds of each ingredient kept in separate bins, and the time it took the fungi culture to compost the mixtures, from the date fed to the date completely composted.

Table A.6   Composting data

Slurry (pounds)	Greens (pounds)	Paper (pounds)	Fed (date)	Composted (date)
86	31	0	13 Jul 90	10 Aug 90
112	79	0	17 Jul 90	13 Aug 90
71	21	0	24 Jul 90	20 Aug 90
203	82	0	27 Jul 90	22 Aug 90
79	28	0	10 Aug 90	12 Sep 90
105	52	0	13 Aug 90	18 Sep 90
121	15	0	20 Aug 90	24 Sep 90
110	32	0	22 Aug 90	8 Oct 90
82	44	9	30 Apr 91	18 Jun 91
57	60	7	2 May 91	20 Jun 91
77	51	7	7 May 91	25 Jun 91
52	38	6	10 May 91	28 Jun 91

1994: The Concrete Slab Problem

The United States Department of Housing and Urban Development (HUD) is considering constructing dwellings of various sizes, ranging from individual houses to large apartment complexes. A principal concern is to minimize recurring costs to occupants, especially the costs of heating and cooling. The region in which the construction is to take place is temperate, with a moderate variation in temperature throughout the year.

With special construction techniques, HUD engineers can build dwellings that do not need to rely on convection—that is, there is no need to rely on opening doors or windows to assist in temperature variation. The dwellings will be single-story, with concrete slab floors as the only foundation. You have been hired as a consultant to analyze the temperature variation in the concrete slab floor to determine whether the temperature averaged over the floor surface can be maintained within a prescribed comfort zone throughout the year. If so, what size/shape of slabs will permit this?

Part 1, Floor Temperature

Consider the temperature variation in a concrete slab given that the ambient temperature varies daily within the ranges given in Table A.7. Assume that the high occurs at noon and the low at midnight. Determine whether slabs can be designed to maintain a temperature averaged over the floor surface within the prescribed comfort zone, considering radiation only. Initially, assume that the heat transfer into the dwelling is through the exposed perimeter of the slab and that the top and bottom of the slabs are insulated. Comment on the appropriateness and sensitivity of these assumptions. If you cannot find a solution that satisfies Table A.7, can you find designs that satisfy a Table A.7 that you propose?

Table A.7   Daily variation in temperature

Ambient temperature		Comfort zone	
High:	85° F	High:	76° F
Low:	60° F	Low:	65° F

Part 2, Building Temperature

Analyze the practicality of the initial assumptions and extend the analysis to temperature variation within the single-story dwelling. Can the house be kept within the comfort zone?

Part 3, Cost of Construction

Suggest a design that considers HUD's objective of reducing or eliminating heating and cooling costs, considering construction restrictions and costs.

1994: The Communications Network Problem

In your company, information is shared among departments on a daily basis. This information includes the previous day's sales statistics and current production guidance. It is important to get this information out as quickly as possible.

Suppose that a communications network is to be used to transfer blocks of data (files) from one computer to another. As an example, consider the graph model in Figure A.13.

Vertices $V_1, V_2, \ldots, V_m$ represent computers, and edges $e_1, e_2, \ldots, e_n$ represent files to be transferred (between computers represented by edge endpoints). $T(e_x)$ is the time that it takes to transfer file e_x, and $C(V_y)$ is the capacity of the computer represented by V_y to transfer files simultaneously. A file transfer involves the engagement of both computers for the entire time it takes to transfer the file. For example, $C(V_y) = 1$ means that computer V_y can be involved in only one transfer at a time.

■ **Figure A.13**

Example of a file transfer network

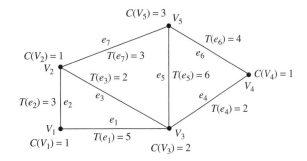

We are interested in scheduling the transfers in an optimal way, to minimize the total time that it takes to complete them all. This minimum total time is called the *makespan*. Consider the following situations for your company.

Situation A

Your corporation has 28 departments. Each department has a computer, each of which is represented by a vertex in Figure A.14. Each day, 27 files must be transferred, represented by the edges in Figure A.14. For this network, $T(e_x) = 1$ and $C(V_y) = 1$ for all x and y. Find an optimal schedule and the makespan for the given network. Can you prove to your supervisor that your makespan is the smallest possible (optimal) for the given network? Describe your approach to solving the problem. Does your approach work for the general case—that is, where $T(e_x)$, $C(V_y)$, and the graph structure are arbitrary?

■ **Figure A.14**

Network for situations A and B

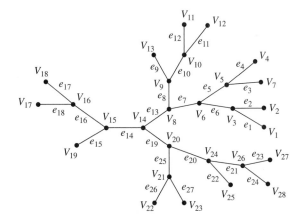

Situation B

Suppose that your company changes the requirements for data transfer. You must now consider the same basic network structure (see Figure A.14) with different types and sizes of files. These files take the amount of time to transfer indicated in Table A.8 by the $T(e_x)$ terms for each edge. We still have $C(V_y) = 1$ for all y. Find an optimal schedule and the makespan for the new network. Can you prove that your makespan is the smallest possible

Table A.8 File transfer time data for situation B

x	1	2	3	4	5	6	7	8	9
$T(e_x)$	3.0	4.1	4.0	7.0	1.0	8.0	3.2	2.4	5.0
x	10	11	12	13	14	15	16	17	18
$T(e_x)$	8.0	1.0	4.4	9.0	3.2	2.1	8.0	3.6	4.5
x	19	20	21	22	23	24	25	26	27
$T(e_x)$	7.0	7.0	9.0	4.2	4.4	5.0	7.0	9.0	1.2

for the new network? Describe your approach to solving this problem. Does your approach work for the general case? Comment on any peculiar or unexpected results.

Situation C

Your corporation is considering expansion. If that happens, there are several new files (edges) that will need to be transferred daily. This expansion will also include an upgrade of the computer system. Some of the 28 departments will get new computers that can handle more than one transfer at a time. All of these changes are indicated in Figure A.15 and Tables A.9 and A.10. What is the best schedule and makespan that you can find? Can you prove that your makespan is the smallest possible for this network? Describe your approach to solving the problem. Comment on any peculiar or unexpected results.

■ **Figure A.15**

Network for situation C

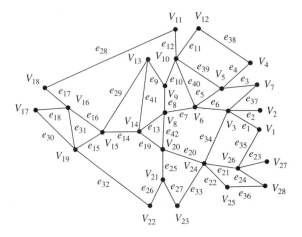

Table A.9 File transfer time data for situation C, for the added transfers

x	28	29	30	31	32	33	34	35
$T(e_x)$	6.0	1.1	5.2	4.1	4.0	7.0	2.4	9.0
x	36	37	38	39	40	41	42	
$T(e_x)$	3.7	6.3	6.6	5.1	7.1	3.0	6.1	

Table A.10   Computer capacity data for situation C

y	1	2	3	4	5	6	7	8	9	10
$C(V_y)$	2	2	1	1	1	1	1	1	2	3

y	11	12	13	14	15	16	17	18	19
$C(V_y)$	1	1	1	2	1	2	1	1	1

y	20	21	22	23	24	25	26	27	28
$C(V_y)$	1	1	2	1	1	1	2	1	1

1995: The Single Helix

A small biotechnical company must design, prove, program, and test a mathematical algorithm to locate in real time all the intersections of a helix and a plane in general positions in space.

Computer-aided geometric design (CAGD) programs enable engineers to view a plane section of an object they design, such as an automobile suspension or a medical device. Engineers may also display on the plane section quantities such as air flow, stress, or temperature, coded by colors or level curves. Plane sections may be rapidly swept through the entire object to gain a three-dimensional visualization of the object and its reactions to motion, forces, or heat. To achieve such results, the computer programs must quickly and accurately locate all the intersections of the viewed plane and every part of the designed object. General equation solvers may in principle compute such intersections, but for specific problems, specific methods may prove faster and more accurate than general methods. In particular, general CAGD software may prove too slow to complete computations in real time or too large to fit in the company's finished medical devices. These considerations have led the company to the following problem.

Problem

Design, justify, program, and test a method to compute all the intersections of a plane and a helix, both in general positions (at any locations and with any orientations) in space. A segment of the helix may represent, for example, a helicoidal suspension spring or a piece of tubing in a chemical or medical apparatus.

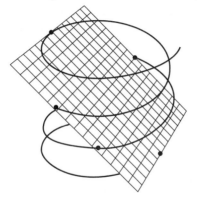

Theoretical justification of the proposed algorithm is necessary to verify the solution from several points of view—for instance, through mathematical proofs of parts of the algorithm and through tests of the final program with known examples. Such documentation and tests will be required by government agencies for medical use.

1995: Aluacha Balaclava College

Aluacha Balaclava College, an undergraduate facility, has just hired a new Provost whose first priority is the institution of a fair and reasonable faculty compensation system. She has hired your consulting team to design a compensation system that reflects the following circumstances and principles.

Faculty are ranked as Instructor, Assistant Professor, Associate Professor, and Professor. Those with Ph.D. degrees are hired at the rank of Assistant Professor. Ph.D. candidates are hired at the rank of Instructor and promoted automatically to Assistant Professor upon completion of their degrees. Faculty may apply for promotion from Associate Professor to Professor after serving at the rank of Associate for 7 or more years. Promotions are determined by the Provost, with recommendations from a faculty committee.

Faculty salaries are for the 10-month period September through June, with raises effective beginning in September. The total amount of money available for raises varies yearly and is generally disclosed in March for the following year.

The starting salary this year for an Instructor with no prior teaching experience was $27,000; $32,000 for an Assistant Professor. Upon hire, faculty can receive credit for up to 7 years of teaching experience at other institutions.

Principles

1. All faculty should get a raise any year that money is available.
2. Promotion should incur a substantial benefit; e.g., promotion in the minimum possible time should result in a benefit roughly equal to 7 years of normal raises.
3. Faculty promoted after 7 or 8 years in rank with careers of at least 25 years should make roughly twice as much at retirement as a starting Ph.D.
4. Experienced faculty should be paid more than less experienced in the same rank. The effect of additional years of experience should diminish over time; that is, if two faculty stay in the same rank, their salaries should equalize over time.

Design a new pay system, first without cost-of-living increases. Incorporate cost-of-living increases, and then design a transition process for current faculty that will move all salaries toward your system without reducing anyone's salary. Existing faculty salaries, ranks, and years of service are shown in Table A.11. Discuss any refinements you think would improve your system.

The Provost requires a detailed compensation system plan for implementation, as well as a brief, clear, executive summary outlining the model, its assumptions, its strengths, its weaknesses, and its expected results, which she can present to the Board and faculty.

Table A.11   Salary data for Aluacha Balaclava College

Case	Years	Rank	Salary	Case	Years	Rank	Salary	Case	Years	Rank	Salary
1	4	ASSO	54,000	2	19	ASST	43,508	3	20	ASST	39,072
4	11	PROF	53,900	5	15	PROF	44,206	6	17	ASST	37,538
7	23	PROF	48,844	8	10	ASST	32,841	9	7	ASSO	49,981
10	20	ASSO	43,549	11	18	ASSO	42,649	12	19	PROF	60,087
13	15	ASSO	38,002	14	4	ASST	30,000	15	34	PROF	60,576
16	28	ASST	44,562	17	9	ASST	30,893	18	22	ASSO	46,351
19	21	ASSO	50,979	20	20	ASST	48,000	21	4	ASST	32,500
22	14	ASSO	38,462	23	23	PROF	53,500	24	21	ASSO	42,488
25	20	ASSO	43,892	26	5	ASST	35,330	27	19	ASSO	41,147
28	15	ASST	34,040	29	18	PROF	48,944	30	7	ASST	30,128
31	5	ASST	35,330	32	6	ASSO	35,942	33	8	PROF	57,295
34	10	ASST	36,991	35	23	PROF	60,576	36	20	ASSO	48,926
37	9	PROF	57,956	38	32	ASSO	52,214	39	15	ASST	39,259
40	22	ASSO	43,672	41	6	INST	45,500	42	5	ASSO	52,262
43	5	ASSO	57,170	44	16	ASST	36,958	45	23	ASST	37,538
46	9	PROF	58,974	47	8	PROF	49,971	48	23	PROF	62,742
49	39	ASSO	52,058	50	4	INST	26,500	51	5	ASST	33,130
52	46	PROF	59,749	53	4	ASSO	37,954	54	19	PROF	45,833
55	6	ASSO	35,270	56	6	ASSO	43,037	57	20	PROF	59,755
58	21	PROF	57,797	59	4	ASSO	53,500	60	6	ASST	32,319
61	17	ASSO	35,668	62	20	PROF	59,333	63	4	ASST	30,500
64	16	ASSO	41,352	65	15	PROF	43,264	66	20	PROF	50,935
67	6	ASST	45,365	68	6	ASSO	35,941	69	6	ASST	49,134
70	4	ASST	29,500	71	4	ASST	30,186	72	7	ASST	32,400
73	12	ASSO	44,501	74	2	ASST	31,900	75	1	ASSO	62,500
76	1	ASST	34,500	77	16	ASSO	40,637	78	4	ASSO	35,500
79	21	PROF	50,521	80	12	ASST	35,158	81	4	INST	28,500
82	16	PROF	46,930	83	24	PROF	55,811	84	6	ASST	30,128
85	16	PROF	46,090	86	5	ASST	28,570	87	19	PROF	44,612
88	17	ASST	36,313	89	6	ASST	33,479	90	14	ASSO	38,624
91	5	ASST	32,210	92	9	ASSO	48,500	93	4	ASST	35,150
94	25	PROF	50,583	95	23	PROF	60,800	96	17	ASST	38,464
97	4	ASST	39,500	98	3	ASST	52,000	99	24	PROF	56,922
100	2	PROF	78,500	101	20	PROF	52,345	102	9	ASST	35,798
103	24	ASST	43,925	104	6	ASSO	35,270	105	14	PROF	49,472
106	19	ASSO	42,215	107	12	ASST	40,427	108	10	ASST	37,021
109	18	ASSO	44,166	110	21	ASSO	46,157	111	8	ASST	32,500
112	19	ASSO	40,785	113	10	ASSO	38,698	114	5	ASST	31,170
115	1	INST	26,161	116	22	PROF	47,974	117	10	ASSO	37,793
118	7	ASST	38,117	119	26	PROF	62,370	120	20	ASSO	51,991
121	1	ASST	31,500	122	8	ASSO	35,941	123	14	ASSO	39,294
124	23	ASSO	51,991	125	1	ASST	30,000	126	15	ASST	34,638
127	20	ASSO	56,836	128	6	INST	35,451	129	10	ASST	32,756
130	14	ASST	32,922	131	12	ASSO	36,451	132	1	ASST	30,000
133	17	PROF	48,134	134	6	ASST	40,436	135	2	ASSO	54,500
136	4	ASSO	55,000	137	5	ASST	32,210	138	21	ASSO	43,160
139	2	ASST	32,000	140	7	ASST	36,300	141	9	ASSO	38,624

(*continued*)

Table A.12 (*Continued*)

Case	Years	Rank	Salary	Case	Years	Rank	Salary	Case	Years	Rank	Salary
142	21	PROF	49,687	143	22	PROF	49,972	144	7	ASSO	46,155
145	12	ASST	37,159	146	9	ASST	32,500	147	3	ASST	31,500
148	13	INST	31,276	149	6	ASST	33,378	150	19	PROF	45,780
151	5	PROF	70,500	152	27	PROF	59,327	153	9	ASSO	37,954
154	5	ASSO	36,612	155	2	ASST	29,500	156	3	PROF	66,500
157	17	ASST	36,378	158	5	ASSO	46,770	159	22	ASST	42,772
160	6	ASST	31,160	161	17	ASST	39,072	162	20	ASST	42,970
163	2	PROF	85,500	164	20	ASST	49,302	165	21	ASSO	43,054
166	21	PROF	49,948	167	5	PROF	50,810	168	19	ASSO	51,378
169	18	ASSO	41,267	170	18	ASST	42,176	171	23	PROF	51,571
172	12	PROF	46,500	173	6	ASST	35,798	174	7	ASST	42,256
175	23	ASSO	46,351	176	22	PROF	48,280	177	3	ASST	55,500
178	15	ASSO	39,265	179	4	ASST	29,500	180	21	ASSO	48,359
181	23	PROF	48,844	182	1	ASST	31,000	183	6	ASST	32,923
184	2	INST	27,700	185	16	PROF	40,748	186	24	ASSO	44,715
187	9	ASSO	37,389	188	28	PROF	51,064	189	19	INST	34,265
190	22	PROF	49,756	191	19	ASST	36,958	192	16	ASST	34,550
193	22	PROF	50,576	194	5	ASST	32,210	195	2	ASST	28,500
196	12	ASSO	41,178	197	22	PROF	53,836	198	19	ASSO	43,519
199	4	ASST	32,000	200	18	ASSO	40,089	201	23	PROF	52,403
202	21	PROF	59,234	203	22	PROF	51,898	204	26	ASSO	47,047

1996: The Submarine Detection Problem

The world's oceans contain an ambient noise field. Seismic disturbances, surface shipping, and marine mammals are sources that, in different frequency ranges, contribute to this field. We wish to consider how this ambient noise might be used to detect large moving objects (e.g., submarines located below the ocean surface). Assuming that a submarine makes no intrinsic noise, develop a method for detecting the presence of a moving submarine, its speed, its size, and its direction of travel, using only information obtained by measuring changes to the ambient noise field. Begin with noise at one fixed frequency and amplitude.

1996: The Contest Judging Problem

When determining the winner of a competition such as the Mathematical Contest in Modeling, there are generally a great many papers to judge. Let's say there are $P = 100$ papers. A group of J judges is collected to accomplish the judging. Funding for the contest constrains both the number of judges that can be obtained and the amount of time that they can judge. For example, if $P = 100$, then $J = 8$ is typical.

Ideally, each judge would read each paper and rank-order them, but there are too many papers for this. Instead, there will be a number of screening rounds in which each judge will read some number of papers and give them scores. Then some selection scheme is used to reduce the number of papers under consideration: If the papers are rank-ordered, then the bottom 30% that each judge rank-orders could be rejected. Alternatively, if the judges do not rank-order the papers, but instead give them numerical scores (e.g., from 1 to 100), then all papers falling below some cutoff level could be rejected.

The new pool of papers is then passed back to the judges, and the process is repeated. A concern is that the total number of papers that each judge reads must be substantially less than P. The process is stopped when there are only W papers left. These are the winners. Typically for $P = 100$, $W = 3$.

Your task is to determine a selection scheme, using a combination of rank-ordering, numerical scoring, and other methods, by which the final W papers will include only papers from among the best $2W$ papers. (By "best," we mean we assume that there is an absolute rank-ordering to which all judges would agree.) For example, the top three papers found by your method will consist entirely of papers from among the best six papers. Among all such methods, the one that requires each judge to read the least number of papers is desired.

Note the possibility of systematic bias in a numerical scoring scheme. For example, for a specific collection of papers, one judge could average 70 points, whereas another could average 80 points. How would you scale your scheme to accommodate changes in the content parameters (P, J, and W)?

1997: The Velociraptor Problem

The velociraptor, *Velociraptor mongoliensis*, was a predator dinosaur that lived during the late Cretaceous period approximately 75 million years ago. Paleontologists think that it was a very tenacious hunter and may have hunted in pairs or even larger packs. Unfortunately, there is no way to observe its hunting behavior in the wild as can be done with modern mammalian predators. A group of paleontologists has approached your team and asked for help in modeling the hunting behavior of the velociraptor. They hope to compare your results with field data reported by biologists studying the behaviors of lions, tigers, and similar predatory animals.

The average adult velociraptor was 3 meters long with a hip height of 0.5 meter and an approximate mass of 45 kg. It is estimated that the animal could run extremely fast, at speeds of 60 km/hr, for about 15 seconds. After that burst of speed, the animal needed to stop and recover from a buildup of lactic acid in its muscles.

Suppose that velociraptor preyed on *Thescelosaurus neglectus*, a bipedal herbivore approximately the same size as the velociraptor. A biomechanical analysis of fossilized thescelosaurus indicates that it could run at a speed of about 50 km/hr almost indefinitely.

Part 1   Assuming the velociraptor is a solitary hunter, design a mathematical model that describes a hunting strategy for a single velociraptor stalking and chasing a single prey, as well as the evasive strategy of the prey. Assume that the thescelosaurus can always detect the velociraptor when it gets within 15 meters but *may* detect this predator at even greater ranges (up to 50 meters) depending upon the nature of the habitat and weather conditions. Additionally, because of its physical structure and strength, the velociraptor has a limited turning radius when running at full speed. This radius is estimated to be three times the animal's hip height. By contrast, the thescelosaurus is extremely agile and has a turning radius of 0.5 meters.

Part 2   Assuming the more realistic situation that the velociraptor hunted in pairs, design a new model that describes a hunting strategy for two velociraptors stalking and chasing a single prey, as well as the evasive strategy of the prey. Use the same other assumptions and limitations as in Part 1.

1997: Mix Well for Fruitful Discussions

Small-group meetings are gaining popularity for the discussion of important issues, particularly long-range planning. It is believed that large groups stymie productive discussion and that a dominant personality usually controls and directs the discussion. Thus in corporation board meetings the Board will meet in small groups to discuss issues before meeting as a whole. These smaller groups still run the risk of control by a dominant personality. In an attempt to reduce this danger, it is common to schedule several sessions with a different mix of people in the groups.

A meeting of An Tostal Corporation will be attended by 29 board members of whom 9 are in-house members (i.e., employees of the corporation). The meeting is to be an all-day affair, with 3 sessions scheduled for the morning and 4 for the afternoon. The sessions will each be 45 minutes, beginning on the hour from 9:00 A.M. to 4:00 P.M., with lunch scheduled at noon. Each morning session will consist of six discussion groups, with each discussion group led by one of the corporation's six senior officers. None of these officers are board members. Thus each senior officer will lead three different discussion groups. The senior officers will not be involved in the afternoon sessions, and each of these sessions will consist of only four different discussion groups.

The president wants a list of board member assignments to discussion groups for each of the seven sessions. The assignments should achieve as much mix of the members as possible. The ideal assignment would have each board member in a discussion group with each other board member the same number of times, while minimizing common membership of groups for the different sessions.

The assignments should also satisfy the following criteria:

1. For the morning sessions, no board member should be in the same senior officer's discussion group twice.

2. No discussion group should contain a disproportionate number of in-house members.

Give a list of assignments for members 1–9 and 10–29 and officers 1–6. Indicate how well the criteria in the previous paragraphs are met. Since it is possible that some board members will cancel at the last minute or that some not scheduled will show up, an algorithm that the secretary can use to adjust the assignments with an hour's notice would be appreciated. It would be ideal if the algorithm could also be used to make assignments for future meetings involving different levels of participation for each type of attendee.

1998: MRI Scanners

Introduction

Industrial and medical diagnostic machines known as magnetic resonance imagers (MRI) scan a three-dimensional object, such as a brain, and deliver their results in the form of a three-dimensional array of pixels. Each pixel consists of one number indicating a color or a shade of gray that encodes a measure of water concentration in a small region of the scanned object at the location of the pixel. For instance, 0 can picture high water concentration in black (ventricles, blood vessels), 128 can picture a medium water concentration in gray (brain nuclei and gray matter), and 255 can picture a low water density in white (lipid-rich white matter consisting of myelinated axons). Such MRI scanners also include facilities to picture on a screen any horizontal or vertical slice through the three-dimensional array (slices are parallel to any of the three Cartesian coordinate axes). Algorithms for picturing slices through oblique planes, however, are proprietary. Current algorithms are limited in terms of the angles and parameter options available; are implemented only on heavily used dedicated workstations; lack input capabilities for marking points in the picture before slicing; and tend to blur and "feather out" sharp boundaries between the original pixels.

A more faithful, flexible algorithm implemented on a personal computer would be useful

1. for planning minimally invasive treatments,
2. for calibrating the MRI machines,
3. for investigating structures oriented obliquely in space, such as post-mortem tissue sections in animal research,
4. for enabling cross sections at any angle through a brain atlas consisting of black-and-white line drawings.

To design such an algorithm, one can access the values and locations of the pixels, but not the initial data gathered by the scanner.

Problem

Design and test an algorithm that produces sections of three-dimensional arrays by planes in any orientation in space, preserving the original gray-scale values as closely as possible.

Data Sets

The typical data set consists of a three-dimensional array A of numbers $A(i, j, k)$ that indicates the density $A(i, j, k)$ of the object at the location $(x, y, z)_{ijk}$. Typically, $A(i, j, k)$ can range from 0 through 255. In most applications, the data set is quite large.

Teams should design data sets to test and demonstrate their algorithms. The data sets should reflect conditions likely to be of diagnostic interest. Teams should also characterize data sets that limit the effectiveness of their algorithms.

Summary

The algorithm must produce a picture of the slice of the three-dimensional array by a plane in space. The plane can have any orientation and any location in space. (The plane can miss some or all data points.) The result of the algorithm should be a model of the density of the scanned object over the selected plane.

1998: Grade Inflation

Background

Some college administrators are concerned about the grading at *A Better Class* (*ABC*) College. On average, the faculty at ABC have been giving out high grades (the average grade now given out is an A−), and it is impossible to distinguish between the good and the mediocre students. The terms of a very generous scholarship allow only the top 10% of the students to be funded, so a class ranking is required.

The dean had the idea of comparing each student to the other students in each class and using this information to build up a ranking. For example, if a student obtains an A in a class in which all students obtain an A, then this student is only "average" in this class. On the other hand, if a student obtains the only A in a class, then that student is clearly "above average." Combining information from several classes might allow students to be placed in deciles (top 10%, next 10%, etc.) across the college.

Problem

Assuming that the grades given out are (A+, A, A−, B+, . . .) can the dean's idea be made to work?

Can any other schemes produce a desired ranking?

A concern is that the grade in a single class could change many student's deciles. Is this possible?

Data Sets

Teams should design data sets to test and demonstrate their algorithms. Teams should characterize data sets that limit the effectiveness of their algorithms.

1999: Deep Impact

For some time, the National Aeronautics and Space Administration (NASA) has been considering the consequences of a large asteroid impact on the earth.

As part of this effort, your team has been asked to consider the effects of such an impact were the asteroid to land in Antarctica. There are concerns that an impact there could have considerably different consequences than one striking elsewhere on the planet.

You are to assume that an asteroid is on the order of 1000 meters in diameter and that it strikes the Antarctic continent directly at the South Pole.

Your team has been asked to provide an assessment of the impact of such an asteroid. In particular, NASA would like an estimate of the probable amount and location of human casualties from this impact, an estimate of the damage done to the food production regions in the oceans of the Southern Hemisphere, and an estimate of possible coastal flooding caused by large-scale melting of the Antarctic polar ice sheet.

1999: Unlawful Assembly

Many public facilities have signs in rooms used for public gatherings which state that it is "unlawful" for the rooms to be occupied by more than a specified number of people. Presumably, this number is based on the speed with which people in the room could be evacuated via the room's exits in case of an emergency. Similarly, elevators and other facilities often have "maximum capacities" posted.

Develop a mathematical model for deciding what number to post on such a sign as being the "lawful capacity." As part of your solution, discuss criteria (other than public safety in the case of a fire or other emergency) that might govern the number of people considered "unlawful" to occupy the room (or space). Also, for the model that you construct, consider the differences between a room with movable furniture such as a cafeteria (with tables and chairs), a gymnasium, a public swimming pool, and a lecture hall with a pattern of rows and aisles. You may wish to compare and contrast what might be done for a variety of different environments: elevator, lecture hall, swimming pool, cafeteria, or gymnasium. Gatherings such as rock concerts and soccer tournaments may present special conditions.

Apply your model to one or more public facilities at your institution (or neighboring town). Compare your results with the stated capacity, if one is posted. If used, your model is likely to be challenged by parties with interests in increasing the capacity. Write an article for the local newspaper defending your analysis.

2000: Air Traffic Control

Dedicated to the memory of Dr. Robert Machol, former chief scientist of the Federal Aviation Agency

To improve safety and reduce air traffic controller workload, the Federal Aviation Agency (FAA) is considering adding, to the air traffic control system, software that would automatically detect potential aircraft flight path conflicts and alert the controller. To that end, an analyst at the FAA has posed the following problems.

Requirement A: Given two airplanes flying in space, when should the air traffic controller consider the objects to be too close and to require intervention?

Requirement B: An airspace sector is the section of three-dimensional airspace that one air traffic controller controls. Given any airspace sector, how do we measure how complex it is from an air traffic workload perspective? To what extent is complexity determined by the number of aircraft simultaneously passing through that sector

1. at any one instant?
2. during any given interval of time?
3. during a particular time of day?

How does the number of potential conflicts arising during those periods affect complexity? Does the presence of additional software tools to automatically predict conflicts and alert the controller reduce or add to this complexity?

In addition to the guidelines for your report, write a summary (no more than two pages) that the FAA analyst can present to Jane Garvey, the FAA administrator, to defend your conclusions.

2000: Radio Channel Assignments

We seek to model the assignment of radio channels to a symmetric network of transmitter locations over a large planar area, so as to avoid interference. One basic approach is to partition the region into regular hexagons in a grid (honeycomb-style), as shown in Figure A.16, where a transmitter is located at the center of each hexagon.

■ **Figure A.16**

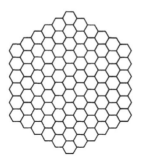

An interval of the frequency spectrum is to be allotted for transmitter frequencies. The interval will be divided into regularly spaced channels, which we represent by integers, 1, 2, 3, ... Each transmitter will be assigned one positive integer channel. The same channel can be used at many locations, provided that interference from nearby transmitters is avoided.

Our goal is to minimize the width of the interval in the frequency spectrum that is needed to assign channels subject to some constraints. This is achieved with the concept of a span. The span is the minimum, over all assignments satisfying the constraints, of the largest channel used at any location. It is not required that every channel smaller than the span be used in an assignment that attains the span. Let s be the length of a side of one of the hexagons. We concentrate on the case that there are two levels of interference.

Requirement A: There are several constraints on frequency assignments. First, no two transmitters within distance $4s$ of each other can be given the same channel. Second, due to spectral spreading, transmitters within distance $2s$ of each other must not be given the same or adjacent channels: Their channels must differ by at least 2. Under these constraints, what can we say about the span in Figure A.16?

Requirement B: Repeat Requirement A, assuming the grid in the example spreads arbitrarily far in all directions.

Requirement C: Repeat Requirements A and B, except now assume, more generally, that channels for transmitters within distance $2s$ differ by at least some given integer k, while those at distance at most $4s$ must still differ by at least 1. What can we say about the span and about efficient strategies for designing assignments, as a function of k?

Requirement D: Consider generalizations of the problem, such as several levels of interference or irregular transmitter placements. What other factors may be important to consider?

Requirement E: Write an article (no more than two pages) for the local newspaper explaining your findings.

2001: Choosing a Bicycle Wheel

Cyclists have different types of wheels they can use on their bicycles. The two basic types of wheels are those constructed using wire spokes and those constructed of a solid disk (Figure A.17). The spoked wheels are lighter, but the solid wheels are more aerodynamic. A solid wheel is never used on the front for a road race but can be used on the rear of the bike.

■ **Figure A.17**

A solid wheel is shown on the left and a spoked wheel is shown on the right.

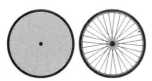

Professional cyclists look at a racecourse and make an educated guess as to what kind of wheels should be used. The decision is based on the number and steepness of the hills, the weather, wind speed, the competition, and other considerations. The director sportif of your favorite team would like to have a better system in place and has asked your team for information to help determine what kind of wheel should be used for a given course.

The director sportif needs specific information to help make a decision and has asked your team to accomplish the tasks listed below. For each of the tasks, assume that the same spoked wheel will always be used on the front but there is a choice of wheels for the rear.

Task 1.   Provide a table giving the wind speed at which the power required for a solid rear wheel is less than for a spoked rear wheel.

The table should include the wind speeds for different road grades ranging from 0% to 10% in 1% increments. (Road grade is defined to be the ratio of the total rise of a hill divided by the length of the road.[1]) A rider starts at the bottom of the hill at a speed of 45 kph, and the deceleration of the rider is proportional to the road grade. A rider will lose about 8 kph for a 5% grade over 100 meters.

Task 2.   Provide an example of how the table could be used for a specific time trial course.

Task 3.   Determine whether the table is an adequate means for deciding on the wheel configuration, and offer other suggestions as to how to make this decision.

2001: Escaping a Hurricane's Wrath

Evacuating the coast of South Carolina ahead of the predicted landfall of Hurricane Floyd in 1999 led to a monumental traffic jam. Traffic slowed to a standstill on Interstate I-26, which is the principal route going inland from Charleston to the relatively safe haven of Columbia in the center of the state. What is normally an easy 2-hour drive took up to 18 hours to complete. Many cars simply ran out of gas along the way. Fortunately, Floyd turned north and spared the state this time, but the public outcry is forcing state officials to find ways to avoid a repeat of this traffic nightmare.

The principal proposal put forth to deal with this problem is the reversal of traffic on I-26, so that both sides, including the coastal-bound lanes, have traffic headed inland from Charleston to Columbia. Plans to carry this out have been prepared (and posted on the Web) by the South Carolina Emergency Preparedness Division. Traffic reversal on principal roads leading inland from Myrtle Beach and Hilton Head is also planned.

Charleston has approximately 500,000 people, Myrtle Beach has about 200,000 people, and another 250,000 people are spread out along the rest of the coastal strip. (More accurate data, if sought, are widely available.)

The interstates have two lanes of traffic in each direction, except in metro areas, where they have three. Columbia, another metro area of around 500,000 people, does not have sufficient hotel space to accommodate the evacuees (including some coming from farther north by other routes), so some traffic continues outbound on I-26 toward Spartanburg; on I-77 north to Charlotte; and on I-20 east to Atlanta. In 1999, traffic leaving Columbia going northwest was moving very slowly.

Construct a model for the problem to investigate what strategies may reduce the congestion observed in 1999. Here are the questions that need to be addressed:

Under what conditions does the plan for turning the two coastal-bound lanes of I-26 into two lanes of Columbia-bound traffic significantly improve evacuation traffic flow?

In 1999, the simultaneous evacuation of the state's entire coastal region was ordered. Would the evacuation traffic flow improve under an alternative strategy that staggers the

---

[1] If the hill is viewed as a triangle, the grade is the sine of the angle at the bottom of the hill.

evacuation, perhaps county-by-county over some time period consistent with the pattern of how hurricanes affect the coast?

Several smaller highways besides I-26 extend inland from the coast. Under what conditions, would it improve evacuation flow to turn around traffic on these?

What effect would it have on evacuation flow to establish more temporary shelters in Columbia, to reduce the traffic leaving Columbia?

In 1999, many families leaving the coast brought their boats, campers, and motor homes. Many drove all of their cars. Under what conditions should there be restrictions on vehicle types or numbers of vehicles in order to guarantee timely evacuation? It has been suggested that in 1999, some of the coastal residents of Georgia and Florida, who were fleeing the earlier predicted landfalls of Hurricane Floyd to the south, came up I-95 and compounded the traffic problems. How big an impact can they have on the evacuation traffic flow?

Clearly identify what measures of performance are used to compare strategies.

Required: Prepare a short newspaper article, not to exceed two pages, explaining the results and conclusions of your study.

2002: Wind and Waterspray

An ornamental fountain in a large open plaza surrounded by buildings squirts water high into the air. On gusty days, the wind blows spray from the fountain onto passersby. The water flow from the fountain is controlled by a mechanism linked to an anemometer (which measures wind speed and direction) located on top of an adjacent building. The objective of this control is to provide passersby with an acceptable balance between an attractive spectacle and a soaking: The harder the wind blows, the lower the water volume and height to which the water is squirted, hence the less spray falls outside the pool area.

Your task is to devise an algorithm that uses data provided by the anemometer to adjust the water flow from the fountain as the wind conditions change.

2002: Airline Overbooking

You're all packed and ready to go on a trip to visit your best friend in New York City. After you check in at the ticket counter, the airline clerk announces that your flight has been overbooked. Passengers need to check in immediately to determine whether they still have a seat.

Historically, airlines know that only a certain percentage of passengers who have made reservations on a particular flight will actually take that flight. Consequently, most airlines *overbook*—that is, they take more reservations than the capacity of the aircraft. Occasionally, however, more passengers will want to take a flight than the capacity of the plane, leading to one or more passengers being *bumped* and thus unable to take the flight for which they had reservations.

Airlines deal with bumped passengers in various ways. Some are given nothing, some are booked on later flights on other airlines, and some are given some kind of cash or airline ticket incentive.

Consider the overbooking issue in light of the current situation:

Fewer flights by airlines from point A to point B

Heightened security at and around airports

Passengers' fear

Loss of billions of dollars in revenue by airlines to date

Build a mathematical model that examines the effects that different overbooking schemes have on the revenue received by an airline company in order to find an optimal overbooking strategy, i.e., the number of people by which an airline should overbook a particular flight so that the company's revenue is maximized. Ensure that your model reflects the issues above, and consider alternatives for handling "bumped" passengers. Additionally, write a short memorandum to the airline's CEO summarizing your findings and analysis.

2003: The Stunt Person

An exciting action scene in a movie is going to be filmed, and you are the stunt coordinator! A stunt person on a motorcycle will jump over an elephant and land in a pile of cardboard boxes to cushion the fall. You need to protect the stunt person, and you must also use relatively few cardboard boxes (lower cost, not seen by camera, etc.).

Your job consists of the following tasks:

- Determine what size boxes to use.
- Determine how many boxes to use.
- Determine how the boxes will be stacked.
- Determine whether any modifications to the boxes would help.
- Generalize to different combined weights (stunt person plus motorcycle) and different jump heights.

Note that in *Tomorrow Never Dies,* the James Bond character on a motorcycle jumps over a helicopter.

2003: Gamma Knife Treatment Planning

Stereotactic radiosurgery delivers a single high dose of ionizing radiation to a radiographically well-defined, small intracranial three-dimensional brain tumor without delivering any significant fraction of the prescribed dose to the surrounding brain tissue. Three modalities are commonly used in this area: the gamma knife unit, heavy charged particle beams, and external high-energy photon beams from linear accelerators.

The gamma knife unit delivers a single high dose of ionizing radiation emanating from 201 cobalt-60 unit sources through a heavy helmet. All 201 beams simultaneously intersect at the isocenter, resulting in an approximately spherical dose distribution at the effective dose levels. Irradiating the isocenter to deliver dose is termed a "shot." Shots can be represented as different spheres. Four interchangeable outer collimator helmets with beam channel diameters of 4, 8, 14, and 18 mm are available for irradiating different size volumes. For a target volume larger than one shot, multiple shots can be used to cover the entire target. In practice, most target volumes are treated with 1 to 15 shots. The target volume is a bounded, three-dimensional digital image that usually consists of millions of points.

The goal of radiosurgery is to deplete tumor cells while preserving normal structures. Because there are physical limitations and biological uncertainties involved in this therapy process, a treatment plan needs to account for all those limitations and uncertainties. In general, an optimal treatment plan is designed to meet the following requirements:

1. Minimize the dose gradient across the target volume.
2. Match specified isodose contours to the target volumes.
3. Match specified dose–volume constraints of the target and critical organ.
4. Minimize the integral dose to the entire volume of normal tissues or organs.
5. Constrain dose to specified normal tissue points below tolerance doses.
6. Minimize the maximum dose to critical volumes.

In gamma unit treatment planning, we have the following constraints:

1. Prohibit shots from protruding outside the target.
2. Prohibit shots from overlapping (to avoid hot spots).
3. Cover the target volume with effective dosage as much as possible, but at least 90% of the target volume must be covered by shots.
4. Use as few shots as possible.

Your tasks are to formulate the optimal treatment planning for a gamma knife unit as a sphere-packing problem and to propose an algorithm to find a solution. While designing your algorithm, you must keep in mind that your algorithm must be reasonably efficient.

2004: Are Fingerprints Unique?

It is a common belief that the thumbprint of every human who has ever lived is different. Develop and analyze a model that will enable you to assess the probability that this is true. Compare the odds (that you found in this problem) of misidentification by fingerprint evidence against the odds of misidentification by DNA evidence.

2004: A Faster QuickPass System

"QuickPass" systems are increasingly appearing to reduce people's time waiting in line, whether it is at tollbooths, amusement parks, or elsewhere. Consider the design of a Quick-Pass system for an amusement park. The amusement park has experimented by offering QuickPasses for several popular rides as a test. The idea is that for certain popular rides, you can go to a kiosk near that ride and insert your daily park entrance ticket, and out will come a slip that states that you can return to that ride at a specific time later. For example, you insert your daily park entrance ticket at 1:15 P.M., and the QuickPass states that you can come back between 3:30 and 4:30 P.M. and use your slip to enter a second, and presumably much shorter, line that will get you to the ride faster. To prevent people from obtaining QuickPasses for several rides at once, the QuickPass machines allow you to have only one active QuickPass at a time.

You have been hired as one of several competing consultants to improve the operation of QuickPass. Customers have been complaining about some anomalies in the test system. For example, customers observed that in one instance, QuickPasses were being offered for a return time as long as 4 hours later. A short time later on the same ride, the QuickPasses were given for times only an hour or so later. In some instances, the lines for people with QuickPasses are nearly as long and slow as the regular lines.

The problem, then, is to propose and test schemes for issuing QuickPasses in order to increase people's enjoyment of the amusement park. Part of the problem is to determine what criteria to use in evaluating alternative schemes. Include in your report a nontechnical summary for amusement park executives who must choose between alternatives from competing consultants.

2005: Flood Planning

Lake Murray in central South Carolina is formed by a large earthen dam, which was completed in 1930 for power production. Model the flooding downstream that would occur if there were a catastrophic earthquake that breached the dam.

Consider two particular questions:

1. Rawls Creek is a year-round stream that flows into the Saluda River a short distance downriver from the dam. How much flooding will occur in Rawls Creek from a dam failure, and how far back will it extend?

2. Could the flood be so massive downstream that water would reach up to the South Carolina State Capitol Building, which is on a hill overlooking the Congaree River?

2005: Tollbooths

Heavily traveled toll roads such as the Garden State Parkway and Interstate 95 are multilane divided highways that are interrupted at intervals by toll plazas. Because collecting tolls is usually unpopular, it is desirable to minimize motorist annoyance by limiting the amount of traffic disruption caused by the toll plazas. Commonly, a much larger number of tollbooths is provided than the number of travel lanes entering the toll plaza. Upon entering the toll plaza, the flow of vehicles fans out to the larger number of tollbooths, and when leaving the toll plaza, the flow of vehicles is required to squeeze back down to a number of travel lanes equal to the number of travel lanes before the toll plaza. Consequently, when traffic is heavy, congestion increases upon departure from the toll plaza. When traffic is very heavy, congestion also builds at the entry to the toll plaza because of the time required for each vehicle to pay the toll.

Make a model to help you determine the optimal number of tollbooths to deploy in a barrier-toll plaza. Explicitly consider the scenario where there is exactly one tollbooth per incoming travel lane. Under what conditions is this more effective than the current practice? Under what conditions is it less effective? Note that the definition of "optimal" is up to you to determine.

2006: Positioning and Moving Sprinkler Systems for Irrigation

There is a wide variety of techniques available for irrigating a field. The technologies range from advanced drip systems to periodic flooding. One approach that is used on smaller ranches is "hand move" irrigation systems. Lightweight aluminum pipes with sprinkler heads are put in place across fields, and they are moved by hand at periodic intervals to ensure that the whole field receives an adequate amount of water. This type of irrigation system is cheaper and easier to maintain than other systems. It is also flexible, allowing for use on a wide variety of fields and crops. The disadvantage is that it requires a great deal of time and effort to move and set up the equipment at regular intervals.

Given that this type of irrigation system is to be used, how can it be configured to minimize the amount of time required to irrigate a field that measures 80 meters by 30 meters? For this task you are asked to find an algorithm to determine how to irrigate the rectangular field in a way that minimizes the amount of time required by a rancher to maintain the irrigation system. One pipe set is used in the field. You should determine the number of sprinklers and the spacing between sprinklers, and you should find a schedule to move the pipes, including where to move them.

A pipe set consists of a number of pipes that can be connected together in a straight line. Each pipe has a 10-centimeter inner diameter with rotating spray nozzles that have a 0.6-centimeter inner diameter. When put together, the resulting pipe is 20 meters long. At the water source, the pressure is 420 kilopascals and has a flow rate of 150 liters per minute. No part of the field should receive more than 0.75 centimeter per hour of water, and each part of the field should receive at least 2 centimeters of water every 4 days. The total amount of water should be applied as uniformly as possible.

2006: Wheelchair Access at Airports

One of the frustrations with air travel is the need to fly through multiple airports, and each stop generally requires each traveler to change to a different airplane. This can be especially difficult for people who are not able to walk easily to a different flight's waiting area. One of the ways that an airline can make the transition easier is to provide a wheelchair and an escort to those people who ask for help. It is generally known well in advance which passengers require help, but it is not uncommon to receive notice when a passenger first registers at the airport. In rare instances, an airline may not receive notice from a passenger until just prior to landing.

Airlines are under constant pressure to keep their costs down. Wheelchairs wear out, are expensive, and require maintenance. There is also a cost for making the escorts available. Moreover, wheelchairs and their escorts must be constantly moved around the airport so that they are available to people when their flight lands. In some large airports, the time required to move across the airport is nontrivial. The wheelchairs must be stored somewhere, but space is expensive and severely limited in an airport terminal. Also, wheelchairs left in high-traffic areas represent a liability risk as people try to move around them. Finally, one of the biggest costs is the cost of holding a plane if someone must wait for an escort and becomes late for his or her flight. The latter cost is especially troubling because it can affect the airline's average flight delay, which can lead to fewer ticket sales as potential customers may choose to avoid that airline.

Epsilon Airlines planners have decided to ask a third party to help them obtain a detailed analysis of the issues and costs of keeping and maintaining wheelchairs and escorts available for passengers. The airline needs to find a way to schedule the movement of wheelchairs throughout each day in a cost-effective way. They also need to find and define the costs for budget planning in both the short and the long term.

Epsilon Airlines has asked your consultant group to put together a bid to help solve this problem. Your bid should include an overview and analysis of the situation to help the planners decide whether you fully understand their problem. They require a detailed description of an algorithm that you would like to implement which can determine where the escorts and wheelchairs should be and how they should move throughout each day. The goal is to keep the total costs as low as possible. Your bid is one of many that the airline will consider. You must make a strong case as to why your solution is the best and show that it will be able to handle a wide range of airports under a variety of circumstances.

Your bid should also include examples of how the algorithm would work for a large (at least four concourses), a medium (at least two concourses), and a small airport (one concourse) under high and low traffic loads. You should determine all potential costs and balance their respective weights. Finally, as populations begin to include a higher percentage of older people who have more time to travel but may require more aid, your report should include projections of potential costs and needs in the future, with recommendations for meeting those future needs.

2007: Gerrymandering

The United States Constitution provides that the House of Representatives shall be composed of some number of individuals (currently 435) who are elected from each state in proportion to the state's population relative to that of the country as a whole. Although this provides a way of determining how many representatives each state will have, it says nothing about how the district represented by a particular representative shall be determined geographically. This oversight has led to egregious (at least some people think so but usually not the incumbent) district shapes that look "unnatural" by some standards.

Hence the following question: Suppose you were given the opportunity to draw congressional districts for a state. How would you do so as a purely "baseline" exercise to create the "simplest" shapes for all the districts in a state? The rules insist only that each district in the state contain the same population. The definition of *simple* is up to you, but you need to make a convincing argument to voters in the state that your solution is fair. As an application of your method, draw geographically simple congressional districts for the state of New York.

2007: The Airplane Seating Problem

Airlines are free to seat passengers waiting to board an aircraft in any order whatsoever. It has become customary to seat passengers with special needs first, followed by first-class passengers (who sit at the front of the plane). Then coach and business-class passengers are seated by groups of rows, beginning with the row at the back of the plane and proceeding forward.

Apart from consideration of the passengers' wait time, from the airline's point of view time is money, and boarding time is best minimized. The plane makes money for the airline only when it is in motion, and long boarding times limit the number of trips that a plane can make in a day.

The development of larger planes, such as the Airbus A380 (800 passengers), reflects efforts to minimize boarding (and deboarding) time.

Devise and compare procedures for boarding and deboarding planes with varying numbers of passengers: small (85–210), midsize (210–330), and large (330–800).

Prepare an executive summary, not to exceed two single-spaced pages, in which you explain your conclusions to an audience of airline executives, gate agents, and flight crews.

Note: The two-page executive summary is to be included in addition to the reports required by the contest guidelines.

An article that appeared in the *New York Times* on November 14, 2006, addressed boarding and deboarding procedures currently being followed and the importance to the airline of finding better solutions. The article can be seen at http://travel2.nytimes.com/2006/11/14/business/14boarding.html.

2008: Take a Bath

Consider the effects on land from the melting of the north polar ice cap due to the predicted increase in global temperatures. Specifically, model the effects on the coast of Florida every 10 years for the next 50 years due to the melting, with particular attention given to large metropolitan areas. Propose appropriate responses to deal with this. A careful discussion of the data used is an important part of the answer.

2008: Creating Sudoku Puzzles

Develop an algorithm to construct Sudoku puzzles of varying difficulty. Develop metrics to define a difficulty level. The algorithm and metrics should be extensible to a varying number of difficulty levels. You should illustrate the algorithm with at least four difficulty levels. Your algorithm should guarantee a unique solution. Analyze the complexity of your algorithm. Your objective should be to minimize the complexity of the algorithm and meet the above requirements.

For further information on the Mathematics Contest in Modeling (MCM), write to COMAP, 57 Bedford Street, Lexington, Massachusetts 02173, or visit www.comap.com.

An Elevator Simulation Algorithm

We will define the terms used in the following algorithm and explain some of its underlying logic. Because the algorithm is complex, this approach should be more revealing than using some hypothetical numbers and taking you step by step through the algorithm. (This is a difficult program to write if you are not using GPSS or another simulation language.)

During the simulation there is a TIME clock that keeps track of the time (given in seconds). Initially, the value of TIME is 0 sec (at 7:50 A.M.), and the simulation ends when TIME reaches 4800 sec (at 9:10 A.M.). Each customer is assigned a number according to the order of his or her arrival: The first customer is labeled 1, the second customer 2, and so forth. Whenever another customer arrives at the lobby, the time between the customer's arrival and the time when the immediately preceding customer arrived is added to the TIME clock. This time between successive arrivals of customers i and $i - 1$ is labeled $between_i$ in the algorithm, and the arrival time of customer i is labeled $arrive_i$. Initially, for the customer arrival submodel, we assume that all values between 0 and 30 have an equal likelihood of occurring.

All four elevators have their own availability times, called $return_j$ for elevator j. If elevator j is currently available at the main floor, its time is the current time, so $return_j =$ TIME. If an elevator is in transit, its availability time is the time at which it will return to the main floor. Passengers enter an available elevator in the numerical order of the elevators: first elevator 1 (if it is available), next elevator 2 (if it is available), and so on. Maximum occupancy of an elevator is 12 passengers.

Whenever another customer arrives in the lobby of the building, two possible situations exist. Either an elevator is available for receiving passengers or no elevator is available and a queue is forming as customers wait for one to become available. Once an elevator becomes available, it is "tagged" for loading, and passengers can enter *only* that elevator until either it is fully occupied (with 12 passengers) or the 15-sec time delay is exceeded before the arrival of another customer. After loading, the elevator departs to deliver all of its passengers. It is assumed that, even if fully loaded with 12 passengers, the elevator waits 15 sec to load the last passenger, allow floor selection, and get under way.

To keep track of which floors have been selected during the loading period of an elevator and the number of times a particular floor has been selected, the algorithm sets up two one-dimensional arrays (having a component for each of the floors 1–12). (Although no one selects floor 1, the indexing is simplified with its inclusion.) These arrays are called $selvec_j$ and $flrvec_j$ for the tagged elevator j. If a customer selects floor 5, for instance, then a 1 is entered into the fifth component position of $selvec_j$ and also into the fifth component of $flrvec_j$. If another customer selects floor 5, then the fifth component of $flrvec_j$ is updated to a 2, and so forth. For example, suppose the passengers in elevator j have selected floor 3 twice, floor 5 twice, and floors 7, 8, and 12 once. Then for this elevator, $selvec_j = (0, 0, 1, 0, 1, 0, 1, 1, 0, 0, 0, 1)$ and $flrvec_j = (0, 0, 2, 0, 2, 0, 1, 1, 0, 0, 0, 1)$. These arrays are then used to calculate the transport time of elevator j so we can determine when it will return to the main floor. As stated previously, that return time is designated $return_j$. The arrays are also used to calculate the delivery times of each passenger in elevator j. Initially, we assume that a customer chooses a floor with equal likelihood. We assume that it takes 10 sec for an elevator to travel between floors, 10 sec to open *and* close its doors, and 3 sec for each passenger to disembark. We also assume that it takes 3 sec for each passenger in a queue to enter the next available elevator.

Summary of Elevator Simulation Algorithm Terms

$between_i$ time between successive arrivals of customers i and $i - 1$ (a random integer varying between 0 and 30 sec)

$arrive_i$ time of arrival from start of clock at $t = 0$ for customer i (calculated only if customer enters a queue waiting for an elevator)

$floor_i$ floor selected by customer i (a random integer varying between 2 and 12)

$elevator_i$ time customer i spends in an elevator

$wait_i$ time customer i waits before stepping into an elevator (calculated only if customer enters a queue waiting for an elevator)

$delivery_i$ time required to deliver customer i to destination floor from time of arrival, including any waiting time

$selvec_j$ binary 0, 1 one-dimensional array representing the floors selected for elevator j, not counting the number of times a particular floor has been selected

$flrvec_j$ integer one-dimensional array representing the number of times each floor has been selected for elevator j for the group of passengers currently being transported to their respective floors

$occup_j$ number of current occupants of elevator j

$return_j$ time from start of clock at $t = 0$ that elevator j returns to the main floor and is available for receiving passengers

$first_j$ an index, the customer number of the first passenger who enters elevator j after it returns to the main floor

$quecust$ customer number of the first person waiting in the queue

$queue$ total length of current queue of customers waiting for an elevator to become available

$startque$ clock time at which the (possibly updated) current queue commences to form

$stop_j$ total number of stops made by elevator j during the entire simulation

$eldel_j$ total time elevator j spends in delivering its current load of passengers

$operate_j$ total time elevator j operates during the entire simulation

$limit$ customer number of the last person to enter an available elevator before it commences transport

max largest index of a nonzero entry in the array $selvec_j$ (highest floor selected)

$remain$ number of customers left in the queue after loading next available elevator

$quetotal$ total number of customers who spent time waiting

$TIME$ current clock time in seconds, starting at $t = 0$

$DELTIME$ average delivery time of a customer to reach destination floor from time of arrival, including any waiting time

$ELEVTIME$ average time a person spends in an elevator

$MAXDEL$ maximum time required for a customer to reach his or her floor of destination from time of arrival

$MAXELEV$ maximum time a customer spends in an elevator

$QUELEN$ number of customers waiting in the longest queue

$QUETIME$ average time a customer who must wait spends in a queue

$MAXQUE$ longest time a customer spends in a queue

Elevator System Simulation Algorithm

Input   None required.

Output   Number of passengers serviced, DELTIME, ELEVTIME, MAXDEL, MAXELEV, QUELEN, QUETIME, MAXQUE, $stop_j$, and the percentage time each elevator is in use.

STEP 1   Initially set the following parameters to zero: DELTIME, ELEVTIME, MAXDEL, MAX-ELEV, QUELEN, QUETIME, MAXQUE, quetotal, remain.

STEP 2   For the first customer, generate time between successive arrivals and floor destination and initialize delivery time:

$$i = 1$$

Generate $between_i$ and $floor_i$

$delivery_i = 15$

STEP 3   Initialize clock time, elevator available clock times, elevator stops, and elevator operating times. Also, initialize all customer waiting times.

$$TIME = between_i$$

For $k = 1\text{--}4 : return_k = TIME$ and $stop_k = operate_k = 0$

For $k = 1\text{--}400 : wait_k = 0$

(The number 400 is an upper-bound guess for the total number of customers)

STEP 4   While TIME ≤ 4800, do Steps 5–32.

STEP 5   Select the first available elevator:

If TIME $\geq return_1$, then $j = 1$ else

If TIME $\geq return_2$, then $j = 2$ else

If TIME $\geq return_3$, then $j = 3$ else

If TIME $\geq return_4$, then $j = 4$

ELSE (no elevator is currently available) GOTO Step 19.

STEP 6   Set as an index the customer number of the first person to occupy tagged elevator, and initialize the elevator occupancy floor selection vectors:

$$\text{first}_j = i, \qquad \text{occup}_j = 0$$
$$\text{For } k = 1\text{–}12\text{: selvec}_j[k] = \text{flrvec}_j[k] = 0$$

STEP 7   Load current customer on elevator j by setting the floor selection vectors and incrementing elevator occupancy:

$$\text{selvec}_j[\text{floor}_i] = 1$$
$$\text{flrvec}_j[\text{floor}_i] = \text{flrvec}_j[\text{floor}_i] + 1$$
$$\text{occup}_j = \text{occup}_j + 1$$

STEP 8   Get next customer and update clock time:

$$i = i + 1$$
Generate between_i and floor_i
$$\text{TIME} = \text{TIME} + \text{between}_i$$
$$\text{delivery}_i = 15$$

STEP 9   Set all available elevators to current clock time:

For $k = 1\text{–}4$:
If $\text{TIME} \geq \text{return}_k$, then $\text{return}_k = \text{TIME}$.
Else leave return_k as is.

STEP 10   If $\text{between}_i \leq 15$ and $\text{occup}_j < 12$, then increase the delivery times for each customer on the tagged elevator j:

For $k = \text{first}_j$ to $i - 1$:
$$\text{delivery}_k = \text{delivery}_k + \text{between}_i$$
and GOTO Step 7 to load current customer on the elevator and get the next customer.
Else (send off the tagged elevator):
Set $\text{limit} = i - 1$ and GOTO Step 11.

The sequence of Steps 11–18 implements delivery of all passengers on the currently tagged elevator j:

STEP 11   For $k = \text{first}_j$ to limit, do Steps 12–16.

STEP 12   Calculate time customer k spends in elevator:

$$N = \text{floor}_k - 1 \text{ (an index)}$$
$\text{elevator}_k = $ travel time up to floor + time to drop off previous
customers + customer k drop off time
+ open/close door times on previous floors
+ open door on current floor
$$= 10N + 3 \sum_{m=1}^{N} \text{flrvec}_j[m] + 3 + 10 \sum_{m=1}^{N} \text{selvec}_j[m] + 5$$

STEP 13   Calculate delivery time for customer k:
$$\text{delivery}_k = \text{delivery}_k + \text{elevator}_k$$

STEP 14   Sum to total delivery time for averaging:
$$\text{DELTIME} = \text{DELTIME} + \text{delivery}_k$$

STEP 15   If $\text{delivery}_k > \text{MAXDEL}$, then $\text{MAXDEL} = \text{delivery}_k$.
Else leave MAXDEL as is.

STEP 16   If $\text{elevator}_k > \text{MAXELEV}$, then $\text{MAXELEV} = \text{elevator}_k$.
Else leave MAXELEV as is.

STEP 17   Calculate total number of stops for elevator j, its time in transit, and the time at which it returns to the main floor:

$$\text{stop}_j = \text{stop}_j + \sum_{m=1}^{12} \text{selvec}_j[m]$$

Max = index of largest nonzero entry in selvec_j (i.e., the highest floor visited)

eldel_j = travel time + passenger drop-off time + door time

$$= 20(\text{Max} - 1) + 3\sum_{m=1}^{12} \text{flrvec}_j[m] + 10\sum_{m=1}^{12} \text{selvec}_j[m]$$

$\text{return}_j = \text{TIME} + \text{eldel}_j$

$\text{operate}_j = \text{operate}_j + \text{eldel}_j$

STEP 18   GOTO Step 5.

The sequence of Steps 19–32 is taken when no elevator is currently available and a queue of customers waiting for elevator service is set up.

STEP 19   Initialize queue:

quecust = i (number for first customer in queue)

startque = TIME (starting time of queue)

queue = 1

arrive_i = TIME

STEP 20   Get the next customer and update clock time:

$i = i + 1$

Generate between_i and floor_i

TIME = TIME + between_i

arrive_i = TIME

queue = queue + 1

STEP 21   Check for elevator availability:

If TIME $\geq \text{return}_1$, then $j = 1$ and GOTO Step 22 else

If TIME $\geq \text{return}_2$, then $j = 2$ and GOTO Step 22 else

If TIME $\geq \text{return}_3$, then $j = 3$ and GOTO Step 22 else

If TIME $\geq \text{return}_4$, then $j = 4$ and GOTO Step 22

ELSE (no elevator is available yet) GOTO Step 20.

STEP 22   Elevator j is available. Initialize floor selection vectors and assess the length of the queue:

For $k = 1$–12: $\text{selvec}_j[k] = \text{flrvec}_j[k] = 0$

remain = queue − 12

STEP 23   If remain ≤ 0, then $R = i$ and occup_j = queue.

Else R = quecust + 11 and $\text{occup}_j = 12$.

STEP 24   Load customers onto elevator j:

For k = quecust to R:

$\text{selvec}_j[\text{floor}_k] = 1$ and $\text{flrvec}_j[\text{floor}_k] = \text{flrvec}_j[\text{floor}_k] + 1$

STEP 25   If queue $\geq$ QUELEN, then QUELEN = queue.

Else leave QUELEN as is.

STEP 26   Update queuing totals:

quetotal = quetotal + occup_j

$$\text{QUETIME} = \text{QUETIME} + \sum_{m=\text{quecust}}^{R} [\text{TIME} - \text{arrive}_m]$$

STEP 27   If $(TIME - startque) \geq MAXQUE$, then $MAXQUE = TIME - startque$.
Else leave MAXQUE as is.

STEP 28   Set index giving number of first customer to occupy tagged elevator:
$first_j = quecust$

STEP 29   Calculate delivery and waiting times for each passenger on the tagged elevator:

For $k = first_j$ to R:
$delivery_k = 15 + (TIME - arrive_k)$
$wait_k = TIME - arrive_k$

STEP 30   If $remain \leq 0$, then set $queue = 0$ and GOTO Step 8 to get next customer.
Else set $limit = R$ and, for $k = first_j$ to $limit$, do Steps 12–17. When finished, GOTO Step 31.

STEP 31   Update queue length and check for elevator availability:

$queue = remain$
$quecust = R + 1$
$startque = arrive_{R+1}$

STEP 32   GOTO Step 20.
The sequence of Steps 33–36 calculates output values for the morning rush-hour elevator simulation.

STEP 33   Output the following values:

$N = i - queue$, the total number of customers served
$DELTIME = DELTIME/N$, average delivery time
MAXDEL, maximum delivery time of a customer

STEP 34   Output the average time spent in an elevator and the maximum time spent in an elevator:

$$ELEVTIME = \sum_{m=1}^{limit} \frac{elevator[m]}{limit} \quad \text{and MAXELEV}$$

STEP 35   Output the number of customers waiting in the longest queue, the average time a customer who waits in line spends in a queue, and the longest time spent in a queue:

QUELEN

$$QUETIME = \frac{QUETIME}{quetotal}$$

MAXQUE

STEP 36   Output the total number of stops for each elevator and the percentage time each elevator is in transport:

For $k = 1-4$: display $stop_k$ and $\dfrac{operate_k}{4800}$
STOP

Note   For ease of presentation in the elevator simulation, TIME is updated only as the next customer arrives in the lobby. Therefore, TIME is not an actual clock being updated every second. It is possible, when a queue has formed, that an elevator returns to the main floor during Step 20, before the next customer arrives. However, loading of the available elevator does not commence until that customer actually arrives. For this reason, the times spent waiting in a queue are slightly on the high side. In Problem 2 you are asked to modify Steps 20–32 in the algorithm so that loading commences immediately upon the return of the first available elevator.

Table B.1   Results of elevator simulation for 15 consecutive days

Simulation number	Numbers of customers serviced	Average delivery time	Maximum delivery time	Average time in elevator	Maximum time in elevator	Number of customers in longest queue	Average time in a queue	Longest time in a queue	Total number of stops for each elevator				Percentage of total time each elevator is in transport			
									1	2	3	4	1	2	3	4
1	328	147	412	89	208	12	40	166	67	76	56	52	84	87	80	75
2	322	146	409	88	211	18	43	176	74	62	52	61	88	80	78	77
3	309	139	385	87	201	12	37	161	62	61	61	62	85	83	80	80
4	331	149	371	89	205	13	42	149	72	69	68	44	85	82	87	73
5	320	146	404	87	208	15	48	178	72	52	58	58	86	78	80	74
6	313	153	405	91	211	13	45	146	69	62	72	47	85	82	82	74
7	328	138	341	88	195	10	35	120	59	66	70	61	86	81	84	82
8	312	147	377	86	198	12	46	163	69	60	61	43	82	82	81	73
9	329	139	352	87	208	11	37	155	58	63	70	57	86	86	82	78
10	314	143	325	88	205	9	35	128	65	68	57	65	83	84	78	83
11	317	137	344	85	202	10	38	129	64	75	64	56	86	85	81	77
12	341	153	396	90	211	18	45	177	83	63	63	53	87	82	78	77
13	318	136	345	80	208	11	36	140	58	64	63	55	91	82	83	73
14	319	140	356	88	208	13	33	135	64	67	58	65	84	83	82	81
15	323	147	386	91	218	15	39	166	76	64	70	54	86	81	87	76
Averages (rounded)	322	144	374	88	206	13	40	153	67	65	63	56	86	83	82	77

Note: All times are measured in seconds and rounded to the nearest second.

Table B.1 gives the results of 15 independent simulations, representing 3 weeks of morning rush hour, according to the preceding algorithm.

B.1   PROBLEMS

1. Consider an intersection of two one-way streets controlled by a traffic light. Assume that between 5 and 15 cars (varying probabilistically) arrive at the intersection every 10 sec in direction 1, and that between 6 and 24 cars arrive every 10 sec going in direction 2. Suppose that 36 cars per 10 sec can cross the intersection in direction 1 and that 20 cars per 10 sec can cross the intersection in direction 2 if the traffic light is green. No turning is allowed. Initially, assume that the traffic light is green for 30 sec and red for 70 sec in

direction 1. Write a simulation algorithm to answer the following questions for a 60-min time period:

a. How many cars pass through the intersection in direction 1 during the hour?

b. What is the average waiting time of a car stopped when the traffic signal is red in direction 1? The maximum waiting time?

c. What is the average length of the queue of cars stopped for a red light in direction 1? The maximum length?

d. What is the average number of cars passing through the intersection in direction 1 during the time when the traffic light is green? What is the maximum number?

e. Answer Problems (a)–(d) for direction 2.

How would you use your simulation to determine the switching period for which the total waiting time in both directions is as small as possible? (You will have to modify it to account for the waiting times in direction 2.)

2. Modify Steps 20–32 in the elevator simulation algorithm so that loading of the first available elevator commences immediately upon its return. Thus, if TIME > return$_j$ so that elevator j is available for loading, then loading commences at time return$_j$ rather than TIME. Consider how you will now process customer i in Step 20 who has not yet quite arrived on the scene.

B.1   PROJECT

1. Find a building in your local area that has from 4 to 12 floors that are serviced by 1–4 elevators. Collect data for the interarrival times (and, possibly, floor destinations) of the customers during a busy hour (e.g., the morning rush hour), and build the inter-arrival and destination submodels based on your data (by constructing the cumulative histograms). Write a computer program incorporating your submodels into the elevator system algorithm to obtain results such as those given in Table B.1.

The Revised Simplex Method

For those of you familiar with matrix algebra, we demonstrate how to accomplish a pivot using matrix techniques. Any desired extreme point can be determined by first inverting a submatrix of the original tableau, followed by premultiplying the original tableau by the inverted submatrix. The method is called the **Revised Simplex Method** and has advantages of both speed and accuracy. In particular, round-off errors can be minimized in subsequent pivots because the Revised Simplex Method can use the original data to perform any desired pivot. We begin by illustrating pivoting by matrix inversion, using the carpenter's problem as presented in Section 7.4 for illustration.

Pivoting by Matrix Inversion and Multiplication

We illustrate how to use matrix inversion and multiplication to move from Tableau 0 to Tableau 1 (see Section 7.4, Example 1, The Carpenter's Problem Revisited). For Tableau 1, we want the set of dependent variables to be (in order)

$$\{x_2, y_2, z\}$$

The corresponding columns from the *original tableau*, in the same order, are selected to form the matrix P:

$$P = \begin{bmatrix} 30 & 0 & 0 \\ 4 & 1 & 0 \\ -30 & 0 & 1 \end{bmatrix}$$

First, compute the inverse of P:

$$P^{-1} = \begin{bmatrix} \frac{1}{30} & 0 & 0 \\ -\frac{2}{15} & 1 & 0 \\ 1 & 0 & 1 \end{bmatrix}$$

Then obtain Tableau 1 by premultiplying the matrix corresponding to Tableau 0 by P^{-1}:

$$T^1 = P^{-1}T^0 = \begin{bmatrix} \frac{1}{30} & 0 & 0 \\ -\frac{2}{15} & 1 & 0 \\ 1 & 0 & 1 \end{bmatrix} \begin{bmatrix} 20 & 30 & 1 & 0 & 0 & 690 \\ 5 & 4 & 0 & 1 & 0 & 120 \\ -25 & -30 & 0 & 0 & 1 & 0 \end{bmatrix}$$

$$= \begin{bmatrix} \frac{2}{3} & 1 & \frac{1}{30} & 0 & 0 & 23 \\ \frac{7}{3} & 0 & -\frac{2}{15} & 1 & 0 & 28 \\ -5 & 0 & 1 & 0 & 1 & 690 \end{bmatrix}$$

The Revised Simplex Method

Note that the columns for the matrix P are selected from the original tableau. Also note that the matrix P^{-1} premultiplies the original data. Any intersection point can be enumerated in this fashion. This method is known as the Revised Simplex Method and is beneficial when solving large linear programs. By returning to the original data to compute P^{-1}, the round-off error (which accumulates in successive pivots using other methods) is reduced. Let's illustrate the idea by computing Tableau 2 by premultiplying the original tableau by an appropriate pivot matrix.

Examining T^1 above, we see that the optimality test determines x_1 as the entering variable. The feasibility test as demonstrated in Section 7.4 determines y_2 as the exiting variable. Thus the new dependent variables are

$$\{x_2, x_1, z\}$$

The columns from the *original tableau* corresponding to the dependent variables, in order, are selected to form the matrix P:

$$P = \begin{bmatrix} 30 & 20 & 0 \\ 4 & 5 & 0 \\ -30 & -25 & 1 \end{bmatrix}$$

We compute the inverse of P:

$$P^{-1} = \begin{bmatrix} \frac{1}{14} & -\frac{2}{7} & 0 \\ -\frac{2}{35} & \frac{3}{7} & 0 \\ \frac{5}{7} & \frac{15}{7} & 1 \end{bmatrix}$$

We now obtain Tableau 2 by premultiplying the matrix corresponding to Tableau 0 by the inverse of the pivot matrix corresponding to the set of dependent variables $\{x_2, x_1, z\}$:

$$T^2 = P^{-1}T^0 = \begin{bmatrix} \frac{1}{14} & -\frac{2}{7} & 0 \\ -\frac{2}{35} & \frac{3}{7} & 0 \\ \frac{5}{7} & \frac{15}{7} & 1 \end{bmatrix} \begin{bmatrix} 20 & 30 & 1 & 0 & 0 & 690 \\ 5 & 4 & 0 & 1 & 0 & 120 \\ -25 & -30 & 0 & 0 & 1 & 0 \end{bmatrix}$$

$$= \begin{bmatrix} 0 & 1 & \frac{1}{14} & -\frac{2}{7} & 0 & 15 \\ 1 & 0 & -\frac{2}{35} & \frac{3}{7} & 0 & 12 \\ 0 & 0 & \frac{5}{7} & \frac{15}{7} & 1 & 750 \end{bmatrix}$$

which is the optimal tableau (see Section 7.4).

The Revised Simplex Method is used (with various enhancements) to solve large problems in which speed and accuracy are important.

APPENDIX

Brief Review of Integration Techniques

u-Substitution

The basic idea underlying *u*-substitution is to perform a simple substitution that converts the intergral into a recognizable form ready for immediate integration. For example, given

$$\int \frac{\cos x}{1 + \sin x}\, dx$$

let $u = 1 + \sin x$ and differentiate to find $du = \cos x\, dx$. Substitution then yields

$$\int \frac{\cos x}{1 + \sin x}\, dx = \int \frac{du}{u} = \ln|u| + C$$

Substituting for u again in this last expression gives

$$\int \frac{\cos x}{1 + \sin x}\, dx = \ln|1 + \sin x| + C$$

Integration by Parts

Recall from calculus that

$$\int u\, dv = uv - \int v\, du$$

In some cases it is necessary to apply the procedure several times before a form is obtained that can easily be integrated. In these and other situations, it is helpful to use the tabular method as follows:

Sign	Derivatives	Integrals
+	u	dv
−	du	v

Diagonal arrows in the table indicate terms to be multiplied (uv in this case). The bottom row in the table has horizontal arrows to indicate the final integral to be evaluated ($\int v\, du$ in the above case). Finally, the sign column is associated with the differentiated term at each stage, beginning with a plus sign and alternating with the minus sign, as suggested by the table format.

Thus the table above would be read as follows:

$$\underbrace{\int u\, dv}_{\text{top row}} = \overset{\overbrace{\quad\text{signs}\quad}}{\underbrace{+\, uv}_{\substack{\text{diagonal}\\\text{arrow}}}} - \underbrace{\int v\, du}_{\substack{\text{horizontal}\\\text{arrow}}}$$

To apply intergration by parts successively, build the table by repeatedly differentiating the derivatives (middle) column and intergrating the integrals (right) column, while the sign (left) column alternates. Terminate the table with a horizontal arrow between the middle and right column when you can readily intergrate the product of the function in the last row or when the last row simply repeats the first row (up to a multiplicative constant). Let us consider several examples.

EXAMPLE 1

Find the integral $\int xe^x\, dx$ by the tabular method.

Solution   We set up the table as follows:

Sign	Derivatives	Integrals
+	x	e^x
−	1	e^x
+	0	e^x

Interpreting the table, we get

$$\int xe^x\, dx = +xe^x - 1\cdot e^x + \int 0\cdot e^x\, dx + C$$
$$= (x-1)e^x + C$$

EXAMPLE 2

Integrate $\int x^2 e^{2x}\, dx$ by the tabular method.

Solution   We set up the table as before:

Sign	Derivatives	Integrals
+	x^2	e^{2x}
−	$2x$	$\dfrac{e^{2x}}{2}$
+	2	$\dfrac{e^{2x}}{4}$
−	0	$\dfrac{e^{2x}}{8}$

Thus

$$\int x^2 e^{2x} dx = +\frac{x^2 e^{2x}}{2} - \frac{2x e^{2x}}{4} + \frac{2e^{2x}}{8} - \int \frac{0 \cdot e^{2x}}{8} dx + C$$

$$= \frac{e^{2x}}{4}(2x^2 - 2x + 1) + C$$

■ ■ ■

EXAMPLE 3

Integrate $\int e^x \sin x \, dx$.

Solution   After filling in the table, we get

Sign	Derivatives	Integrals
+	$\sin x$	e^x
−	$\cos x$	e^x
+	$-\sin x$	e^x

Thus

$$\int e^x \sin x \, dx = e^x \sin x - e^x \cos x + \int (-\sin x) e^x dx + C$$

or

$$\int e^x \sin x \, dx = \frac{e^x(\sin x - \cos x)}{2} + C_1$$

■ ■ ■

Examples 1–3 illustrate the two basic strategies of integration by parts: (1) Choose a term to differentiate whose successive derivatives eventually become zero or repeat, and (2) continue to differentiate by parts until the integrand (up to a multiplicative constant) is repeated in the bottom row, as in Example 3. In choosing the term dv to integrate, you may find the following mnemonic "detail ladder" useful:

dv

exponential

trigonometric

algebraic

inverse trigonometric

logarithmic

To use the ladder, choose the term dv to integrate in order of priority from the top to the bottom. Conversely, the term u to differentiate is chosen from bottom to top. For example, when integrating

$$\int x^2 e^x dx$$

which involves a polynomial and an exponential, integrate the exponential $dv = e^x dx$ and differentiate the polynomial $u = x^2$. The above mnemonic device is a rule of thumb only and may not work in some cases.

Rational Functions

Given an algebraic fraction with a polynomial in both the numerator and the denominator (that is, **a rational function**), division may lead to a simpler form. If the highest power in the numerator is equal to or greater than the highest power in the denominator, first perform polynomial division and then integrate the result. For example,

$$\frac{y+1}{y-1} = 1 + \frac{2}{y-1}$$

so

$$\int \frac{y+1}{y-1}\,dy = \int \left(1 + \frac{2}{y-1}\right)dy = y + 2\ln|y-1| + C$$

Partial Fractions

In algebra you learned to sum fractional expressions by finding a common denominator. For example,

$$\frac{2}{x-1} + \frac{4}{x+3} = \frac{2(x+3) + 4(x-1)}{(x-1)(x+3)}$$

$$= \frac{6x+2}{x^2+2x-3}$$

For purposes of integration we need to reverse this procedure. That is, given the integral

$$\int \frac{6x+2}{x^2+2x-3}\,dx$$

we use partial fraction decomposition to obtain a new expression that is readily integrable:

$$\int \left(\frac{2}{x-1} + \frac{4}{x+3}\right)dx = 2\ln|x-1| + 4\ln|x+3| + C$$

This process of splitting a fraction $f(x)/g(x)$ into a sum of fractions with linear or quadratic denominators is called **partial fraction decomposition**. For the method to work, the degree of the numerator $f(x)$ must be less than the degree of the denominator $g(x)$; otherwise, you must first perform polynomial long division. To use the method, the denominator must be factored into linear and quadratic factors. In Examples 4–6 we review three cases that may exist for the factored denominator:

1. Distinct linear factors
2. Repeated linear factors
3. Quadratic factors

EXAMPLE 4 *Distinct Linear Factors*

Find the integral $\displaystyle\int \frac{2x^2 - x + 1}{(x+1)(x-3)(x+2)}\, dx$.

Solution   We must find constants A, B, and C such that

$$\frac{2x^2 - x + 1}{(x+1)(x-3)(x+2)} = \frac{A}{x+1} + \frac{B}{x-3} + \frac{C}{x+2} \tag{1}$$

Algebraic Method   In this method you multiply through by the factored denominator to obtain

$$2x^2 - x + 1 = A(x-3)(x+2) + B(x+1)(x+2) + C(x+1)(x-3)$$

Then expand the right-hand side and combine like powers of x:

$$2x^2 - x + 1 = (A + B + C)x^2 + (-A + 3B - 2C)x + (-6A + 2B - 3C)$$

Next equate the coefficients of like powers of x on both sides of this last equation. This procedure results in a system of linear algebraic equations involving our three unknowns:

$$
\begin{aligned}
A + B + C &= 2 \\
-A + 3B - 2C &= -1 \\
-6A + 2B - 3C &= 1
\end{aligned}
$$

Solution of this system by elimination or by the method of determinants yields

$$A = -1, \qquad B = \frac{4}{5}, \qquad \text{and} \qquad C = \frac{11}{5}$$

Thus

$$
\begin{aligned}
\int \frac{2x^2 - x + 1}{(x+1)(x-3)(x+2)}\, dx &= -\int \frac{dx}{x+1} + \frac{4}{5}\int \frac{dx}{x-3} + \frac{11}{5}\int \frac{dx}{x+2} \\
&= -\ln|x+1| + \frac{4}{5}\ln|x-3| + \frac{11}{5}\ln|x+2| + C
\end{aligned}
$$

Heaviside Method   There is a shortcut method for finding the constants in the partial fraction decomposition of $f(x)/g(x)$. First, write the rational function with $g(x)$ completely factored into its linear terms:

$$\frac{f(x)}{g(x)} = \frac{f(x)}{(x - r_1)(x - r_2) \cdots (x - r_n)} \tag{2}$$

To find the constant A_i associated with the term

$$\frac{A_i}{x - r_i}$$

in the partial fraction decomposition, cover the factor $x - r_i$ in the denominator of the right-hand side of Equation (2) and replace all the uncovered x's with the number r_i. For instance, to find the constant A in Equation (1), cover the factor $x + 1$ in the denominator and replace all the uncovered x's with $x = -1$.

$$A = \frac{2 - (-1) + 1}{(\boldsymbol{x + 1})(-1 - 3)(-1 + 2)} = \frac{4}{(-4)(1)} = -1$$
$$\uparrow$$
$$\text{covered}$$

Likewise, we find B by covering the factor $x - 3$ and replacing all the uncovered x's with $x = 3$.

$$B = \frac{2(9) - 3 + 1}{(3 + 1)(\boldsymbol{x - 3})(3 + 2)} = \frac{16}{4(5)} = \frac{4}{5}$$
$$\uparrow$$
$$\text{covered}$$

Finally, C is determined when $x = -2$.

$$C = \frac{2(4) - (-2) + 1}{(-2 + 1)(-2 - 3)(\boldsymbol{x + 2})} = \frac{11}{(-1)(-5)} = \frac{11}{5}$$
$$\uparrow$$
$$\text{covered}$$

The integration is the same as before. We emphasize that *the Heaviside method can be used only with distinct linear factors.* In the next example, we present another method for finding the constants when the linear factors are repeated. Of course, you can always resort to the more tedious algebraic method.   ▪ ▪ ▪

EXAMPLE 5   *A Repeated Linear Factor*

Find the integral $\displaystyle\int \frac{3P}{(P + 4)^2(P + 1)}\, dP.$

Solution   We need to find constants A, B, and C such that

$$\frac{3P}{(P + 4)^2(P + 1)} = \frac{A}{P + 4} + \frac{B}{(P + 4)^2} + \frac{C}{P + 1}$$

or

$$3P = A(P + 4)(P + 1) + B(P + 1) + C(P + 4)^2 \tag{3}$$

Substitution Method   Since Equation (3) is an identity, it holds for every value of P. Thus, to obtain three equations for finding the unknowns A, B, and C, we simply substitute convenient values for P:

$$P = -4: \quad -12 = -3B$$
$$P = -1: \quad -3 = 9C$$
$$P = 0: \quad 0 = 4A + B + 16C$$

to give the solutions $A = \frac{1}{3}$, $B = 4$, $C = -\frac{1}{3}$. Thus

$$\int \frac{3P}{(P+4)^2(P+1)}\, dP = \int \left[\frac{1}{3(P+4)} + \frac{4}{(P+4)^2} - \frac{1}{3(P+1)} \right] dP$$

$$= \frac{1}{3} \ln|P+4| - \frac{4}{P+4} - \frac{1}{3} \ln|P+1| + C \quad \blacksquare \blacksquare \blacksquare$$

EXAMPLE 6   *A Quadratic Factor*

Find the integral $\displaystyle\int \frac{dP}{(P+1)(P^2+1)}$.

Solution   We must find constants A, B, and C such that

$$\frac{1}{(P+1)(P^2+1)} = \frac{A}{P+1} + \frac{BP+C}{P^2+1}$$

Thus

$$1 = A(P^2+1) + (BP+C)(P+1)$$

Since this expression is to hold for all P, the coefficients of like powers of P on both sides of the equation must be equal. After collecting like powers of P on the right-hand side, we get

$$0P^2 + 0P^1 + 1P^0 = (A+B)P^2 + (B+C)P + (A+C)$$

which yields the linear system

$$0 = A + B$$
$$0 = B + C$$
$$1 = A + C$$

The solution is $A = \frac{1}{2}$, $B = -\frac{1}{2}$, and $C = \frac{1}{2}$. Thus

$$\int \frac{dP}{(P+1)(P^2+1)} = \int \left[\frac{1}{2(P+1)} + \frac{-\dfrac{P}{2} + \dfrac{1}{2}}{P^2+1} \right] dP$$

$$= \frac{1}{2} \ln|P+1| - \frac{1}{4} \ln|P^2+1| + \frac{1}{2} \tan^{-1} P + C \quad \blacksquare \blacksquare \blacksquare$$

Index